Food Process Modeling and Control

Topics in Chemical Engineering

A series edited by R. Hughes, University of Salford, UK

Volume 1	HEAT AND MASS TRANSFER IN PACKED BEDS by N. Wakao and S. Kaguei
Volume 2	THREE-PHASE CATALYTIC REACTORS by P.A. Ramachandran and R.V. Chaudhari
Volume 3	DRYING: Principles, Applications and Design by Cz. Strumillo and T. Kudra
Volume 4	THE ANALYSIS OF CHEMICALLY REACTING SYSTEMS: A Stochastic Approach by L.K. Doraiswamy and B.K. Kulkarni
Volume 5	CONTROL OF LIQUID–LIQUID EXTRACTION COLUMNS by K. Najim
Volume 6	CHEMICAL ENGINEERING DESIGN PROJECT: A Case Study Approach by M.S. Ray and D.W. Johnston
Volume 7	MODELLING, SIMULATION AND OPTIMIZATION OF INDUSTRIAL FIXED BED CATALYTIC REACTORS by S.S.E.H. Elnashaie and S.S. Elshishini
Volume 8	THREE-PHASE SPARGED REACTORS edited by K.D.P. Nigam and A. Schumpe
Volume 9	DYNAMIC MODELLING, BIFURCATION AND CHAOTIC BEHAVIOUR OF GAS-SOLID CATALYTIC REACTORS by S.S.E.H. Elnashaie and S.S. Elshishini
Volume 10	THERMAL PROCESSING OF BIO-MATERIALS edited by T. Kudra and Cz. Strumillo
Volume 11	COMPOSITION MODULATION OF CATALYTIC REACTORS by P.L. Silveston
Volume 12	FOOD PROCESS MODELING AND CONTROL: Chemical Engineering Applications by M. Özilgen

Food Process Modeling and Control: Chemical Engineering Applications

Mustafa Özilgen

CRC Press
Taylor & Francis Group
Boca Raton London New York

CRC Press is an imprint of the
Taylor & Francis Group, an **informa** business

Amsteldijk 166
1st Floor
1079 LH Amsterdam
The Netherlands

British Library Cataloguing in Publication Data

A catalogue record for this book is available from the British Library.

ISBN: 90-5699-143-4 (softcover)

To Sibel and Arda

Contents

Introduction to the Series

The subject matter of chemical engineering covers a very wide spectrum of learning and the number of subject areas encompassed in both undergraduate and graduate courses is inevitably increasing each year. This wide variety of subjects makes it difficult to cover the whole subject matter of chemical engineering in a single book. The present series is therefore planned as a number of books covering areas of chemical engineering which, although important, are not treated at any length in graduate and postgraduate standard texts. Additionally, the series will incorporate recent research material which has reached the stage where an overall survey is appropriate, and where sufficient information is available to merit publication in book form for the benefit of the profession as a whole.

Inevitably, with a series such as this, constant revision is necessary if the value of the texts for both teaching and research purposes is to be maintained. I would be grateful to individuals for criticisms and for suggestions for future editions.

R. HUGHES

Preface

A chemical engineer working in the food process industry needs to have an interdisciplinary education including biology, microbiology, biochemistry, statistics, etc., in addition to the standard chemical engineering curricula based on differential and numerical equations, unit operations, kinetics, transport phenomena, rheology, etc. Although the book covers topics from the indicated classes, it does not aim to substitute any of them. A chemical engineer working in the food process industry is expected to design a process starting with biological raw materials to produce an edible product for human consumption. This book goes beyond the scope of a typical food engineering text and includes bioprocesses, i.e., microbial growth, product formation, etc., associated with foods, food additives and beverage fermentations.

Although the fundamental principles of chemical engineering are applied to food processing and preservation, the nature of the raw materials requires special interest:

i) Food raw materials have a cellular structure, while the typical raw materials of the other chemical process industries do not have a cellular structure. Also, more than 70% of the raw materials of the foods is water, therefore different models are used to describe the unit operations than those of the other chemical process industries.
ii) There are almost no gas phase reactions in food processing but microbial interactions are important. The sensory aspects of food processing are modeled with analogy models, therefore the kinetics of the processes require special treatment.
iii) Some unit operations involving crystallization, freezing time calculations, thermal process calculations, extraction processes, etc. require special attention in the food industry and are not commonly used in the same way as in the other chemical process industries.
iv) Statistical process analysis recently became a hot topic in general chemical engineering research, whereas such methods have been used in the food process industries for many years. The hazards involved in food processing are substantially different than in the other chemical process industries.

This book is authored for junior and senior level engineering undergraduates or beginner graduate students. The basic purpose is to familiarize students with the fundamental techniques of mathematical modeling rather than exposing them to theoretical analysis, mathematical derivations or proving theorems. Readers are expected to have an intermediary level of mathematics background including calculus and ordinary differential equations. Important intermediary level mathematical prerequisites are presented in tables to make the fundamental information readily available in a concise manner, while focusing attention on modeling techniques rather than the mathematical details. Reviewing the whole literature concerning a specific process was not among the purposes of the book. The examples were selected from landmark studies, but they were also required to be easily understandable by readers without much background knowledge. In addition to chemical engineers, I expect professional people who work with the engineers but do not have an engineering degree, i.e., food scientists, or agricultural, food, bioprocessing/bio-resource, industrial and mechanical engineers who work in the food industry, to benefit from this book.

Computer flowcharts to present the modeling details are given in the book. Readers are expected to have an intermediary level of computer skills; a spreadsheet program and knowledge of a computer language, i.e., BASIC, will be sufficient to do most of the computer work.

The material covered in this book was collected and tested during nine years of classroom teaching and research in the Food Engineering Department of the Middle East Technical University. An earlier version of Chapter 5 was also taught in the University of California at Davis. I do appreciate all of my former students' comments and help, especially from Gürdal Atıcı, Sibel Babacan, Hülya Kahraman-Doğan, Kilichan Kaynak, Murat Özdemir and Gülüm Şumnu, to improve the text. I received tremendous professional help from my wife Dr Sibel Özilgen in every stage of this laborious work. She prepared most of the computer drawings, and she is actually the secret co-author of the book.

I appreciate the permissions granted by ACS (American Chemical Society), Academic Press, AFRC Silsoe Research Institute, AIChE (American Institute of Chemical Engineers), Akadémiai Kiadó És Nyomda (Budapest), American Society of Enology and Viticulture, ASAE (American Society of Agricultural Engineers), ASME (American Society of Mechanical Engineers), AT&T Bell Labs, Biometrika Trustees, Canadian Institute of Food Science and Technology, Chapman & Hall, EMAP Maclaren Ltd, Elsevier Science Ltd, Elsevier Science Publishers B.V., Food & Nutrition Press, Inc., IAMFES (International Association of Milk, Food and Environmental Sanitarians), Inc., IFT (Institute of Food Technologists), IARC (International Agency for Research on Cancer), Harwood Academic Publishers GmbH, John Wiley & Sons Inc., Marcel Dekker, Inc., McGraw-Hill, Inc., Oxford University Press, Inc., Pergamon Press, Ltd, Plenum Press, SCI (Society of Chemical Industries), Springer-Verlag, Techpress Publishing Co. Ltd, University of Tokyo Press, and Verlag Chemie GmbH to use their copyrighted material in the book. I also appreciate permissions granted or help offered by Drs Leslie Bluhm, Felix Franks, Kan-Ichi Hayakawa, Michael F. Kozempel, Kunihiro Kuwajima, Bart M. Nicolai, R. Larry Merson, Güner Özay, David S. Reid, James F. Steffe and Thomas B. Whitaker.

Mustafa Özilgen

CHAPTER 1

What is Process Modeling?

In the context of food processing a *mathematical model* means an approximate representation of a process in mathematical terms. Fundamental principles of mathematical modeling have been reviewed in numerous studies (Rand, 1983; Meyer, 1985; Riggs, 1988; Teixeira and Shoemaker, 1989; Cleland, 1990 and Luyben, 1990). A mathematical model can never be an actual representation of a process, since it would be very difficult, confusing or impossible to describe the whole system with mathematical formulations. In a typical process, inputs $x_1, x_2 \dots x_N$ may generate the outputs $y_1, y_2 \dots y_M$. With a model the cause-and-result relation between the major process inputs $(x_1, x_2 \dots x_n)$ and outputs $(y_1, y_2 \dots y_m)$ may be formulated in mathematical terms after simplification. Negligible inputs $x_{n+1} \dots x_N$ and outputs $y_{m+1} \dots y_M$ are not included in the model. One of the common mistakes made by engineers inexperienced in modeling is to get lost in the complexity of the process. The best rule of thumb in process modeling is the 80–20 rule, which means you get 80% of the benefit with the first 20% of the model complexity (Glasscock and Hale, 1994).

Mathematical modeling is usually done after obtaining the data in tabular and graphical forms. The model is a short-hand description of the data and estimates the values of the outputs $(y_1, y_2 \dots y_m)$ when the values of the inputs $(x_1, x_2 \dots x_m)$ are entered. The model may help to explain the details of the relation between the inputs and the outputs, which may not be understood by plotting the data only, and may explain the mechanism of the events. In the following pages we will frequently use the sentence

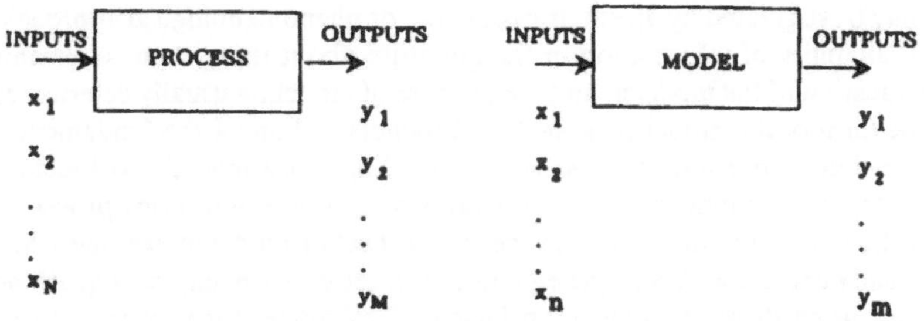

FIGURE 1.1 Comparison of input/output for a process and its model.

"*comparison of the model with the experimental data is shown in Figure*".... This sentence actually means that the two boxes given in Figure 1.1 are compared. Experimental data will be presented usually with symbols and represent the process; the mathematical model will be obtained after going through pertinent steps of Figure 1.2, will consist of the other box and will be presented with solid, dashed or dotted lines.

A good mathematical model should be general (apply to a wide variety of situations), realistic (based on correct assumptions), precise (its estimates should be finite numbers, or definite mathematical entities), accurate (its estimates should be correct or very near to correct) and there should be no trend in the deviations of the model from the experimental data. A good model should be robust (relatively immune to errors in the input data), and fruitful (its conclusions are useful or point the way to other good models).

Mathematical models may be categorized as *empirical*, *analog* or *phenomenological* models depending on what basis the functional relation is suggested. An empirical model assumes the form of the functional relation between the input and the output variables. There is usually no theoretical background sought when suggesting this relation. Empirical models are best when used within the range of the experimental data they are based on. An analogy model may be suggested for a relatively less known process by considering its similarity to a well known process, i.e., electrical circuit analogs may be used for modeling heat transfer, or stress/strain relations. Phenomenological models use a theoretical approach based on conservation of mass, energy, momentum, etc. to suggest the form of the mathematical model. They may include many different types including microscopic (distributed parameter) or macroscopic (lumped parameter) models.

The first step in building a mathematical model is the definition of the system. The answer to the question "What is going to be predicted by the model by what input data?" should be given whilst defining the system. Controlling factors of the system should also be identified, and the data should show the effects of the individual controlling factors. The system may be simplified by neglecting the effects of the marginal inputs, i.e., $x_{n+1} \ldots x_N$ and outputs $y_{m+1} \ldots y_M$. The form of the mathematical model may be suggested by an empirical, analog or phenomenological approach. Availability of information in the literature about the system, skills and education of the modeler, and the purpose of modeling usually determines the form of the model suggested. In Chapters 2, 3 and 4 the fundamental principles of phenomenological model building with application of kinetics, transport phenomena and unit operations to Food Engineering processes will be discussed in detail with theoretical background and examples. We usually end up with a single equation or a set of mathematical equations after using these fundamental principles. Techniques for solving of these equations will be discussed in Chapter 2. Empirical model building will be discussed in Chapter 5.3. Application of mathematical modeling to process control will be discussed in Chapter 5.9. Comparison of the mathematical model, i.e., solution of the equations, with the experimental data is the final stage of modeling. The model is validated if it agrees with the data. If such an agreement should not be obtained, all the steps of modeling, starting with the definition of the system, are repeated until satisfactory representation is obtained (Fig. 1.2).

Example 1.1. Death kinetics of microorganisms in dough In this example the modeling principles described in Figure 1.2 are explained with data. Where the process is death of microorganisms, the input to the model is time after the beginning of the death phase, and the output of the model is the colony count of surviving microorganisms at any time after the begin-

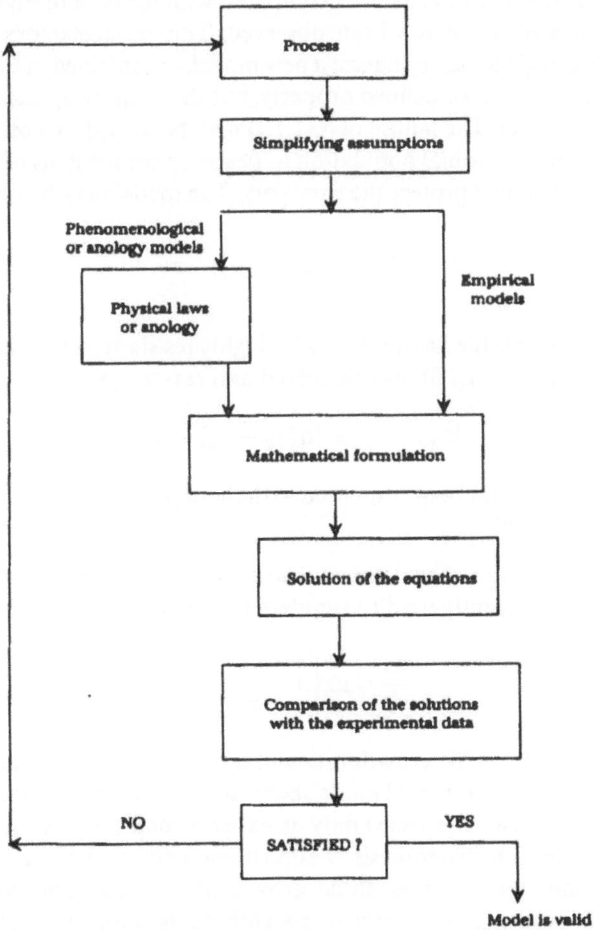

FIGURE 1.2 Schematic description of modeling.

ning of the death phase. Generally a microbial death model presumes that the microorganisms are equally labile, do not affect each other and that a constant fraction of the viable population dies in unit time:

$$\frac{dx}{dt} = -kx \tag{E.1.1.1}$$

where x is the concentration of the viable microorganisms at time t and k is the death rate constant. After integration (E.1.1.1) may be expressed as:

$$\ln x = \ln x_0 - kt \tag{E.1.1.2}$$

where x_0 is the viable microbial concentration at the beginning of the death phase. Equation (E.1.1.2) has been compared with the data in Figure E.1.a and clear disagreement has been observed. This unsatisfactory attempt with (E.1.1.1) requires us to suggest a new model, as explained in Figure 1.2. The process seems to be defined properly, but the simplifying assumptions may not be correct. The failure of (E.1.1.1) may be caused by non-uniform resistance of the microbial population to death, or constituents of the dead microorganisms may protect the survivors. The model may be revised as:

$$\frac{dx}{dt} = -k(x - x_{\infty}) \tag{E.1.1.3}$$

where x_{∞} represents the finally attainable highly resistant microbial concentration. Equation (E.1.1.3) may be solved and rearranged as:

$$\ln(x - x_{\infty}) = \ln(x_0 - x_{\infty}) - kt \tag{E.1.1.4}$$

Equation (E.1.1.4) has been compared with the experimental data in Figure E.1.b and validated.

Example 1.2. Kinetics of galactose oxidase production The logistic model (Chapter 3) is frequently used to simulate microbial growth:

$$\frac{dx}{dt} = \mu x\left(1 - \frac{x}{x_{max}}\right) \tag{E.1.2.1}$$

where μ is initial specific growth rate and x_{max} is the maximum attainable value of x. This is an empirical model, because it simulates the data without any theoretical basis. It is based only on experimental observations: When $x \ll x_{max}$, the term in parenthesis is almost one and can be neglected, the equation simulates the exponential growth ($dx/dt = \mu x$), and when x is comparable with x_{max}, the term in parenthesis becomes important and simulates the inhibitory effect of over-crowding on microbial growth. When $x = x_{max}$ the term in parenthesis becomes zero, and the equation will predict

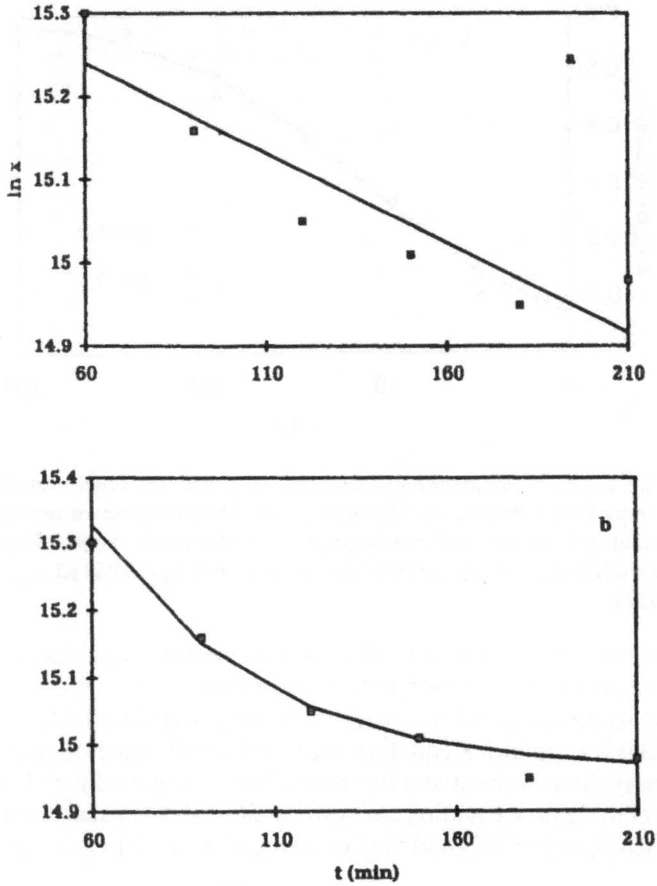

FIGURE E.1 a) Equation (E.1.1.2) ($\ln x = 15.37 - 0.00216t$, correlation coefficient $r = -0.92$) does not agree with the data; b) Equation (E.1.1.4) ($\ln(x - x_\infty) = 15.64 - 0.025\,t$, correlation coefficient $= -0.999$) simulates the data, when $x_\infty = 3.15 \times 10^6$ cfu/g and $k = 0.025\ \text{min}^{-1}$. Models are represented with lines (—) and the data are shown in symbols (■). Experimental data were taken from Yönden *et al.* (1992).

no growth ($dx/dt = 0$). The logistic equation may be integrated as:

$$x = \frac{x_0 e^{\mu t}}{1 - (x_0/x_{max})(1 - e^{\mu t})} \qquad \text{(E.1.2.2)}$$

Comparison of (E.1.2.2) with the experimental data is given in Figure E.1.2.1. Galactose oxidase production may be simulated with the Luedeking-Piret model (Ögel and Özilgen, 1995) (Chapter 3):

$$\frac{dc_p}{dt} = \alpha x + \beta \frac{dx}{dt} \qquad \text{(E.1.2.3)}$$

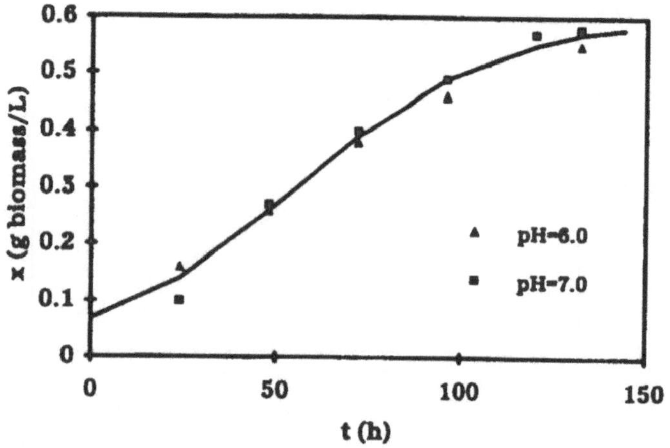

FIGURE E.1.2.1 Comparison of (E.1.2.2)(———) with the experimental data (■ and ▲) reported by Shatzman and Kosman (1977). The growth curves were almost the same at both pH values, therefore a single growth model may be suggested by using the whole data (Ögel and Özilgen, 1995). Model parameters: $x_0 = 0.068$ g/l, $x_{max} = 0.6$ g/l $\mu = 0.037$ h^{-1}

where α and β are constants. The term αx represents galactose oxidase production by the microorganisms regardless of their growth; $\beta dx/dt$ represents the additional enzyme production in proportion with the growth rate. This is an empirical equation, because it simply relates the experimental observations without any theoretical basis. Integrated and differential forms of the logistic equation were used to simulate biomass concentration x and dx/dt, respectively and the Luedeking-Piret model was integrated as:

$$c_P = c_{P0} + \alpha A + \beta B \tag{E.1.2.4}$$

where c_{P0} is the amount of the product in the fermentor when $t = 0$,

$$A = \frac{x_{max}}{\mu}\ln\left(1 - \frac{x_0}{x_{max}}(1 - e^{\mu t})\right) \text{ and } B = x_0\left(\frac{e^{\mu t}}{1 - \frac{x_0}{x_{max}}(1 - e^{\mu t})} - 1\right).$$

Comparison of (E.1.2.4) with the experimental data is shown in Figure E.1.2.2.

Example 1.3. Kinetics of lipid oxidation in foods (Özilgen and Özilgen, 1990) Lipid oxidation is one of the major reasons for spoilage of the lipid foods and occurs stagewise:

Initiation: $RH \rightarrow R\bullet + H\bullet$

Propagation: $R\bullet + O_2 \rightarrow RO_2\bullet$

$RO_2\bullet + RH \rightarrow RO_2H + R\bullet$

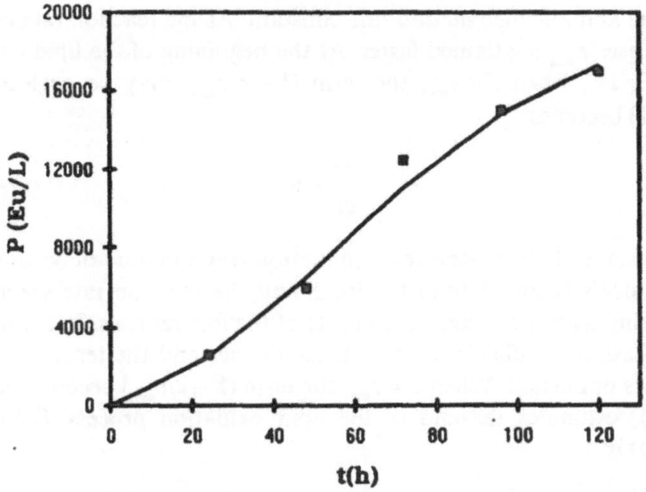

FIGURE E.1.2.2 Comparison of (E.1.2.4)(———) with the experimentally determined galactose oxidase activity (■) at pH = 6.7 (original data are reported by Shatzman and Kosman, 1977). Model parameters: α = 43 Eu/g biomass *h* and *b* = 32 10^3 Eu/g biomass.

Termination: $R\bullet + R\bullet \rightarrow RR$

$$R\bullet + RO_2\bullet \rightarrow RO_2R$$

$$RO_2\bullet + RO_2\bullet \rightarrow RO_2R + O_2$$

where RH represents the lipids, R•, H• and $RO_2\bullet$ the free radicals, and RO_2H and RO_2R are the oxidation products. Parameter R actually represents different chemical species, since too many reactions, involving many intermediary products, occur simultaneously during lipid oxidation. Such a complex reaction mechanism makes it impossible to monitor concentration of all the individual chemicals and it is difficult to use the conventional chemical kinetic rate expressions for modeling.

There is an analogy between lipid oxidation and microbial growth: Both have initiation, propagation and termination phases. Equation (E.1.2.1) simulates microbial growth successfully. Microbial growth, like lipid oxidation, occurs through a large number of complex reactions. Therefore a logistic equation may be used to simulate lipid oxidation:

$$\frac{dc}{dt} = kc\left(1 - \frac{c}{c_{max}}\right) \tag{E.1.3.1}$$

where c is the total amount of the oxidation products, dc/dt is the lipid oxidation rate, c_{max} is the total amount of the lipids available for oxidation,

t is time and k is the reaction rate constant. As the reaction rate constant k increases c_{max} is attained faster. At the beginning of the lipid oxidation process, i.e., when $c \ll c_{max}$ the term $(1 - c/c_{max})$ may be neglected and (E.1.3.1) becomes

$$\frac{dc}{dt} = kc$$

At this stage of the process the termination reactions do not occur and the free radicals accumulate in the food, and the reaction rate seems to be increasing with the accumulation rate of the free radicals. In the termination phase free radicals react with each other and the term $(1 - c/c_{max})$ becomes important. When $c = c_{max}$ the term $(1 - c/c_{max})$ becomes zero and (E.1.3.1) estimates the end of the lipid oxidation process. Solution to (E.1.3.1) is:

$$c = \frac{c_0 e^{kt}}{1 - (c_0/c_{max})(1 - e^{kt})} \quad \text{(E.1.3.2)}$$

where c_0 is the concentration of the oxidized products at the beginning of oxidation. A sample set of simulations is shown in Figure E.1.3.

Example 1.4. Analogy between cake filtration and ultrafiltration processes In a cake filtration process (Chapter 4.6) the total pressure drop through the filter (ΔP) is:

$$\Delta P = \Delta P_c + \Delta P_m \quad \text{(E.1.4.1)}$$

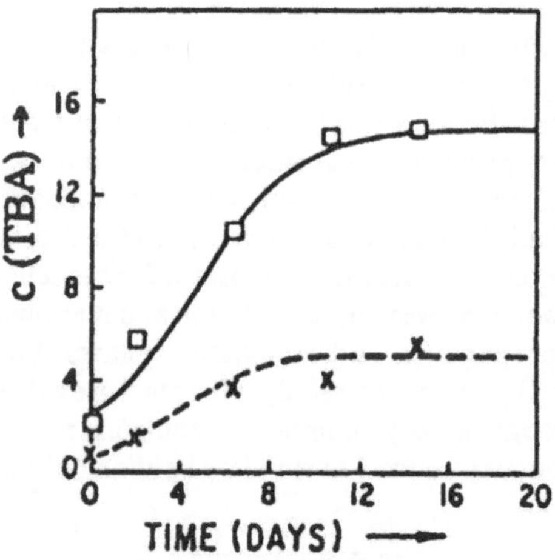

FIGURE E.1.3 Lipid oxidation in ground raw poultry meat at 2–4°C. Model: ---- and ———, data: □ pH 4.70, × pH 6.25 (Özilgen and Özilgen, 1990; ©IFT, reproduced by permission).

where ΔP_c and ΔP_m are the pressure drops through the cake and the medium, respectively and are expressed theoretically as:

$$\Delta P_c = \frac{\alpha \mu c V}{A^2} \frac{dV}{dt} \tag{E.1.4.2}$$

and

$$\Delta P_m = \frac{R_m \mu}{A} \frac{dV}{dt} \tag{E.1.4.3}$$

where α = specific resistance of the cake, μ = viscosity of the filtrate, c = mass of the solids per unit volume of the feed, A = filter area, V = volume of the filtrate, t = time and R_m = filter medium resistance. Equations (E.1.4.2) and (E.1.4.3) may be substituted in (E.1.4.1) to obtain Sperry's Equation:

$$\frac{dV}{dt} = \frac{A\Delta P}{\mu} \frac{1}{R_m + \alpha c V/A} \tag{E.1.4.4}$$

Equation (E.1.4.4) has been based on theoretical considerations, therefore it may be referred to as a phenomenological model. Foods are usually highly complex systems. Theoretical expressions are frequently found inefficient to simulate the processes involving foods and biological materials. DeLagarza and Boulton (1984) modified (E.1.4.4) to model wine filtrations:

$$\frac{dV}{dt} = \frac{A\Delta P}{\mu} \frac{1}{R_m + \alpha c (V/A)^n} \tag{E.1.4.5}$$

$$\frac{dV}{dt} = \frac{A\Delta P}{\mu} \frac{1}{R_m + \exp(\kappa V/A)} \tag{E.1.4.6}$$

Equation (E.1.4.6) may be integrated as:

$$V = \frac{A}{\kappa}\left(\ln\left(\frac{\kappa \Delta P}{\mu R_m} t + 1\right)\right) \tag{E.1.4.7}$$

Comparison of (E.1.4.7) with the data is shown in Figure E.1.4.1.

In an ultrafiltration process a fraction of the solute is permitted to pass through the membrane and the remaining solutes are rejected. Use of batch ultrafiltration is limited to laboratory scale separations only, due to accumulation of the rejected particles on the membrane. Although (E.1.4.4)–(E.1.4.6) are suggested for cake filtrations only, they may be used to simulate batch ultrafiltration based on the analogy between the processes: Resistance to flow by the membrane, and the rejected solids in an ultrafiltration process are analogous to the medium and the cake resistance in cake filtration. Equation (E.1.4.5) with $n = 2$ was used by Bayindirli *et al.* (1988) to model sequential batch ultrafiltration of red beet extract (Fig. E.1.4.2).

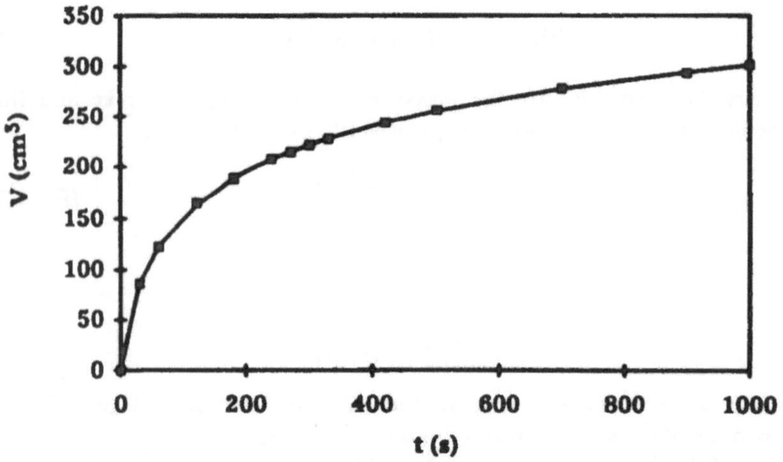

FIGURE E.1.4.1 Comparison of (E.1.4.7) (——) with the experimental data (■). Experimental data were obtained during cake filtration of apple juice ($\mu = 1.65\ 10^{-3}$ Pa s) in a filter with $A = 30.2$ cm^2, $\Delta P = 6.5\ 10^4$ Pa, pre coating = 0.25 g/cm^2, filter aid dose = 0.005 g/cm^3 juice. Model constants were $\kappa = 0.447$ cm^{-1} and $R_m = 2.05\ 10^8$ cm^{-1} (Bayindirli *et al.*, 1989).

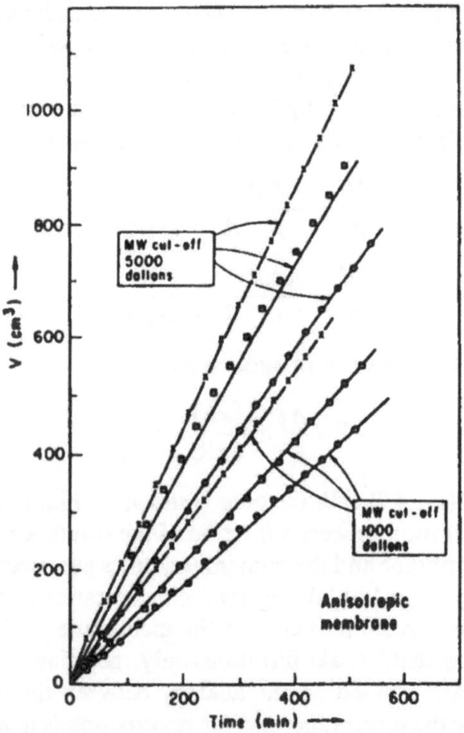

FIGURE E.1.4.2 Comparison of (E.1.4.5) ($n = 2$) (——) with the experimental data (●: $\Delta P = 3 \times 10^5$ Pa, ■: $\Delta P = 4 \times 10^5$ Pa, x: $\Delta P = 5 \times 10^5$ Pa) (Bayindirli *et al.*, 1988; ©IFT, reproduced by permission).

REFERENCES

Bayindirli, A., Yildiz, F. and Özilgen, M. (1988) Modeling of sequential batch ultrafiltration of red beet extract. *Journal of Food Science*, **53**, 1418–1422.

Bayindirli, L., Özilgen, M. and Ungan, S. (1989) Modeling of apple juice filtrations. *Journal of Food Science*, **54**, 1003–1006.

Cleland, A. C. (1990) Food Refrigeration Process Analysis, Design and Simulation. Elsevier Applied Science, London.

Glasscock, D. A. and Hale, J. C. (1994) Process simulation: The art and science of modeling. *Chemical Engineering*, **101**(11), 82–89.

De LaGarza, F. and Boulton, R. The modeling of wine filtrations. *American Journal of Enology and Viticulture*, **35**, 189–195

Luyben, W. L., (1990) *Process Modeling, Simulation and Control for Chemical Engineers*, McGraw-Hill, Singapore, pp. 15–17.

Meyer, W. J. (1985) Concepts of Mathematical Modeling. McGraw-Hill, Singapore, pp. 139–222.

Ögel, B. Z. and Özilgen, M. (1995) Regulation and kinetic modeling of galactose oxidase secretion. *Enzyme and Microbial Technology*, **17**, 870–876.

Özilgen, S. and Özilgen, M. (1990) Kinetic model of lipid oxidation in Foods. *Journal of Food Science*, **55**, 498–536.

Rand, W. M. (1983) Development and analysis of empirical mathematical kinetic models pertinent to food processing and storage, in *Computer-Aided Techniques in Food Technology*, Saguy, I. (ed.) Marcel Dekker, New York, pp. 49–70.

Riggs, J. B. (1988) A systematic approach to modeling. *Chemical Engineering Education*, **22**(1), 26–29.

Shatzman, A. R. and Kosman, D. J. (1977) Regulation of galactose oxidase synthesis and secretion, in *Dactylium dendroides*: Effects of pH and culture density. *Journal of Bacteriology*, **130**, 455–463.

Teixeira, A. A. and Shoemaker, C. F. (1989) *Computerized Food Processing Operations*, Avi, New York, pp. 135–168.

Yöndem, F., Özilgen, M. and Bozoglu, T. F. (1992) Kinetic aspects of leavening with mixed cultures of *Lactobacillus plantarum* and *Saccharomyces cerevisiae*. *Lebensmittel-Wissenschaft und Technologie*, **25**, 162–167.

CHAPTER 2

Transport Phenomena Models

2.1. THE GENERAL PROPERTY BALANCE EQUATION

Phenomenological models are mostly based on conservation principles. Conservation of mass, energy and momentum are mathematically analogous; discussion of either one equally applies to the others. A microscopic property (ψ) balance around a differential volume element gives:

$$\frac{\partial \psi}{\partial t} + \nabla \cdot (\psi v) = \psi_G + (\nabla \cdot \delta \nabla \psi) \tag{2.1}$$

with constant δ (2.1) becomes

$$\frac{\partial \psi}{\partial t} + \nabla \cdot (\psi v) = \psi_G + \delta \nabla^2 \psi \tag{2.2}$$

where δ is diffusivity (Tab. 2.1) ∇ and ∇^2 are del and Laplacian operators.

$\partial\psi/\partial t$	= accumulation rate of the property at a fixed point in the volume element
$\nabla \cdot (\psi v)$	= net input rate of the property to the point with convection
ψ_G	= generation rate of the property at the point
$\delta\nabla^2\psi$ or $\nabla \cdot \delta\nabla\psi$	= net input rate of the property to the point with molecular diffusion

Detailed discussion of derivation of the transport equations may be found in the classical Transport Phenomena Books (Bird *et al.*, 1960; Brodkey and Hershey, 1988, Whitaker, 1977).

Table 2.1 List of ψ and δ for use in general property balance (Brodkey and Hershey, 1988)

transport phenomena	flux	flux units	diffusivity ($m^{-2}s^{-1}$)	concentration of property ψ	units of ψ
heat	q	$J m^{-2} s^{-1}$	α	$\rho c_p T$	$J m^{-3}$
mass	J	$kg m^{-2} s^{-1}$	D	c	$kg m^{-3}$
momentum	σ	$N m^{-2}$	ν	ρv	$kg m^{-2} s^{-1}$

Parameters α and ν are called the thermal diffusivity and kinematic viscosity (or viscous diffusivity), respectively and defined as

$$\nu = \frac{\mu}{\rho} \tag{2.3}$$

and

$$\alpha = \frac{k}{\rho c_p} \tag{2.4}$$

where c_p, k, ρ and μ are specific heat, thermal conductivity, density and viscosity, respectively. Parameter D is the diffusion coefficient for mass transfer.

Equation (2.2) is the starting point of almost all the transport phenomena problems. We choose appropriate ψ from Table 2.1, and expand ∇ or ∇^2 in appropriate coordinate systems, then solve the differential equation. Equations for the flux expressions are given in Table 2.2:

Equations (2.5)–(2.7) imply that the fluxes J_z, q_z and σ_{zy} are proportional with the gradients dc/dz, dT/dz and dv_y/dz, respectively. Parameters D, k and μ are actually the proportionality constants. It should be noticed that mass and heat transfer occurs in the same direction with the gradient. The negative signs imply that the mass, heat and momentum transfers occur from higher to lower concentration, temperature or momentum regions, respectively. Liquid molecules are in contact with each other. When a layer of these molecules are set in motion in the y direction, momentum is transferred to the other liquid layers in the z direction due to the molecular interactions (σ_{zy} = viscous flux of y momentum in the z direction).

Choosing an appropriate coordinate system (Fig. 2.1) with an appropriate origin may facilitate the modeling process substantially. A cylindrical coordinate system with z axis located on the center line of the cylinder is preferred when the equation of continuity is used to describe drying behavior of a cylindrical rise grain (Example 2.12); a spherical coordinate system with the origin located to the center of the tuber is preferred when the equation of energy is used to evaluate temperature profiles along a spherical potato (Example 2.9).

Table 2.2 Empirical laws for one directional fluxes

mass transfer	Fick's law	$J_z = -D\frac{dc}{dz}$	(2.5)
heat transfer	Fourier's law	$q_z = -k\frac{dT}{dz}$	(2.6)
momentum transfer	Newton's law	$\sigma_{zy} = -\mu\frac{dv_y}{dz}$	(2.7)

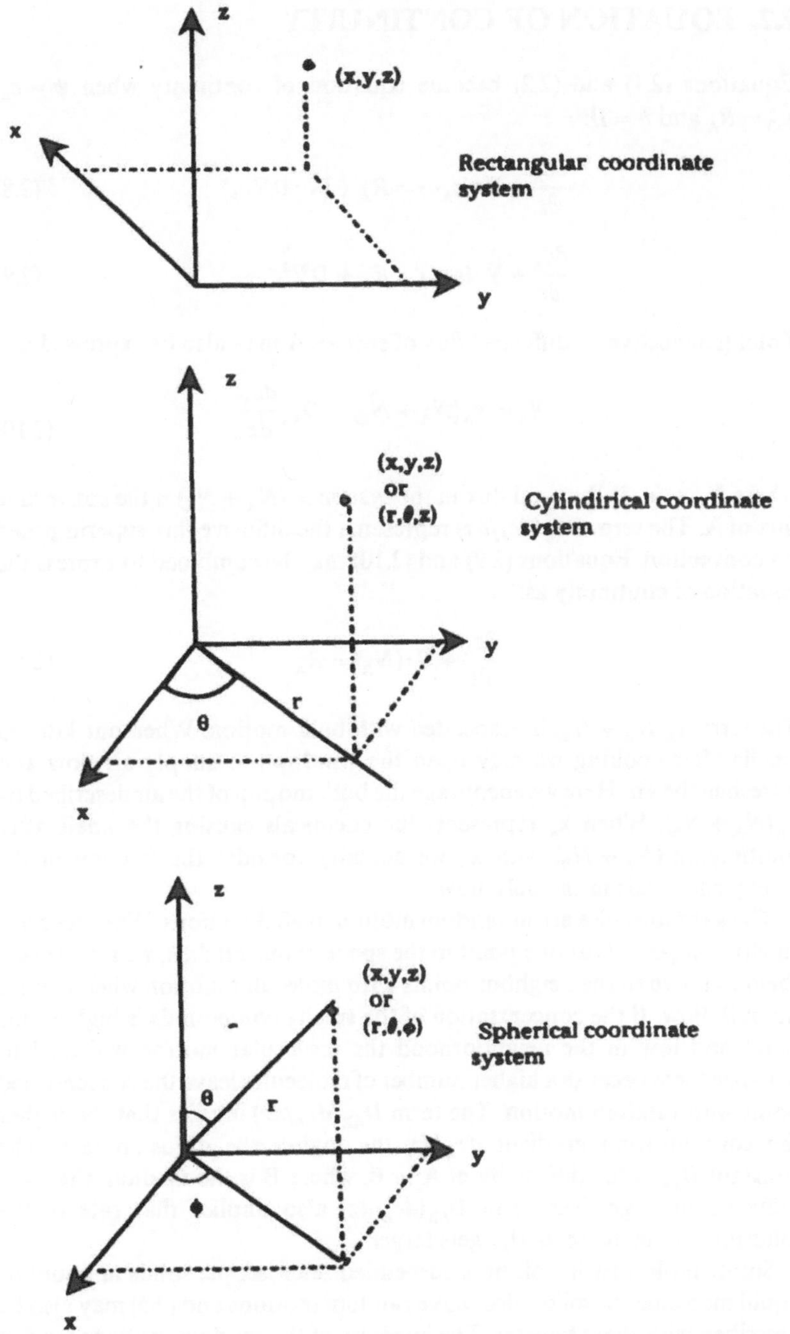

FIGURE 2.1 Coordinate systems employed with transport phenomena models.

2.2. EQUATION OF CONTINUITY

Equations (2.1) and (2.2) become equation of continuity when $\psi = c_A$, $\psi_G = R_A$ and $\delta = D$:

$$\frac{\partial c_A}{\partial t} + \nabla \cdot (c_A v) = R_A + (\nabla \cdot D \nabla c_A) \tag{2.8}$$

$$\frac{\partial c_A}{\partial t} + \nabla \cdot (c_A v) = R_A + D \nabla^2 c_A \tag{2.9}$$

Total (convective + diffusive) flux of species A may also be expressed as:

$$N_A = x_A(N_A + N_B) - D_{AB}\frac{dc_A}{dz} \tag{2.10}$$

where $N_A + N_B$ is the total flux in the system $x_A(N_A + N_B)$ is the convective flux of A. The term $D_{AB}(dc_A/dz)$ represents the diffusive flux superimposed on convection. Equations (2.9) and (2.10) may be combined to express the equation of continuity as:

$$\frac{\partial c_A}{\partial t} + \nabla \cdot (N_A) = R_A \tag{2.11}$$

The term $x_A(N_A + N_B)$ is associated with bulk motion. When our kitchen smells after cooking we may open the windows to supply air flow and refreshen the air. Here we encourage the bulk motion of the air described by $x_A(N_A + N_B)$. When x_A represents the chemicals causing the smell, after multiplying $(N_A + N_B)$ with x_A we actually consider the fraction of the smelly chemicals in the bulk flow.

The gas molecules are in random motion in all directions. When we have smelly compounds at one point in the space in our kitchen, we may expect them to move to the neighbor points with molecular motion when there is no bulk flow. If the concentration of the smelly compounds is high at one point and low in the neighborhood the molecular motion will tend to equalize them because a higher number of molecules leave the concentrated point with random motion. The term $D_{AB}(dc_A/dz)$ implies that the higher the concentration gradient dc_A/dz, the higher the diffusion rate. The constant D_{AB} is the diffusivity of A in B, where B is the medium that A is diffusing through. The term $D_{AB}(dc_A/dz)$ also implies that rate of the diffusion will increase as D_{AB} gets larger.

Solute molecules in solution, suspended microscopic solids in liquid or liquid molecules in solids also make random motions and (2.5) may also be described with these systems. The intensity of the random motions tend to decrease with liquid and solid systems, as the molecular matrix of the diffusion medium gets more rigid.

Table 2.3 Equation of continuity for different coordinate systems

i) Rectangular coordinates:

$$\frac{\partial c_A}{\partial t}+\left(\frac{\partial N_{Ax}}{\partial x}+\frac{\partial N_{Ay}}{\partial y}+\frac{\partial N_{Az}}{\partial z}\right)=R_A \tag{2.12}$$

ii) Cylindrical coordinates:

$$\frac{\partial c_A}{\partial t}+\left(\frac{1}{r}\frac{\partial rN_{Ar}}{\partial r}+\frac{1}{r}\frac{\partial N_{A\theta}}{\partial \theta}+\frac{\partial N_{Az}}{\partial z}\right)=R_A \tag{2.13}$$

iii) Spherical coordinates:

$$\frac{\partial c_A}{\partial t}+\left(\frac{1}{r^2}\frac{\partial r^2 N_{Ar}}{\partial r}+\frac{1}{r\,\mathrm{Sin}\theta}\frac{\partial N_{A\theta}\mathrm{Sin}\theta}{\partial \theta}+\frac{1}{r\,\mathrm{Sin}\theta}\frac{\partial N_{A\phi}}{\partial \phi}\right)=R_A \tag{2.14}$$

Example 2.1. Diffusion of water through a wine barrel during aging During aging of red wine in oak barrels (radius $=R$, length $=L$, thickness of the wood $=\lambda$) small amounts of water diffuses through the wood and evaporates. Assuming that all of the barrel's surface is available for mass transfer, obtain an expression for the evaporation rate. Fraction of water at the inner surface of the wood is x_{in} and at the outer surface is $x_{out}=\beta\mathscr{H}$, where β is a constant and $\mathscr{H}$ is the humidity of air in the cellar.

Solution It is assumed that the barrel is a perfect cylinder and water diffuses out from all the surfaces. Since the thickness of the wood is much smaller than that of the barrel we may use the rectangular coordinates to simulate water transport through the wood with equation of continuity

$$c\frac{\partial x}{\partial t}+\frac{\partial N_w}{\partial z}=R_w.$$

Since $c(\partial x/\partial t)=0$ at steady state and $R_w=0$ (water does not react in the wood) the equation of continuity becomes $(dN_w/dz)=0$, implying that N_w is a constant along z direction. Flux of water in wood is $N_w=x(N_w+N_{wood})+D_w(dc/dz)$, which becomes $N_w=-cD_w(dx/dz)$ since only small amounts of water are available in the wood ($x\ll 1$). After combining the final forms of the flux expression and the equation of continuity we obtain $(d^2x/dz^2)=0$. Upon integration, the moisture profile along the wood becomes $x=\mathscr{K}_1 z+\mathscr{K}_2$. Constants $\mathscr{K}_1$ and $\mathscr{K}_2$ may be evaluated from the boundary conditions as: $\mathscr{K}_1=(\beta\mathscr{H}-x_{in})/\lambda$ and $\mathscr{K}_2=x_{in}$. We may substitute $dx/dz=(\beta\mathscr{H}-x_{in})/\lambda$ in $N_w=-c\,D_w(dx/dz)$ and obtain $N_w=cD_w(x_{in}-\beta\mathscr{H})/\lambda$. Total water loss through the barrel is $(A_{barrel})(N_w)=2\pi R(L+R)cD_w(x_{in}-\beta\mathscr{H})/\lambda$ where A_{barrel} is the total mass transfer area through the barrel. We should recognize that as the humidity of the cellar increases, moisture loss through the wood decreases.

2.3. EQUATION OF ENERGY

After substituting $\psi = \rho c_p T$ (2.1) becomes

$$\frac{\partial(\rho c_p T)}{\partial t} + \nabla \cdot (\rho c_p T v) = \dot{\psi}_G + (\nabla . \delta \nabla \rho c_p T) \tag{2.15}$$

The generation term $\dot{\psi}_G$ accounts for heat generation by viscous dissipation in a microwave field, radiation, etc. Equation (2.15) for incompressible fluids may be expanded for various coordinate systems as given in Table 2.4.

Table 2.4 Equation of energy for different coordinate systems

i) Rectangular coordinates:

$$\frac{\partial T}{\partial t} + v_x \frac{\partial T}{\partial x} + v_y \frac{\partial T}{\partial y} + v_z \frac{\partial T}{\partial z} = \frac{\dot{\psi}_G}{\rho c_p} + \frac{\partial}{\partial x}\left(\alpha \frac{\partial T}{\partial x}\right) + \frac{\partial}{\partial y}\left(\alpha \frac{\partial T}{\partial y}\right) + \frac{\partial}{\partial z}\left(\alpha \frac{\partial T}{\partial z}\right) \tag{2.16}$$

ii) Cylindrical coordinates:

$$\frac{\partial T}{\partial t} + v_r \frac{\partial T}{\partial r} + \frac{v_\theta}{r}\frac{\partial T}{\partial \theta} + v_z \frac{\partial T}{\partial z} = \frac{\dot{\psi}_G}{\rho c_p} + \frac{1}{r}\frac{\partial}{\partial r}\left(r\alpha \frac{\partial T}{\partial r}\right) + \frac{1}{r^2}\frac{\partial}{\partial \theta}\left(\alpha \frac{\partial T}{\partial \theta}\right) + \frac{\partial}{\partial z}\left(\alpha \frac{\partial T}{\partial z}\right) \tag{2.17}$$

iii) Spherical coordinates:

$$\frac{\partial T}{\partial t} + v_r \frac{\partial T}{\partial r} + \frac{v_\theta}{r}\frac{\partial T}{\partial \theta} + \frac{v_\phi}{r \sin\theta}\frac{\partial T}{\partial \phi} = \frac{\dot{\psi}_G}{\rho c_p} + \frac{1}{r^2}\frac{\partial}{\partial r}\left(r^2 \alpha \frac{\partial T}{\partial r}\right) + \frac{1}{r^2 \sin\theta}\frac{\partial}{\partial \theta}\left(\alpha \sin\theta \frac{\partial T}{\partial \theta}\right) + \frac{1}{r^2 \sin^2\theta}\frac{\partial}{\partial \phi}\left(\alpha \frac{\partial T}{\partial \phi}\right) \tag{2.18}$$

2.4. EQUATION OF MOTION

The equation of motion is more complicated than the equations of energy and continuity, because in momentum balance ψ is a vector composed of components ψ_x, ψ_y, and ψ_z. When ρ is constant, after substituting $\psi_x = \rho v_x$ and $\delta = \nu$ in (2.1) and (2.2) the x component of the momentum balance becomes

$$\frac{\partial v_x}{\partial t} + (\nabla \cdot v) v_x = \frac{\dot{\psi}_{Gx}}{\rho} + (\nabla \cdot \nu \nabla v_x) \tag{2.19}$$

and

$$\frac{\partial v_x}{\partial t}+(\nabla\cdot v)\,v_x=\frac{\psi_{Gx}}{\rho}+\nu\nabla^2 v_x \tag{2.20}$$

It should be noticed that $\nabla\cdot v=0$ when ρ is constant. Equations for the y and z components of the momentum balance may be written similarly as (2.19) and (2.20). When the rate of momentum generation is proportional to the pressure gradient and x component of the gravitational acceleration, i.e., $\psi_{Gx}=-(\partial p/\partial x)+\rho g_x$, (2.19) becomes the Navier-Stokes equation and its expansion for different coordinate systems is given in Table 2.5.

Table 2.5 Navier-Stokes equation for different coordinate systems

i) x component in rectangular coordinates:

$$\frac{\partial v_x}{\partial t}+v_x\frac{\partial v_x}{\partial x}+v_y\frac{\partial v_x}{\partial y}+v_z\frac{\partial v_x}{\partial z}=-\frac{1}{\rho}\frac{\partial p}{\partial x}+g_x+\nu\left(\frac{\partial^2 v_x}{\partial x^2}+\frac{\partial^2 v_x}{\partial y^2}+\frac{\partial^2 v_x}{\partial z^2}\right) \tag{2.21}$$

ii) r component in cylindrical coordinates:

$$\frac{\partial v_r}{\partial t}+v_r\frac{\partial v_r}{\partial r}+\frac{v_\theta}{r}+\frac{\partial v_r}{\partial \theta}+v_z\frac{\partial v_r}{\partial z}-\frac{v_\theta^2}{r}=-\frac{1}{\rho}\frac{\partial p}{\partial r}g_r+\nu\frac{\partial^2 v_r}{\partial r^2}+\frac{\nu}{r}\frac{\partial v_r}{\partial r}-\frac{\nu v_r}{r^2}+\frac{\nu}{r^2}\frac{\partial^2 v_r}{\partial \theta^2}-\frac{2\nu}{r^2}\frac{\partial^2 v_\theta}{\partial \theta}+\nu\frac{\partial^2 v_r}{\partial z^2} \tag{2.22}$$

iii) r component in spherical coordinates:

$$\frac{\partial v_r}{\partial t}+v_r\frac{\partial v_r}{\partial r}+\frac{v_\theta}{r}\frac{\partial v_r}{\partial \theta}+\frac{v_\phi}{r\sin\theta}\frac{\partial v_r}{\partial \phi}-\frac{v_\theta^2}{r}-\frac{v_\phi^2}{r}=-\frac{1}{\rho}\frac{\partial p}{\partial r}+g+\frac{\nu}{r^2}\frac{\partial}{\partial r}\left(\frac{r^2\partial v_r}{\partial r}\right)+\frac{\nu}{r^2\sin\theta}\frac{\partial}{\partial \theta}\left(\sin\theta\frac{\partial v_r}{\partial \theta}\right)+\frac{\nu}{r^2\sin^2\theta}\frac{\partial^2 v_r}{\partial \phi^2}-\frac{2\nu v_r}{r^2} \tag{2.23}$$

2.5. THEORIES FOR LIQUID TRANSPORT COEFFICIENTS

Although viscosity, thermal conductivity and mass diffusion coefficients are defined by empirical relations in (2.5)–(2.7), there are theoretical explanations to these equations for simple systems. Foods are either liquid, or solid or their combination. Theoretical models for viscosity, thermal conductivity and mass diffusion coefficients are available for pure liquids or suspensions of fine particles. Theoretical models are rarely used to evaluate values of μ, k or D, empirical models or the actual experimental data are preferred in process calculations.

i) Eyring's theory of liquid viscosity: In a pure liquid at rest the individual molecules are considered to be confined in a large *cage* formed by its nearest neighbors (Fig. 2.2). This cage represents an energy barrier of $\Delta G^{\#}/N_{Av}$ ($\Delta G^{\#}$ = energy barrier, N_{Av} = Avagadro number). The theory (Glasstone *et al.*, 1942) suggests that the liquid at rest undergoes rearrangements in which one molecule at a time is transferred to an adjoining *hole*. The rate (frequency) of jumps in each direction is:

$$k = \left(\frac{\kappa T}{h}\right) \exp\left(-\frac{\Delta G_{o}^{\#}}{RT}\right) \tag{2.24}$$

where h is the Planck constant, κ is the Boltzman constant, T is temperature $\Delta G_{o}^{\#}$ is the energy barrier (molar free energy of activation) and R is gas constant.

When liquid moves in the x direction with velocity gradient dv_x/dy the energy barrier is distorted and the frequency of molecular rearrangements increases. The distorted energy barrier is

$$-\Delta G^{\#} = -\Delta G_{o}^{\#} \pm \frac{a}{\delta}\frac{\sigma_{yx} V_m}{2} \tag{2.25}$$

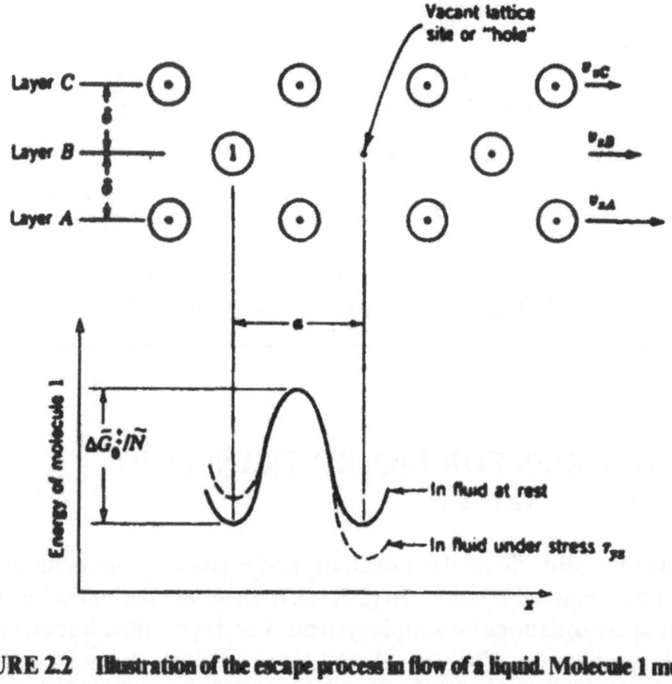

FIGURE 2.2 Illustration of the escape process in flow of a liquid. Molecule 1 must pass through a *bottle neck* to reach the vacant site (Reprinted by permission of John Wiley & Sons Inc. from *Transport Phenomena*, Bird *et al.*, © 1960 John Wiley & Sons Inc.).

where ΔG^{*} is distorted mean free energy of activation, a is length of molecular jump, δ is the distance between the molecular layers, V_m is the volume of one mole of liquid, σ_{xy} is the applied shear stress and $(a/\delta)(\partial_{yx} V_m/2)$ is the approximation of the work done on the molecules as they move to the top of the energy barriers with the applied shear stress. The minus sign is employed to describe the movement of the molecules against the shear stress. When $(\sigma_{xy} a V_m/2\delta RT) \ll 1$ and $\delta/a = 1$, the product of the net frequency of jumps and a/δ is the shear rate which is:

$$-\frac{dv_x}{dy} = \frac{\kappa T}{h}\exp(-\Delta G_e^{*}/RT)\frac{\sigma_{xy} V_m}{2RT} \tag{2.26}$$

where h = Planck constant. Equation (2.26) is Newton's law of viscosity with (Bird *et al.*, 1960);

$$\mu = \frac{N_{Av} h}{V_m} e^{\Delta G_e^{*}/RT} \tag{2.27}$$

Temperature effects on liquid viscosity may be expressed with the Arrhenius expression:

$$\mu = \mu_0 e(E_a/RT) \tag{2.28}$$

where μ_0 is pre-exponential constant, E_a is the activation energy and R is the gas constant.

Example 2.2. Kinetic compensation relations for the viscosity of fruit juices (Özilgen and Bayindirli, 1992) Gibbs free energy of activation at rest may be expressed as:

$$\Delta G_e^{*} = \Delta H^{*} - T\Delta S^{*}$$

where ΔH^{*} = activation enthalpy and ΔS^{*} = activation entropy. After substituting this expression (2.27) becomes

$$\mu = \frac{N_{Av} h}{V_m}\exp(\Delta H^{*}/RT)\exp(-\Delta S^{*}/R)$$

Comparing this equation with (2.28) requires

$$E_a = \Delta H^{*} + RT$$

and

$$\mu_0 = \frac{N_{Av} h}{2.72\, V_m}\exp(-\Delta S^{*}/R)$$

In experiments performed under slightly different experimental conditions, generally linear relationships are observed between the activation energy E_a and the frequency factor μ_0; and the activation enthalpy ΔH^{*} and the activation entropy ΔS^{*}:

$$\ln \mu_0 = \alpha E_a + \beta$$

and

$$\Delta S^{\#} = \delta \Delta H^{\#} + \phi$$

These equations are called the compensation relations. Since μ_0 and $\Delta S^{\#}$; E_a and $\Delta H^{\#}$ are interrelated, validity of either equation implies that the other one is also valid. Although E_a, μ_0, $\Delta H^{\#}$ and $\Delta S^{\#}$ varies with the experimental conditions, α, β, δ and ϕ are constants through out the range of the experiments. A sample set of compensation relations are given in Figure E.2.2:

The compensation relations may be used for interpolation in process design when data is available in the close range, but not under the exactly required conditions (Özilgen and Özilgen, 1996).

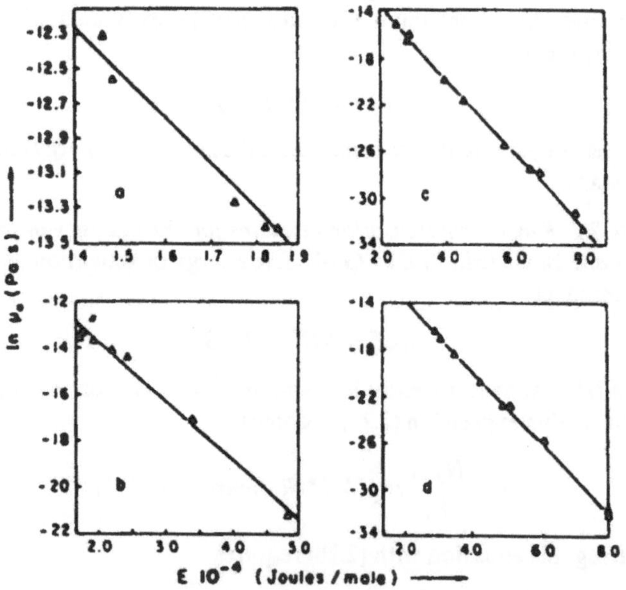

FIGURE E.2.2 Variation of parameter $\ln \mu_0$ with activation energy: a) Sour cherry juice; Equation of the line: $\ln \mu_0 = -8.64 - 2.61 \times 10^{-4} E_a$ ($r = -0.98$) b) Clarified apple juice; Equation of the line: $\ln \mu_0 = -8.80 - 2.51 \times 10^{-4} E_a$ ($r = -0.99$) c) Apple juice; Equation of the line: $\ln \mu_0 = -7.09 - 3.19 \times 10^{-4} E_a$ ($r = -0.96$) d) Grape juice; Equation of the line: $\ln \mu_0 = -7.31 - 3.10 \times 10^{-4} E_a$ ($r = -0.99$) Reprinted from Journal of Food Engineering 17(2), Özilgen and Bayindirli, Frequency factor-activation energy compensation relations for viscosity of the fruit juices, pages 143–151, © 1992 with kind permissions from Elsevier Science Ltd, The Boulevard, Langford Lane, Kidlington 0X5 1GB, UK.

ii) Thermal Conductivity of Liquids: Thermal conductivity of monatomic low density gases is $k = 1/3\ \rho c_v u \lambda$ (c_v = specific heat, u = mean molecular velocity, λ = collision length). This equation may be modified for liquids after assuming that pure liquid molecules are arranged in a cubic lattice, energy is transferred from one lattice plane to the next, the heat capacity of monatomic liquids at constant volume is about the same as for a solid at high temperature and the mean molecular speed is the same as the sonic velocity v_s and the distance that energy travels per single collision is the same as the lattice spacing $(\tilde{V}/N_{Av})^{1/2}$ as (Bird *et al.*, 1960):

$$k = 3\left(\frac{N_{Av}}{\tilde{V}}\right)^{2/3} \kappa v_s \qquad (2.29)$$

where $\tilde{V}$ = specific volume.

iii) Hydrodynamic theory of diffusion in liquids: Diffusion of a single particle or solute molecule through a stationary medium is calculated by balancing the forces around the particle:

$$D_{AB} = \kappa T \frac{v}{F} \qquad (2.30)$$

where v is the velocity of the particle, F is the drag force on the particle and v/F is the mobility of the particle. When $N_{Re} \ll 1$ and there is no tendency for fluid to slip at the surface of the diffusing particle:

$$F_A = 6\pi\mu v d_p \qquad (2.31)$$

where d_p is the effective particle diameter. Substituting (2.31) into (2.30) gives

$$D_{AB} = \frac{\kappa T}{6\pi\mu d_p} \qquad (2.32)$$

Equation (2.32) is called Stokes – Einstein equation.

When there is no tendency for the fluid to stick to the surface of the diffusing particle then (2.31) and (2.32) become

$$F_A = 4\pi\mu v d_p \qquad (2.33)$$

and

$$D_{AB} = \frac{\kappa T}{4\pi\mu v d_p} \qquad (2.34)$$

iv) Eyring's theory of liquid diffusion (Glasstone *et al.*, 1944): As explained for viscosity, Eyring's theory assumes that the liquid molecules are entrapped into a *cage* made by neighbor molecules. The molecule is expected to exceed an activation barrier to jump into the neighboring *holes* with a first order reaction. A constant average distance is involved into each jump. Derivation of the expression for diffusivity i.e. diffusion coefficient, is very similar to that of the viscosity. The resulting equation is:

$$D_{AB} = \frac{\kappa T}{\mu}\left(\frac{N_{Av}}{V}\right)^{1/3} \tag{2.35}$$

Temperature effects on diffusivity may be described with the Arrhenius expression:

$$D = D_0 \exp\left\{-\frac{E_a}{R\,T}\right\} \tag{2.36}$$

where D_0 = pre-exponential constant.

Example 2.3. Temperature effects on diffusivity of water in starch Parameters E_a and D_0 were given for diffusion in hydrated amylo pectin (fraction of starch) and potatoes as (Özilgen, 1993):

diffusion system	D_0 (m^2/s)	E_a (J/mol)
water in hydrated amylo pectin	3.37×10^{-6}	2.55×10^{4}
glucose in potato tissue	1.78×10^{-6}	2.03×10^{4}
potassium in potato tissue	4.08×10^{-6}	2.09×10^{4}

a) Compare the diffusivities of water in amylo pectin, glucose in potato tissue and potassium in potato tissue at 65°C.

Solution $D = D_0 \exp\left\{-\frac{E_a}{R\,T}\right\}$ and $R = 8.314$ J/mol K after substituting the numbers we will have:

diffusion system	D (m^2/s)
water in hydrated amylo pectin	2.04×10^{-9}
glucose in potato tissue	1.34×10^{-9}
potassium in potato tissue	2.40×10^{-9}

Water molecules are smaller than the potassium molecules and both of them are smaller than the glucose molecules; potato tissue is cellular, amylopectin has a simpler structure. This results indicate that the diffusivity of smaller molecules is greater in structurally simpler systems.

b) Compare the diffusivity of potassium in potato tissue at 65°C and 90°C.

Solution After substituting the numbers in (2.36) we will have:

diffusion system	$D(m^2/s)$
potassium in potato tissue at 65°C	2.04×10^{-9}
potassium in potato tissue at 90°C	4.00×10^{-9}

The results show 1.67 folds increase in D with 25°C increase in T.

Example 2.4. Compensation relations for diffusivity of water in starch The compensation relations, similarly as those of viscosity, applies to diffusivity:

$$\ln D_0 = \alpha E_a + \beta$$

These equations may also be used for interpolation in process design (Özilgen and Özilgen, 1996). Figure E.2.4 describes variation of $\ln D_0$ with the activation energy in different media.

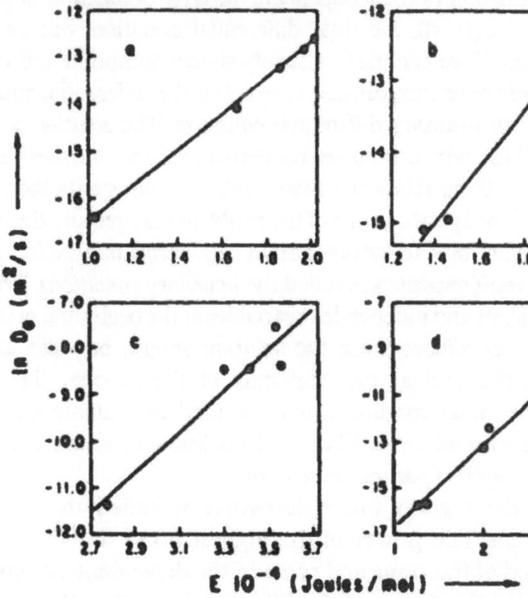

FIGURE E.2.4 Variation of parameter $\ln D_0$ with activation energy. a) Diffusion of water in hydrated amioca, equation of the line: $\ln D_0 = -20.25 + 3.75 \times 10^{-4} E_a$ ($r = 0.99$). b) Diffusion of water in hydrated hygon, equation of the line: $\ln D_0 = -20.51 + 3.92 \times 10^{-4} E_a$ ($r = 0.99$). c) Diffusion of water in 75% hygon and 25% sucrose, equation of the line: $\ln D_0 = -23.31 + 4.37 \times 10^{-4} E_a$ ($r = 0.96$). d) Diffusion of glucose, potassium, magnesium and ascoric acid in potato tissue, equation of the line: $\ln D_0 = -20.20 + 3.54 \times 10^{-4} E_a$ ($r = 0.99$). Reprinted from *Journal of Food Engineering*, 17(2), Özilgen and Bayindirli, Frequency factor-activation energy compensation relations for viscosity of the fruit juices, pages 143–151, © 1992 with kind permissions from Elsevier Science Ltd, The Boulevard, Langford Lane, Kidlington 0X5 1GB, UK.

2.6. ANALYTICAL SOLUTIONS TO ORDINARY DIFFERENTIAL EQUATIONS

In most transport phenomena problems we choose an equation from Tables 2.3–2.5, to describe the system, then simplify it by eliminating the negligible terms. The next step is solving this differential equation. ***Differential equations*** involve derivatives or differentials of one or more dependent variables with respect to one or more independent variables. If the dependent variable is a function of one independent variable, the differential equation involves only ordinary derivatives and called an ***ordinary differential equation***. If the dependent variable is a function of two or more independent variables, its derivatives with respect to each independent variable are called partial derivatives, and the equation is called a ***partial differential equation***. When we simplify the equation of continuity (Tab. 2.3), if concentration of species A remains as the single dependent variable as a function of the independent variable of location in x direction and time, (i.e., $c_A = c_A(t, x)$), the final differential equation will be a partial differential equation. If we can make a steady state assumption and neglect the time dependence of concentration, (i.e., $c_A = c_A(x)$), the differential equation will be converted into an ordinary differential equation. The ***solution*** is a relation between the variables, which involves no derivatives and satisfies the original differential equation. If this relation contains unknown constants, then it is called a general solution. The specifications of the problem may require the dependent and independent variables to satisfy certain requirements at some geometric boundaries, these requirements are called the ***boundary conditions***. The relation between the dependent and independent variables at the beginning of the process is called the ***initial condition***. Since the solution should be valid also at the geometric boundaries and at the beginning of the process, the boundary conditions and the initial conditions may be used to evaluate the unknown constants of the general solution. When all the unknown constants are obtained we will have the ***complete (particular) solution***.

The order of the highest order derivative is called the ***order of the differential equation***. The power of the highest order derivative after the equation is cleared of fractions and roots in the dependent variable and its derivatives, is called the ***degree of the differential equation***. If every dependent variable and every derivative are of first degree and there are no products of the dependent variable with the derivatives, the differential equation is called a ***linear differential equation***. If all the terms in the equation contains the dependent variable and its derivatives the differential equation is ***homogeneous***.

Example 2.5. Characterization of a differential equation Characterize the differential equation

$$\frac{d^2y}{dx^2} = f(x)\left[1 + \left(\frac{dy}{dx}\right)^2\right]^{3/2}$$

Solution The highest order derivative is d^2y/dx^2, therefore this is a second order differential equation. We may clear this differential equation from the fractions in the dependent variable and its derivatives as $(d^2y/dx^2)=f(x)(1+(dy/dx)^2)^3$ power of the highest order derivative is 2, therefore this is a second degree differential equation. The equation is non-linear because it contains second order term $(d^2y/dx^2)^2$, and another non-linear term $(1+(dy/dx)^2)^3$. This is a non-homogeneous equation, because an $f(x)$ term will appear after expanding the right hand side of the equation and it does not have a dependent variable, or its derivatives. This is an ordinary differential equation, because dependent variable y is a function of a single independent variable x.

Techniques for solving the ordinary differential equations are discussed in the classical differential equations books, i.e., Ross (1989). The solutions are usually time consuming and the techniques for solving aie easily forgotten if they are not used for a while. We are very lucky in the sense that only limited types of differential equations arise after simplifying the general transport equations in most of the food engineering analysis, and we may find the general solutions to those differential equations in Table 2.6.

Example 2.6. Heat transfer in a continuous plug flow sterilization reactor In a continuous plug flow (Chapter 3.6) sterilization process milk enters into a heating tube at a constant temperature T_0 and the temperature changes along the flow direction only.

a) Obtain the governing ordinary differential equation of heat transfer

Solution Equation of energy in cylindrical coordinates is:

$$\frac{\partial T}{\partial t}+v_r\frac{\partial T}{\partial r}+\frac{v_\theta}{r}\frac{\partial T}{\partial \theta}+v_z\frac{\partial T}{\partial z}=\psi_G+\frac{1}{r}\frac{\partial}{\partial r}\left(r\alpha\frac{\partial T}{\partial r}\right)+\frac{1}{r^2}\frac{\partial}{\partial \theta}\left(\frac{\alpha\,\partial T}{\partial \theta}\right)$$
$$+\frac{\partial}{\partial z}\left(\alpha\frac{\partial T}{\partial z}\right) \tag{2.17}$$

The following simplifications may be made:

$\frac{\partial T}{\partial t}=0$ (steady state)

$v_r=v_\theta=0$ (no flow in r and θ directions)

$\frac{\partial T}{\partial \theta}=0$ (no temperature profile in θ direction)

$\psi_G=0$ (no heat generation)

Table 2.6 General solutions to common ordinary differential equations

Equation	Solution
i) First order, first degree	
separable $g(y)\,dx + f(x)\,dy = 0$	$\int_{x_1}^{x_2} \frac{dx}{f(x)} = -\int_{y_1}^{y_2} \frac{dy}{g(y)}$
exact $M(x, y)\,dx + N(x, y)\,dy = 0$ with $\frac{\partial M}{\partial y} = \frac{\partial N}{\partial x}$	$\left(\int^{x} M dx\right)_y + \left(\int^{y} R dy\right)_x = C$ with $R = N - \frac{\partial}{\partial y}\int^{x} M dx$ or $-\left(\int^{y} N dy\right)_x + \left(\int^{x} S dx\right)_y = C$ with $S = M - \frac{\partial}{\partial x}\int^{x} N dy$
homogeneous $M(x, y)\,dx + N(x, y)\,dy = 0$ with $M(\lambda x, \lambda y) = \lambda^n M(x, y)$ and $N(\lambda x, \lambda y) = \lambda^n N(x, y)$ where λ is a constant, then M and N are homogeneous of degree n	Substitute $y = vx$, then equation becomes separable
linear $\frac{dy}{dx} + P(x)y = Q(x)$	$y = \exp\left\{-\int^{x} P dx\right\}\int^{x} Q \exp\left\{\int^{x} P dx\right\} dx$ $+ C \exp\left\{-\int^{x} P dx\right\}$
ii) First order, special ordinary differential equations	
Bernoulli's equation $\frac{dy}{dx} + P(x)y = Q(x)y^n$ $n \neq 1$	Substitute $v = y^{n-1}$, then equation becomes linear $\frac{dv}{dx} + P_1(x)v = Q_1(x)$ with $P_1(x) = (1 - n)\,P(x)$ and $Q_1(x) = (1 - n)\,Q(x)$
Riccatic equation $\frac{dy}{dx} + P(x)y + Q(x)y^2 = R(x)$	Substitute $y = \frac{du}{dx}\frac{1}{Q(x)u}$ equation becomes linear $\frac{d^2u}{dx^2} + P(x)\frac{du}{dx} - R(x)Q(x)u = 0$

iii) Simultaneous linear equations

$$(\alpha_{11}D+\beta_{11})x_1+(\alpha_{12}D+\beta_{12})x_2+(\alpha_{13}D+\beta_{13})x_3=\gamma_1(t)$$
$$(\alpha_{21}D+\beta_{21})x_1+(\alpha_{22}D+\beta_{22})x_2+(\alpha_{23}D+\beta_{23})x_3=\gamma_2(t)$$
$$(\alpha_{31}D+\beta_{31})x_1+(\alpha_{32}D+\beta_{32})x_2+(\alpha_{33}D+\beta_{33})x_3=\gamma_3(t)$$

where $D=\frac{d}{dt}$, α and β's are constants

equations are transformed into single linear ordinary differential equations as:

$$\Delta x_1=\mathcal{D}x_1$$
$$\Delta x_3=\mathcal{D}x_2$$
and $$\Delta x_3=\mathcal{D}x_3$$

where

$$\Delta=\begin{vmatrix}\alpha_{11}D+\beta_{11} & \alpha_{12}D+\beta_{12} & \alpha_{13}D+,\\ \alpha_{21}D+\beta_{21} & \alpha_{22}D+\beta_{22} & \alpha_{23}D+,\\ \alpha_{31}D+\beta_{31} & \alpha_{32}D+\beta_{32} & \alpha_{33}D+,\end{vmatrix}$$

$$\mathcal{D}x_1=\begin{vmatrix}\gamma_1(t) & \alpha_{12}D+\beta_{12} & \alpha_{13}D+\beta_{13}\\ \gamma_2(t) & \alpha_{22}D+\beta_{22} & \alpha_{23}D+\beta_{23}\\ \gamma_3(t) & \alpha_{32}D+\beta_{32} & \alpha_{33}D+\beta_{33}\end{vmatrix}$$

$$\mathcal{D}x_2=\begin{vmatrix}\alpha_{11}D+\beta_{11} & \gamma_1(t) & \alpha_{13}D+\beta_{13}\\ \alpha_{21}D+\beta_{21} & \gamma_2(t) & \alpha_{23}D+\beta_{23}\\ \alpha_{31}D+\beta_{31} & \gamma_3(t) & \alpha_{33}D+\beta_{33}\end{vmatrix}$$

$$\mathcal{D}x_3=\begin{vmatrix}\alpha_{11}D+\beta_{11} & \alpha_{12}D+\beta_{12} & \gamma_1(t)\\ \alpha_{21}D+\beta_{21} & \alpha_{22}D+\beta_{22} & \gamma_2(t)\\ \alpha_{31}D+\beta_{31} & \alpha_{32}D+\beta_{32} & \gamma_3(t)\end{vmatrix}$$

iv) First order, higher degree differential equations

solvable for y

$$y=f\left(x,\frac{dy}{dx}\right) \qquad \frac{dy}{dx}P=\frac{\partial f}{\partial x}+\frac{\partial f}{\partial P}\frac{\partial P}{\partial x} \qquad y=\int^x P\,dx+C$$

solvable for x

$$y=f\left(y,\frac{dy}{dx}\right) \qquad \frac{dy}{dx}=\frac{1}{P}=\frac{\partial f}{\partial y}+\frac{\partial f}{\partial P}\frac{\partial P}{\partial y} \qquad x=\int^y \frac{1}{P}dy+C$$

v) Second order, linear differential equations

homogeneous with constant coefficients

$$\frac{d^2y}{dx^2}+\alpha_1\frac{dy}{dx}+\alpha_2 y=0$$

$$\lambda_{1,2}=\frac{-\alpha_1\pm\sqrt{\alpha_1^2-4\alpha_2}}{2}$$

i) $\lambda_2\neq\lambda_2$ real numbers $y=C_1\exp(\lambda_1 x)+C_2\exp(\lambda_2$

ii) $\lambda_1=\lambda_2$ real numbers $y=(C_1+C_2x)\exp(\lambda_2 x)$

iii) $\lambda_{1,2}=a\pm ib$ imaginary numbers
$y=e^{ax}(C_1\sin bx+C_2\cos bx)$

iv) with higher order equations assume a solution $y=Ce^{\lambda x}$, substitute in the differential equation and obtain the characteristic equation to solve λ

Table 2.6 (continued)

	Special cases a) $\alpha_1 = 0,\ \alpha_2 = \lambda^2,\ y = C_1 \sin \lambda x + C_2 \cos \lambda x$ b) $\alpha_1 = 0,\ \alpha_2 = -\lambda^2,\ y = C_1 \sinh \lambda x + C_2 \cos \lambda x$
non-homogeneous with constant coefficients $\frac{d^2y}{dx^2} + \alpha_1 \frac{dy}{dx} + \alpha_2 y = r(x)$	$y = y_h + y_p$ y_h = homogeneous solution is found as shown above y_p = particular solution, may be found with the method of undetermined coefficients: Assume particular solution $y_p(x)$ according to $r(x)$, then substitute $y_p(x)$ in the equation and determine the unknown coefficients (α) after equating the coefficients of the equal power terms on both sides. $r(x) y_p(x)$ (assumed solution) κx^n — $\alpha_1 x^n + \alpha_2 x^{n-1} + \cdots + \alpha_{n-1} x + \alpha_n$ κe^{px} — αe^{px} $\kappa \cos qx$ or $\kappa \sin qx$ — $\alpha_1 \sin qx + \alpha_1 \cos qx$
Cauchy-Euler equation $\alpha_0 x^2 \frac{d^2y}{dx^2} + \alpha_1 x \frac{dy}{dx} + \alpha_2 y = h(x)$	after substituting $x = e^t$ may be converted into a linear differential equation $\frac{d^2y}{dt^2} + \beta_1 \frac{dy}{dt} + \beta_2 y = H(t)$, where $\beta_1 = \frac{\alpha_1 - \alpha_0}{\alpha_0}$, $\beta_2 = \frac{\alpha_2}{\alpha_0}$, $H(t) = \frac{h(e^t)}{\alpha_0}$
contains no dy/dx and no x $\frac{d^2y}{dx^2} = f(y)$	substitute $\frac{dy}{dx} = P$, $\frac{d^2y}{dx^2} = \frac{dP}{dx}$, $\frac{dP}{dx} = P\frac{dP}{dy} = f_1(y)$ original differential equation becomes $\int^P P\,dP - \int^y f_1(y)\,dy = C_1$, after integration obtain $P = f_2(y)$ integrate once more to obtain the solution $\int^y \frac{dy}{f_2(y)} - y = C_2$
contains no x and no y $\frac{d^2y}{dx^2} = f\left(\frac{dy}{dx}\right)$	substitute $\frac{dy}{dx} = P$, $\frac{d^2y}{dx^2} = \frac{dP}{dx} = f_1(P)$ original differential equation becomes $\int^P \frac{dP}{f_1(P)} - x = C_1$, after integration obtain $P = f_2(x)$, since $\frac{dy}{dx} = P$ integrate once more to obtain the solution

vi) Second order, non-linear differential equations

$$y=\int^{y} f_2(x)\,dx + C_2$$

contains no y

$f\left(\dfrac{d^2y}{dx^2},\dfrac{dy}{dx},x\right)=0$

substitute $\dfrac{dy}{dx}=P,\ \dfrac{d^2y}{dx^2}=\dfrac{dP}{dx},\ f_1\left(\dfrac{dP}{dx},P,x\right)=0$ rearrange to obtain $\dfrac{dP}{dx}=f_2(P,x)$ solve for P to obtain, $P=f_3(x)$. Since $\dfrac{dy}{dx}=P$ integrate once more to obtain the final solution

$$y=\int^{x} f_3(x)\,dx + C$$

contains no x

$f\left(\dfrac{d^2y}{dx^2},\dfrac{dy}{dx},y\right)=0$

substitute $\dfrac{dy}{dx}=P,\ \dfrac{d^2y}{dx^2}=P\dfrac{dP}{dy},\ f_1\left(\dfrac{dP}{dx},P,y\right)=0$ solve for P to obtain $P=f_2(y)$. Since $P=\dfrac{dy}{dx}$ the final solution is $\displaystyle\int^{y}\frac{dy}{f_2(y)}-x=C$

$v_z\dfrac{\partial T}{\partial z}\gg\dfrac{\partial}{\partial z}\left(\alpha\dfrac{\partial T}{\partial z}\right)$ (heat transfer with conduction is much smaller than heat transfer with convection in z direction)
Equation (2.17) becomes:

$$v_z\frac{\partial T}{\partial z}=\frac{1}{r}\frac{\partial}{\partial r}\left(r\alpha\frac{\partial T}{\partial r}\right)$$

or

$$v_z\frac{\partial T}{\partial z}=\frac{\alpha}{r}\frac{\partial T}{\partial r}+\alpha\frac{\partial^2 T}{\partial r^2}$$

We may average each term along the r direction as:

$$\frac{1}{\pi R^2}\int_0^R v_z\frac{\partial T}{\partial z}2\pi r\,dr=\frac{1}{\pi R^2}v_z\frac{\partial T}{\partial z}(\pi R^2-0)=v_z\frac{\partial T}{\partial z}$$

$$\frac{1}{\pi R^2}\int_0^R\frac{\alpha}{T}\frac{\partial T}{\partial z}2\pi r\,dr=\frac{2\alpha}{R^2}\int_0^R\frac{\partial T}{\partial r}dr=\frac{2\alpha}{R^2}\int_T^{T_w}dT=\frac{2\alpha}{R^2}(T_w-T)$$

$\frac{1}{\pi R^2}\int_0^R \alpha\frac{\partial^2 T}{\partial r^2}2\pi r dr = \frac{2\alpha}{R^2}\int_0^R r\frac{\partial^2 T}{\partial r^2}dr$. We may use integration by parts as

$$\frac{2\alpha}{R^2}\int_0^R r\frac{\partial^2 T}{\partial r^2}dr = \frac{2\alpha}{R^2}\left[r\frac{dT}{dr}\Big|_{r=0}^{r=R} - \int_0^R \frac{\partial T}{\partial r}dr\right] \text{ then } -k\frac{dT}{dr}\Big|_{r=R} = h(T-T_w)$$

to obtain $\frac{1}{\pi R^2}\int_0^R \alpha\frac{\partial^2 T}{\partial r^2}2\pi r dr = \frac{2\alpha h}{Rk}(T-T_w) - \frac{2\alpha}{R^2}(T_w - T)$. After substituting all the averaged terms and $K = (Rkv_z/2\alpha h)$ into the simplified energy equation we obtain $(dT/dz) + (1/K)T - (1/K)T_w = 0$

b) Solve the differential equation when T_w = constant. Show that the solution confirms the boundary conditions $T = T_0$ at $z = 0$ and $T = T_w$ when $z \to \infty$.

Solution The governing equation $(dT/dz) + (1/K)T - (1/K)T_w = 0$ is separable and its solution is given in Table 2.6:

Equation	Solution
$g(y)dx + f(x)dy = 0$	$\int_{x_1}^{x_2}\frac{dx}{f(x)} = -\int_{y_1}^{y_2}\frac{dy}{g(y)}$
with the following substitutions	
dx	dT
$f(x)$	$(T-T_w)/K$
dy	dz
$g(y)$	1

therefore $\int_{T_w-T_0}^{T_w-T}\frac{KdT}{(T-T_w)} = -\int_0^z dz$ Upon integration we obtain an expression for variation of T along the z direction:

$$T = T_w + (T_0 - T_w)\exp(-z/K)$$

When we substitute $z = 0$ and $z \to \infty$ the model predicts $T = T_0$ and, $T = T_w$ respectively.

c) Solve the differential equation when the wall temperature increases with distance $T_{wall} = T_{w0} + \alpha Z$. Where T_{w0} and α are constants.

Solution The governing equation $(\partial T/dz) + (1/K)T = (\alpha z/K) + (T_{w0}/K)$. The solution may be assumed to be $T = T^c + T^p$, where T^c is the complementary and T^p is the particular solution.

i) Complementary solution: $(dT^c/dz) + (1/K)T^c = 0$, this is a simple separable equation and its solution is: $T^c = Ce^{-z/K}$ where C is a constant.

ii) Particular solution may be obtained from the equation $(dT^p/dz) + (1/K)T^p = (\alpha z/K) + (T^p_{w0}/K)$. This equation has constant coefficients and may be solved with the method of undetermined coefficients. According to the right hand side of the equation T^p may be assumed with the help of Table 2.6 as: $T^p = A_1 z + A_2$, therefore $(dT^p/dz) = A_1$. After substituting equivalents of T^p and (dT^p/dz) into the equation we will have $A_1 + (1/K)(A_1 z + A_2) = (\alpha z/K) + (T^p_{w0}/K)$.

Coefficients of the same powers of z should be the same on both sides of the equation:

power of z	equal coefficients
z^0	$A_1 + \frac{A_2}{K} = \frac{T_{w0}}{K}$
z^1	$A_1 = \alpha$

The unknown constants are: $A_1 = \alpha$ and $A_2 = T_{w0} - \alpha K$ and the particular solution is: $T^p = \alpha(z - K) + T_{w0}$.

The model for variation of the milk temperature along the tube is: $T = T^p + T^c = \alpha(z - K) + T_{w0} + Ce^{-z/K}$. The boundary condition $T = T_0$ at $z = 0$ requires $C = T_0 - \alpha K - T_{w0}$, therefore the model is:

$$T = T_{w0} + \alpha(z - K) + (T_0 - \alpha K - T_{w0})\, e^{-z/K}$$

d) Solve the differential equation when $T_{wall} = T_{w0}e^{aZ} + \beta$.

Solution The governing equation $(dT/dz) + (1/K)T = (T_{w0}e^{aZ}/K) + (\beta/K)$. The solution may be assumed to be $T = T^c + T^p$ where T^c is the complementary and T^p is the particular solution.

i) Complementary solution: $(dT^c/dz) + (1/K)T^c = 0$, therefore the complementary solution is the same as in the previous case: $T^c = Ce^{-z/K}$.

ii) Particular solution may be obtained from the equation $(dT^p/dz) + (1/K)T^p = (T_{w0}e^{aZ}/K) + (\beta/K)$. This equation has constant coefficients and may be solved by following the same procedure as in the previous case. According to the right hand side of the equation T^p may be assumed with the help of Table 2.6 as: $T^p = A_1 e^{az} + A_2$, therefore $dT^p/dz = A_1 \alpha e^{aZ}$. After substituting equivalents of T^p and dT^p/dz into the equation we will have $A_1 \alpha e^{aZ} + (1/K)(A_1 e^{aZ} + A_2) = (T_{w0}e^{aZ}/K) + (\beta/K)$. This equation may be rearranged as: $(A_1\alpha + A_1/K)\, e^{aZ} + (A_2/K) = (T_{w0}e^{aZ}/K) + (\beta/K)$.

Coefficients of the same powers of z should be the same on both sides of the equation:

term	equal coefficients
z^0	$\frac{A_2}{K} = \frac{\beta}{K}$
e^{az}	$A_1\alpha + \frac{A_1}{K} = \frac{T_{w0}}{K}$

The unknown constants are: $A_1 = T_{wo}/(\alpha K + 1)$ and $A_2 = \beta$ and the particular solution is: $T^p = (T_{wo}/(\alpha K + 1))e^{\alpha z} + \beta$.
The model for variation of the milk temperature along the tube is:
$T = T^p + T^c = (T_{wo}/(K + 1))\, e^{\alpha z} + \beta + Ce^{-z/K}$. The boundary condition $T = T_0$ at $z = 0$ requires $C = T_0 - (T_{wo}/(\alpha K + 1)) - \beta$, therefore the model is:

$$T = \frac{T_{w0}}{\alpha K + 1} e^{\alpha z} + \beta + \left(T_0 - \frac{T_{w0}}{\alpha K + 1} - \beta\right) e^{-z/K}$$

e) Solve the differential equation when $T_{wall} = 3Z^2 + e^Z + 2Ze^Z + 3Z + 3e^{5Z}$

Solution The governing equation $(dT/dz) + (1/K)T = (1/K)(3Z^2 + e^Z + 2Ze^Z + 3Z + 3e^{5Z})$. The solution may be assumed to be $T = T^c + T^p$ where T^c is the complementary and T^p is the particular solution.

i) Complementary solution: $(dT^c/dz) + (1/K)T^c = 0$, therefore the complementary solution is the same as in the previous case: $T^c = Ce^{-z/K}$.

ii) Particular solution may be obtained from the equation $(dT^p/dz) + (1/K)\, T^p = (1/K)(3Z^2 + e^Z + 2Ze^Z + 3Z + 3e^{5Z})$. This equation has constant coefficients and may be solved by following the same procedure as in the previous case. According to the right hand side of the equation T^p may be assumed by the help of Table 2.6 as:

term	T^p
z^2	$A_1 z^2 + A_2 z + A_3$
e^z	$A_4 e^z$
ze^z	$A_5 ze^z$ (z^0 terms are the same as that of e^z)
z	solution is included in that of z^2
e^{5z}	$A_6 e^{5z}$
	$T^p = A_1 z^2 + A_2 z + A_3 + A_4 e^z + A_5 ze^z + A_6 e^{5z}$

therefore

$$\frac{dT^p}{dz} = 2A_1 z + A_2 + A_4 e^z + A_5 ze^z + A_5 e^z + 5A_6 e^{5z}.$$

After substituting equivalents of T^p and dT^p/dz into the equation we will have

$$\left(A_2 + \frac{A_3}{K}\right) + \left(2A_1 + \frac{A_2}{K}\right) z + \left(\frac{A_1}{K}\right) z^2 + \left(A_4 + A_5 + \frac{A_4}{K}\right) e^z$$
$$+ \left(5A_6 + \frac{A_6}{K}\right) e^{5z} + \left(\frac{A_5}{K}\right) ze^z = \frac{1}{K}(3Z^2 + e^Z + 2Ze^Z + 3Z + 3e^{5z})$$

Coefficients of the same powers of z should be the same on both sides of the equation:

term	equal coefficients
z^0	$A_2 + \frac{A_3}{K} = 0$
z^1	$2A_1 + \frac{A_2}{K} = \frac{3}{K}$
z^2	$\frac{A_1}{K} = \frac{3}{K}$
e^z	$A_4 + A_5 + \frac{A_4}{K} = \frac{1}{K}$
e^{5z}	$5A_6 + \frac{A_6}{K} = \frac{3}{K}$
ze^z	$\frac{A_5}{K} = 0$

The unknown constants are: $A_1 = 3$, $A_2 = 3(1-2K)$, $A_3 = 3K(2K-1)$, $A_4 = 1/(1+K)$, $A_5 = 0$ and $A_6 = 3/(5K+1)$

The model for variation of the milk temperature along the tube is:

$T = T^c + T^p = Ce^{-z/K} + 3Kz^2 + 3(1-2K)z + 3K(2K-1) + (1/(1+K))e^z + (3/(5K+1))e^{5z}$.

BC: $T = T_0$ at $z = 0$, requires $C = T_0 - (30K^4 + 21K^3 - 12K^2 + 5K + 4)/(5K^2 + 6K + 1)$, therefore the model is:

$$T = \left\{T_0 - \left(\frac{30K^4 + 21K^3 - 12K^2 + 5K + 4}{5K^2 + 6K + 1}\right)\right\}e^{-z/K}$$

$$+ 3Kz^2 + 3(1-2K)z + 3K(2K-1) + \frac{1}{1+K}e^z + \frac{3}{5K+1}e^{5z}$$

Example 2.7. Diffusion of the modified atmosphere gas mixtures through a polymer film (*Ilter et al.*, 1991) Gas mixtures of 3–15% oxygen, 5–20% carbon dioxide, and balance nitrogen are referred to as modified atmospheres and are used to replace normal air in storage processes to extend the shelf life of fresh produce. Diffusion of oxygen and carbon dioxide from a low density polyethylene package was studied by using a diffusion cell associated with the experimental set-up shown in Figure E.2.7.1.

The diffusion cell has two major compartments: upstream and downstream sides. The upstream side of the cell was maintained at a constant gas composition. The downstream side of the diffusion cell is a closed system, separated from the upstream side by means of a polymer film and filled initially with nitrogen. The gas composition changes in the downstream side of the cell due to the permeation of the gases through the film, and this data is used to evaluate the permeation characteristics of the film to the gas

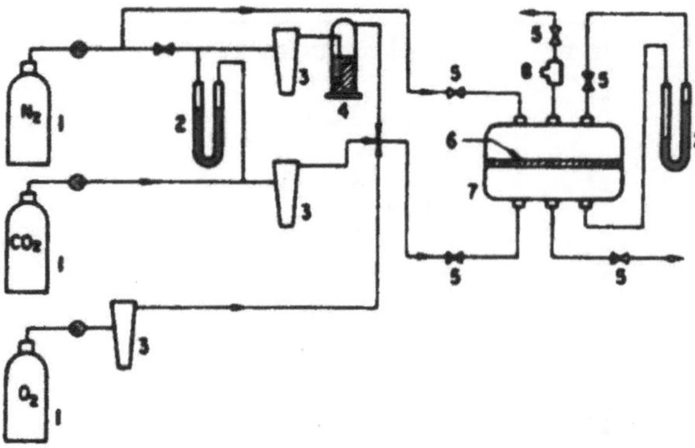

FIGURE E.2.7.1 Experimental set-up. 1. Gas tanks, 2. manometers, 3. rotameters, 4. water reservoir, 5. valves, 6. polymer film, 7. diffusion cell, 8. sampling port (Ilter *et al.*, 1991, © SCI, reproduced by permission).

mixtures. Carbon dioxide and oxygen balances within the downstream side of the diffusion cell are:

$$N_c S = V\frac{dc_c}{dt}$$

and

$$N_o S = V\frac{dc_o}{dt}$$

where N_c, N_o = carbon dioxide and oxygen fluxes through the film, S = mass transfer area, V = volume of the downstream side of the cell, c_c, c_o = carbon dioxide and oxygen concentrations in the down stream side of the cell, t = time. We may use the equation of continuity in a rectangular coordinate system to simulate carbon dioxide flux through the film:

$$\frac{\partial c_c}{\partial t} + \left(\frac{\partial N_c}{\partial x} + \frac{\partial N_c}{\partial y} + \frac{\partial N_c}{\partial z}\right) = R_c \tag{2.12}$$

This expression may be simplified under steady state conditions, i.e., $dc_c/dt = 0$, with one dimensional flux, i.e., $dN_c/dx = dN_c/dy = 0$, and when carbon dioxide does not involve any chemical reactions, i.e., $R_c = 0$ as:

$$\frac{dN_c}{dz} = 0$$

A similar expression may be obtained for the oxygen flux:

$$\frac{dN_o}{dz} = 0$$

Flux was previously expressed for non interfering species as:

$$N_A = x_A(N_A + N_B) - D_{AB}\frac{dc_A}{dz} \tag{2.10}$$

In the present system carbon dioxide and oxygen fluxes may interfere with each other, therefore equation (2.10) may be modified as:

$$N_c = x_c(N_c + N_o + N_n) - D_c\frac{d\bar{c}_c}{dz} - \beta_o N_o$$

where $\bar{c}_c$ is the carbon dioxide concentration in the film, N_n = nitrogen flux through the film, D_c = apparent diffusivity of carbon dioxide through the film and β_o = constant. Since pressure is constant within the diffusion cell we may state that the total flux of carbon dioxide and oxygen into the diffusion cell equals the nitrogen flux from the cell to upstream gas mixture, implying that

$$N_c + N_o + N_n = 0$$

then the flux expression for carbon dioxide will be

$$N_c = -D_c\frac{d\bar{c}_c}{dz} - \beta_o N_o$$

A similar flux expression may be obtained for oxygen as:

$$N_o = -D\frac{d\bar{c}_o}{dz} - \beta_c N_c$$

We may substitute the simplified flux expressions in the final forms of the equation of continuity and obtain:

$$\frac{d^2\bar{c}_c}{dz^2} = 0$$

and

$$\frac{d^2\bar{c}_o}{dz^2} = 0$$

with the BCs:

$$\bar{c}_c = H_c c_{cu} \quad \text{at } z = 0 \text{ when } t > 0$$

$$\bar{c}_c = H_c c_c \quad \text{at } z = L \text{ when } t > 0$$

and

$$\bar{c}_o = H_o c_{ou} \quad \text{at } z = 0 \text{ when } t > 0$$

$$\bar{c}_o = H_o c_o \quad \text{at } z = L \text{ when } t > 0$$

where H_c and H_o are Henry's law constants for carbon dioxide and oxygen, respectively, c_{cu} and c_{ou} are the upstream gas concentrations and L is the thickness of the film. We may integrate these equations twice to obtain the concentration profiles along the film:

$$\bar{c}_c = \frac{H_c}{L}(c_c - c_{cu})z + H_c c_{cu}$$

and

$$\bar{c}_o = \frac{H_o}{L}(c_o - c_{ou})z + H_c c_{ou}$$

Then the fluxes of carbon dioxide and oxygen become (assumption $\beta_c \beta_o \ll 1$):

$$N_c = -D_c \frac{d\bar{c}_c}{dz} - \beta_o N_o = -\frac{D_c H_c}{L}(c_c - c_{cu}) + \beta_o \frac{D_o H_o}{L}(c_o - c_{ou})$$

$$N_o = -D_o \frac{d\bar{c}_o}{dz} - \beta_c N_c = -\frac{D_o H_o}{L}(c_o - c_{ou}) + \beta_c \frac{D_c H_c}{L}(c_c - c_{cu})$$

Carbon dioxide and oxygen balances around the diffusion cells were and $N_c S = V(dc_c/dt)$ and $N_o S = V(dc_o/dt)$. After substituting the expressions for N_c and N_o we will have:

$$\frac{dc_c}{dt} = -\phi_1(c_c - c_{cu}) + \phi_2(c_o - c_{ou})$$

$$\frac{dc_o}{dt} = -\phi_3(c_o - c_{ou}) + \phi_4(c_c - c_{cu})$$

where

$$\phi_1 = \frac{D_c S H_c}{VL}, \quad \phi_2 = \beta_o \frac{D_o S H_o}{VL}, \quad \phi_3 = \frac{D_o S H_o}{VL} \quad \text{and} \quad \phi_4 = \beta_c \frac{D_c S H_c}{VL}$$

We will re arrange these simultaneous equations as:

$$(D + \phi_1)c_c - \phi_2 c_o = \phi_1 c_{cu} - \phi_2 c_{ou}$$

$$-\phi_4 c_c + (D + \phi_3)c_o = -\phi_4 c_{cu} + \phi_3 c_{ou}$$

We will use Table 2.6 to solve these equations simultaneously:

$$\Delta c_c = \mathscr{D} c_c$$

$$\Delta c_o = \mathscr{D} c_o$$

and

$$\Delta = \begin{vmatrix} D+\phi_1 & -\phi_2 \\ -\phi_4 & D+\phi_3 \end{vmatrix} \quad \mathscr{D}c_c = \begin{vmatrix} \phi_1 c_{cu} - \phi_2 c_{ou} & -\phi_2 \\ -\phi_4 c_{cu} + \phi_3 c_{ou} & D+\phi_3 \end{vmatrix}$$

and

$$\mathscr{D}c_o = \begin{vmatrix} D+\phi_1 & \phi_1 c_{cu} - \phi_2 c_{ou} \\ -\phi_4 & -\phi_4 c_{cu} + \phi_3 c_{ou} \end{vmatrix}$$

Then we will have

$$\frac{d^2 c_c}{dt^2} + (\phi_1 + \phi_3)\frac{dc_c}{dt} + (\phi_1\phi_3 - \phi_2\phi_4)c_c = (\phi_1\phi_3 - \phi_2\phi_4)c_{cu}$$

and

$$\frac{d^2 c_o}{dt^2} + (\phi_1 + \phi_3)\frac{dc_o}{dt} + (\phi_1\phi_3 - \phi_2\phi_4)c_o = (\phi_1\phi_3 - \phi_2\phi_4)c_{ou}$$

These are second order ordinary differential equations with constant coefficients and their solutions can be obtained by using Table 2.6 and the ICs:

$$c_c = 0 \quad \text{at } t = 0$$

$$c_o = (c_{ou})_{\text{initial}} \quad \text{at } t = 0$$

$$c_c = r_3\{\exp(r_1 t) - \exp(r_2 t)\} + c_{cu}\{1 - \exp(r_2 t)\}$$

$$c_o = r_4\{\exp(r_1 t) - \exp(r_2 t)\} + c_{ou}\{1 - \exp(r_2 t)\} + (c_{ou})_{\text{initial}}\exp(r_2 t)$$

Adjustable constants $r_1 - r_4$ were obtained from the data with regression, and variation of c_c and c_o with time under different experimental conditions are plotted in Figure E.2.7.2.

2.7. TRANSPORT PHENOMENA MODELS INVOLVING PARTIAL DIFFERENTIAL EQUATIONS

We will solve problems involving partial differential equations by using three different techniques: i) Separation of variables, ii) Combination of variables and iii) The Laplace transformations. All of these techniques help us to reduce the problem into ordinary differential equations and make it possible to obtain the solution through the use of simple techniques.

Example 2.8. Temperature profiles in a steak during frying and in meat analog during cooking A steak ($\alpha = 8\ 10^{-8}\ \text{m}^2/\text{s}$, dimensions = 10 cm × 10 cm × 1 cm, $T_0 = 4°\text{C}$) is dipped into vegetable oil ($T = 177°\text{C}$) for frying. We may assume that the surfaces attain the oil temperature immediately. If heat transfer within the steak may be simulated with conduction only, draw

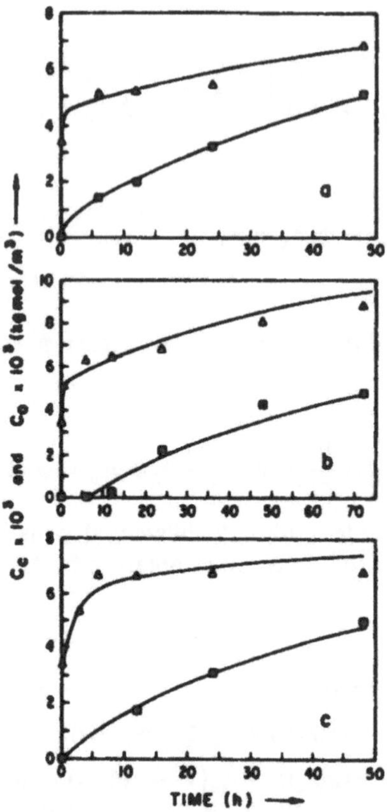

FIGURE E.2.7.2 Comparison of the model (—) with the experimental CO_2 (■) and O_2 (▲) concentrations in the downstream side of the diffusion cell. Upstream consisted of a) 13.0% CO_2, 14.3% O_2 and balance nitrogen, b) 18.8% CO_2, 12.0% O_2 and balance nitrogen, c) 14.1% CO_2, 13.0% O_2, balance nitrogen (on dry basis) and water vapor to obtain 95% relative humidity (Ilter *et al.*, 1991, © SCI, reproduced by permission).

the temperature profiles of the steak 5 and 10 minutes after beginning frying.

Solution Equation of energy in rectangular coordinate system is:

$$\frac{\partial T}{\partial t}+v_x\frac{\partial T}{\partial x}+v_y\frac{\partial T}{\partial y}+v_z\frac{\partial T}{\partial z}=\dot{\psi}_G+\frac{\partial}{\partial x}\left(\alpha\frac{\partial T}{\partial x}\right)+\frac{\partial}{\partial y}\left(\alpha\frac{\partial T}{\partial y}\right)+\frac{\partial}{\partial z}\left(\alpha\frac{\partial T}{\partial z}\right) \tag{2.16}$$

The process conditions imply $v_x=v_y=v_z=0$ and $\dot{\psi}_G=0$. The steak is a thin sheet therefore conduction through the longer dimensions may be neglected: $(\partial/\partial x)(\alpha(\partial T/\partial x))=0$, $(\partial/\partial y)(\alpha(\partial T/\partial y)=0$, then (2.16) becomes

$$\frac{\partial T}{\partial t}=\frac{\partial}{\partial z}\left(\alpha\frac{\partial T}{\partial z}\right) \tag{E.2.8.1}$$

Since α is constant it may be rearranged as

$$\frac{\partial^2 T}{\partial z^2}-\frac{1}{\alpha}\frac{\partial T}{\partial t}=0$$

This is a partial differential equation with

BC1: $T=T_1$ at $z=-L$ when $t>0$ ($2L$ = thickness of the steak)

BC2: $T=T_1$ at $z=L$ when $t>0$

BC3: $\dfrac{dT}{dz}$ at $z=0$ when $t>0$

BC4: $T=T_1$ for all z when $t\to\infty$

IC: $T=T_0$ when $t=0$ for all z

where T_1 = surface temperature = 177°C and L = half of the thickness of the steak = 0.5 cm. We may introduce the dimensionless variables

$$\theta=\frac{T-T_1}{T_0-T_1},\quad \eta=\frac{z}{L}\quad\text{and}\quad \tau=\frac{\alpha t}{L^2}$$

We may write the equation, BCs and the IC in dimensionless variables:

$$\frac{\partial^2\theta}{\partial\eta^2}-\frac{\partial\theta}{\partial\tau}=0$$

BC1: $\theta=0$ at $\eta=-1$

BC2: $\theta=0$ at $\eta=1$

BC3: $\dfrac{d\theta}{d\eta}=0$ at $\eta=0$

BC4: $\theta=0$ for all η when $\tau\to\infty$

IC: $\theta=1$ when $\tau=0$

Solution of this partial differential equation was also described by Bird *et al.* (1960, pages 354–356). We may use the method of *separation of variables* for solution since both the partial differential equation, boundary conditions and the initial condition are homogeneous; the boundaries are fixed and the boundary and the initial conditions are convenient to use with the orthogonality principle, i.e., they are equal to 0, 1 or -1.

The method of separation of variables requires separation of θ into two functions $f(\eta)$ and $g(\tau)$:

$$\theta = f(\eta)\, g(\tau)$$

where f is function of η, and g is function of τ.
The partial differential equation may be rewritten in terms of the new functions as:

$$g\frac{\partial^2 f}{\partial \eta^2} - f\frac{\partial g}{\partial \tau} = 0$$

We may rearrange this equation as:

$$\frac{1}{f}\frac{\partial^2 f}{\partial \eta^2} = \frac{1}{g}\frac{\partial g}{\partial \tau} = \lambda$$

Different variables appear on both sides of the equation. This equation may be satisfied if both sides equals a constant λ.

Solution to $\frac{1}{g}\frac{\partial g}{\partial \tau} = \lambda$ is $g = C_1 e^{\lambda\tau}$

We may determine the sign of λ by considering the boundary conditions:

Assumption	consequences of the BCs	conclusions
$\lambda > 0$	$g \to \infty$ when $\tau \to \infty$, therefore $T \to \infty$	impossible
$\lambda = 0$	$g = C_1$, therefore $\theta = f(\eta)$ only	impossible
$\lambda < 0$	$g = 0$ when $\tau \to \infty$, therefore $\theta = 0$	satisfies BC4

These results require $\lambda = -C^2$ and consequently $g = C_1 e^{-C^2\tau}$. Square of any real number C is always positive, but the preceding negative sign makes it negative.

Solution to $\frac{1}{f}\frac{\partial^2 f}{\partial \eta^2} = \lambda$ is $f(\eta) = C_2 \sin(C_\eta) + C_3 \cos(C_\eta)$ (Table 2.6).

Original solution was

$$\theta = f(\eta)\, g(t) = C_1 e^{-C^2\tau}(C_2 \sin(C_\eta) + C_3 \cos(C_\eta))$$

Temperature profiles within the steak should be symmetrical. The cos function is symmetrical ($\cos(C_\eta) = \cos(-C_\eta)$), but the sin function is not ($\sin(C_\eta) \neq \sin(-C_\eta)$), therefore we may not have a sin function in the solution, and $C_2 = 0$, then the solution becomes:

$$\theta = C_n \exp(-C^2\tau) \cos(C_\eta)$$

where $C_n = C_1 C_3$.

Constants C_n and C are still not determined. We may use BC2 to determine C:
$\theta = 0$ at $\eta = 1$ implying $0 = C_n \exp(-C^2\tau)\cos(C)$. This equation holds if $C = (n + (1/2))\pi$ and the solution becomes

$$\theta_n = \sum_{i=0}^{\infty} C_n \exp\left\{-\left(n+\frac{1}{2}\right)^2 \pi^2 \tau\right\} \cos\left[\left(n+\frac{1}{2}\right)\pi\eta\right]$$

Constant C_n still needs to be determined. We may use the IC to determine C_n:

$$\text{IC: when } \tau = 0\ \theta = 1 \text{ and } 1 = \sum_{i=0}^{\infty} C_n \cos\left[\left(n+\frac{1}{2}\right)\pi\eta\right]$$

cos function is an orthogonal function and the orthogonality principle may be applied as:

$$\int_a^b \cos\left[\left(n+\frac{1}{2}\right)\pi\eta\right]\cos\left[\left(m+\frac{1}{2}\right)\pi\eta\right] d\eta = 0$$

if $m \neq n$ within the range of $-1 \leqslant \eta \leqslant 1$

After multiplying both sides of the final expression for θ_n with $\cos((m + (1/2))\pi\eta)$ and applying the orthogonality principle we may obtain $C_n = (2(-1)^n/(n + (1/2)\pi))$ and the solution becomes

$$\frac{T - T_1}{T_0 - T_1} = \sum_{n=0}^{\infty} \frac{2(-1)^n}{\left(n+\frac{1}{2}\right)\pi} \exp\left\{-\left(n+\frac{1}{2}\right)^2 \pi^2 \alpha t/L^2\right\} \cos\left[\left(n+\frac{1}{2}\right)\pi z/L\right]$$

Temperatures along the steak 5 and 10 minutes after beginning frying are given in Table E.2.8 and plotted in Figure E.2.8.1.

Table E.2.8. Variation of temperature with distance from the center of the steak after 5 and 10 minutes from the beginning of the frying process

$z\,(m)$	0	0.001	0.002	0.003	0.004	0.005
T(°C) when $t = 5$ min	156.4	157.5	160.3	164.9	170.6	177.0
T(°C) when $t = 10$ min	175.1	175.2	175.5	175.9	176.4	177.0

The partial differential equation used in this example constitutes the basis to scale-up a meat analog cooking process (Altomare, 1988). The analogs may be defined as engineered foods, fabricated to offer some perceptible or tangible benefits over the natural version. Typical benefits include opportunities for reduced cost, improved nutrition and efficiencies in the food cycle, controlled composition, shelf stability, and convenience. A schematic drawing of a bacon analog production process is depicted in

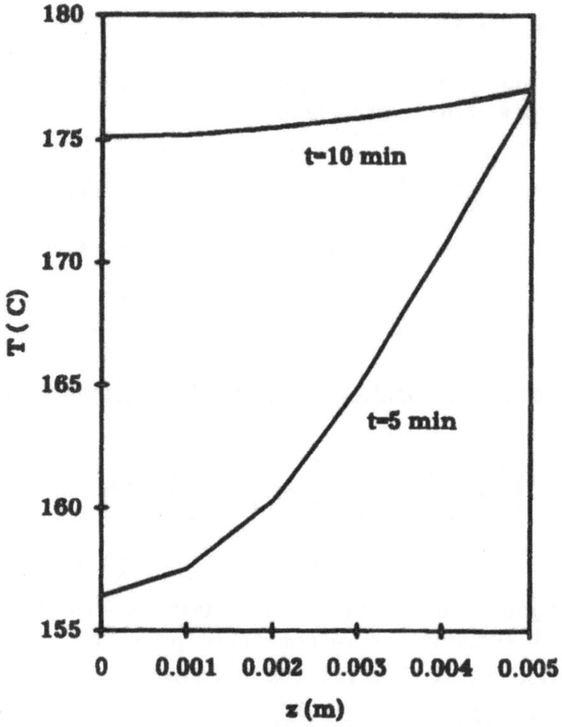

FIGURE E.2.8.1 **Temperature profiles along the steak after 5 and 10 minutes after the beginning of the frying process.**

Figure E.2.8.2. After alternately layering red, pink and white proteinaceous slurries through specially designed nozzles on to a stainless steel belt, the slurry layers are transported through a steam heated chamber, where humid heat coagulates the slurries into a solid mass; then the product is sliced, pre-fried and packaged.

In the early stages of the product development process, the primary objective is to use creative culinary and scientific talents to prepare an attractive, marketable new product. Once the decision is made to commercialize the product efforts shift toward understanding each individual operation to insure successful scale-up.

It was reported that the cooked slab may be sliced without smearing only when the center temperature reaches about 90.5°C. Since the slab is cooked in the autoclave in a saturated steam atmosphere the surface temperature, i.e., the boundary condition BC1, is the saturated steam temperature. Solutions to (E.2.8.1) are depicted in Figure E.2.8.3, which clearly indicates that the scale-up criterion for the bacon analog slab is maintaining the same dimensionless time - dimensionless temperature correlation to achieve the

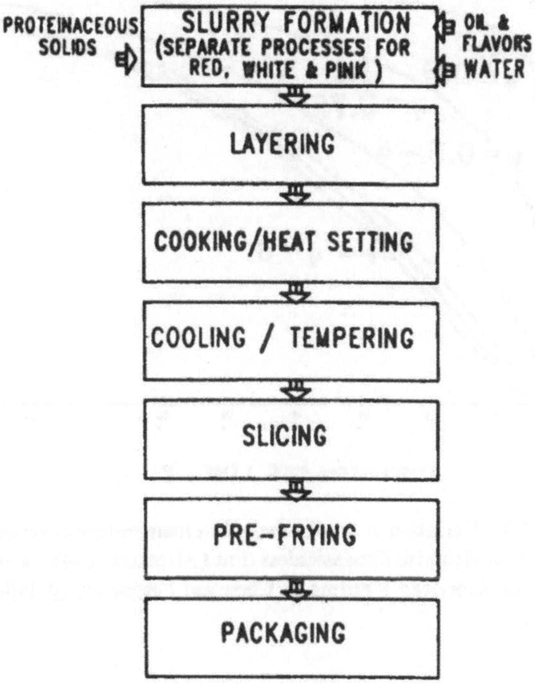

FIGURE E.2.8.2 Bacon analog process (Altomare, 1988). Reproduced with permission of the American Institute of Chemical Engineers. © 1988 AIChE. All rights reserved.

cooking criterion of "about 90.5°C of center temperature". The actual cooking time varies with the loaf thickness as depicted in Figure E.2.8.4.

Example 2.9. Temperature profiles in a spherical potato tuber during blanching (Sarikaya and Özilgen, 1991) A spherical potato tuber ($\alpha = 0.19\ 10^{-6}\,\mathrm{m^2/s}$, radius $R = 4.1$ cm, $T_0 = 18$°C) is dipped into blanching water where its surface temperature rises to T_1°C. If heat transfer within the potato tuber may be simulated with conduction only, plot variation of the temperature with time at $r = 3.1$ cm and $r = 0$ cm from the center.

Solution Equation of energy in spherical coordinate system is:

$$\frac{\partial T}{\partial t} + v_r \frac{\partial T}{\partial r} + \frac{v_\theta}{r}\frac{\partial T}{\partial \theta} + \frac{v_\phi}{r\sin\theta}\frac{\partial T}{\partial \phi} = \psi_G + \frac{1}{r^2}\frac{\partial}{\partial r}\left(r^2 \alpha \frac{\partial T}{\partial r}\right)$$
$$+ \frac{1}{r^2 \sin\theta}\frac{\partial}{\partial \theta}\left(\alpha \sin\theta \frac{\partial T}{\partial \theta}\right) + \frac{1}{r^2 \sin^2\theta}\frac{\partial}{\partial \phi}\left(\alpha \frac{\partial T}{\partial \phi}\right) \tag{2.18}$$

The process conditions imply $v_\theta = v_\phi = v_r = 0$ and $\psi_G = 0$. The potato tuber is symmetrical, therefore conduction through the radial directions

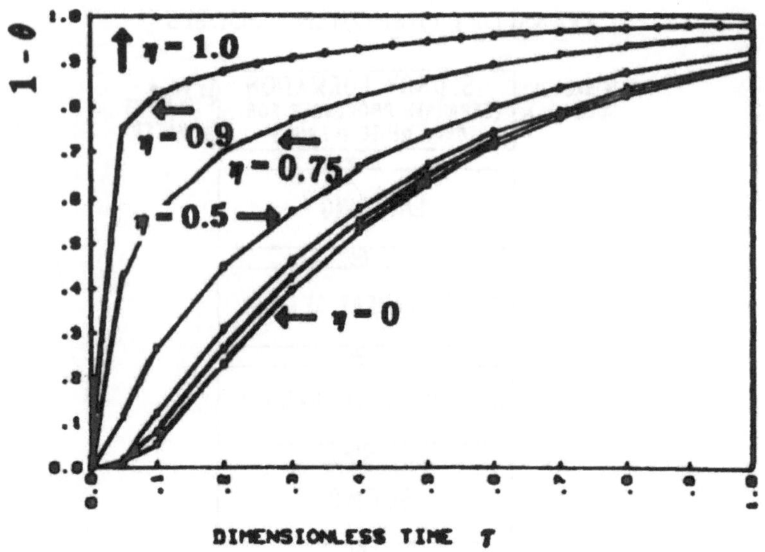

FIGURE E.2.8.3 Variation of the dimensionless temperature at various dimensionless locations of the slab with dimensionless time (Altomare, 1988). Reproduced with permission of the American Institute of Chemical Engineers.

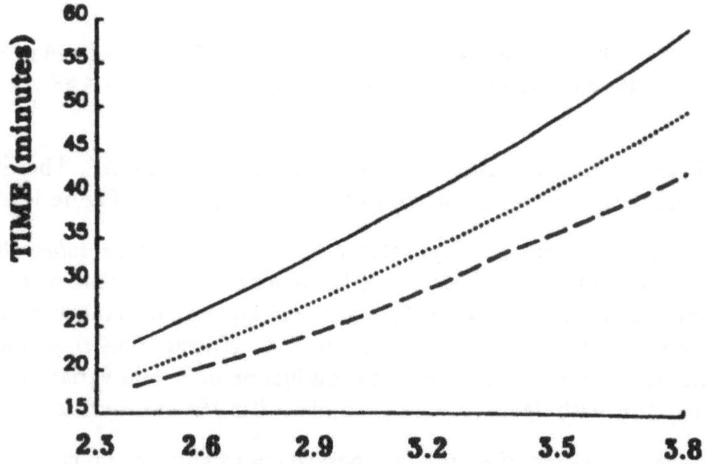

FIGURE E.2.8.4 Variation of the cooking time of the bacon analog with the loaf thickness. Initial slab temperature was 29.4°C, surface temperature 96.7°C, thermal diffusivity and cooking criteria were described with different lines: (—) $\alpha = 3.9 \times 10^{-4}\ \mathrm{m^2/h}$, center temperature 90.5°C; (···) $\alpha = 4.7 \times 10^{-4}\ \mathrm{m^2/h}$, center temperature 90.5°C; (---) $\alpha = 4.7 \times 10^{-4}\ \mathrm{m^2/h}$ and center temperature 87.8°C (Altomare, 1988). Reproduced with permission of the American Institute of Chemical Engineers.

may be neglected:

$$\frac{1}{r^2\sin\theta}\frac{\partial}{\partial\theta}\left(\alpha\sin\theta\frac{\partial T}{\partial\theta}\right)=0, \frac{1}{r^2\sin^2\theta}\frac{\partial}{\partial\phi}\left(\alpha\frac{\partial T}{\partial\phi}\right)=0,$$

then (2.18) becomes

$$\frac{\partial T}{\partial t}=\frac{1}{r^2}\frac{\partial}{\partial r}\left(r^2\alpha\frac{\partial T}{\partial r}\right)$$

Since α is constant it may be rearranged as

$$\frac{\partial T}{\partial t}=\alpha\left[\frac{2}{r}\frac{\partial T}{\partial r}+\frac{\partial^2 T}{\partial r^2}\right]$$

This is a partial differential equation with

BC1: $T = T_1$ at $r = R$ when $t > 0$

BC2: $\frac{dT}{dr}=0$ at $r = 0$ when $t > 0$

BC3: $T = T_1$ for all r when $t \to \infty$

IC: $T = T_0$ when $t = 0$ for all r

We may introduce the dimensionless variables

$$\theta=\frac{T-T_1}{T_0-T_1}, \quad \eta=\frac{r}{R} \quad \text{and} \quad \tau=\frac{t\alpha}{R^2}.$$

We may write the equation, BCs and the IC in dimensionless variables:

$$\frac{\partial\theta}{\partial\tau}=\frac{2}{\eta}\frac{\partial\theta}{\partial\eta}+\frac{\partial^2\theta}{\partial\eta^2}$$

BC1: $\theta = 0$ at $\eta = 1$ when $\tau > 0$

BC2: $\frac{d\theta}{d\eta}=0$ (θ is finite) at $\eta = 0$ when $\tau > 0$

BC3: $\theta = 0$ for all η when $\tau \to \infty$

IC: $\theta = 1$ when $\tau = 0$

The method of separation of variables requires separation of θ into two functions $f(\eta)$ and $g(\tau)$:

$$\theta(\eta, \tau) = f(\eta)g(\tau)$$

where f is function of η, and g is function of τ.

The partial differential equation may be rewritten in terms of the new functions as:

$$\frac{1}{f}\left[\frac{d^2f}{d\eta^2}+\frac{2}{\eta}\frac{df}{d\eta}\right]=\frac{1}{g}\frac{dg}{d\tau}=\lambda$$

After following the same principle with the previous example we will find $\lambda=-C^2$ and consequently $g=C_1e^{-C^2\tau}$.
The other equation

$$\frac{1}{f}\left[\frac{d^2f}{d\eta^2}+\frac{2}{\eta}\frac{df}{d\eta}\right]=-C^2$$

may be rearranged as:

$$\eta^2\frac{d^2f}{d\eta^2}+2\eta\frac{df}{d\eta}+\eta^2C^2f=0$$

We may define a new variable: $u=f\eta$, then this differential equation may be written as: $(d^2u/d\eta^2)+C^2u=0$ and its solution is (Tab. 2.6):

$$u=C_2\cos(C\eta)+C_3\sin(C\eta) \text{ or } f=(C_2/\eta)\cos(C\eta)+(C_3/\eta)\sin(C\eta)$$

BC2 implies θ is finite when $\eta=0$, but $(C_2/\eta)\cos(C\eta)\rightarrow\infty$, therefore $(C_2/\eta)\cos(C\eta)$ may not be a solution, and we conclude that $C_2=0$, therefore the solution is: $\theta=(C_n/\eta)e^{-C^2\tau}\sin(C\eta)$. BC1 requires $\theta=0$ at $\eta=1$ when $\tau>0$, therefore $0=C_ne^{-C^2\tau}\sin(C)$. The BC is satisfied when $C=\pi n$, and the solution becomes

$$\theta=\frac{1}{\eta}\sum_{n=0}^{\infty}\{C_ne^{-(\pi n)2\tau}\sin(\pi n\eta)\}$$

The IC requires $\theta=1$ when $\tau=0$, i.e., $1=\frac{1}{\eta}\sum_{n=1}^{\infty}C_n\sin(\pi n\eta)$ we may determine C_n after applying the orthogonality principle as:

$$C_n=2\frac{(-1)^{n+1}}{\pi n},$$

then the solution is $\theta=\frac{1}{\eta}\sum_{n=0}^{\infty}\left\{2\frac{(-1)^{n+1}}{\pi n}e^{-(\pi n)^2\tau}\sin(\pi n\eta)\right\}$ or

$$\frac{T-T_1}{T_0-T_1}=\frac{R}{r}\left(\frac{2}{\pi}\right)\sum_{n=0}^{\infty}\left\{\frac{(-1)^{n+1}}{\pi n}e^{-(\pi n)^2\tau}\sin\left(\frac{\pi n r}{R}\right)\right\}$$

Comparison of the model with the data at $r = 3.1$ cm and $r = 0$ cm from the center is shown in Figure E.2.9.

Example 2.10. Temperature profiles in a sausage during cooking A sausage ($\alpha = 6.5\,10^{-8}$ m^2/s, length = 10 cm, radius = 0.8 cm, $T_0 = 4$°C) is dipped into boiling water where its surface temperature rises to $T_1 = 102$°C. If heat transfer within the sausage may be simulated with conduction only, plot the temperature profiles along the radius 3 and 7 minutes after the beginning of the process.

Solution Equation of energy in cylindrical coordinate system is:

$$\frac{\partial T}{\partial t} + v_r \frac{\partial T}{\partial r} + \frac{v_\theta}{r}\frac{\partial T}{\partial \theta} + v_z \frac{\partial T}{\partial z} = \dot{\psi}_G + \frac{1}{r}\frac{\partial}{\partial r}\left(r\alpha \frac{\partial T}{\partial r}\right) + \frac{1}{r^2}\frac{\partial}{\partial \theta}\left(\alpha \frac{\partial T}{\partial \theta}\right) + \frac{\partial}{\partial z}\left(\alpha \frac{\partial T}{\partial z}\right) \quad (2.17)$$

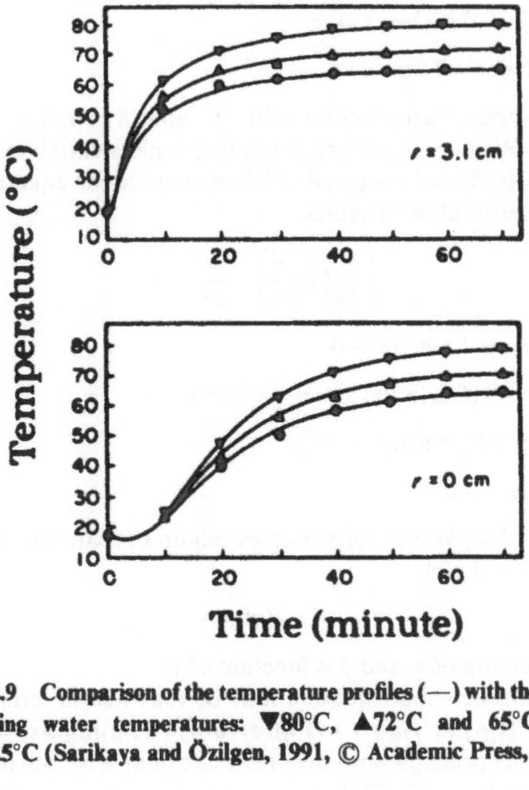

FIGURE E.2.9 Comparison of the temperature profiles (—) with the experimental data. Blanching water temperatures: ▼80°C, ▲72°C and 65°C; $R = 4.1$ cm, $T_0 = 17.5 \pm 1.5$°C (Sarikaya and Özilgen, 1991, © Academic Press, reproduced by permission).

The process conditions imply $v_\theta = v_r = v_z = 0$ and $\dot{\psi}_G = 0$. The sausage is a thin cylinder therefore conduction through the longer dimension may be neglected:

$$\frac{1}{r^2}\frac{\partial}{\partial\theta}\left(\alpha\frac{\partial T}{\partial\theta}\right)=0,\ \frac{\partial}{\partial z}\left(\alpha\frac{\partial T}{\partial z}\right)=0,$$

then (2.17) becomes

$$\frac{\partial T}{\partial t}=\frac{1}{r}\frac{\partial}{\partial r}\left(r\alpha\frac{\partial T}{\partial r}\right)$$

Since α is constant it may be rearranged as

$$\frac{1}{\alpha}\frac{\partial T}{\partial t}-\frac{1}{r}\frac{\partial}{\partial r}\left(r\frac{\partial T}{\partial r}\right)=0$$

This is a partial differential equation with

BC1: $T = T_1$ at $r = R$ when $t > 0$

BC2: $(dT/dr) = 0$ at $r = 0$ when $t > 0$

BC3: $T = T_1$ for all r when $t \to \infty$

IC: $T = T_0$ when $t = 0$ for all r

where T_1 = surface temperature = 102 °C and R = radius of the sausage = 0.8cm. We may introduce the dimensionless variables $\theta = (T - T_1)/(T_0 - T_1)$, $\eta = (r/R)$ and $\tau = (t\alpha/R^2)$. We may write the equation, BCs and the IC in dimensionless variables:

$$\frac{1}{\eta}\frac{\partial}{\partial\eta}\left(\eta\frac{\partial\theta}{\partial\eta}\right)-\frac{\partial\theta}{\partial\tau}=0$$

BC1: $\theta = 0$ at $\eta = 1$ when $\tau > 0$

BC2: $(d\theta/d\eta) = 0$ (θ is finite) at $\eta = 0$ when $\tau > 0$

BC3: $\theta = 0$ for all η when $\tau \to \infty$

IC: $\theta = 1$ when $\tau = 0$

The method of separation of variables requires separation of θ into two functions $f(\eta)$ and $g(\tau)$:

$$\theta = f(\eta)\,g(\tau)$$

where f is function of η, and g is function of τ.

The partial differential equation may be rewritten in terms of the new functions as: $(1/f)((d^2f/d\eta^2) + (1/\eta)(df/d\eta)) = (1/g)(dg/d\tau) = \lambda$ After following the same principle with the previous example we will find $\lambda = -C^2$ and consequently $g = C_1 e^{-C^2\tau}$.

The other equation $(1/f)((d^2f/d\eta^2)+(1/\eta)(df/d\eta)) = -C^2$ may be rearranged as:

$$\eta^2\frac{d^2f}{d\eta^2}+\eta\frac{df}{d\eta}+\eta^2C^2f=0$$

This is Bessel's equation with $v=0$ and its solution is (Tabs. 2.8 and 2.7):

$$f=C_2J_0(C_\eta)+C_3Y_0(C_\eta)$$

Plots of J_0 and Y_0 versus η were given in Figure 2.3 and shows that at $\eta=0$ $Y_0\to\infty$. Presence of the term $C_3Y_0(C_\eta)$ in the solution will cause f (and θ) $\to\infty$ at $\eta=0$, but contradict the fact that $\theta=$ finite at $\eta=0$; therefore $C_3Y_0(C_\eta)$ can not be a solution and $C_3=0$. The combined solution is:

$$\theta=f(\eta)\,g(\tau)=C_ne^{-C^2\tau}J_0(C\eta)$$

where $C_n=C_1C_2$.
BC1: $\theta=0$ at $\eta=1$ when $\tau>0$ requires $0=C_ne^{-C^2\tau}J_0(C)$. The requirement is satisfied when $C=B_n=$ eigen values of J_0. (Definition: Values of x which make $J_0(x)=0$ are called the eigen values of J_0 and given in Tab. 2.14), then the solution becomes

$$\theta=\sum_{n=1}^{\infty}C_n\exp(-B_n^2\tau)J_0(B_n\eta)$$

IC: $\theta=1$ when $\tau=0$ requires

$$1=\sum_{n=1}^{\infty}C_nJ_0(B_n\eta)$$

J_0 is an orthogonal function and requires

$$\int_0^1 J_0(B_n\eta)J_0(B_m\eta)\,\eta\,d\eta=0 \text{ if } n\neq m.$$

We may apply the orthogonality principle to the IC equation as:

$$\int_0^1 J_0(B_n\eta)\,\eta\,d\eta=C_n\int_0^1(J_0(B_n\eta))^2\,\eta\,d\eta$$

We use identities $\int\alpha x^kJ_{k-1}(\alpha x)dx=x^kJ_k(\alpha x)$ (see Tab. 2.8)
$J_{-n}(x)=(-1)^nJ_n(x)$ (see Tab. 2.7, for $n=$ positive integer)
and

$$\int_{x=0}^{\infty}(J_k(\alpha x))^2x\,dx=\frac{1}{2}x^2(J_k^2(\alpha x)-J_{k-1}(\alpha x)J_{k+1}(\alpha x))$$

(see Tab. 2.8)

then calculate $C_n = (2/B_n J_1(B_n))$ The final solution is:

$$\theta = \sum_{n=1}^{\infty} \frac{2}{B_n J_1(B_n)} \exp(-B_n^2 \tau) J_0(B_n \eta)$$

Temperature profiles along the radius 3 and 7 minutes after the beginning of the process were calculated by using Table 2.9 and plotted in Figure E.2.10.

Table E.2.10 Variation of temperature with distance from the center of the sausage 3 and 7 minutes after dipping into boiling water

r (m)	0	0.002	0.004	0.006	0.008
T(°C) when $t = 3$ min	47.7	52.0	65.5	83.6	102.0
T(°C) when $t = 7$ min	88.7	89.8	93.1	97.5	102.0

Example 2.11. Determination of the temperature profile in a spherical potato tuber by using the generalized Bessel's equation (Sarikaya and Özilgen,

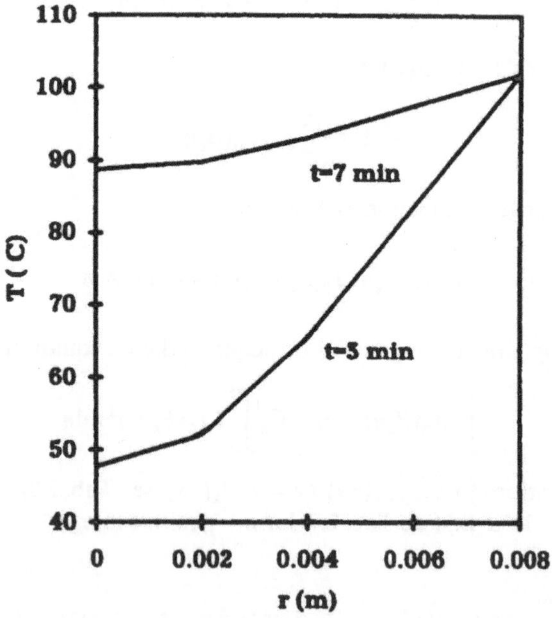

FIGURE E.2.10 Temperature profiles along the sausage 3 and 7 minutes after dipping into boiling water.

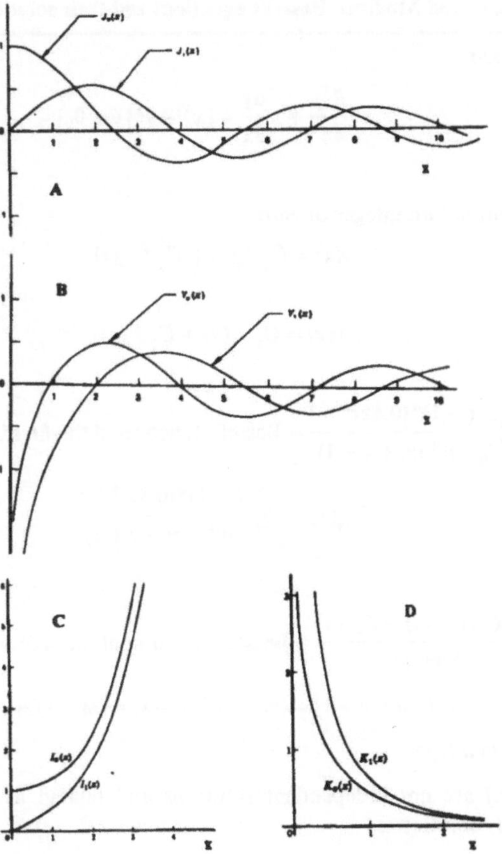

FIGURE 2.3. Variation of Bessel functions of (A) the first and (B) the second kinds; and the modified Bessel functions of (C) the first and (D) the second kinds with the independent variable x.

1991) It was shown that the temperature profile in a spherical potato tuber (Example 2.9) may be described as

$$\frac{\partial\theta}{\partial\tau}=\frac{2}{\eta}\frac{\partial\theta}{\partial\eta}+\frac{\partial^2\theta}{\partial\eta^2}$$

with

BC1: $\theta=0$ at $\eta=1$ when $\tau>0$
BC2: $(d\theta/d\eta)=0$ (θ is finite) at $\eta=0$ when $\tau>0$
BC3: $\theta=0$ for all η when $\tau\rightarrow\infty$

Table 2.7 Bessel's and Modified Bessel's equations and their solutions

i) Bessel's equation

$$x^2\frac{d^2y}{dx^2}+x\frac{dy}{dx}+(x^2-\nu^2)y=0$$

Solutions:

a) ν is positive but not an integer or zero

$$y(x)=C_1\,J_\nu(x)+C_2J_{-\nu}(x)$$

or

$$y(x)=C_1\,J_\nu(x)+C_2J_\nu(x)$$

where

$$J_\nu(x)=\sum_{m=0}^{\infty}\frac{(-1)^m(0.5x)^{2m+\nu}}{m!\,\Gamma(m+\nu+1)}=\text{Bessel's function of the first kind of order }\nu$$

$$J_{-\nu}(x)=\sum_{m=0}^{\infty}\frac{(-1)^m(0.5x)^{2m-\nu}}{m!\,\Gamma(m-\nu+1)}$$

and

$$Y_\nu(x)=\frac{\text{Cos}(\nu\pi)\;J_\nu(x)-J_{-\nu}(x)}{\text{Sin}(\nu\pi)}=\text{Bessel's function of the second kind of order }\nu$$

Γ = Gamma function, $\Gamma(m+\nu+1)=(m+\nu)\Gamma(m+\nu)=(m+\nu)(m+\nu-1)\,\Gamma(m+\nu-1)$

b) $\nu=n=$ positive integer

$J_{-n}(x)$ and $J_n(x)$ are not independent solutions and related as $J_{-n}(x)=(-1)^nJ_n(x)$, therefore the only solution is

$$y(x)=C_1\,J_n(x)+C_2J_n(x)$$

where

$$J_n(x)=\frac{x^n}{2^n n!}\left\{1-\frac{x^2}{2^2\,1!(n+1)}+\frac{x^4}{2^4\,2!(n+1)(n+2)}-\frac{x^6}{2^6\,3!(n+1)(n+2)(n+3)}+\cdots\right\}$$

$$Y_n(x)=\frac{2}{\pi}\left\{(\ln(0.5x)+0.5772)J_n(x)-0.5\sum_{m=0}^{\infty}\left((-1)^{m+1}[\phi(m)+\phi(m+n)]\frac{(0.5x)^{2m+n}}{m!(m+n)!}\right)\right\}$$

$$\phi(m)=\sum_{k=1}^{\infty}\frac{1}{k}\text{ with }\phi(0)\equiv 0$$

ii) Modified Bessel's equation

$$x^2\frac{d^2y}{dx^2}+x\frac{dy}{dx}-(x^2+\nu^2)y=0$$

Solutions:

a) When ν is positive but not an integer or zero $y(x)=C_1I_\nu(x)+C_2I_{-\nu}(x)$ or $y(x)=C_1I_\nu(x)+C_2K_\nu(x)$

Table 2.7 (Continued)

where

$$I_v(x)=\sum_{m=0}^{\infty}\frac{(0.5x)^{2m+v}}{m!\Gamma(m+v+1)} = \text{Modified Bessel's function of the first kind of order } v$$

$$I_{-v}(x)=\sum_{m=0}^{\infty}\frac{(0.5x)^{2m-v}}{m!\Gamma(m-v+1)}$$

$$K_v(x)=(0.5\pi)\frac{I_{-v}(x)-I_v(x)}{\text{Sin}(v\pi)} = \text{Modified Bessel's function of the second kind of order } v$$

b) When $v=n=$ positive integer or zero $I_{-n}(x)$ and $I_n(x)$ are not independent solutions and related as $I_{-n}(x)=I_n(x)$, therefore the only solution is $y(x)=C_1I_n(x)+C_2K_n(x)$

where

$$I_n(x)=\frac{x^n}{2^n n!}\left\{1+\frac{x^2}{1^2\,1!\,(n+1)}+\frac{x^4}{2^4\,2!\,(n+1)(n+2)}+\frac{x^6}{2^6\,3!\,(n+1)(n+2)(n+3)}+\cdots\right\}$$

$$K_n(x)=(-1)^{n+1}(\ln(0.5x)+0.5772)I_n(x)+0.5\sum_{m=0}^{n-1}\left(\frac{(-1)^m(n-m-1)!\,2^{n-2m}}{m!\,x^{n-2m}}\right)+$$

$$\frac{(-1)^n}{2}\sum_{m=0}^{\infty}\left\{\frac{n+2m}{m!(n+m)!\,2^{n+m}}\left(\left[1+\frac{1}{2}+\cdots+\frac{1}{m}\right]+\left[1+\frac{1}{2}+\cdots+\frac{1}{n+m}\right]\right)\right\}$$

Table 2.8 Important properties for Bessel's equation; Bessel and modified Bessel functions and their derivatives

i) $x^2\frac{d^2y}{dx^2}+x\frac{dy}{dx}+(C^2x^2-v^2)\,y=0$ ($C=$ constant) becomes Bessel's equation

$t^2\frac{d^2y}{dt^2}+x\frac{dy}{dt}+(t^2-v^2)\,y=0$ after substituting $t=Cx$

ii) Half order functions

$$J_{1/2}(x)=\sqrt{\frac{2}{\pi x}}\sin x \qquad J_{-1/2}(x)=\sqrt{\frac{2}{\pi x}}\cos x$$

$$I_{1/2}(x)=\sqrt{\frac{2}{\pi x}}\sinh x \qquad I_{-1/2}(x)=\sqrt{\frac{2}{\pi x}}\cosh x$$

iii) Differential properties

$$J_{n-1}(x)+J_{n+1}(x)=\left(\frac{2n}{x}\right)J_n(x) \qquad Y_{n-1}(x)+Y_{n+1}(x)=\left(\frac{2n}{x}\right)Y_n(x)$$

$$J_{n-1}(x)-J_{n+1}(x)=2\frac{d(J_n(x))}{dx} \qquad Y_{n-1}(x)-Y_{n+1}(x)=2\frac{d(Y_n(x))}{dx}$$

$$nJ_n(x)+x\frac{d(J_n(x))}{dx}=J_{n-1}(x) \qquad nY_n(x)+x\frac{d(Y_n(x))}{dx}=x\,Y_{n-1}(x)$$

$$nJ_n(x)-x\frac{d(J_n(x))}{dx}=J_{n+1}(x) \qquad nY_n(x)-x\frac{d(Y_n(x))}{dx}=x\,Y_{n+1}(x)$$

Table 2.8 (Continued)

$$I_{n-1}(x) - I_{n+1}(x) = \left(\frac{2n}{x}\right) I_n(x) \qquad K_{n-1}(x) + K_{n+1}(x) = -\left(\frac{2n}{x}\right) K_n(x)$$

$$I_{n-1}(x) + I_{n+1}(x) = 2\frac{d(I_n(x))}{dx} \qquad K_{n-1}(x) + K_{n+1}(x) = -2\frac{d(K_n(x))}{dx}$$

$$nI_n(x) + x\frac{d(I_n(x))}{dx} = xI_{n-1}(x) \qquad nK_n(x) + x\frac{d(K_n(x))}{dx} = -xK_{n-1}(x)$$

$$nI_n(x) - x\frac{d(I_n(x))}{dx} = x - I_{n+1}(x) \qquad nK_n(x) - x\frac{d(K_n(x))}{dx} = xK_{n+1}(x)$$

iv) Integral properties

$$\int \alpha x^k J_{k-1}(\alpha x)dx = x^k J_k(\alpha x) \qquad \int \alpha x^k Y_{k-1}(\alpha x)dx = x^k Y_k(\alpha x)$$

$$\int \alpha x^k I_{k-1}(\alpha x)dx = x^k I_k(\alpha x) \qquad \int \alpha x^k K_{k-1}(\alpha x)dx = -x^k K_k(\alpha x)$$

$$\int_{x=0}^{\infty} [J_k(\alpha x)]^2\, xdx = \frac{1}{2}x^2 [J_k^2(\alpha x) - J_{k-1}(\alpha x) J_{k+1}(\alpha x)]$$

$$\int_{x=0}^{\infty} [I_k(\alpha x)]^2\, xdx = \frac{1}{2}x^2 [I_k^2(\alpha x) - I_{k-1}(\alpha x) I_{k+1}(\alpha x)]$$

Table 2.9 Generalized form of the Bessel's equation

$$x^2\frac{d^2y}{dx^2} + x(a + 2bx^r)\frac{dy}{dx} + [c + dx^{2s} - b(1 - a - r)x^r + b^2x^{2r}]\, y = 0$$

is called the generalized Bessel's equation and its solution is:

$$y = x^{(1-a)/2} \exp\left\{-\frac{bx^r}{r}\right\}\left[C_1 Z_p\left(\frac{\sqrt{|d|}}{s}x^s\right) + C_2 Z_{-p}\left(\frac{\sqrt{|d|}}{s}x^s\right)\right]$$

where $p = \frac{1}{s}\sqrt{\left(\frac{1-a}{2}\right)^2 - c}$

if $\frac{\sqrt{d}}{s}$ = real, $p \neq 0$ or $p \neq$ integer $\qquad Z_p = J_p,\ Z_{-p} = J_{-p}$

if $\frac{\sqrt{d}}{s}$ = real, $p = 0$ or $p =$ integer $\qquad Z_n = J_p,\ Z_{-p} = Y_n$

if $\frac{\sqrt{d}}{s}$ = imaginary, $p \neq 0$ or $p \neq$ integer $\qquad Z_p = J_p,\ Z_{-p} = I_{-p}$

if $\frac{\sqrt{d}}{s}$ = imaginary, $p = 0$ or $p =$ integer $\qquad Z_p = I_p,\ Z_{-p} = K_n$

Table 2.10 Bessel functions $J_0(x)$ and $J_1(x)$ (see the note in Table 2.14)

x	$J_0(x)$	$J_1(x)$	x	$J_0(x)$	$J_1(x)$	x	$J_0(x)$	$J_1(x)$	x	$J_0(x)$	$J_1(x)$
0.0	1.0000	0.0000	4.0	−0.3971	−0.0660	8.0	0.1717	0.2346	12.0	0.0477	−0.2234
0.1	0.9975	0.0499	4.1	−0.3887	−0.1033	8.1	0.1475	0.2476	12.1	0.0697	−0.2157
0.2	0.9900	0.0995	4.2	−0.3765	−0.1386	8.2	0.1222	0.2580	12.2	0.0908	−0.2050
0.3	0.9776	0.1483	4.3	−0.3610	−0.1719	8.3	0.0960	0.2657	12.3	0.1108	−0.1943
0.4	0.9604	0.1960	4.4	−0.3423	−0.2028	8.4	0.0692	0.2708	12.4	0.1296	−0.1807
0.5	0.9385	0.2423	4.5	−0.3205	−0.2311	8.5	0.0419	0.2731	12.5	0.1469	−0.1655
0.6	0.9120	0.2867	4.6	−0.2961	−0.2566	8.6	0.0146	0.2728	12.6	0.1626	−0.1487
0.7	0.8812	0.3290	4.7	−0.2693	−0.2791	8.7	−0.0125	0.2697	12.7	0.1766	−0.1307
0.8	0.8463	0.3688	4.8	−0.2404	−0.2985	8.8	−0.0392	0.2641	12.8	0.1887	−0.1114
0.9	0.8075	0.4059	4.9	−0.2097	−0.3147	8.9	−0.0653	0.2559	12.9	0.1988	−0.0912
1.0	0.7652	0.4401	5.0	−0.1776	−0.3276	9.0	−0.0903	0.2453	13.0	0.2069	−0.0703
1.1	0.7196	0.4709	5.1	−0.1443	−0.3371	9.1	−0.1142	0.2324	13.1	0.2129	−0.0489
1.2	0.6711	0.4983	5.2	−0.1103	−0.3432	9.2	−0.1367	0.2174	13.2	0.2167	−0.0271
1.3	0.6201	0.5220	5.3	−0.0758	−0.3460	9.3	−0.1577	0.2004	13.3	0.2183	−0.0052
1.4	0.5669	0.5419	5.4	−0.0412	−0.3453	9.4	−0.1768	0.1816	13.4	0.2177	0.0166
1.5	0.5118	0.5579	5.5	−0.0068	−0.3414	9.5	−0.1939	0.1613	13.5	0.2150	0.0380
1.6	0.4554	0.5699	5.6	0.0270	−0.3343	9.6	−0.2090	0.1395	13.6	0.2101	0.0590
1.7	0.3980	0.5778	5.7	0.0599	−0.3241	9.7	−0.2218	0.1166	13.7	0.2032	0.0791
1.8	0.3400	0.5815	5.8	0.0917	−0.3110	9.8	−0.2323	0.0928	13.8	0.1943	0.0984
1.9	0.2818	0.5812	5.9	0.1220	−0.2951	9.9	−0.2403	0.0684	13.9	0.1836	0.1165
2.0	0.2239	0.5767	6.0	0.1506	−0.2767	10.0	−0.2459	0.0435	14.0	0.1711	0.1334
2.1	0.1666	0.5683	6.1	0.1773	−0.2559	10.1	−0.2490	0.0184	14.1	0.1570	0.1488
2.2	0.1104	0.5560	6.2	0.2017	−0.2329	10.2	−0.2496	−0.0066	14.2	0.1414	0.1626
2.3	0.0555	0.5399	6.3	0.2238	−0.2081	10.3	−0.2477	−0.0313	14.3	0.1245	0.1747
2.4	0.0025	0.5202	6.4	0.2433	−0.1816	10.4	−0.2434	−0.0555	14.4	0.1065	0.1850

Table 2.10 (Continued)

x	$J_0(x)$	$J_1(x)$	x	$J_0(x)$	$J_1(x)$	x	$J_0(x)$	$J_1(x)$	x	$J_0(x)$	$J_1(x)$
2.5	−0.0484	0.4971	6.5	0.2601	−0.1538	10.5	−0.2366	−0.0789	14.5	0.0875	0.1934
2.6	−0.0968	0.4708	6.6	0.2740	−0.1250	10.6	−0.2276	−0.1012	14.6	0.0679	0.1999
2.7	−0.1424	0.4416	6.7	0.2851	−0.0953	10.7	−0.2164	−0.1224	14.7	0.0476	0.2043
2.8	−0.1850	0.4097	6.8	0.2931	−0.0652	10.8	−0.2032	−0.1422	14.8	0.0271	0.2066
2.9	−0.2243	0.3754	6.9	0.2981	−0.0349	10.9	−0.1881	−0.1603	14.9	0.0064	0.2069
3.0	−2.601	0.3391	7.0	0.3001	−0.0047	11.0	−0.1712	−0.1768			
3.1	−0.2921	0.3009	7.1	0.2991	0.0252	11.1	−0.1528	−0.1913			
3.2	−0.3202	0.2613	7.2	0.2951	0.0543	11.2	−0.1330	−0.2039			
3.3	−0.3443	0.2207	7.3	0.2882	0.0826	11.3	−0.1121	−0.2143			
3.4	−0.3643	0.1792	7.4	0.2786	0.1096	11.4	−0.0902	−0.2225			
3.5	−0.3801	0.1374	7.5	0.2663	0.1352	11.5	−0.0677	−0.2284			
3.6	−0.3918	0.0955	7.6	0.2516	0.1592	11.6	−0.0446	−0.2320			
3.7	−0.3992	0.0538	7.7	0.2346	0.1813	11.7	−0.0213	−0.2333			
3.8	−0.4026	0.0128	7.8	0.2154	0.2014	11.8	−0.0020	−0.2323			
3.9	−0.4018	−0.0272	7.9	0.1944	0.2192	11.9	−0.0250	−0.2290			

Table 2.11 Bessel functions $Y_0(x)$ and $Y_1(x)$ (see the note in Table 2.14)

x	$Y_0(x)$	$Y_1(x)$	x	$Y_0(x)$	$Y_1(x)$	x	$Y_0(x)$	$Y_1(x)$
0.0	$-\infty$	$-\infty$	1.50	0.382	−0.412	3.50	0.189	0.410
0.05	−1.979	−12.790	1.60	0.420	−0.347	3.60	0.148	0.415
0.10	−1.534	−6.459	1.70	0.452	−0.285	3.70	0.106	0.417
0.15	−1.271	−4.364	1.80	0.477	−0.224	3.80	0.064	0.414
0.20	−1.081	−3.324	1.90	0.497	−0.164	3.90	0.023	0.408
0.25	−0.932	−2.704	2.00	0.510	−0.107	4.00	−0.017	0.398
0.30	−0.807	−2.293	2.10	0.518	−0.052	4.10	−0.056	0.384
0.35	−0.700	−2.000	2.20	0.521	0.0015	4.20	−0.094	0.368
0.40	−0.606	−1.781	2.30	0.518	0.052	4.30	−0.130	0.348
0.45	−0.521	−1.610	2.40	0.510	0.100	4.40	−0.163	0.326
0.50	−0.444	−1.471	2.50	0.498	0.146	4.50	−0.195	0.301
0.60	−0.308	−1.260	2.60	0.481	0.188	4.60	−0.223	0.274
0.70	−0.191	−1.103	2.70	0.460	0.228	4.70	−0.249	0.244
0.80	−0.087	−0.978	2.80	0.436	0.263	4.80	−0.272	0.213
0.90	0.0056	−0.873	2.90	0.408	0.296	4.90	−0.292	0.181
1.00	0.088	−0.781	3.00	0.377	0.325	5.00	−0.308	0.148
1.10	0.162	−0.698	3.10	0.343	0.350			
1.20	0.228	−0.621	3.20	0.307	0.371			
1.30	0.286	−0.548	3.30	0.269	0.338			
1.40	0.338	−0.479	3.40	0.230	0.401			

Table 2.12 Modified Bessel functions $I_0(x)$ and $I_1(x)$ (see the note in Table 2.14)

x	$I_0(x)$	$I_1(x)$	x	$I_0(x)$	$I_1(x)$	x	$I_0(x)$	$I_1(x)$
0.0	1.000	0.0000	1.80	1.990	1.317	3.60	8.028	6.793
0.10	1.002	0.050	1.90	2.128	1.448	3.70	8.739	7.436
0.20	1.010	0.100	2.00	2.280	1.591	3.80	9.517	8.140
0.30	1.023	0.152	2.10	2.446	1.745	3.90	10.369	8.913
0.40	1.040	0.204	2.20	2.629	1.914	4.00	11.30	9.759
0.50	1.063	0.258	2.30	2.830	2.098	4.10	12.32	10.69
0.60	1.092	0.314	2.40	3.049	2.298	4.20	13.44	11.70
0.70	1.126	0.372	2.50	3.290	2.517	4.30	14.67	12.82
0.80	1.166	0.433	2.60	3.553	2.755	4.40	16.01	14.04
0.90	1.213	0.497	2.70	3.842	3.016	4.50	17.48	15.39
1.00	1.266	0.565	2.80	4.157	3.301	4.60	19.09	16.86
1.10	1.326	0.637	2.90	4.503	3.613	4.70	20.86	18.48
1.20	1.394	0.715	3.00	4.881	3.953	4.80	22.79	20.25
1.30	1.469	0.797	3.10	5.294	4.326	4.90	24.91	22.20
1.40	1.553	0.886	3.20	5.747	4.734	5.00	27.24	24.34
1.50	1.647	0.982	3.30	6.243	5.181			
1.60	1.750	1.085	3.40	6.785	5.670			
1.70	1.864	1.196	3.50	7.378	6.206			

Table 2.13 Modified Bessel functions $K_0(x)$ and $K_1(x)$ (see the note in Table 2.14)

x	$K_0(x)$	$K_1(x)$	x	$K_0(x)$	$K_1(x)$	x	$K_0(x)$	$K_1(x)$
0.0	$-\infty$	$-\infty$	1.50	0.214	0.277	3.50	0.0196	0.0222
0.05	3.114	19.910	1.60	0.188	0.241	3.60	0.0175	0.0198
0.10	2.427	9.854	1.70	0.165	0.209	3.70	0.0156	0.0176
0.15	2.030	6.477	1.80	0.146	0.183	3.80	0.0140	0.0157
0.20	1.753	4.776	1.90	0.129	0.160	3.90	0.0125	0.0140
0.25	1.541	3.747	2.00	0.114	0.140	4.00	0.0112	0.0125
0.30	1.327	3.056	2.10	0.101	0.123	4.10	0.0100	0.0111
0.35	1.233	2.559	2.20	0.0893	0.108	4.20	0.0089	0.0099
0.40	1.114	2.184	2.30	0.0791	0.0950	4.30	0.0080	0.0089
0.45	1.013	1.892	2.40	0.0702	0.0837	4.40	0.0071	0.0079
0.50	0.924	1.656	2.50	0.0623	0.0739	4.50	0.0064	0.0071
0.60	0.778	1.303	2.60	0.0554	0.0653	4.60	0.0057	0.0063
0.70	0.660	1.050	2.70	0.0492	0.0577	4.70	0.0051	0.0056
0.80	0.565	0.862	2.80	0.0438	0.0511	4.80	0.0046	0.0050
0.90	0.487	0.716	2.90	0.0390	0.0453	4.90	0.0041	0.0045
1.00	0.421	0.602	3.00	0.0347	0.0402	5.00	0.0037	0.0040
1.10	0.366	0.510	3.10	0.0310	0.0356			
1.20	0.318	0.434	3.20	0.0276	0.0316			
1.30	0.278	0.372	3.30	0.0246	0.0281			
1.40	0.244	0.321	3.40	0.0220	0.0250			

Table 2.14 Values of B_n for which $J_0(B_n) = 0$ and corresponding values of $J_1(B_n)$

n	B_n	$J_1(B_n)$
1	2.4048	0.5191
2	5.5201	-0.3404
3	8.6537	0.2715
4	11.7915	-0.2324
5	14.9309	0.2065

Note: For extensive list of the mathematical functions given in Tables 2.10–2.14 an interested reader may refer to *Mathematical Tables*, Volume 6, Part 1, *Bessel Functions of orders zero and unity*, British Association of Advancement of Science, Cambridge University Press, 1958.

IC: $\theta = 1$ when $\tau = 0$
where $\theta = (T - T_1)/(T_0 - T_1)$, $\eta = (r/R)$ and $\tau = (t\alpha/R^2)$.

After using the method of separation of variables

$$\theta(\eta, \tau) = f(\eta)\, g(\tau)$$

Table 2.15 Values of $K_{1/4}(x)$ and its derivatives $(d(K_{1/4}(x))/dx)$ (Carsten and McKerrow, 1944)

x	$K_{1/4}(x)$	$\frac{d(K_{1/4}(x))}{dx}$	x	$K_{1/4}(x)$	$\frac{d(K_{1/4}(x))}{dx}$
0.1	2.69	12.3	3.1	0.031229	0.036030
0.2	1.878	5.50	3.2	0.027833	0.031983
0.3	1.448	3.39	3.3	0.024817	0.028409
0.4	1.1651	2.370	3.4	0.022137	0.025251
0.5	0.9603	1.772	3.5	0.019755	0.022457
0.6	0.80399	1.3795	3.6	0.017635	0.019983
0.7	0.68058	1.1037	3.7	0.015749	0.017791
0.8	0.58086	0.90026	3.8	0.014069	0.015847
0.9	0.49893	0.74502	3.9	0.012572	0.014121
1.0	0.37342	0.62346	4.0	0.011238	0.012589
1.1	0.32486	0.52635	4.1	0.010049	0.011228
1.2	0.28345	0.44754	4.2	0.0089877	0.010018
1.3	0.28345	0.38278	4.3	0.0080407	0.0089420
1.4	0.24794	0.32900	4.4	0.0071953	0.0079841
1.5	0.21736	0.28396	4.5	0.0064404	0.0071312
1.6	0.19091	0.24597	4.6	0.0057660	0.0063714
1.7	0.16797	0.21372	4.7	0.0051633	0.0056943
1.8	0.14801	0.18621	4.8	0.0046247	0.0050906
1.9	0.13060	0.16262	4.9	0.0041430	0.0045522
2.0	0.11538	0.14233	5.0	0.0037123	0.0040718
2.1	0.10204	0.12480	6.0	0.0012500	0.0013514
2.2	0.090341	0.10961	8.0	0.0001470	0.0001560
2.3	0.080054	0.096426	10.0	0.0000178	0.0000187
2.4	0.070999	0.084941			
2.5	0.063017	0.074919			
2.6	0.055973	0.066156			
2.7	0.049750	0.058480			
2.8	0.044246	0.051744			
2.9	0.039374	0.045826			
3.0	0.035057	0.040618			

it was found that

$$g = C_1 e^{-C^2 \tau}$$

and the other equation was rearranged as

$$\eta^2 \frac{d^2 f}{d\eta^2} + 2\eta \frac{df}{d\eta} + \eta^2 C^2 f = 0$$

Use the generalized Bessel's equation to obtain the same solution.

Solution The generalized Bessel's equation and its solution are (Tab. 2.9):

$$\eta^2\frac{d^2f}{d\eta^2}+\eta(a+2b\eta^r)\frac{df}{d\eta}+(c+d\eta^{2s}-b(1-a-r)\eta^r+b^2\eta^{2r})f=0$$

and

$$f=\eta^{(1-a)/2}\exp\left\{-\frac{b\eta^r}{r}\right\}\left[C_1Z_p\left(\frac{\sqrt{|d|}}{s}\eta^s\right)+C_2Z_{-p}\left(\frac{\sqrt{|d|}}{s}\eta^s\right)\right]$$

where $p=(1/s)(\sqrt{((1-a)/2)^2-c}$

In the given example $a=2, b=0, c=0, d=C^2$ and $s=1$, therefore $p=1/2$

Since $C^2>0\,\frac{\sqrt{d}}{s}=$ real $Z_p=J_p\ Z_{-p}=J_{-p}$ then the solution is

$$f=\eta^{(1-a)/2}\exp\left\{-\frac{b\eta^r}{r}\right\}\left[C_1Z_p\left(\frac{\sqrt{|d|}}{s}\eta^s\right)+C_2Z_{-p}\left(\frac{\sqrt{|d|}}{s}\eta^s\right)\right.$$

$$f=\eta^{-1/2}(C_1J_{1/2}(C\eta)+C_2J_{-p}(C\eta))$$

Table 2.8 shows that $J_{1/2}(x)=\sqrt{(2/\pi x)}\sin x$ and $J_{-1/2}(x)=\sqrt{(2/\pi x)}\cos x$, therefore

$$f=\frac{1}{\eta}\left[C_1\sqrt{\frac{2}{\pi}}\sin(C\eta)+C_2\sqrt{\frac{2}{\pi}}\cos(C\eta)\right]$$

BC2 implies θ is finite when $\eta=0$, but

$$\frac{C_2}{\eta}\sqrt{\frac{2}{\pi}}\cos(C\eta)\rightarrow\infty,$$

therefore $\frac{C_2}{\sqrt{\eta}}\cos(C\eta)$ may not be a solution, and we conclude that $C_2=0$, therefore the solution is:

$$\theta=fg=\frac{C_n}{\eta}e^{-C^2\tau}\sin(C\eta)$$

Constants C and C_n will be evaluated the same way as in the previous solution and we will obtain

$$\frac{T-T_1}{T_0-T_1}=\frac{R}{r}\left(\frac{2}{\pi}\right)\sum_{n=0}^{\infty}\left\{\frac{(-1)^{n+1}}{n}e^{-(\pi n)^2\tau}\tau\sin\left(\frac{\pi nr}{R}\right)\right\}$$

Example 2.12. Liquid diffusion model for drying rough rice The rice grain may be represented by a finite cylinder of radius R and length $2L$. The moisture removal is assumed to be in liquid state and described by the equation of continuity in cylindrical coordinate system:

$$\frac{\partial c_A}{\partial t}+\left(\frac{1}{r}\frac{\partial rN_{Ar}}{dr}+\frac{1}{r}\frac{\partial N_{A\theta}}{\partial\theta}+\frac{\partial N_{Az}}{\partial z}\right)=R_A \tag{2.13}$$

We may neglect the diffusion in angular direction $(1/r)(\partial N_{A\theta}/\partial\theta)=0$, since water does not involve any reactions $R_A=0$, then (2.13) becomes:

$$\frac{\partial c_A}{\partial t}+\left(\frac{1}{r}\frac{\partial r\dot{N}_{Ar}}{dr}+\frac{\partial N_{Az}}{\partial z}\right)=0$$

Total flux of water in z direction is:

$$N_{Az}=x_A(N_A+N_B)-D_{AB}\frac{dc_A}{dz} \tag{2.10}$$

Total flux of water in r direction may be stated as:

$$N_{Ar}=x_A(N_A+N_B)-D_{AB}\frac{dc_A}{dr}$$

we may assume that the fraction of water in rice x_A is small and neglect the term $x_A(N_A+N_B)$ we will obtain the expressions for N_{Az} and N_{Ar} as:

$$N_{Az}=-D_{AB}\frac{dc_A}{dz}$$

$$N_{Ar}=-D_{AB}\frac{dc_A}{dr}$$

rough rice has 22–24 % water. Neglecting the term $x_A(N_A+N_B)$ simplifies the solution of the problem, but exaggerates effect of diffusion described by the terms $D_{AB}(dc_A/dz)$ or $D_{AB}(dc_A/dz)$. Since the process is treated like pure diffusion in contrast to the actual mechanism, we will refer the constant D_{AB} as the *apparent diffusivity*. Apparent constants are frequently used in modeling. After substituting the final forms of N_{Az} and N_{Ar} in the simplified form of the equation of continuity and neglecting the subscripts we will have:

$$\frac{\partial c}{\partial t}=D\left[\frac{1}{r}\frac{\partial}{\partial r}\left(r\frac{\partial c}{\partial r}\right)+\frac{\partial}{\partial z}\left(\frac{\partial c}{\partial z}\right)\right]$$

This equation may be simplified further:

$$\frac{\partial c}{\partial t}=D\left[\left(\frac{\partial^2 c}{\partial r^2}\right)+\frac{1}{r}\frac{\partial c}{\partial r}+\frac{\partial^2 c}{\partial z^2}\right]$$

The boundary and the initial conditions are:

BC1 $c(t,R,z)=c_e$

BC2,3 $c(t, r, \pm L) = c_e \; t > 0$

BC4 $\left.\dfrac{dc}{dr}\right|_{r=0} = 0$

BC5 $\left.\dfrac{dc}{dz}\right|_{z=0} = 0$

IC: $c(0, r, z) = c_0$

Dimensionless variables $c = (c - c_e)/(c_0 - c_e)$, $\zeta = (r/R)$, $\eta = (z/L)$, $\tau = (tD/R^2)$ and $\ell = (L/R)$ may be introduced, then the partial differential equation, BCs and IC becomes:

$$\frac{\partial c}{\partial \tau} = \frac{\partial^2 c}{\partial \zeta^2} + \frac{1}{\zeta}\frac{\partial c}{\partial \zeta} + \frac{1}{\ell^2}\frac{\partial^2 c}{\partial \eta^2}$$

BC1 $c(\tau, 1, \eta) = 0$

BC2,3 $c t(\tau, \zeta, \pm 1) = 0 \; \tau > 0$

BC4 $\left.\dfrac{dc}{d\zeta}\right|_{\zeta=0} = 0$

BC5. $\left.\dfrac{dc}{d\eta}\right|_{\eta=0} = 0$

IC: $c(0, \zeta, \eta) = 1$

A new dependent variable may be defined to apply the method of the separation of variables:

$$c(\tau, \zeta, \eta) = \mathscr{R}(\tau, \zeta)\,\mathscr{Z}(\tau, \eta)$$

The dimensionless partial differential equation may be arranged as:

$$\mathscr{R}\left(\frac{1}{\ell^2}\frac{\partial^2 \mathscr{Z}}{\partial \eta^2} - \frac{\partial \mathscr{Z}}{\partial \tau}\right) + \mathscr{Z}\left(\frac{\partial^2 \mathscr{R}}{\partial \zeta^2} + \frac{1}{\zeta}\frac{\partial \mathscr{R}}{\partial \zeta} - \frac{\partial \mathscr{R}}{\partial \tau}\right) = 0$$

This equation is satisfied if:

$$\frac{\partial^2 \mathscr{R}}{\partial \zeta^2} + \frac{1}{\zeta}\frac{\partial \mathscr{R}}{\partial \zeta} - \frac{\partial \mathscr{R}}{\partial \tau} = 0$$

(unsteady state diffusion in radial direction)

with the BCs and the IC as

BC $\mathscr{R}(\tau, 1) = 0$

BC $\left.\dfrac{d\mathscr{R}}{d\zeta}\right|_{\zeta=0} = 0$

IC $\mathscr{R}(0, \zeta) = 1$

and

$$\frac{1}{\ell^2}\frac{\partial^2 \mathscr{Z}}{\partial \eta^2} - \frac{\partial \mathscr{Z}}{\partial \tau} = 0 \text{ (unsteady state diffusion in longitudinal direction)}$$

BC $\mathscr{Z}(\tau, 1) = 0$

BC $\left.\frac{d\mathscr{Z}}{d\eta}\right|_{\eta=0} = 0$

IC $\mathscr{Z}(0, \eta) = 1$

We may use the method of separation variables to solve the unsteady state diffusion in longitudinal direction: $\mathscr{Z}(\tau, \eta) = f(\eta)\,g(\tau)$ and the equation is rewritten as: $(1/\ell^2)(1/f)(d^2f/d\eta^2) = (1/g)(dg/d\tau) = -C^2$, individual equations and their solutions are

differential equation	solution
$\frac{dg}{d\tau} + C^2 g = 0$	$g = C_1 \exp(-C^2\tau)$ (see the previous examples)
$\frac{d^2f}{d\eta^2} + C^2\ell^2 f = 0$	$f = C_2 \sin(C\ell\eta) + C_3 \cos(C\ell\eta)$ (see Tab. 2.6)

The solution to our problem should be symmetric, but the sin function is not symmetric, i.e., $\sin(\alpha x) \neq \sin(-\alpha x)$, therefore $C_2 = 0$ and

$$\mathscr{Z}(\tau, \eta) = C_m \exp(-C^2\tau)\cos(C\ell\eta)$$

where $C_m = C_1 C_3$
The BC is $\mathscr{Z}(\tau, 1) = 0$, i.e., $0 = C_4 \exp(-C^2\tau)\cos(C\ell)$ and satisfied when

$C = \beta_m = \frac{(2m-1)\pi}{2\ell}$ and the final equation is:

$$\mathscr{Z}(\tau, \eta) = C_m \exp\{-(\beta_m)^2\tau\}\cos(\beta_m \ell\eta)$$

The IC is $\mathscr{Z}(0, \eta) = 1$, i.e., $1 = C_m \cos(\beta_m \ell\eta)$
after applying the orthogonality principle within the limits of $-1 \leqslant \eta \leqslant 1$ we obtain the remaining constant and the solution becomes

$$\mathscr{Z}(\tau, \eta) = \frac{2}{\ell}\sum_{m=1}^{\infty}\left[\frac{(-1)^{m+1}}{\beta_m}\right]\cos(\beta_m \ell\eta)\exp(-\beta_m^2\tau)$$

We may use the method of separation variables to solve the unsteady state diffusion in radial direction:

$$\mathscr{R}(\tau, \zeta) = u(\zeta)\, v(\tau)$$

After using similar procedures as in the previous examples we will get

$$v = C_4 \exp(-C^2 \tau)$$

and

$$u = C_5 J_0(C\xi) + C_6 Y_0(C\zeta)$$

BC $\left.\frac{d\mathscr{R}}{d\zeta}\right|_{\zeta=0} = 0$ requires $u =$ finite at $\zeta = 0$, but $Y_0(C\zeta) \to -\infty$, therefore $C_6 = 0$ and the solution becomes $\mathscr{R}(\tau, \zeta) = C_n \exp(-C^2\tau)\, J_0(C\zeta)$, where $C_n = C_4 C_5$.
BC $\mathscr{R}(\tau, 1) = 0$ requires $0 = C_n \exp(-C^2\tau)\, J_0(C)$, therefore $C = B_n =$ eigen values of $J_0(\zeta)$ and the solution becomes

$$\mathscr{R}(\tau, \zeta) = \sum_{n=1}^{\infty} C_n \exp(-B_n^2\tau) J_0(B_n\zeta)$$

We may use the orthogonality principle to determine C_n as in the previous examples: $C_n = \frac{2}{B_n J_1(B_n)}$, then the solution is:

$\mathscr{R}(\tau, \zeta) = \sum_{n=1}^{\infty} \frac{2}{B_n J_1(B_n)} \exp(-B_n^2\tau) J_0(B_n\zeta)$ and the overall solution to the problem is:

$$c(\tau, \zeta, \eta) = \mathscr{Z}(\tau, \eta)\, \mathscr{R}(\tau, \zeta) = \frac{c - c_e}{c_0 - c_e} = \frac{4}{\ell} \sum_{n=1}^{\infty} \sum_{m=1}^{\infty}$$

$$\left[\frac{(-1)^{m+1}}{B_n J_1(B_n) \beta_m}\right] J_0(B_n\zeta) \cos(\beta_m \ell \eta) \exp(-(\beta_m^2 + B_n^2)\tau)$$

Ece and Cihan (1993) suggested an empirical expression apparent diffusivity:

$$D(T, \nu) = 0.05915\, \nu^{-0.85} \exp\left\{-\frac{4706}{\nu^{0.076} T}\right\}$$

where $D =$ apparent diffusivity (m^2/h), $T =$ temperature (K) and $\nu =$ air velocity (m/s). In a diffusion controlled process diffusivity is not affected by air velocity, but since D is the apparent diffusivity, such a dependence is not surprising. Variation of the diffusivity with temperature is shown in Figure E.2.12.

Example 2.13 Salting of semi-hard ripened cheese During production of semi-hard ripened white cheese (feta cheese) salting is one of the principle

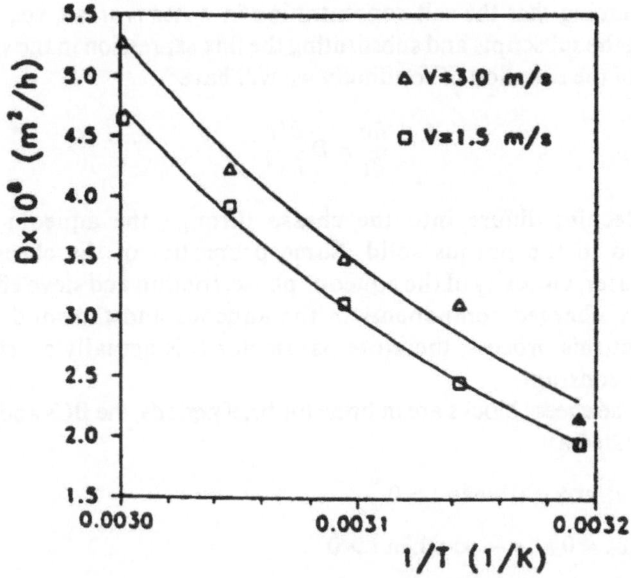

FIGURE E.2.12 Variation of the diffusion coefficient with temperature and air speed (Ece and Cihan, 1993, © ASAE, reproduced by permission).

steps. Cheese blocks are dipped into brine during the salting process, where salt diffuses into the cheese. The equation of continuity in the rectangular coordinate system may describe this process:

$$\frac{\partial c_A}{\partial t}+\left(\frac{\partial N_{Ax}}{\partial x}+\frac{\partial N_{Ay}}{\partial y}+\frac{\partial N_{Az}}{\partial z}\right)=R_A \tag{2.12}$$

When mass transfer is in the x direction only, i.e.,

$$\frac{\partial N_{Ay}}{\partial y}=\frac{\partial N_{Az}}{\partial z}=0$$

and salt does not involve into any chemical reaction, i.e., $R_A=0$ (2.12) is simplified as:

$$\frac{\partial c_A}{\partial t}+\frac{\partial N_{Ax}}{\partial x}=0$$

Salt flux in the x direction was expressed as:

$$N_{Ax}=x_A(N_A+N_B)-D_A\frac{dc_A}{dx} \tag{2.10}$$

After assuming that the salt concentration in water is small, i.e., $x_A \cong 0$, dropping the subscripts and substituting the flux expression in the simplified form of the equation of continuity we will have

$$\frac{\partial c}{\partial t} = D\frac{\partial^2 c}{\partial x^2}$$

Salt molecules diffuse into the cheese through the aqueous phase entrapped in the porous solid. Some properties of the cheese, i.e., bound water, viscosity of the aqueous phase, friction and sieve effects of the pores, charged components in the aqueous and the solid phases also affect this process, therefore parameter D is actually an effective diffusion constant.

When the cheese blocks are in brine for brief periods, the BCs and the IC may be stated as:

BC1. $c = c_1$ at $x = 0$ when $t > 0$

BC2. $c = c_0 = 0$ at $x \rightarrow \infty$ when $t > 0$

IC. $c = c_0 = 0$ at $x > 0$ when $t = 0$

We may define dimensionless concentration $\mathfrak{c} = (c - c_0)/(c_1 - c_0)$ and form a new variable $\eta = (x/\sqrt{4Dt})$. We may anticipate that the solution will be in the form of $\mathfrak{c} = \phi(\eta)$. This is called the *method of combination of variables.* There are many examples in the literature of this partial differential equation with the given boundary and the initial conditions which may be solved with this substitution. In terms of the new variables we will have $d\mathfrak{c}/dt = -(1/2)(\eta/t)\phi'$ and $d^2\mathfrak{c}/dx^2 = (\eta^2/x^2)\phi''$ where primes indicate differentiation with respect to η, then the partial differential equation becomes:

$$\phi'' + 2\eta\phi' = 0$$

with the boundary conditions

BC1 $\eta = 0$ $\phi = 1$

BC2 $\eta \rightarrow \infty$ $\phi = 0$

If ϕ' is replaced with ψ, then we get a first order separable equation $\psi' + 2\eta\psi = 0$ and the solution is:

$$\phi' = \psi = C_1 e^{-\eta^2}$$

Integrating once more gives

$$\phi = C_1 \int_0^{\eta} e^{-\eta^2} d\eta + C_2$$

After using BC1 we will obtain $C_2 = 1$. Constant C_1 may be evaluated after using BC2: $C_1 = (-1/\int_0^\infty e^{-\eta^2} d\eta) = (-(2/\sqrt{\pi})$ then the solution becomes

$$\phi = 1 - \frac{2}{\sqrt{\pi}} \int_0^\eta e^{-\eta^2} d\eta.$$

The error function is defined as: $\mathrm{erf}(\eta) = 2/\sqrt{\pi} \int_0^\eta \exp(-\eta^2) d\eta$ (Tab. 2.15), then the final form of the solution is $(c - c_0)/(c_1 - c_0) = 1 - \mathrm{erf}(x/2\sqrt{Dt})$ or $(c_1 - c)/(c_1 - c_0) = \mathrm{erf}\,(x/2\sqrt{Dt})$

Turhan and Kaletunç (1992) reported that the value of the effective diffusivity was affected by the duration of the experiments and found $D = 0.33\ 10^{-9}$ m^2/s during 3 days of brining, but the diffusivity became $0.31\ 10^{-9}$ m^2/s when the brining time was 9 days. Comparison of their experimentally determined salt concentration measurements with the model are shown in Figure E.2.13. It should be noticed that considering the cheese block as an infinite body may be valid during the initial stages of the brining process, but this assumption is not valid in the later stages of the experiments, where salt profiles also develop at the depths of the cheese.

2.8. CHART SOLUTIONS TO UNSTEADY STATE CONDUCTION PROBLEMS

In most thermal processes solid foods are heated in a liquid medium, i.e., blanching of potatoes. The ratio of conductive resistance in a solid to convective resistance in a fluid is described by the Biot number ($Bi = hR/k$, $Bi = h\,L/k$, etc.). Biot number may have values between 0 and ∞. When $k \to \infty$ ($Bi \to 0$) internal resistance to heat transfer is negligible, therefore the solid may be assumed to be at uniform temperature and the heat transfer rate to the solid food may be expressed as:

$$\rho c V \frac{dT}{dt} = hA(T_1 - T) \tag{2.37}$$

where ρ, c A and V are density, specific heat, effective heat transfer area and volume of the food, respectively, h is convective heat transfer coefficient, T and T_1 are the food and the medium temperatures, respectively. When $Bi \to \infty$ ($h \to \infty$) the solid surface attains the temperature of the heating medium immediately with the contact of the phases. Under these conditions the external heat transfer resistance disappears and the temperature

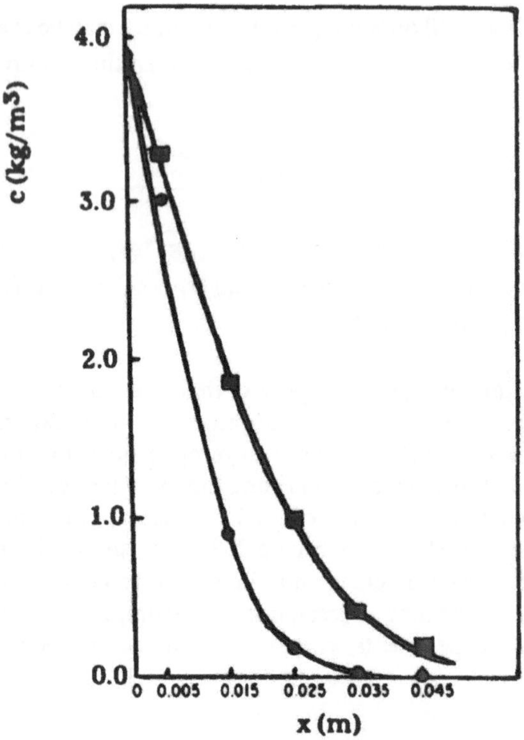

FIGURE E.2.13 Experimental and theoretical salt concentration profiles in white cheese. -■-: concentration profile after 9 days of brining, -●-: concentration profile after 3 days of brining, (Turhan and Kaletunç, 1992; © IFT, redrawn by permission).

Table 2.16 Definition of Error function and its important properties

error function erf(η) of the argument η	$\operatorname{erf}(\eta) = \frac{2}{\sqrt{\pi}} \int_{\zeta=0}^{\zeta=\eta} \exp(-\zeta^2)\,d\zeta$
complimentary error function erfc(η) of the argument η	$\operatorname{erfc}(\eta) = 1 - \operatorname{erf}(\eta) = 1 - \frac{2}{\sqrt{\pi}} \int_{\zeta=\eta}^{\zeta=\infty} \exp(-\zeta^2)\,d\zeta$
	$\frac{d}{d\eta}[\operatorname{erf}(\eta)] = \frac{2}{\sqrt{\pi}} \exp(-\eta^2)$
	$\operatorname{erf}(\infty) = 1, \operatorname{erf}(-\eta) = -\operatorname{erf}(\eta)$

Table 2.17 Values of erf(η), erfc(η) and $2/\sqrt{\pi}\exp(-\eta^2)$ (For extensive list of the mathematical functions given in Table 2.17 an interested reader may refer to *Handbook of Mathematical Functions*, Abrahamovitz, M. and Stegun, I. (editors). Dover Publications, New York, 1965)

η	erf(η)	erfc(η)	$\frac{2}{\sqrt{\pi}}\exp(-\eta^2)$	η	erf(η)	erfc(η)	$\frac{2}{\sqrt{\pi}}\exp(-\eta^2)$
0	0	1.00	1.128	0.80	0.742	0.258	0.595
0.05	0.056	0.944	1.126	0.85	0.771	0.229	0.548
0.10	0.112	0.888	1.117	0.90	0.797	0.203	0.502
0.15	0.168	0.832	1.103	0.95	0.821	0.179	0.458
0.20	0.223	0.777	1.084	1.00	0.843	0.157	0.415
0.25	0.276	0.724	1.060	1.10	0.880	0.120	0.337
0.30	0.329	0.671	1.031	1.20	0.910	0.090	0.267
0.35	0.379	0.621	0.998	1.30	0.934	0.066	0.208
0.40	0.428	0.572	0.962	1.40	0.952	0.048	0.159
0.45	0.475	0.525	0.922	1.50	0.966	0.034	0.119
0.50	0.520	0.480	0.879	1.60	0.976	0.024	0.087
0.55	0.563	0.437	0.834	1.70	0.984	0.016	0.063
0.60	0.604	0.396	0.787	1.80	0.989	0.011	0.044
0.65	0.642	0.378	0.740	1.90	0.993	0.007	0.030
0.70	0.678	0.322	0.691	2.00	0.995	0.005	0.021
0.75	0.711	0.289	0.643				

profile established only in the solid phase. Qualitative temperature profiles in a liquid/solid system are shown in Figure 2.4.

Chart solutions to the conduction problems are frequently used in Bioprocess and Food Engineering analysis when both the internal and the external heat transfer resistances are not negligible in the following geometries:

i) Infinite plate of thickness $2L$, for which $T = T(x, t)$ and x is measured from the plate centerline
ii) Infinitely long cylinder with outside radius R, for which $T = T(r, t)$ and r is measured from the centerline
iii) Sphere with radius R, for which $T = T(r, t)$ and r is measured from the center.

The boundary conditions for all three geometries are similar. The first boundary condition specifies the minimum in the temperature profile at the midplane location of the infinite plane, centerline of the infinitely long cylinder or center of the sphere:

$$\left.\frac{dT}{dx}\right|_{x=0} = \left.\frac{dT}{dr}\right|_{r=0} = 0 \tag{2.38}$$

The second boundary condition requires that the heat transferred from the exterior surface of the solid is removed by a fluid with an ambient

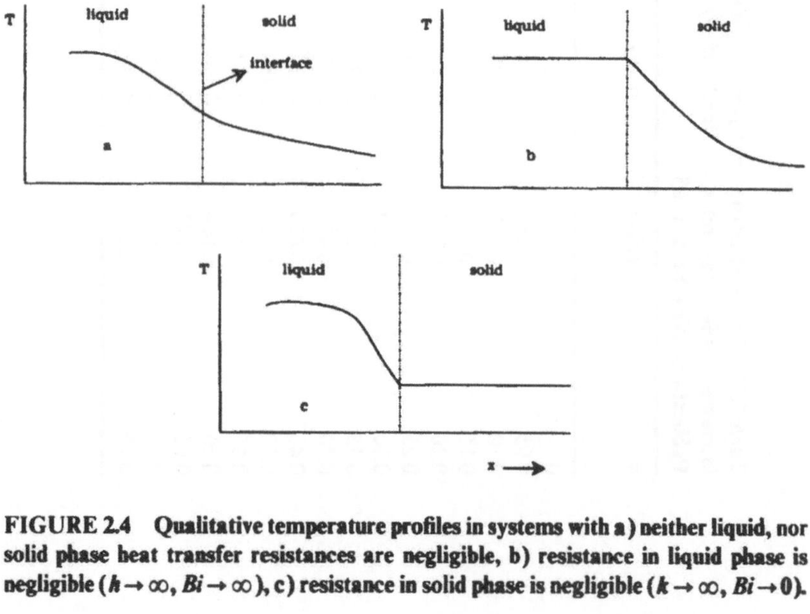

FIGURE 2.4 Qualitative temperature profiles in systems with a) neither liquid, nor solid phase heat transfer resistances are negligible, b) resistance in liquid phase is negligible ($h \to \infty$, $Bi \to \infty$), c) resistance in solid phase is negligible ($k \to \infty$, $Bi \to 0$).

temperature T_1 and a heat transfer coefficient h. This boundary condition may be expressed mathematically for a slab as:

$$h(T_s - T_1) = -k\left.\frac{dT}{dx}\right|_s \tag{2.39.a}$$

and for sphere or cylinder as:

$$h(T_s - T_1) = -k\left.\frac{dT}{dr}\right|_s \tag{2.39.b}$$

where subscript s indicates the quantity evaluated at the surface conditions. There is a second possible boundary condition to substitute (2.39.b) if $h \to \infty$, i.e., thermal resistance of the convection layer is negligible and the surface temperature of the solid equals that of the surrounding fluid.

$$T_s = T_1 \tag{2.39.c}$$

The initial conditions for these geometries may be

$$T = T_0 \text{ when } t = 0 \text{ for all } x \text{ or } r \tag{2.40}$$

Temperature distribution is presented in graphical form (Heisler, 1947) in terms of the dimensionless temperature $Y = (T_1 - T)/(T_1 - T_0)$, Fourier number $(\alpha t/L^2)$ (or $\alpha t/R^2$) and the Biot number (hL/k) or (hR/k).

Relations between these dimensionless groups are presented for the midplate of the slab (Fig. 2.5.a), centerline of the cylinder (Fig. 2.6.a) and center of the sphere (Fig. 2.7.a) or for any location as a function of the mid plate temperature (T_c) of the slab (Fig. 2.5.b), centerline temperature (T_c) of the cylinder (Fig. 2.6.b) and the center temperature (T_c) of the sphere (Fig 2.7.b). In these cases an additional dimensionless parameter $(T_1 - T)/(T_1 - T_c)$ is used.

The use of one dimensional Heisler charts may be extended to two and three dimensional problems. This method involves using the product of the

Table 2.18 Dimensionless positions and groups used to construct the Heisler charts

Geometry	Position	Biot number	Fourier number
infinite plate with thickness $2L$	$\frac{x}{L}$	$\frac{hL}{k}$	$\frac{\alpha t}{L^2}$
infinite cylinder with radius R	$\frac{r}{R}$	$\frac{hR}{k}$	$\frac{\alpha t}{R^2}$
sphere with radius R	$\frac{r}{R}$	$\frac{hR}{k}$	$\frac{\alpha t}{R^2}$

Y values from the one dimensional charts. The basis for obtaining two or three dimensional solutions from one dimensional charts is similar to the reasoning of the method of separation of variables. When the external heat transfer resistance is negligible we may use the chart given in Example 4.46 to calculate the average temperature of the solids immersed in a heating medium.

Example 2.14. Estimation of processing time and local temperatures during thermal processing of a conduction heating food by using charts A conduction heating food with $T_0 = 4°C$, $k = 0.63$ W/mK, $c_p = 4.19$ kJ/kg K, $\rho = 978$ kg/m^3 is sterilized in 307 x 306 cans at retort temperature $T_1 =$ 110°C. For complete sterilization the critical point temperature must reach 100°C. The heat transfer coefficient between can and steam is $h_{outside} = 250$ W/m^2 K. i) Estimate the process time, ii) Estimate the temperature of the food at the end of the process at the point where $r = 0.5\,R$, $x = 0.5\,L$ (radius of the can = R, height of the can = $2L$)

Solution
$$L = \left(\frac{1}{2}\right)\left(3 + \frac{7}{16}\right)(2.54) = 4.36 \text{ cm},$$

$$R = \left(\frac{1}{2}\right)\left(3 + \frac{6}{16}\right)(2.54) = 4.29 \text{ cm}, \quad \alpha = \frac{k}{\rho c_p} = 9.25\,10^{-7}\ \text{m}^2/\text{min}$$

(See example 4.6 for the size conversion)

i) Determination of the process time (solutions at the can center)

We will first calculate the Biot number for the infinite slab and the infinite cylinder, then estimate the process time, then calculate the Fourier numbers and use the charts to calculate Y_{slab} and $Y_{cylinder}$. Since $Y_{can} = Y_{slab}\,Y_{cylinder}$ we may calculate $Y_{can} = (T_1 - T)/(T_1 - T_0)$ and T. If $T = 100°C$, the estimated process time is correct. We will repeat the trial and error solution until we obtain the correct processing time.

Geometry	Position	inverse Biot number	Fourier number ($t = 85$ min estimated)	Y (from charts)
infinite slab (thickness = $2L$)	$\frac{x}{L} = 0$	$\frac{k}{hL} = 0.058$	$\frac{\alpha t}{L^2} = 0.041$	0.3 (Figure 2.5.a)
infinite cylinder (radius = R)	$\frac{r}{R} = 0$	$\frac{k}{hR} = 0.058$	$\frac{\alpha t}{R^2} = 0.041$	0.3 (Figure 2.6.a)

$Y_{can} = Y_{slab}\, Y_{cylinder} = 0.09, \quad Y_{can} = \frac{T_1 - T}{T_1 - T_0},$ therefore $0.09 = \frac{110 - T}{110 - 4} =$ 100.4 °C ≅ 100 °C, therefore the processing time should be 85 minutes.

ii) Calculations at the point where $r = 0.5\,R$, $x = 0.5\,L$ at the end of the process

Geometry	Position	inverse Biot number	$\frac{T_1 - T}{T_1 - T_c}$ (from charts)
infinite slab (thickness = $2L$)	$\frac{x}{L} = 0.5$	$\frac{k}{hL} = 0.058$	0.74 (Figure 2.5.b)
infinite cylinder (radius = R)	$\frac{r}{R} = 0.5$	$\frac{k}{hR} = 0.058$	0.72 (Figure 2.6.b)

$$\left.\frac{T_1 - T}{T_1 - T_c}\right|_{can} = \left.\frac{T_1 - T}{T_1 - T_c}\right|_{slab} \times \left.\frac{T_1 - T}{T_1 - T_c}\right|_{cylinder} = 0.53$$

$\frac{110 - T}{110 - 100} = 0.53$, therefore $T = 104.7$°C.

2.9. INTERFACIAL MASS TRANSFER

Mass transfer usually occurs between two phases in Food and Bioprocess Engineering operations. Water transport from food materials to air during drying, diffusion of oxygen through the protective packaging to food materials during storage, oxygen transfer from a bubble to broth during fermentation are among the common examples. A typical plot describing the variation of the mass fractions of the diffusing compound around the interface is shown in Figure 2.8.

During interfacial mass transfer the driving force is the concentration gradient within each phase, with no resistance to mass transfer at the interface. There is usually a concentration jump at the interface, but the interfacial concentrations are in local equilibrium, which may be described with Henry's law or another equilibrium relation:

$$y_i = \mathscr{H}\, x_i \tag{2.41}$$

where y_i and x_i are mole fractions at the gas and liquid sides of the interface, respectively and $\mathscr{H}$ is the partition coefficient;

$$y_i = \frac{H}{P_{tot}}\, x_i \tag{2.42}$$

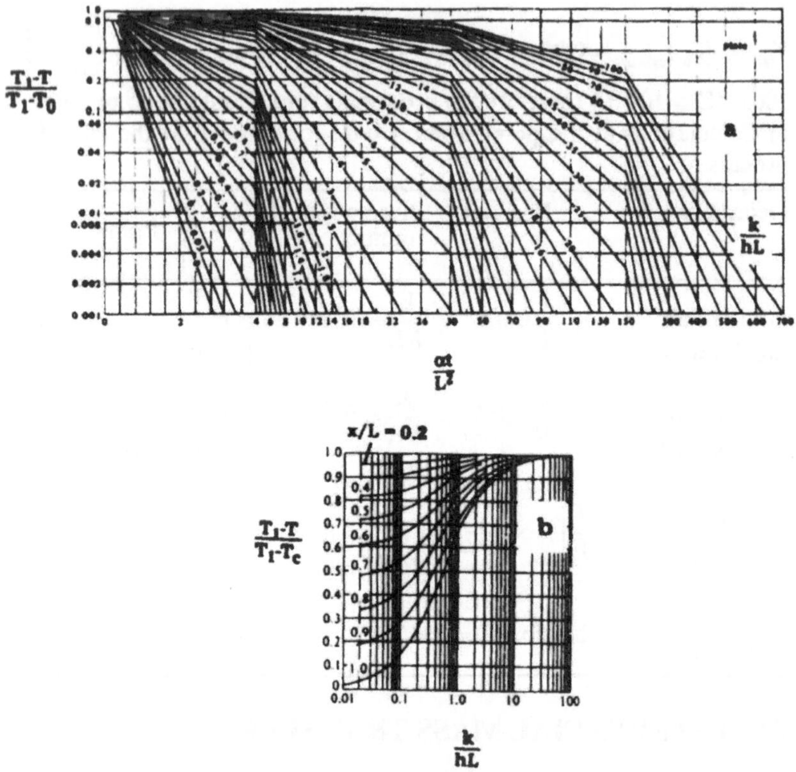

FIGURE 2.5 (a) Dimensionless midplane temperature for an infinite plane with thickness 2L (Heisler, 1947, © ASME, reproduced by permission). (b) Dimensionless local temperature for an infinite plane with thickness 2L as a function of midplane temperature (Heisler, 1947, © ASME, reproduced by permission).

where H is the Henry's Law constant and P_{tot} is the total pressure in the gas side. Although there is local equilibrium at the interface, this does not imply that the bulk average mole fractions of the two phases are in equilibrium. Interfacial mass transfer may be simulated with the following empirical equations:

$$N_A = k_G(y_{A\infty} - y_{Ai}) \qquad (2.43.a)$$

$$N_A = k_L(x_{Ai} - x_{A\infty}) \qquad (2.43.b)$$

$$N_A = \mathscr{K}_G(y_{A\infty} - y^*) \qquad (2.43.c)$$

$$N_A = \mathscr{K}_L(x_A^* - x_{A\infty}) \qquad (2.43.d)$$

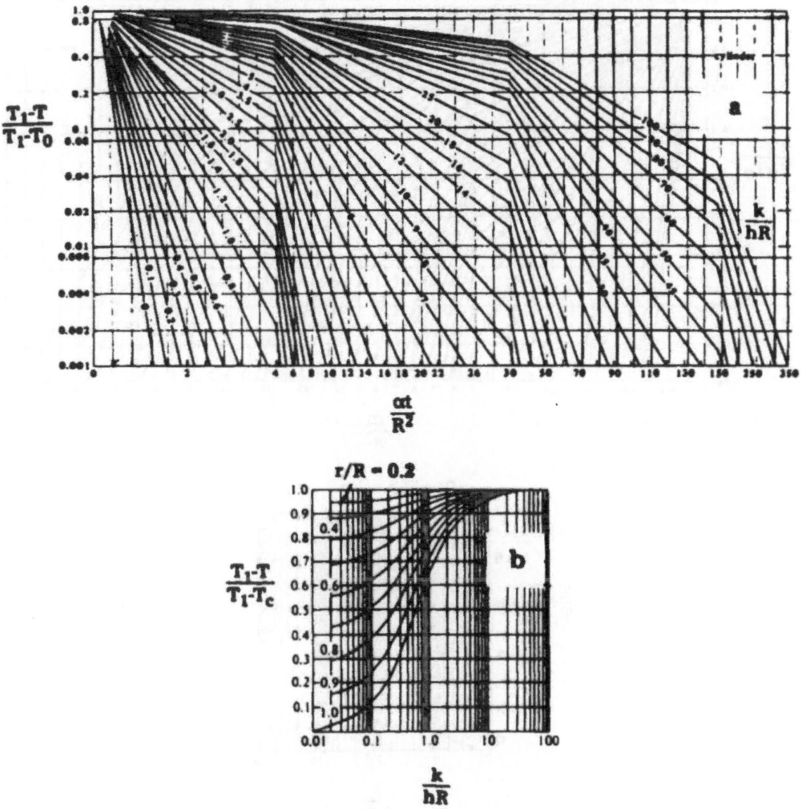

FIGURE 2.6 **(a) Dimensionless axis temperature for an infinite cylinder with radius *R* (Heisler, 1947, © ASME, reproduced by permission). (b) Dimensionless local temperature for an infinite cylinder with radius *R* as a function of axis temperature (Heisler, 1947, © ASME, reproduced by permission).**

$$N_A = \mathscr{k}_c(c_{Ai} - c_{A\infty}) \tag{2.43.e}$$

$$N_A = \mathscr{K}_c(c_A^* - c_{A\infty}) \tag{2.43.f}$$

where $y_{A\infty}$ and $x_{A\infty}$ are the bulk mole fractions of species *A* far from the interface in the gas and the liquid phases, respectively; x_A^* is the mole fraction of species *A* in equilibrium with $y_{A\infty}$ and y_A^* is the mole fraction of species *A* in equilibrium with $x_{A\infty}$ (Figs. 2.8 and 2.9). Equations (2.43.e) and (2.43.f) are essentially the same as (2.43.b) and (2.43.d), but written in terms of concentrations, instead of the mole fractions. Equations (2.43.a) – (2.43.f) are empirical equations, because they consider the mass transfer rate a constant fold of the driving force. Mass transfer coefficients $\mathscr{k}_G$, $\mathscr{k}_L$, $\mathscr{K}_G$, $\mathscr{K}_L$, $\mathscr{k}_c$ and $\mathscr{K}_c$ are actually the proportionality constants. The equilibrium curve and the driving forces shown in Figure 2.9.

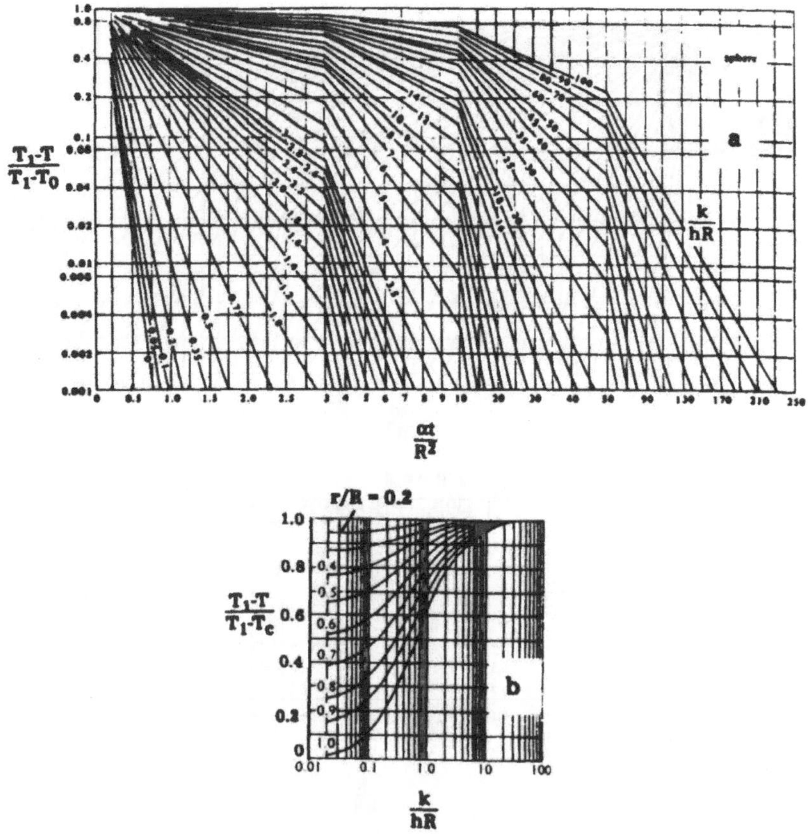

FIGURE 2.7 (a) Dimensionless center temperature for a sphere with radius *R* (Heisler, 1947, © ASME, reproduced by permission). (b) Dimensionless local temperature for a sphere with radius *R* as a function of center temperature (Heisler, 1947, © ASME, reproduced by permission).

2.10. CORRELATIONS FOR PARAMETERS OF THE TRANSPORT EQUATIONS

Density, specific heat, thermal conductivity, viscosity, diffusivity and mass transfer coefficient are the constants appearing in the transport equations (2. 4) – (2.7), (2.37), (2.39) and (2.43). The density, specific heat, thermal conductivity and diffusivity are usually correlated to the food composition. Convective heat transfer coefficient h and mass transfer coefficient k are usually expressed in terms of the dimensionless numbers. Some examples to these correlations are presented here. A discussion of all the correlations available in the literature is beyond the scope of the book; more information may be found in the references.

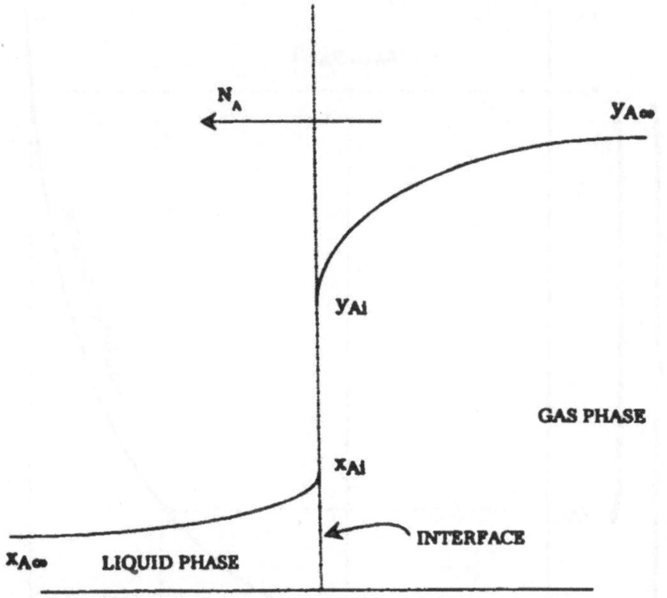

FIGURE 2.8 Variation of the mass fractions of a diffusing compound around the interface.

i) Density of dried vegetables (Lozano *et al.*, 1983)

A correlation relating the density of the bulk, sliced and particle carrots, pear, potato, sweet potato and garlic to the moisture content is presented as:

$$\rho = h + \ell \frac{x}{x_0} + p \exp\left(-q \frac{x}{x_0}\right) \tag{2.44}$$

where ρ = density (kg/m^3), x = moisture content (kg water/kg dry matter), x_0 = initial moisture content (kg water/kg dry matter); h, ℓ, p and q are constants. Values of these constants for two typical examples are:

Foodstuff	density	h	ℓ	p	q
carrot	bulk	0.984	0	0.224	1.800
carrot	particle	1.497	−0.294	−0.253	39.793
garlic	bulk (sliced)	1.130	−0.567	0.187	−0.866
garlic	particle	2.694	0	−1.316	−1.1638

ii) Specific heat

Siebel or Choi and Okos equations may be used to predict the specific heats of the foods. Siebel's equation (ASHRAE, 1965) may be stated above

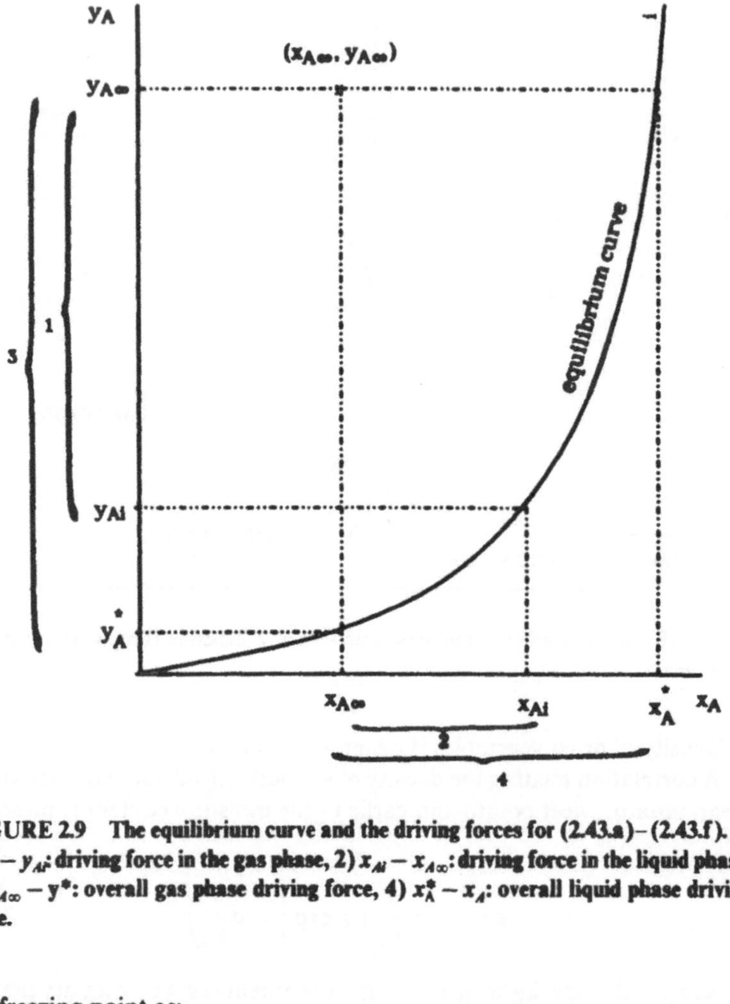

FIGURE 2.9 The equilibrium curve and the driving forces for (2.43.a)–(2.43.f). 1) $y_{A\infty} - y_{Ai}$: driving force in the gas phase, 2) $x_{Ai} - x_{A\infty}$: driving force in the liquid phase, 3) $y_{A\infty} - y^*$: overall gas phase driving force, 4) $x_A^* - x_A$: overall liquid phase driving force.

the freezing point as:

$$c = 1674.72\,\omega_f + 837.36\,\omega_s + 4186.8\,\omega_w \tag{2.45.a}$$

and below the freezing point as

$$c = 1674.72\,\omega_f + 837.36\,\omega_s + 2093.4\,\omega_w$$

where c = specific heat of the food (J/kg K), ω_f = mass fraction of fat, ω_s = mass fraction of non-fat solids, ω_w = mass fraction of water in the food.

Choi and Okos (1987) gives the correlations for the specific heats of the foods above the freezing point as:

$$c = \omega_p c_p + \omega_f c_f + \omega_c c_c + \omega_{fi} c_{fi} + \omega_a c_a + \omega_w c_{waf} \tag{2.46.a}$$

where c, c_p, c_f, c_c, c_{fi}, c_a and c_{waf} are specific heats of the food, proteins, fat, carbohydrates, fibers, ash and water (above the freezing point), respectively (all are in J/kg K); ω_p, ω_f, ω_c, ω_{fi}, ω_a and ω_w represents the mass fractions of proteins, fat, carbohydrates, fiber, ash and moisture, respectively. The following equations are used to predict the individual specific heats:

$$c_p = 2008.2 + 1208.9 \times 10^{-3} T - 1312.9 \times 10^{-6} T^2 \quad (2.46.b)$$

$$c_f = 1984.2 + 1473.3 \times 10^{-3} T - 4800.8 \times 10^{-6} T^2 \quad (2.46.c)$$

$$c_c = 1548.8 + 1962.5 \times 10^{-3} T - 5939.9 \times 10^{-6} T^2 \quad (2.46.d)$$

$$c_{fi} = 1845.9 + 1930.6 \times 10^{-3} T - 4650.9 \times 10^{-6} T^2 \quad (2.46.e)$$

$$c_a = 1092.6 + 1889.6 \times 10^{-3} T - 3681.7 \times 10^{-6} T^2 \quad (2.46.f)$$

and

$$c_{waf} = 4176.2 - 9.0862 \times 10^{-5} T - 5473.1 \times 10^{-6} T^2 \quad (2.46.g)$$

Siebel's equation assumes that all types of nonfat solids have the same specific heat and, all water is frozen below the freezing point. However these assumptions may not always be correct and lead to inaccurate results. Choi-Okos equation generally predicts higher values than Siebel's equation at high moisture contents. Siebel's equation has been found to agree closely with the experimental data when $\omega_w > 0.7$ and when no fat is present. Choi-Okos equation is more accurate at low moisture contents and a wider range of product composition, since it is based on published experimental values for wide variety of foods. Siebel's equation is simpler than the Choi-Okos equation, therefore it is preferred in calculations where error tolerance is not too low (Toledo, 1991).

iii) Thermal conductivity of meat

Thermal conductivity is commonly related to the moisture content of non-fat meat when heat flow is non-perpendicular to the fibers as (Pham and Willix, 1989):

$$k = 0.080 + 0.52x \quad (0 < T < 60\ ^\circ\mathrm{C}) \quad (2.47.a)$$

or

$$k = -0.28 + 1.9x - 0.0092T \quad (-40 < T < -5\ ^\circ\mathrm{C}) \quad (2.47.b)$$

where x = moisture content (kg water/kg dry matter), k is thermal conductivity (W/mK). There are also correlations relating the thermal conductivity to structure (Pham and Willix, 1989):

Model	Equation
parallel model	$k_{eff} = \sum_{i=1}^{n} v_i k_i \qquad (2.48)$
series model	$\frac{1}{k_{eff}} = \sum_{i=1}^{n} \frac{v_i}{k_i} \qquad (2.49)$
Maxwell-Eucken model (used with food dispersions) Water/ice or fat = continuous phase; other constituents = dispersed phase	$k_{eff} = k_c \frac{1 - 2rv_d}{1 + rv_d} \qquad (2.50)$ where $r = \frac{k_c - k_d}{2k_c + k_d}$
Levy's model (replace v_d in Maxwell-Eucken model with function F). Proposed to obtain the same numerical result when the continuous phase becomes the dispersed phase and vice versa	$2F = \frac{2}{s} - 1 + 2v_d - \left[\left(\frac{2}{s} - 1 + 2v_d\right)^2 - \frac{8v_d}{s}\right]^{1/2}$ where $s = \frac{(k_d - k_c)^2}{(k_d + k_c)^2 + \frac{k_d k_c}{2}}$
Hill-Leitman-Sunderland model (Simulates heat transfer through a network in a continuous phase with simultaneous parallel and series conduction)	$k_{eff} = (2\phi - \phi^2)k_d + (1 - 4\phi + 3\phi^2)k_c + 8(\phi - \phi^2)\frac{k_c k_d}{\phi k_c + (4 - \phi)k_d} \qquad (2.51)$ where $\phi = 2 - \sqrt{4 - 2v_d}$

k_{eff} = effective thermal conductivity (W/mK)
k_c = thermal conductivity of the continuous phase (W/mK)
k_d = thermal conductivity of the dispersed phase (W/mK)
k_i = thermal conductivity of ith component (W/mK)
v_d = volume fraction of the dispersed phase
v_i = volume fraction of ith component

An interested reader is referred to Miles *et al.* (1983) for extensive coverage of the equations for thermal conductivity in foods.

iv) Viscosity of microbial suspensions

Most foods and biological solutions contain suspended solids. Atkinson and Mavituna (1991) recommend using the following equations to predict viscosity of the microbial suspensions:

Model	Equation	
Kunitz equation	$\mu_m = \mu_1 \left\{ \frac{1 + 0.5\phi_s}{(1 - \phi_s)^4} \right\}$ used when $\phi_s < 0.4$	(2.52)
Mori and Ototake equation	$\mu_m = \mu_1 \left\{ 1 + \frac{1.56\phi_s}{0.52 - \phi_s} \right\}$ used when $\phi_s \leqslant 0.1$	(2.53)
Mooney's correlation (used when experimental data are available to determine the adjustable parameter *c*)	$\ln \frac{\mu_m}{\mu_l} = 1 + \frac{2.5\phi_s}{1 - c\phi_s}$	(2.54)
Einstein equation (used with yeast suspensions at low volume fractions)	$\mu_m = \mu_l + (1 + \phi_s)$	(2.55)
Vand equation (used to estimate viscosity of yeast suspensions up to $\phi_s = 0.14$)	$\mu_m = \mu_l + (1 + 2.5\phi_s + 7.25\phi_s^2)$	(2.56)

ϕ_s = fraction of the suspended solids
μ_l = viscosity of the liquid phase
μ_s = viscosity of the solid suspension

v) Moisture diffusivity in granular starch (Vagenas and Karathanos, 1991)

There are some equations available for porous solids which relate the effective (apparent) diffusivity to the diffusivities in each phase, these are empirical equations which include:

Model	Equation	
parallel model	$D_{eff} = (1 - \varepsilon) D_s + \varepsilon D_g$	(2.57)
series model	$\frac{1}{D_{eff}} = \frac{(1 - \varepsilon)}{D_s} + \frac{\varepsilon}{D_g}$	(2.58)
random model	$D_{eff} = D_s^{1-\varepsilon} D_g^{\varepsilon}$	(2.59)
mixed model	$\frac{1}{D_{eff}} = \frac{(1 - f)}{(1 - \varepsilon) D_s + \varepsilon D_g} + f \left[\frac{1 - \varepsilon}{D_s} + \frac{\varepsilon}{D_g} \right]$	(2.60)
Topper model	$\frac{1}{D_{eff}} = \left[\frac{1 - \varepsilon^{3/4}}{D_s} + \frac{\varepsilon^{1/3}}{\varepsilon^{1/3} D_g + D_s(1 - \varepsilon^{1/3})} \right]$	(2.61)
Bahren's model	$D_{eff} = D_s \frac{(p + 1) + (p - 1)\varepsilon}{(p + 1) - (p - 1)\varepsilon}$	(2.62)
Maxwell model	$D_{eff} = D_s \frac{D_g + 2D_s - 2\varepsilon(D_s - D_g)}{D_g + 2D_s + \varepsilon(D_s - D_g)}$	(2.63)

$p = D_g/D_s$
D_{eff} = effective diffusivity
D_p = diffusivity in the solid
D_g = diffusion in the gas phase
f = constant
ε = porosity

vi) Convective heat transfer coefficients during heat transfer to canned foods in steritort (Rao *et al.*, 1985)

System	correlation and range of the dimensionless groups	
heat transfer to canned Newtonian liquids in steritort (Rao *et al.*, 1985)	$Nu_D = 0.135\,(Gr\,Pr)^{0.323} + 0.391 \times 10^{-3} \left(Re_D\,Pr\frac{D}{L}\right)^{1.369}$ $24 \leqslant Nu_D \leqslant 272$ $0.4 \leqslant Re_D \leqslant 458$ $2.8 \leqslant Pr \leqslant 476$ $2.0 \times 10^4 \leqslant Gr \leqslant 2.9 \times 10^9$ $1.13 \leqslant L/D \leqslant 1.37$	(2.64.a)
heat transfer to canned non-Newtonian liquids in steritort (Rao *et al.*, 1985) (see 2.11, Rheological models)	$Nu_D = 2.60\,(Gr^*\,Pr^*)^{0.205} + 7.15 \times 10^{-7} \left(Re_D^*\,Pr^*\frac{D}{L}\right)^{1.837}$ $24 \leqslant Nu_D \leqslant 160$ $0.4 \leqslant Re_D^* \leqslant 96$ $205 \leqslant Pr^* \leqslant 5100$ $100 \leqslant Gr_D^* \leqslant 5.2 \times 10^5$ $1.13 \leqslant L/D \leqslant 1.37$	(2.64.b)

D = can diameter (m)
g = gravitational acceleration (m/s^2)

$Gr = \frac{gD^3\rho^2\beta\Delta T}{\eta^2}$ = Grashof number based on can diameter (dimensionless)

$Gr^* = \frac{gD^3\rho^2\beta\Delta T}{\eta_{aM}^2}$ = Generalized Grashof number for the non-Newtonian fluids (based on can diameter, dimensionless)

h = convective heat transfer coefficient (W/m^2 K)
k = thermal conductivity (W/m K)
K = consistency index of power law fluids (N s^n/m^2) (defined in (2.71))
L = length of can (m)
n = Flow behavior index of power law fluids (defined in (2.71))
N = Revolutions per second

$Nu_D = \frac{hD}{k}$ = Nusselt number (based on can diameter, dimensionless)

$Pr = \frac{c\eta}{k}$ = Prandl number (dimensionless)

$Pr^* = \frac{c\eta_a}{k}$ = Generalized Prandl number for the non-Newtonian fluids (dimensionless)

$Re_D = \frac{D^2 N \rho}{\eta}$ = Reynolds number (based on can diameter, dimensionless)

$Re_D^* = \frac{D^2 N^{2-n} \rho}{8^{n-1} K \left(\frac{3n+1}{4n}\right)^n}$ = Generalized Reynolds number for the non-Newtonian fluids (based on can diameter, dimensionless)

β = coefficient of volumetric expansion (1/°C)

η = viscosity (Pa/s)

$\eta_a = \frac{K\, 8^{n-1}}{N^{1-n}} \left(\frac{3n+1}{4n}\right)^n$ = apparent viscosity for the non-Newtonian liquids

ρ = density (Kg/m^3)

vii) Mass transfer coefficient k for oxygen transfer in fermenters

Oxygen transfer from gas phase, i.e., bubbles, to the fermentation broth, is among the most important phenomena determining the biomass and product yield in fermentation processes. Atkinson and Mavituna (1991) presented a detailed review of the correlations available in literature. Bailey and Ollis (1986) recommended the following correlations for oxygen transfer from the air bubbles to fermentation broth:

Model		Range of the dimensionless groups
$Sh = 1.01\, Pe^{1/3}$	(2.65)	$Re \ll 1$, $Pe \gg 1$
$Sh = 2.0 + 0.60\, Re^{1/2} Sc^{1/3}$	(2.66)	$Re \gg 1$
$Sh = 0.31\, Gr^{1/3}\, Sc^{1/2}$	(2.67)	$d_b < 2.5$ mm
$Sh = 0.42\, Gr^{1/3}\, Sc^{1/2}$	(2.68)	$d_b > 2.5$ mm

d_b = bubble diameter (m)

D = molecular diffusivity of oxygen in fermentation broth (m^2/s)

g = gravitational acceleration (m/s^2)

Gr = Grash of number $= \frac{d_b^3 \rho_c(\rho_c - \rho_g) g}{\mu_c^2}$ (dimensionless)

Pe = Peclet number $= \frac{d_b v}{D}$ (dimensionless)

Re = Reynolds number $= \frac{d_b v \rho_c}{\mu}$ (dimensionless)

Sc = Schmidt number $= \frac{k\, d_b}{D}$ (dimensionless)

Sh = Sherwood number $= \frac{\mu_c}{\rho_c D}$ (dimensionless)

v = approaching velocity far away from the bubble (m/s)
μ_c = viscosity of the continuous phase (N s/m^2)
ρ_c = density of the continuous phase (kg/m^3)
ρ_g = density of the gas phase (kg/m^3)

Example 2.15. Drying behavior of frozen beef (*Tütüncü et al.*, 1990) Meat is stored in a refrigerated, as well as in a frozen state, or aged at refrigeration temperatures for one or two weeks to permit tenderization. When a refrigerated or frozen meat slab is stored without an adequate moisture barrier, it loses weight by evaporation of water or sublimation of ice. Equation (2.43.d) describes interfacial mass transfer:

$$N_A = \mathcal{K}_L (x^* - x) \tag{2.43.d}$$

Total mass flux N_A was expressed calculated as:

$$N_A = \frac{1}{A} V \frac{dx}{dt}$$

where A = interfacial area, $V = A\,h$ = volume of the beef slab, h = thickness of the slab, x = average fraction of water in the beef. After combining these expressions we obtain

$$\frac{dx}{dt} = -K(x - x^*)$$

where $K = \mathcal{K}/h$. The water content of lean beef is very high (about 76%), and the equilibrium water content may be assumed negligible. The term "equilibrium water content" refers to the water or ice content of beef in equilibrium with the ambient air. Since drying rates are very slow at the refrigeration or freezing temperatures, numerical values of x do not change much and remain always substantially greater than x^*, thus mass transfer equation may be rewritten as:

$$\frac{dx}{dt} = -Kx$$

upon integration we will obtain

$$\ln x = \ln x_0 - Kt$$

This is an equation of a line with independent variable t, dependent variable $\ln x$, intercept (with $1/t = 0$ axis) $\ln x_0$ and slope K. Figure E.2.15.1 shows

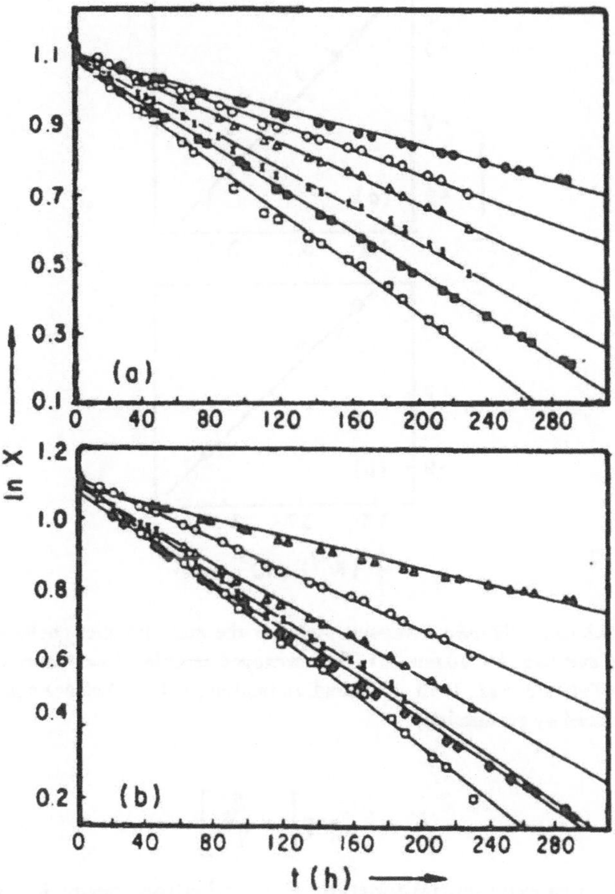

FIGURE E.2.15.1 Sample plots for comparison of the drying model to the data (Tütüncü *et al.*, 1990, © Canadian Institute of Food Science and Technology, reproduced by permission). a) Unwrapped meat and ground beef samples at −5°C. b) Cloth wrapped meat and ground beef samples at −5°C. Symbols: ○ Meat cut in parallel direction with the fibers, $h = 20$ mm, △ Meat cut in perpendicular direction with the fibers, $h = 20$ mm x, Meat cut in parallel direction with the fibers, $h = 10$ mm □ Meat cut in perpendicular direction with the fibers, $h = 10$ mm, ● Unwrapped ground beef, $h = 20$ mm, ■ Unwrapped ground beef, $h = 10$ mm, ▲ Cloth wrapped ground beef, $h = 20$ mm, ◆ Cloth wrapped ground beef, $h = 10$ mm —— best fitting line h = thickness of the slabs.

comparison of this equation with the experimental data obtained under different conditions.

The effect of temperature on the overall mass transfer coefficient, K, was described by the Arrhenius expression:

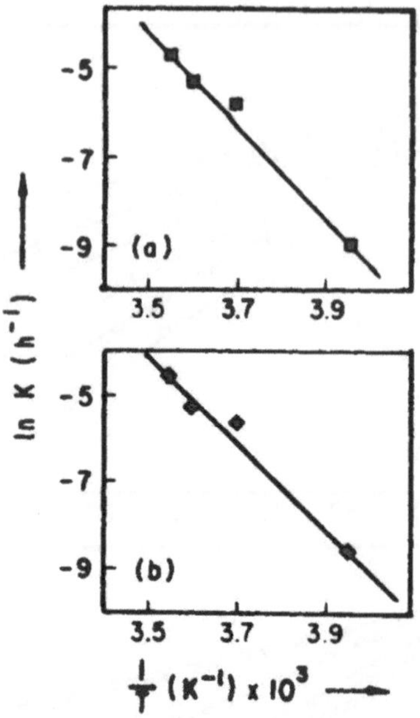

FIGURE E.2.15.2 Sample Arrhenius plots for the mass transfer coefficient K. a) Unwrapped samples, $h = 10$ mm. b) Cloth wrapped samples, $h = 10$ mm, (—) Best fitting line (Tütüncü *et al.*, 1990, © Canadian Institute of Food Science and Technology, reproduced by permission).

$$K = K_0 \exp\left[-\frac{E_a}{RT}\right]$$

where K_0 = pre-exponential constant, E_a = activation energy, R = gas constant and T = absolute temperature. The Arrhenius expression may be linearized as:

$$\ln K = \ln K_0 - \frac{E_a}{R}\frac{1}{T}$$

when $\ln K$ is plotted versus $1/T$, $\ln K_0$ is the intercept with $1/T = 0$ axis, and E_a/R is the slope. Comparison of the linearized Arrhenius equation with the experimental data obtained under various conditions is given in Figure E.2.15.2.

Example 2.16. Theoretical expressions for interfacial mass transfer coefficients Theoretical expressions may be obtained for interfacial mass transfer coefficients under limited conditions. Total mass flux N_A is expressed empirically as:

$$N_A = k_c(c_{Ai} - c_{A\infty}) \qquad (2.43.e)$$

Where c_{Ai} = concentration of A at the interface and $c_{A\infty}$ = concentration of A away from the interface, k_c = mass transfer coefficient. In stationary systems *film models* are used, which assume that the concentration profile development is confined within an imaginary thin film with thickness = δ, through which mass transfer occurs under steady state conditions. We may use the following BCs with such systems:

BC1: $c_A = c_{Ai}$ at $x = 0$ when $t > 0$

BC2: $c_A = c_{A\infty}$ at $x = \delta$ when $t > 0$

In agitated systems *penetration and surface renewal models* are used and based on unsteady state mass transfer to pockets of liquid exposed to the interface ($x = 0$) for a short time, then swept away and remixed with the bulk of the liquid. We may use the following IC and BCs with such systems:

IC $c_A = c_{A\infty}$ for all x when $t = 0$

BC1: $c_A = c_{Ai}$ at $x = 0$ when $t > 0$

BC2: $c_A = c_{A\infty}$ as $x \to \infty$ when $t > 0$

Obtain an expressions for k_c in terms of diffusivity of A and the contact time t.

Solution Equation of continuity in rectangular coordinate system is:

$$\frac{\partial c_A}{\partial t} + \left(\frac{\partial N_{Ax}}{\partial x} + \frac{\partial N_{Ay}}{\partial y} + \frac{\partial N_{Az}}{\partial z}\right) = R_A \tag{2.12}$$

When mass transfer is in the x direction only, i.e. $(\partial N_{Ay}/\partial y) = (\partial N_{Az}/\partial z) = 0$ there are no chemical reactions, i.e., $R_A = 0$, (2.12) is simplified as:

$$\frac{\partial c_A}{\partial t} + \frac{\partial N_{Ax}}{\partial x} = 0$$

Flux in the x direction is:

$$N_{Ax} = x_A(N_A + N_B) - D_A \frac{dc_A}{dx} \tag{2.10}$$

After assuming that concentration of A in water is small, i.e., $x_A \cong 0$ the flux expression will be

$$N_A = -D_A \frac{dc_A}{dx}$$

We may substitute the flux expression in the simplified form of the equation of continuity to obtain

$$\frac{\partial c_A}{\partial t} = D_A \frac{\partial^2 c_A}{\partial x^2}$$

i) Film model

Since the steady state conditions prevail $(\partial c_A/dt)=0$ and we will have $(\partial^2 c_A/dx^2)=0$. After integrating this expression twice we obtain the concentration profile $c_A=K_1x+K_2$. The BCs require $K_1=(c_{A\infty}-c_{Ai})/\delta$, $K_2=c_{Ai}$, therefore the concentration profile along the film is $c_A=(c_{A\infty}-c_{Ai})(x/\delta)+c_{Ai}$. Flux through the interface is: $N_A=-D_A\left.\frac{dc_A}{dx}\right|_{x=0}$ Therefore $N_A=(D_A/\delta)(c_{Ai}-c_{A\infty})$. Comparing this equation with $N_A=k(c_{Ai}-c_{A\infty})$ requires $k_c=\frac{D_A}{\delta}$

ii) Penetration and surface renewal models

Solution of this partial differential equation with the method of combination of variables was previously given in Example 2.13:

$$c_A(t)=(c_{Ai}-c_{A\infty})\,\mathrm{erfc}\left(\frac{x}{2\sqrt{D_At}}\right)+c_{A\infty}$$

therefore

$$\frac{dc_A}{dx}=\frac{(c_{Ai}-c_{A\infty})}{\sqrt{\pi D_At}}\exp\left(-\frac{x}{4D_At}\right)$$

Flux through the interface is:

$$N_A=-D_A\left.\frac{dc_A}{dx}\right|_{x=0}$$

Therefore

$$N_A=\sqrt{\frac{D_A}{\pi t}}(c_{Ai}-c_{A\infty})$$

Comparing this equation with $N_A=k(c_{Ai}-c_{A\infty})$ requires $k_c=\sqrt{\frac{D_A}{\pi t}}$

2.11. RHEOLOGICAL MODELING

A simple classification of the rheological models has been presented in Figure 2.10.

Since the raw materials are highly variable, maintaining uniform consistency of the processed foodstuffs is one of the major challenges in the industry, and rheological measurements are among the powerful tools for quality assurance (Race, 1991). Flow behavior of orange juice, salad dressing or ketchup and spreadability of mayonnaise affect consumer preference, therefore characterizing, i.e., modeling, rheological behavior of such products is important for the manufactures. Rheological behavior of fermentation media affects the heat and oxygen transfer, which in turn

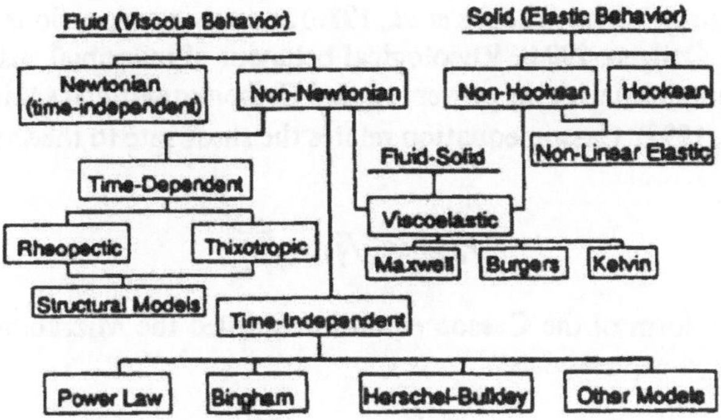

FIGURE 2.10 Simple classification of rheological behavior (Steffe, 1992, © Steffe J. F., reproduced by permission).

contributes significantly to bioprocess design and economics. An extensive review of the rheological models used in bio-processes has been given by Atkinson and Mavituna (1991).

Rheological models are empirical and valid only within the range of the experimental work that their constants are determined. The Herschel-Bulkley model is among the most general models:

$$\sigma = K\dot{\gamma}^n + \sigma_0 \tag{2.69}$$

where σ = shear stress (Pa), $\dot{\gamma}$ = shear rate = $(d\gamma/dt)$ (s^{-1}), K = consistency index (Pa s^n), and σ_0 = yield stress (Pa). The yield stress may be defined as the minimum shear stress required to initiate flow. Special cases of the Herschel-Bulkley model may be described as:

Special case	model	
$n = 1$	Bingam plastic: $\sigma = K\dot{\gamma} + \sigma_0$	(2.70)
$\sigma_0 = 0$	Power law: $\sigma = K\dot{\gamma}^n$	(2.71)
	Pseudo plastic: $n < 1$	
	Dilatant: $n > 1$	
$\sigma_0 = 0$, $n = 1$	Newtonian: $\sigma = \mu\dot{\gamma}$	(2.72)
	(flow behavior index K is substituted with viscosity μ)	

Rheological behavior of serum samples of low-pulp concentrated orange juice (Vitali and Rao, 1984 a, b), apple sauce and apple sauce serum (Rao *et al.*, 1986), apple juice–bentonite suspensions (Dik and Özilgen, 1994) and tomato juice concentrates (Tanglertpaibul and Rao, 1987) were described by the power law model. The Casson model is widely used to describe the rheological behavior of liquid foods, e.g., cocoa and hot chocolate (Steffe,

1992); yogurt (Parnell-Clunies *et al.*, 1986) and raisin suspensions (Pekyardimci and Özilgen, 1994). Rheological behavior of microbial suspensions were also modeled with the power law and Casson equations (Atkinson and Mavituna, 1992). Casson equation relates the shear rate to the shear stress as:

$$\sqrt{\sigma} = K\sqrt{\dot{\gamma}} + \sqrt{\sigma_0} \tag{2.73}$$

A modified form of the Casson equation is called the Mizrahi and Berk equation and expressed as:

$$\sqrt{\sigma} = K\dot{\gamma}^n + \sqrt{\sigma_0} \tag{2.74}$$

Constants of the rheological models are generally evaluated by curve fitting and the yield stress is determined by extrapolation to zero shear rate. Such a model may represent the relation between the shear stress and the shear rate within the range of the experiments, but the shear rate may not represent the actual physical properties of the fluid. Detailed discussion of the relations between the fluid properties and the yield stress has been discussed many authors, i.e., Qui and Rao (1988). Michaels and Bolger (1962) proposed a structural model for flow behavior of kaolin suspensions. This model considers small clusters of particles (and enclosed water) as the basic flow units of the flocculated suspensions. At low shear rates those flow units group into aggregates to form a "network" extending through the container. The "network" may give a finite yield stress to the suspension at low shear rates. Hunter and Nicol (1968) improved this theory to correlate the rheological behavior of kaolinite sol with its surface properties. Metz *et al.* (1979) and Berkman-Dik *et al.* (1992) suggested that this concept may be useful for studying the rheological behavior of the mold suspensions. The "network" concept may also describe the physical structure of the food or biological dispersions.

Temperature effects on the consistency index K of (2.71) described with the Arrhenius expression with many fluids, i.e., tomato concentrates (Harper and El-Sahrigi, 1965), concentrated orange juice (Vitali and Rao, 1984b), apple juice bentonite suspensions (Dik and Özilgen, 1994):

$$K = K_0 e^{E/RT} \tag{2.75}$$

where K_0 was the pre-exponential coefficient, R was gas constant and E was the activation energy. Consistency index K of (2.71) may also be related to the solids concentration of the medium with an exponential, i.e., (2.76.a), (Vitali and Rao, 1984b) or a power type expression, i.e., (2.76.b), (Dik *et al.*, 1995):

$$K = K_0 \exp(\alpha_1 c) \tag{2.76.a}$$

$$K = K_0 + K_1 c_1^{\alpha 1} + K_2 c_2^{\alpha 2} + K_3 c_3^{\alpha 3} \tag{2.76.b}$$

where α_i and K_i are constants; c and c_i are solids concentration.

Apparent viscosity is defined as the ratio of the shear stress to the shear rate:

$$\eta_{app} = \frac{\sigma}{\dot{\gamma}} \tag{2.77}$$

Temperature effects on apparent viscosity may also be described with the Arrhenius expression (Vitali and Rao, 1984b):

$$\eta_{app} = \eta_0 \, e^{E/RT} \tag{2.78}$$

where η_0 is a pre-exponential constant. Concentration dependence of the apparent viscosity may be expressed as (Harper and El-Sahrigi, 1965):

$$\eta_{app} = \kappa c^{\alpha} \tag{2.79}$$

or (Vitali and Rao, 1984b):

$$\eta_{app} = \kappa \exp(\beta c) \tag{2.80}$$

where α, β and κ are constants and c is concentration.

Time dependent rheological models are observed when the shear stress-shear-rate relation varies with time under shear. Thixotropic and rheopectic materials exhibit, respectively, decreasing and increasing shear stress (and apparent viscosity) over time at a fixed shear rate. In other words, thixotropy is time-dependent thinning and rheopecty is time-dependent thickening.

Example 2.17. Rheological behavior of the orange juice concentrate (Vitali and Rao, 1984b) Equation (2.71) may be linearized as $\log \sigma = \log K + n \log \dot{\gamma}$

Stress-strain data obtained with Perna orange juice at 65° Brix is compared with the power law model between − 18.8 °C and 29.2 °C in Figure E.2.17.1. Model parameters obtained under these experimental conditions were reported in Table E.2.17.

Table E.2.17 Power law parameters for Perna orange juice at 65°Brix (Vitali and Rao, 1984b, Table 1)

T(°C)	K ($N\ s^n/m^2$)	n
− 18.8	24.45	0.763
− 14.5	15.59	0.763
− 9.9	8.80	0.791
− 5.4	6.49	0.770
− 0.8	4.74	0.759
9.5	2.06	0.781
19.4	1.25	0.774
29.2	0.68	0.799

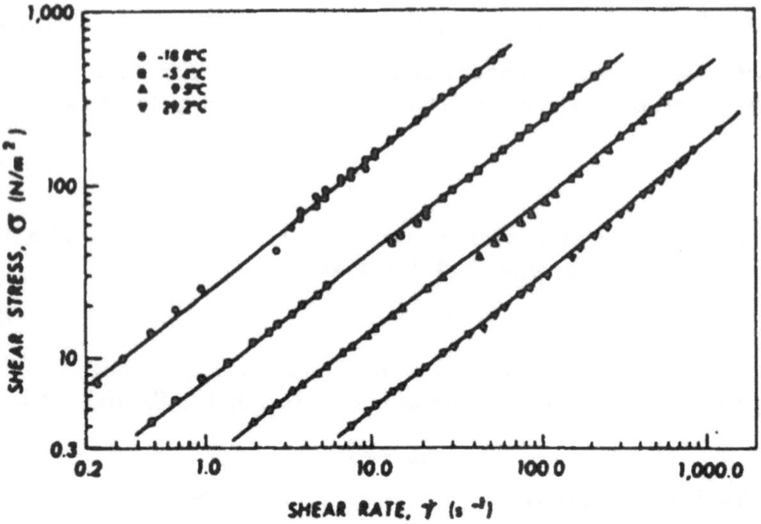

FIGURE E.2.17.1 Applicability of the power law model to Perna orange juice at 65° Brix between − 18.8°C and 29.2°C (Vitali and Rao, 1984b, © IFT, reproduced by permission). (°Brix is identical to per cent by weight and used specifically for soluble solids or sugar contents of solutions).

Table E.2.17 shows that the average value of the flow behavior index*n* is 0.775, therefore the power law model may be stated as $\sigma = K\dot{\gamma}^{0.775}$.

Vitali and Rao's data (1984b) showed that temperature and concentration effects on apparent viscosity may be described with (2.78) and (2.80), respectively (Figs. E.2.17.2 and 3).

Example 2.18. Rheological aspects of extrusion processes A slurry of raw materials, i.e., pre gelatinized flour, etc., is forced by single or multiple screws to flow along a barrel and pass through a die. The temperature may vary along the barrel and a processed or semi-processed product is obtained. Mathematical modeling of the overall extrusion process is beyond the scope of this text and an interested reader is referred to the comprehensive discussion by Levine (1992). The food extrudates are highly non-Newtonian and exhibit shear thinning behavior, normally described by the power law (pseudo plastic n < 1) model:

$$\sigma = K\dot{\gamma}^{n} \tag{2.71}$$

where $0.05 \leqslant n \leqslant 0.78$ for the common processes (Levine, 1992). In addition to the viscosity being a strong function of the shear environment, the

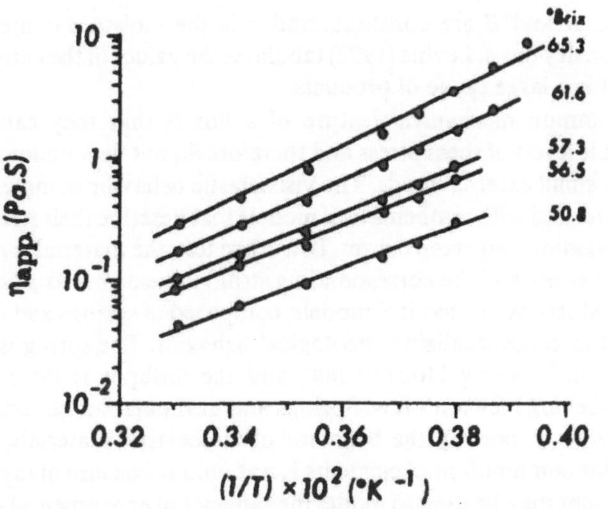

FIGURE E.2.17.2 Arrhenius plots for Perna orange juice concentrates between 50 and 65°Brix (Vitali and Rao, 1984b, © IFT, reproduced by permission). Apparent viscosity was determined at $\dot{\gamma} = 100\,s^{-1}$.

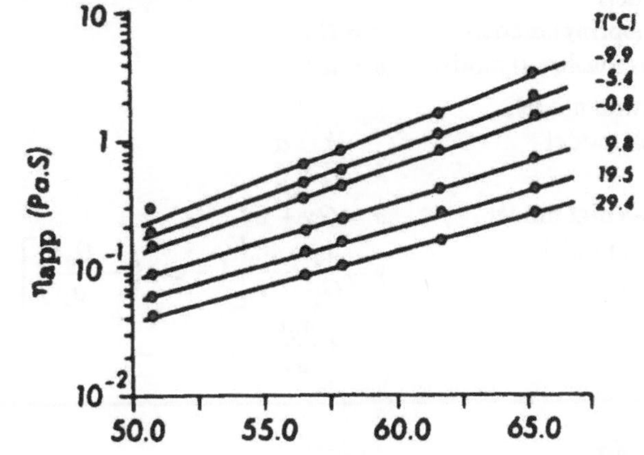

FIGURE E.2.17.3 Exponential relationships between°Brix and apparent viscosity at $100\,s^{-1}$ (Vitali and Rao, 1984b, © IFT, reproduced by permission). Apparent viscosity was determined at $\dot{\gamma} = 100\,s^{-1}$.

formulation and temperature have important roles in the viscous behavior of the extrudates. To account these effects (2.71) is usually modified:

$$\sigma = m_0 e^{A/T} e^{Bx} \dot{\gamma}^n \tag{E.2.18.1}$$

where m_0, A and B are constants and x is the moisture content of the product on dry basis. Levine (1992) tabulates the values of the constants m_0, A and B for a large range of products.

The common mechanical feature of solids is that they can support appreciable levels of shear stress and therefore do not flow under their own weight or small external loads. The viscoelastic behavior of materials may be characterized with mathematical models that describe their stress-strain, stress-relaxation and creep curves. In a creep test, the material is subject to a constant stress and the corresponding strain is measured as a function of time $\gamma(t)$. Massless mechanical models, composed of springs and dashpots, are useful in conceptualizing rheological behavior. The spring is an ideal solid element obeying Hooke's law; and the dashpot is the ideal fluid element obeying Newton's law. Springs and dashpots can be connected in various ways to portray the behavior of viscoelastic materials; however a particular combination of elements is not unique because many different combinations may be used to model the same set of experimental data. The most common mechanical analogs of rheological behavior are Maxwell and Kelvin-Voigt models (Figure 2.11) (Steffe, 1992):

Basic models

i) Elastic (spring) models $\sigma = G\gamma$ (2.83)

ii) Viscous (dashpot) model $\sigma = \mu\dot{\gamma}$ (2.84)

Combination models

i) Maxwell model

$$\dot{\gamma} = \frac{\dot{\sigma}}{G} + \frac{\sigma}{\mu} \quad (2.85)$$

ii) Kelvin-Voigt model

$$\sigma = G\gamma + \mu\dot{\gamma} \quad (2.86)$$

iii) Burgers model

$$\gamma = \frac{\sigma_0}{G_1} + \frac{\sigma_0}{G_2}\left[1 - \exp\left(-\frac{G_2 t}{\mu_2}\right)\right] + \frac{\sigma_0 t}{\mu_3} \quad (2.87)$$

where $\dot{\sigma} = \dfrac{d\sigma}{dt}$

Most food materials exhibit non-linear viscoelasticity and rheological memory. In special cases there are also specific modes of behavior characterized by discontinuities in the stress-strain relationships, i.e., a bio-yield point in some fruits, stress generation in muscle tissues during rigor mortis, etc. To account for non-linearity the linear models may be modified (Peleg, 1983):

$$\sigma = C_1\gamma + C_2\dot{\gamma} + C_3\gamma\dot{\gamma} \text{ (modified Kelvin-Voigt model)} \quad (2.87)$$

$$\sigma = C_1 + C_2\dot{\gamma} + C_3\dot{\gamma}^2 + C_4\dot{\gamma}^3 + \cdots \text{ (Green-Rivlin model)} \quad (2.88)$$

Example 2.19. Maxwell, Kelvin-Voigt and Burgers models of the stress-strain relations Obtain mathematical equations describing the variation of the strain as a function of time under constant stress with the materials following a) Maxwell b) Kevin-Voigt c) Burgers models.

Solution a) The Maxwell model described in Figure 2.11 is composed of a spring and a dashpot in series. The spring and the dashpot experience the same stress, but the total strain is the sum of the individual strains:

$$\gamma = \gamma_{spring} + \gamma_{dashpot}$$

differentiating this equation with respect to time yields

$$\dot{\gamma} = \dot{\gamma}_{spring} + \dot{\gamma}_{dashpot}$$

Since $\sigma = G\gamma_{spring}$ with the spring and $\sigma = \mu\,\dot{\gamma}_{dashpot}$ with the dashpot; and $\gamma_{spring} = \sigma/G$ and $\dot{\gamma}_{dashpot} = \sigma/\mu$ we will have

$$\dot{\gamma} = \frac{1}{G}\frac{d\sigma}{dt} + \frac{\sigma}{\mu}$$

or

$$\sigma + \tau_{ret}\frac{d\sigma}{dt} = \mu\dot{\gamma}$$

where $\tau_{ret} = \mu/G \cong$ time takes out a molecule to be stretched when deformed.
Under constant stress σ_0 this equation becomes

$$\frac{d\gamma}{dt} = \frac{\sigma_0}{\mu}$$

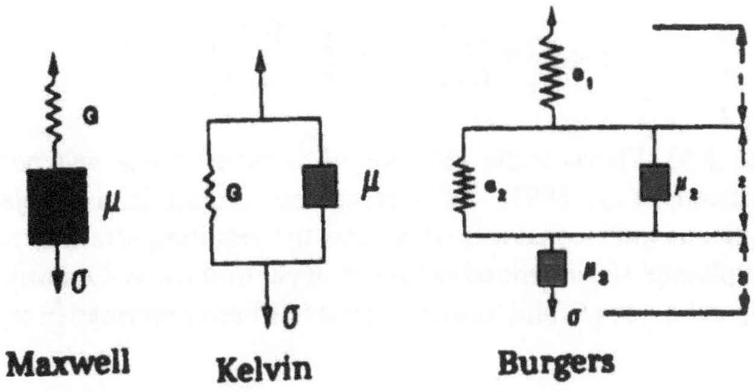

FIGURE 2.11 **Maxwell, Kelvin-Voigt and Burgers models.**

The initial strain in the system is given as

$$\text{IC: } \gamma = \frac{\sigma_0}{G} \text{ at } t = 0$$

The spring responds to the stress immediately, while the dashpot delays, therefore we observe the response of the spring, i.e., the first element in the IC. The solution to the differential equation with the given IC is

$$\gamma = \frac{\sigma_0}{G} + \frac{\sigma_0}{\mu} t$$

b) The Kelvin-Voigt model was described in Figure 2.11 is composed of a spring and a dashpot in parallel. The spring and the dashpot are strained equally, but the total stress is the sum of the individual stresses:

$$\sigma = G\gamma + \mu\dot{\gamma} \tag{2.86}$$

which may be rewritten under constant stress σ_0 as

$$\frac{d\gamma}{dt} + \frac{\gamma}{\tau_{ret}} = \frac{\sigma_0}{\mu}$$

where τ_{ret} = retardation time is unique for the material in question. The system was not strained initially:

$$\text{IC: } \gamma = 0 \text{ at } t = 0$$

Solution of the differential equation with the given IC is:

$$\gamma = \frac{\sigma_0}{G}\left(1 - \exp\left(-\frac{t}{\tau_{ret}}\right)\right)$$

c) The Burgers model was described in Figure 2.11 is composed of a Maxwell and a Kelvin-Voigt elements in series. The total strain is the sum of the individual Maxwell and Kevin-Voigt element strains:

$$\gamma = \frac{\sigma_0}{G_1} + \frac{\sigma_0}{G_2}\left\{1 - \exp\left(-\frac{G_2 t}{\mu_2}\right)\right\} + \frac{\sigma_0}{\mu_3} t$$

Example 2.20. Visco-elastic behavior of cheddar cheese and potato flesh (Purkayastha, et al., 1985) In a creep test a material is subject to an instantaneous (and constant) stress and the resulting strain is measured. The compliance $J(t)$ is defined as the strain per unit stress. Quantification of the creep behavior of solid food materials has been described in the general form:

$$J(t) = K_0 + K_1 t + \sum_{i=2}^{N} K_i\left\{1 - \exp\left(-\frac{t}{\tau_i}\right)\right\} \tag{E.2.20.1}$$

where K_0, K_1 and K_i's are constants, N = number of the Kelvin-Voigt elements in the system, and τ_i = time characteristics (retardation times). The relation between $J(t)$, its individual terms, and t is shown in Figure E.2.20.1. The creep curve can always be fitted by the given mathematical model provided that there is no restriction on N. It was demonstrated by various researchers that the total number of constants (K's and τ_i's) is generally between 3 and 12.

Puskayastha *et al.* (1985) demonstrated that $J(t)$ may be expressed in simpler form:

$$J(t) = K_0^\bullet + K_1^\bullet t + \frac{t}{K_2^\bullet + K_3^\bullet t} \tag{E.2.20.2}$$

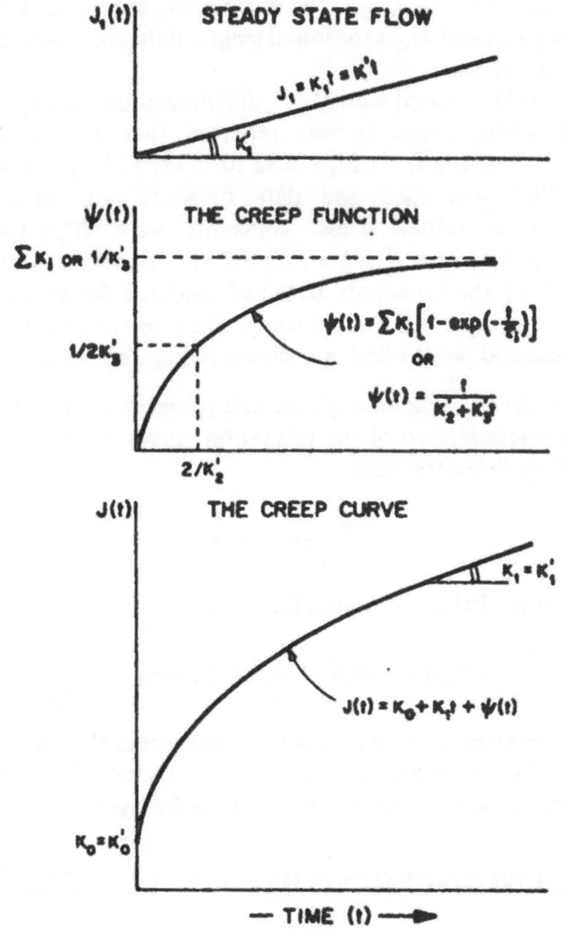

FIGURE E.2.20.1 Schematic diagram of creep compliance curve and its three components as described in (E.2.20.1) (Puskayastha *et al.*, 1985, © IFT, reproduced by permission).

Puskayastha *et al.* (1985) calculated the compliance versus time relations as:

$$J(t) = \frac{\gamma_E(t)}{\sigma_0} = \frac{A_0\,\Delta H(t)}{F\,H_0}$$

and

$$J_c(t) = \frac{\gamma_H(t)}{\sigma(t)} = \frac{H_0 A_0 \ln\left(\dfrac{H_0}{H_0 - \Delta H}\right)}{F(H_0 - \Delta H)}$$

where $J_c(t)$ is the corrected compliance, σ_0 is the initial stress and calculated by dividing the imposed force (F) by the original cross sectional area of the undeformed specimen (A_0); γ_E is the engineering strain, γ_H is the natural strain (Hencky's strain), H_0 is the initial length of the specimen, $\Delta H(t)$ is the absolute deformation.

When (E.2.20.1) was used with $N = 1$, and data obtain with potato under 64.8 kPa of initial stress, it was reported that $K_0 = 0.405\ \text{MPa}^{-1}$. $K_1 = 1.65 \times 10^{-4}\ (\text{MPa s})^{-1}$), $K_2 = 3.32\ 10^{-2}\ \text{MPa}^{-1}$, $\tau_2 = 3.94 \times 10^{-2}$ s. When (E.2.20.2) was used and data obtained with cheddar chees under 40 kPa of initial stress, constants were $K_0^* = 1.40\ \text{MPa}^{-1}$, $K_1^* = 8.82 \times 10^{-3} (\text{MPa s})^{-1}$, $K_2^* = 7.53\ 10^{-2}$ MPa s and $K_3^* = 0.474$ MPa. Complete list of the constants obtained under different experimental conditions are given in the original study. The creep compliance curves of cheddar cheese and potato flesh are shown in Figure E.2.20.2.

Example 2.21. Stress relaxation of fruit gels (*Gamero, et al.*, 1993) Mathematical charcacterization of the relaxation curves of the fruit gel were made according Peleg equation

$$\frac{\sigma_0 t}{\sigma_0 - \sigma(t)} = k_1 + k_2 t \qquad \text{(E.2.21.1)}$$

and Nussinovitch, Peleg, Normand Equation

$$\frac{\sigma(t)}{\sigma_0} = C_0 + C_1 \exp\left(-\frac{t}{10}\right) + C_2 \exp\left(-\frac{t}{100}\right) \qquad \text{(E.2.21.2)}$$

where $\sigma(t)$ is the stress at time t, σ_0 is the initial stress at time $t = 0$; C_0, C_1 and C_2, k_1 and k_2 are constants. Figure E.2.21 shows the variation of the model parameters with the composition of the fruit gels.

2.12. THE ENGINEERING BERNOULLI EQUATION

Under steady state conditions, on a unit mass basis and for an incompressible fluid the Engineering Bernoulli equation is commonly written as (Heldman and Singh, 1981):

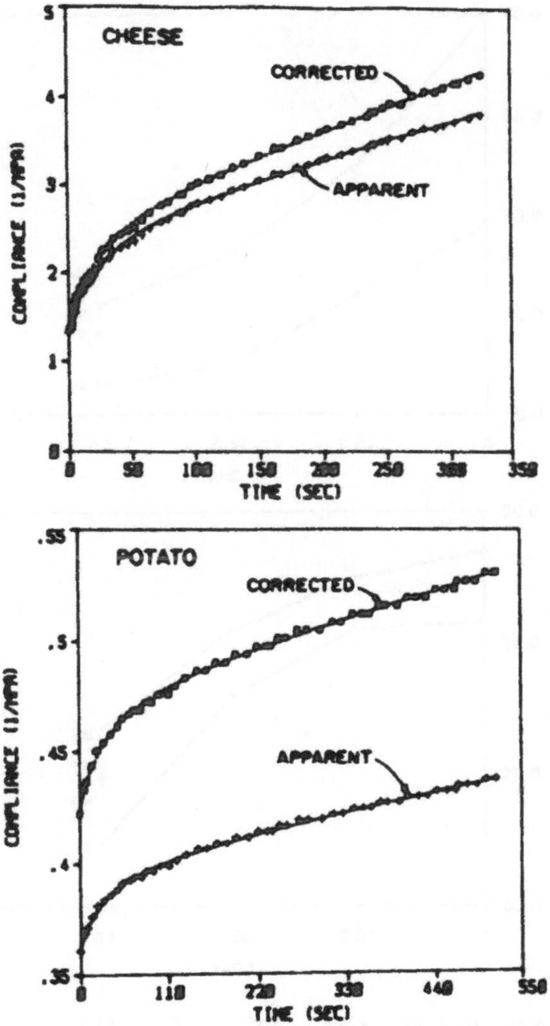

FIGURE E.2.20.2 Creep compliance curves of cheddar cheese and potato flesh fitted by (E.2.20.2). Experimental data are shown in symbols, the model is represented with solid lines (Puskayastha *et al.*, 1985, © IFT, reproduced by permission).

$$g h_1 + \frac{P_1}{\rho} + \frac{\bar{v}_1^2}{2\alpha} - W = g h_2 + \frac{P_2}{\rho} + \frac{\bar{v}_2^2}{2\alpha} + \Sigma F \tag{2.89}$$

where subscripts refer to points 1 and 2 in a fluid handling system, Z is the elevation from a reference, g is gravitational acceleration, P is pressure; ρ and $\bar{v}$ are the density and the average velocity of the liquid, respectively;

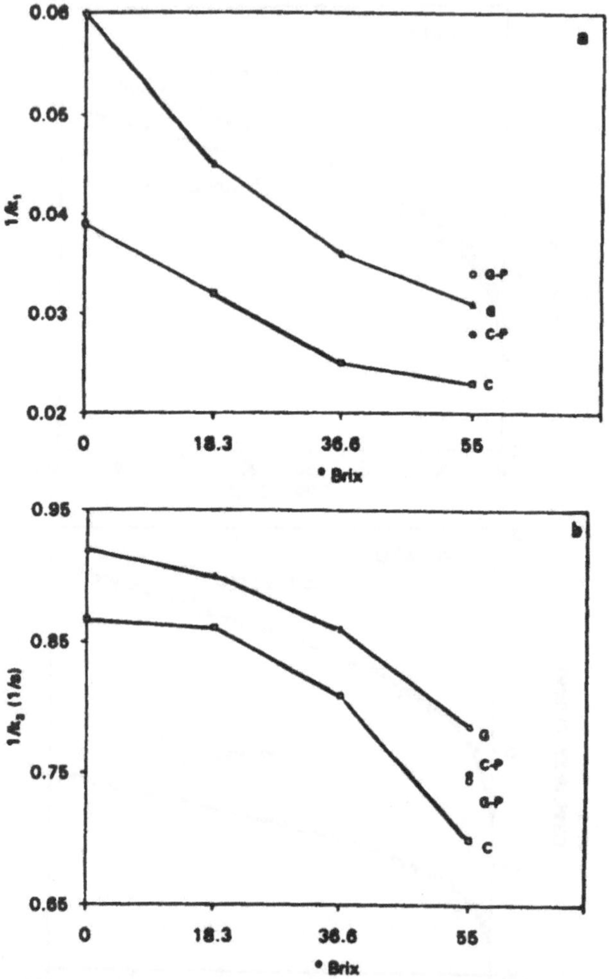

FIGURE E.2.21 **Variation of the model parameters with the composition of the fruit gels. Sample compositions were abbreviated as: *C* = carrageenan only, *G* = carrageenan plus locust bean gum, *C-P* = carrageenan with fruit pulp, *G-P* = carrageenan plus locust bean gum with fruit pulp (Gamero, *et al.*, 1993, © IFT, reproduced by permission).**

α is the kinetic energy correction factor, W is the work output per unit mass of the liquid, ΣF is frictional loss of energy per unit mass of the liquid. The gh and $\bar{v}^2/2$ terms are the potential and the kinetic energy per unit mass of the liquid. The term P/ρ is the pressure energy, i.e., ability of doing pressure volume work by unit mass of liquid. Equation (2.89) may be used for modeling flow in foods or pipeline design.

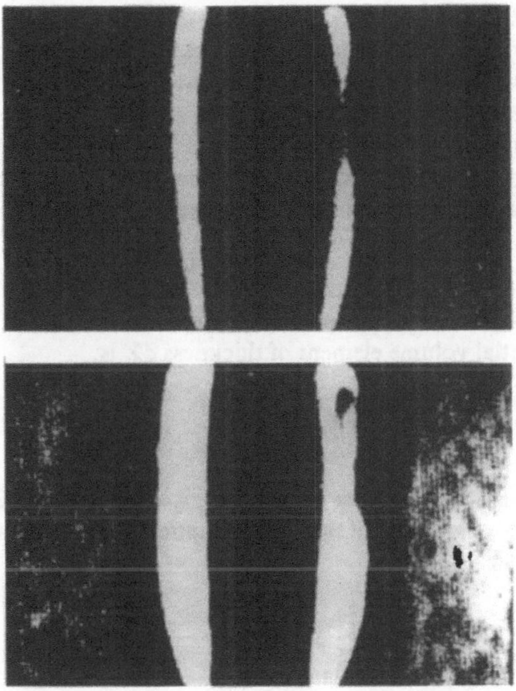

FIGURE E.2.22.1 Transverse NMR images showing the widening of the cut surfaces due to shrinkage of the curd soon after cutting (top) and at the end of the experiments (bottom). Reprinted from *Process Biochemistry* 29 (5), Özilgen and Kauten, NMR analysis and modeling of shrinkage and whey expulsion in rennet curd, pages 373–379, © 1994 with kind permissions from Elsevier Science Ltd, The Boulevard, Langford Lane, Kidlington 0X5 1GB, UK.

Example 2.22. Shrinkage and whey expulsion in rennet curd (*Özilgen and Kauten*, 1994) The casein micelles of milk have hydrophobic cores and hydrophilic external layers. Paracaseins are produced via removal of casein macro peptide residues and the hydrophobic core is exposed during renetting. The paracasein micelles first flocculate, and then form a gel. The free surfaces of the gel micelles are reactive, and approach others with Brownian motion or deformation to form a network. The network increases the internal stress and urges the gel to shrink. It applies pressure on whey and forces it to synerese. Syneresis is delayed until the gel is disturbed. The surfaces of the undisturbed gel are not permeable for water since they stick on the vessel walls and are covered with lipids at the milk/air interfaces. When the undisturbed gel is cut, it contracts to expel the whey.

Syneresis is observed with newly cut curd during cheese making and is critical to the yields, texture, moisture content, flavor and quality of the final cheese. Expulsion of whey from the rennet curd is coupled with rearrangement of the network (Fig. E.2.22.1).

Total pump work, W, associated with expelling water from a curd with thickness Z under constant pressure, P, upon differential volume change, dV, via shrinkage is (work is done by the curd on the fluid):

$$W = P\,dV = PAv_s \tag{1.34}$$

Pump work associated with expelling a differential amount of water, dM, from a differential volume element of thickness dZ is:

$$W = PAv_s\frac{dZ}{Z_i} \tag{E.2.22.1}$$

where A = cross sectional area of the curd, v_s = velocity of the shrinking curd surface, Z_i = thickness of the curd. Equation (2.89) was stated for unit mass of water as:

$$gh_1 + \frac{P_1}{\rho} + \frac{\bar{v}_1^2}{2\alpha} - W = gh_2 + \frac{P_2}{\rho} + \frac{\bar{v}_2^2}{2\alpha} + \Sigma F \tag{2.89}$$

since the elevation of the curd is the same everywhere, i.e., $gh_1 = gh_2$, pressure drop along the curd is negligible, i.e., $P_1 = P_2$, and when $\alpha = 1$ for expulsion of differential amount of water dM (2.89) becomes:

$$-PAv_s\frac{dZ}{Z_i} = dM\left(\frac{\bar{v}_2^2}{2} - \frac{\bar{v}_1^2}{2}\right) + dM\,\Sigma F \tag{E.2.22.2}$$

Total resistance to flow in the curd may be expressed as:

$$\Sigma F = K(Z_i - Z)\bar{v}_2^2 \tag{E.2.22.3}$$

where $Z_i - Z$ is the distance over which differential amount of water dM is pumped and K is a constant. After substituting this expression for ΣF and $\bar{v}_1^2 = 0$ the final form of the Bernoulli equation was integrated to obtain an expression for the amount of water removed from the curd as a function of distance from the cut surface:

$$M = M_i + \frac{C}{K}\ln(1 + 2K(Z_i - Z)) \tag{E.2.22.4}$$

where

$$C = \frac{PAv_s}{\bar{v}_2^2 Z_i}$$

Profiles of the remaining water are calculated by subtracting this equation from the constant initial water content. The intensity of the NMR signal, S, along the curd is proportional with the amount of entrapped water and expressed as:

$$S = S_i + \frac{C}{K}\ln(1 + 2K(Z_i - Z)) \qquad \text{(E.2.22.5)}$$

where S_i is the intensity of the NMR signal on the cutting surface, C and K are constants. This model was compared with the experimental data after substituting values of S_i. Values of parameter C were 0.21 ± 0.05, values of the parameter K were 0.09 ± 0.01. The minimal variation of these parameters with time confirms the consistency of the model and the measurements. Comparison of the model with the data is exemplified in Figure E.2.22.2.

In equation (2.89) energy losses due to friction were shown with ΣF, which include losses in straight pipe, valves and fittings and may be expressed as:

$$\Sigma F = \frac{2f\bar{v}^2 L}{D} + \sum_{i=1}^{m}\left[\frac{k_f \bar{v}^2}{2}\right]_m \qquad (2.90)$$

where f = Fanning friction factor, L = pipe length, D = inner diameter of the pipe, k_f = friction loss coefficient (unique for any particular valve or fitting) and m is the number of the fittings in the flow system. Different values of $\bar{v}$, k_f and f may be required when the pumping system includes pipes having different diameters.

A procedure was outlined by Steffe and Morgan (1986) to calculate the energy losses for non-Newtonian fluids under laminar flow conditions. We may calculate f from

$$f = \frac{16}{\psi Re} \qquad (2.91)$$

where

$$Re = \frac{D^n(\bar{v}^{2-n})\rho}{8^{n-1}K}\left(\frac{4n}{1+3n}\right)^n \qquad (2.92)$$

$$\psi = (1+3n)^n(1-\xi_0)^{n+1}\left[\frac{(1-\xi_0)^2}{1+3n} + \frac{2\xi_0(1-\xi_0)}{1+2n} + \frac{\xi_0^2}{1+n}\right]^2 \qquad (2.93)$$

and

$$\xi_0 = \frac{2\tau_0}{f\rho\bar{v}^2}$$

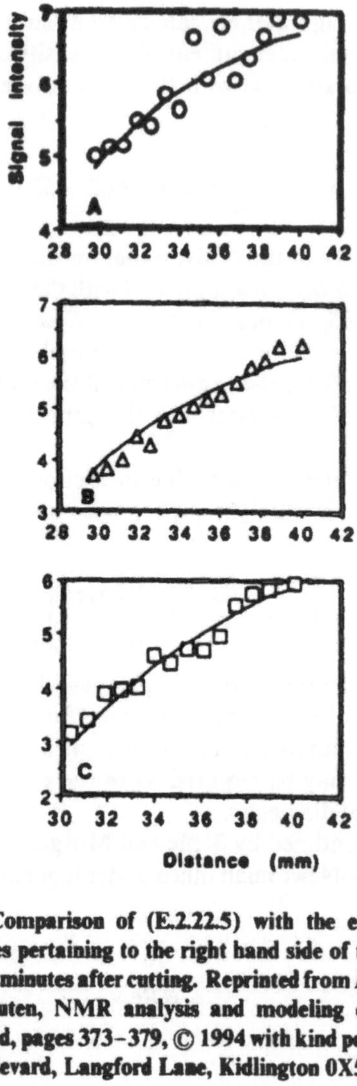

FIGURE E.2.22.2 Comparison of (E.2.22.5) with the experimentally obtained NMR signal intensities pertaining to the right hand side of the incubation chamber: (A) 0, (B) 76, (C) 136 minutes after cutting. Reprinted from *Process Biochemistry* 29 (5), Özilgen and Kauten, NMR analysis and modeling of shrinkage and whey expulsion in rennet curd, pages 373–379, © 1994 with kind permissions from Elsevier Science Ltd, The Boulevard, Langford Lane, Kidlington 0X5 1GB, UK.

For power law and Newtonian liquids $\xi_0 = 0$ and $\psi = 1$, therefore f may be calculated easily using (2.91) and (2.92). Reynolds number is also given as

$$Re = 2\,He\left(\frac{n}{1+3n}\right)^2\left(\frac{\psi}{\xi_0}\right)^{(2/n)-1} \qquad (2.94)$$

where He is the Hedstrom number defined as

$$He = \frac{D^2 \rho}{K}\left(\frac{\tau_0}{K}\right)^{(2/n)-1} \tag{2.95}$$

To calculate the friction factor for Herschel-Bulkey fluids, ξ_0 is estimated through iteration of (2.94) using equations (2.92), (2.93) and (2.94), then f can be calculated using (2.91) and (2.92). Non-Newtonian foods rarely flow under turbulent conditions; however it is recommended to check the flow behavior using the Hanks and Ricks (1974) diagram (Fig. 2.12), which gives the value of the critical number, Re_c, where laminar flow ends.

The kinetic energy correction factor α is 2 in turbulent flow. When the flow is laminar, α may be calculated from the Osorio-Steffe equation (1984):

$$\alpha = \frac{2(1+3n+2n^2+2n^2\xi_0+2n\xi_0+2n^2\xi_0^2)^3(2+3n)(3+5n)(3+4n)}{(1+2n)^2(1+3n)^2(18+nf_1(\xi_0)+n^2f_2(\xi_0)+n^3f_3(\xi_0)+n^4f_4(\xi_0)+n^5f_5(\xi_0))} \tag{2.96}$$

where

$f_1(\xi_0) = 105 + 66\,\xi_0$

$f_2(\xi_0) = 243 + 306\,\xi_0 + 85\,\xi_0^2$

$f_3(\xi_0) = 279 + 522\,\xi_0 + 350\xi_0^2$

$f_4(\xi_0) = 159 + 390\,\xi_0 + 477\,\xi_0^2$

$f_5(\xi_0) = 36 + 108\,\xi_0 + 216\xi_0^2$

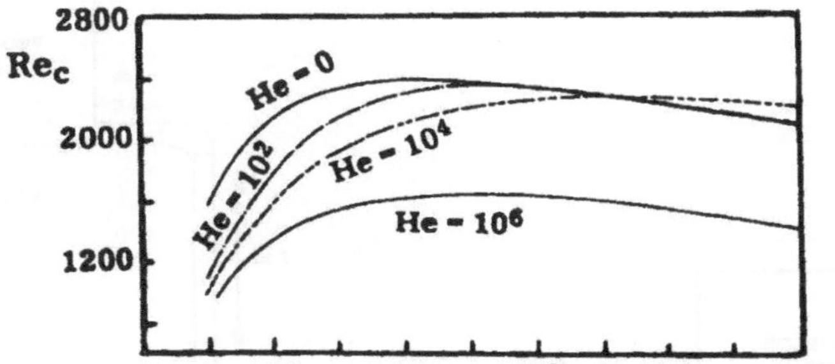

FIGURE 2.12 Variation of the critical Reynolds number as a function of Hedstrom number and the flow behavior index. An interested reader may refer to Hanks and Ricks (1974) for wider range plots.

Example 2.23. Pump power requirement with a Herschel-Bulkley fluid A Herschel-Bulkley fluid ($\tau_0 = 150$ Pa, $K = 5.20$ Pa s, $n = 0.45$, $\rho = 1300$ kg/m^3) is pumped in the following system (inside diameter of the steel pipe was 0.04089 m before the pump and 0.05250 m after the pump) with a flow rate of 3×10^{-3} m^3/s (Fig. E.2.23). Friction loss factors for elbows and the valves are $k_f = 0.45$ and $k_f = 0.90$, respectively. Calculate the power requirement for the pump. You may neglect sudden contraction and sudden expansion friction losses at the exit of the storage tank and at the entrance of the process unit.

Solution Major steps in calculations:

		before the pump	after the pump
$v = \dfrac{\text{flow rate of the fluid}}{\text{cross sectional area of the pipe}}$		$v = 1.39$ m/s	$v = 2.28$ m/s
$Re = \dfrac{D^n(\bar{v}^{2-n})\rho}{8^{n-1}K}\left(\dfrac{4n}{1+3n}\right)^n$	(2.92)	$Re = 592.3$	$Re = 304.4$
$He = \dfrac{D^2\rho}{K}\left(\dfrac{\tau_0}{K}\right)^{(2/n)-1}$	(2.95)	$He = 44038.9$	$He = 72597.4$

Re was also expressed as:

$$Re = 2\,He\left(\frac{n}{1+3n}\right)^2\left(\frac{\psi}{\xi_0}\right)^{(2/n)-1} \tag{2.94}$$

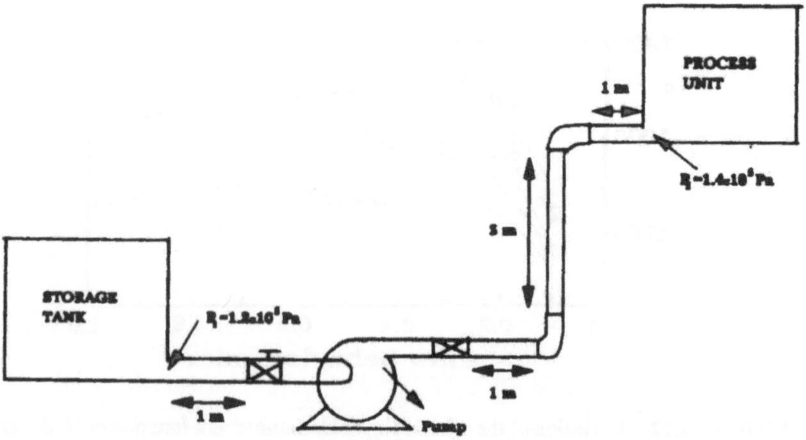

FIGURE E.2.23 Schematic diagram of the flow system.

After substituting the numbers in (2.94) we will have

$$0.0567 = \left(\frac{\psi}{\xi_0}\right)^{3.44} \quad \text{after the pump} \qquad \text{(E.2.23.1)}$$

and

$$0.182 = \left(\frac{\psi}{\xi_0}\right)^{3.44} \quad \text{before the pump} \qquad \text{(E.2.23.2)}$$

We also have

$$\psi = (1+3n)^n (1-\xi_0)^{n+1}\left[\frac{(1-\xi_0)^2}{1+3n} + \frac{2\xi_0(1-\xi_0)}{1+2n} + \frac{\xi_0^2}{1+n}\right]^2 \qquad (2.93)$$

We will combine (E.2.23.1) with (2.93) and (E.2.23.2) with (2.93) then apply trial and error solution to obtain $\zeta_0 = 0.42$ after the pump and $\zeta_0 = 0.33$ before the pump. We will substitute these numbers in (2.93), (2.91), (2.96) and (2.90):

equation		before pump	after pump
$\psi = (1+3n)^n(1-\xi_0)^{n+1} \times \left[\frac{(1-\xi_0)^2}{1+3n} + \frac{2\xi_0(1-\xi_0)}{1+2n} + \frac{\xi_0^2}{1+n}\right]^2$	(2.93)	$\psi = 0.205$	$\psi = 0.18$
$f = \frac{16}{\psi Re}$	(2.91)	$f = 0.132$	$f = 0.292$
$\alpha = \frac{2(1+3n+2n^2+2n^2\xi_0+2n\xi_0+2n^2\xi_0^2)^3}{(1+2n)^2(1+3n)^2(18+nf_1(\xi_0)}$		$\alpha = 1.43$	$\alpha = 1.38$
$\frac{(2+3n)(3+5n)(3+4n)}{+n^2f_2(\xi_0)+n^3f_3(\xi_0)+n^4f_4(\xi_0)+n^5f_5(\xi_0))}$	(2.96)		
$\Sigma F = \frac{2f\bar{v}^2L}{D} + \sum_{i=1}^{m}\left[\frac{k_f\bar{v}^2}{2}\right]_m$	(2.90)	$\Sigma F = 35.9$ J/kg	$\Sigma F = 109.19$ J/kg

Since $gh_1 = gh_2$, (2.89) was simplified as:

$$\frac{P_1}{\rho} + \frac{\bar{v}_1^2}{2\alpha} - W = \frac{P_2}{\rho} + \frac{\bar{v}_2^2}{2\alpha} + \Sigma F \qquad \text{(E.2.23.3)}$$

where subscripts 1 and 2 refer to before and after the pump, respectively. After substituting the numbers in (E.2.23.3) we will obtain $W = -188.7$ J/kg.

2.13. LAPLACE TRANSFORMATION IN MATHEMATICAL MODELING

Most Food and Bioprocess Engineering problems are solved in time domain. Transformation into a Laplace domain (Tabs. 2.19 and 2.20)

Table 2.19 Laplace transformation and its important properties

Laplace transformation $\mathscr{L}[f(t)]$ of a function $f(t)$	$\mathscr{L}[f(t)] = \int_0^\infty e^{-st} f(t)\,dt$ comments: i) If the integral on the right hand side of the definition can not be calculated, the Laplace transform of the function $f(t)$ would not be available. ii) The Laplace transform involves no information for the behavior of $f(t)$ when $t < 0$.
The Laplace transform is linear	$\mathscr{L}[\alpha f_1(t) + \beta f_2(t)] = \alpha\mathscr{L}[f_1(t)] + \beta\mathscr{L}[f_2(t)]$
Shifting theorem	$\mathscr{L}[e^{-at} f(t)] = \int_0^\infty e^{-(s+a)t} f(t)\,dt = f(s+a)$
Heaviside's expansion theorem 1 $Q(s)$ has higher degree than $P(s)$	$\mathscr{L}^{-1}\left\{\frac{P(s)}{Q(s)}\right\} = \sum_{m=1}^{\infty} \frac{P(s_m)}{Q'(s_m)} \exp(s_m t)$ where s_m = roots of $Q(s) = 0$ $Q'(s) = \frac{dQ}{ds}$
Heaviside's expansion theorem 2 $Q(s) = (s - s_1)^{m_1}(s - s_2)^{m_2}\ldots$ $(s - s_n)^{m_n}$ degree of $P(s)$ is less than $\left\{\sum_{j=1}^{n} m_j\right\} - 1$	$\mathscr{L}^{-1}\left\{\frac{P(s)}{Q(s)}\right\} = \sum_{k=1}^{n} \sum_{\ell=1}^{m_k} \frac{\phi_{k\ell}(s_k)}{(m_k - \ell)!(\ell - 1)!} t^{m_k - \ell} \exp(s_k t)$ where $\phi_{k\ell}(s) = \frac{d^{\ell-1}}{ds^{\ell-1}}\left[\frac{P(s)}{Q_k(s)}\right]$ $Q_k(s) = \frac{Q(s)}{(s - s_k)^{m_k}}$

converts an ordinary differential equation into an algebraic equation, or a partial differential equation into an ordinary differential equation and makes it easier to obtain the solution. The procedure of using Laplace transformations is described in Figure 2.13.

Example 2.24. Solution of a linear ordinary differential equation with Laplace transformations An ordinary differential equation

$$\frac{d^2x(t)}{dt^2} + \frac{dx(t)}{dt} + 0.125\,x(t) = 1 \tag{E.2.24.1}$$

Table 2.20 Laplace transforms of simple functions ($f(t) = 0$ when $t < 0$)

	$f(t)$ when $t \geqslant 0$	$\mathscr{L}[f(t)]$
step function	1	$\frac{1}{s}$
ramp function	t	$\frac{1}{s^2}$
	t^n	$\frac{n!}{s^{n+1}}$
exponential function	$e^{-\alpha t}$	$\frac{1}{s+\alpha}$
	$t^n e^{-\alpha t}$	$\frac{n!}{(s+\alpha)^{n+1}}$
sine function	$\sin \kappa t$	$\frac{\kappa}{s^2+\kappa^2}$
cosine function	$\cos \kappa t$	$\frac{s}{s^2+\kappa^2}$
	$e^{-\alpha t} \sin \kappa t$	$\frac{\kappa}{(s+\alpha)^2+\kappa^2}$
	$e^{-\alpha t} \cos \kappa t$	$\frac{s+\alpha}{(s+\alpha)^2+\kappa^2}$
hyperbolic sine function	$\sinh \kappa t$	$\frac{\kappa}{s^2-\kappa^2}$
hyperbolic cosine function	$\cosh \kappa t$	$\frac{s}{s^2-\kappa^2}$
unit impulse	$f(t)=0 \quad t<0$ $f(t)=\delta(t) \quad t=0$ $f(t)=0 \quad t>0$	1
the derivative	$\frac{df(t)}{dt}$	$sf(s)-f(0)$
higher order derivatives	$\frac{d^n f(t)}{dt^n}$	$s^n f(s)-s^{n-1}f(0)-s^{n-2}\left(\frac{df(t)}{dt}\right)_{t=0}-$ $s^{n-3}f\left(\frac{d^2 f(t)}{dt^2}\right)_{t=0}-\ldots-\left(\frac{d^{n-1}f(t)}{dt^{n-1}}\right)_{t=0}$
integral	$\int_0^t f(t)\,dt$	$\frac{1}{s}f(s)$
complimentary error function	$\mathrm{erfc}\left(\frac{k}{2\sqrt{t}}\right)$	$\frac{1}{s}\exp(-k\sqrt{s})$ where $k \geqslant 0$

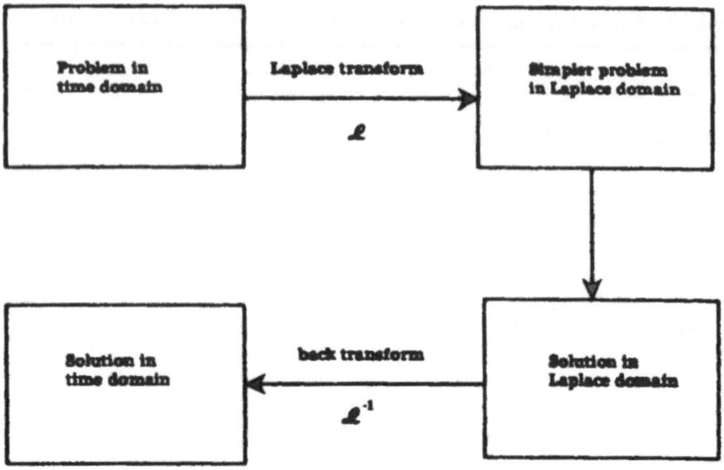

FIGURE 2.13 Schematic description of problem solving by using Laplace transformation.

with the initial conditions

$$x(0) = 0 \tag{E.2.24.2}$$

and

$$\frac{dx(0)}{dt} = 0 \tag{E.2.24.3}$$

is converted into the Laplace domain with transformation of the individual terms as:

$$\mathscr{L}\left\{\frac{d^2x(t)}{dt^2}\right\} = s^2x(s) - s\,x(0) - \frac{dx(0)}{dt} = s^2\,x(s) \tag{E.2.24.4}$$

$$\mathscr{L}\left\{\frac{dx(t)}{dt}\right\} = s\,x(s) - x(0) = s\,x(s) \tag{E.2.24.5}$$

$$\mathscr{L}\{x\} = x(s) \tag{E.2.24.6}$$

$$\mathscr{L}\{1\} = \frac{1}{s} \tag{E.2.24.7}$$

After substituting the transformations (E.2.24.4)–(E.2.24.7) in (E.2.24.1) we will obtain the Laplace transformation of the original ordinary differential equation as:

$$s^2\,x(s) + s\,x(x) + 0.125\,x(s) = \frac{1}{s} \tag{E.2.24.8}$$

Equation (E.2.24.8) may be rearranged as:

$$x(s)=\frac{1}{(s^2+s+0.125)(s)} \tag{E.2.24.9}$$

We may apply the partial fractions as:

$$\frac{1}{(s^2+s+0.125)(s)}=\frac{A}{s}+\frac{Bs+C}{s^2+s+0.125} \tag{E.2.24.10}$$

It should be noticed that the second term has a polynomial at the denominator and its numerator is also a polynomial such that (power of the numerator) = (power of the denominator) − 1.
Equation (E.2.24.10) requires

$$A(s^2+s+0.125)+(Bs+C)s=1 \tag{E.2.24.11}$$

Equation (E.2.24.11) may be rearranged as:

$$(A+B)s^2+(A+C)s+0.125\,A=1 \tag{E.2.24.12}$$

Coefficients of the equal powers of s should be the same on both sides of (E.2.24.12), therefore

$$A=8 \tag{E.2.24.13}$$

$$A+C=0 \tag{E.2.24.14}$$

and

$$A+B=0 \tag{E.2.24.15}$$

Equations (E.2.24.13)–(E.2.24.15) requires $B=-8$ and $C=-8$, then (E.2.24.9) may be rewritten as:

$$x(s)=8\left\{\frac{1}{s}-\frac{s+1}{s^2+s+0.125}\right\} \tag{E.2.24.15}$$

We should transform (E.2.23.15) into the time domain to obtain the solution to the problem:

$$\mathcal{L}^{-1}\left\{\frac{1}{s}\right\}=1 \tag{E.2.24.16}$$

We may apply the Heaviside's expansion theorem (Tab. 2.19) to obtain the back transform of the second term:

$$\mathcal{L}^{-1}\left\{\frac{P(s)}{Q(s)}\right\}=\sum_{m=1}^{\infty}\frac{P(s_m)}{Q'(s_m)}\exp(s_m t) \tag{E.2.24.17}$$

$$P(s)=s+1 \tag{E.2.24.18}$$

$$Q(s)=s^2+s+0.125 \tag{E.2.24.19}$$

$$Q'(s) = 2s + 1$$

The roots of $Q(s) = s^2 + s + 0.125 = 0$ are $s_{1,2} = \dfrac{-1 \pm \sqrt{1-(4)(1)(0.125)}}{(2)(1)}$,
therefore $s_1 = -0.854$ and $s_2 = -0.146$

$$\mathscr{L}^{-1}\left\{\frac{P(s)}{Q(s)}\right\} = \sum_{m=1}^{\infty} \frac{P(s_m)}{Q'(s_m)} \exp(s_m t) = \frac{s_1+1}{2s_1+1}\exp(s_1 t) + \frac{s_2+1}{2s_2+1}\exp(s_2 t)$$

$$= -0.206\exp(-0.854t) + 1.206\exp(-0.146t) \quad \text{(E.2.24.20)}$$

Example 2.25. Diffusion of salt into semi hard white cheese In Example 2.13. diffusion of salt into semi hard white cheese was described with the following partial differential equation Boundary and initial conditions:

$$\frac{\partial c}{\partial t} = D\frac{\partial^2 c}{\partial x^2}$$

BC1. $c = c_1$ at $x = 0$ when $t > 0$

BC2. $c = c_0 = 0$ at $x \to \infty$ when $t > 0$

IC. $c = c_0 = 0$ at $x > 0$ when $t = 0$

We used the method of combination of variables to solve this problem. The Laplace transformations may also be used to obtain this solution. Transformation of each term into a Laplace domain is:

$$\mathscr{L}\left\{D\frac{\partial^2 c}{\partial x^2}\right\} = D\frac{\partial^2 c(s)}{\partial x^2}$$

$$\mathscr{L}\left\{\frac{\partial c}{\partial t}\right\} = s\,c(s) - c_0$$

and that of the partial differential equation is:

$$D\frac{\partial^2 c(s)}{\partial x^2} = s\,c(s) - c_0$$

and may be rearranged as:

$$\frac{\partial^2 c(s)}{\partial x^2} - \frac{s}{D}c(s) - \frac{c_0}{D} = 0$$

The boundary conditions are transformed into the Laplace domain as:

BC1. $c(s) = \dfrac{c_1}{s}$ at $x = 0$

BC2. $c(s) = \dfrac{c_0}{s} = 0$ at $x \to \infty$

This is a second order non-homogeneous ordinary differential equation with constant coefficients. Its solution may be obtained by using Table 2.6 as:

$$c(s) = K_1 \exp(-\sqrt{s/D}\,x) + K_2 \exp(\sqrt{s/D}\,x) + \frac{c_0}{s}$$

BC2 requires $c(s) = \text{finite}$ at $x \to \infty$, but $\exp(\sqrt{s/D}\,x) \to \infty$, therefore $K_2 = 0$. We may use BC1 to evaluate $K_1 = (c_1 - c_0)/s$, then the solution is

$$c(s) = \frac{c_1 - c_0}{s} \exp(-\sqrt{s/D}\,x) + \frac{c_0}{s}$$

We may use Table 2.18 to make back transformations as:

$$\mathscr{L}^{-1}\left\{\frac{c_0}{s}\right\} = c_0$$

and

$$\mathscr{L}^{-1}\left\{\frac{1}{s}\exp(-\sqrt{s/D}\,x)\right\} = \operatorname{erfc}\left(\frac{x}{2\sqrt{Dt}}\right)$$

the back transformation of the solution is

$$c(t) = (c_1 - c_0)\operatorname{erfc}\left(\frac{x}{2\sqrt{Dt}}\right) + c_0$$

upon rearrangement we obtain

$$\frac{c_1 - c}{c_1 - c_0} = \operatorname{erfc}\left(\frac{x}{2\sqrt{Dt}}\right)$$

This is the same solution as we obtained in Example 2.13.

Example 2.26. Temperature profiles in a steak during frying In Example 2.8 a steak was dipped into vegetable oil for frying, where heat transfer was simulated with the following partial differential equation, BCs and the IC:

$$\frac{\partial^2 T}{\partial z^2} - \frac{1}{\alpha}\frac{\partial T}{\partial t} = 0$$

BC1: $T = T_1$ at $z = -L$ when $t > 0$ ($2L$ = thickness of the steak)

BC2: $T = T_1$ at $z = L$ when $t > 0$

BC3: $\frac{dT}{dz}$ at $z=0$ when $t>0$

BC4: $T=T_1$ for all z when $t\to\infty$

IC: $T=T_0$ when $t=0$ for all z

We may transform the partial differential equation and the BCs into the Laplace domain:

$$\frac{d^2T(s)}{dz^2}-\frac{s}{\alpha}T(s)+\frac{T_0}{\alpha}=0$$

BC1: $T(s)=\frac{T_1}{s}$ at $z=-L$

BC2: $T(s)=\frac{T_1}{s}$ at $z=L$

BC3: $\frac{dT(s)}{dz}=0$ at $z=0$

Solution of the ordinary differential equation may be obtained by using Table 2.6:

$$T(s)=K_1\exp\left(\sqrt{\frac{s}{\alpha}}z\right)+K_2\exp\left(-\sqrt{\frac{s}{\alpha}}z\right)+\frac{T_0}{s}$$

After using BCs 2 and 3 we will find

$$K_1=K_2=\frac{(T_1-T_0)/s}{\exp\left(\sqrt{\frac{s}{\alpha}}L\right)+\exp\left(-\sqrt{\frac{s}{\alpha}}L\right)},$$

then the solution will be

$$T(s)=\frac{T_1-T_0}{s}\frac{\exp\left(\sqrt{\frac{s}{\alpha}}z\right)+\exp\left(-\sqrt{\frac{s}{\alpha}}z\right)}{\exp\left(\sqrt{\frac{s}{\alpha}}L\right)+\exp\left(-\sqrt{\frac{s}{\alpha}}L\right)}+\frac{T_0}{s}$$

After using the identity $\cosh(u)=\frac{e^u+e^{-u}}{2}$ we will obtain

$$T(s)=\frac{T_1-T_0}{s}\frac{\cosh\left(\sqrt{\frac{s}{\alpha}}z\right)}{\cosh\left(\sqrt{\frac{s}{\alpha}}L\right)}+\frac{T_0}{s}$$

We may use Heaviside's Theorem 1 (Tab. 2.19) to obtain the back transform:

$$\mathscr{L}^{-1}\left\{\frac{P(s)}{Q(s)}\right\}=\sum_{m=1}^{\infty}\frac{P(s_m)}{Q'(s_m)}\exp(s_m t)$$

We may also use the identity $\cosh(u)=\cos(i\,u)$, then we will have

$$P(s)=\cosh\left(\sqrt{\frac{s}{\alpha}}z\right)=\cos\left(i\sqrt{\frac{s}{\alpha}}z\right)$$

$$Q(s)=s\cosh\left(\sqrt{\frac{s}{\alpha}}L\right)=s\cos\left(i\sqrt{\frac{s}{\alpha}}L\right)$$

$Q(s)=0$ when $s_{m1}=0$ and $s_{m2}=-\left(n+\frac{1}{2}\right)^2\left(\frac{\pi}{L}\right)^2\alpha$

$$Q'(s)=\cosh\left(\sqrt{\frac{s}{\alpha}}L\right)+\frac{1}{2}\left(\sqrt{\frac{s}{\alpha}}L\right)\sinh\left(\sqrt{\frac{s}{\alpha}}L\right)=\cos\left(i\sqrt{\frac{s}{\alpha}}L\right)$$

$$+\frac{1}{2}\sqrt{\frac{s}{\alpha}}L\sin\left(i\sqrt{\frac{s}{\alpha}}L\right)\sum_{m=1}^{\infty}\frac{P(s_m)}{Q'(s_m)}\exp(s_m t)$$

$$=\frac{\cos(0)}{\cos(0)+0}\exp(0)$$

$$+\sum_{n=0}^{\infty}\frac{\cos\left[-\left(n+\frac{1}{2}\right)\pi\frac{z}{L}\right]}{\cos\left[-\left(n+\frac{1}{2}\right)\pi\right]+\frac{1}{2}\left(n+\frac{1}{2}\right)\pi\sin\left[-\left(n+\frac{1}{2}\right)\pi\right]}$$

$$\times\exp\left[-\left(n+\frac{1}{2}\right)^2\left(\frac{\pi}{L}\right)^2\alpha t\right]$$

after substituting $\cos(-\beta)=\cos(\beta)$, $\sin(-\beta)=-\sin(\beta)$ and $\sin((n+(1/2))\pi)=(-1)^n$ we will have

$$\sum_{m=1}^{\infty}\frac{P(s_m)}{Q'(s_m)}\exp(s_m t)=1-2\sum_{n=0}^{\infty}(-1)^n\frac{\cos\left[\left(n+\frac{1}{2}\right)\pi\frac{z}{L}\right]}{\left(n+\frac{1}{2}\right)\pi}$$

$$\times\exp\left[-\left(n+\frac{1}{2}\right)^2\left(\frac{\pi}{L}\right)^2\alpha t\right]$$

We also have $\mathscr{L}^{-1}\{T(s)\}=T$ and $\mathscr{L}^{-1}\left\{\frac{T_0}{s}\right\}=T_0$

After substituting the back transformations we will have:

$$\frac{T-T_1}{T_0-T_1}=2\sum_{n=0}^{\infty}(-1)^n\frac{\cos\left[\left(n+\frac{1}{2}\right)\pi\frac{z}{L}\right]}{\left(n+\frac{1}{2}\right)\pi}$$
$$\times\exp\left[-\left(n+\frac{1}{2}\right)^2\left(\frac{\pi}{L}\right)^2\alpha t\right]$$

This is the same solution as we had in Example 2.8.

2.14. NUMERICAL METHODS IN MATHEMATICAL MODELING

There are numerous Food and Bioprocess Engineering problems involving ordinary or partial differential equations which may not be solved with the analytical solution techniques. The numerical solution techniques are usually easier than the analytical techniques and may be preferred when computers are available. The analytical solutions are usually represented as continuous functions within the range of the independent variables. A differential equation may be transformed into a numerical form using the differentiation formulas given in Table 2.21. The numerical solutions are presented as an array of numbers with some space between them. There is usually no error involved into the analytical solution techniques, but the numerical methods are subject to inherent error of the preferred technique (Tab. 2.21).

Numerical differentiation is usually subject to larger error than numerical integration, therefore modeling techniques requiring integration are preferred over the ones involving differentiation. There are truncation, round off and propagation errors involved in application of the numerical methods. The truncation error arises when an infinite series expansion of a function is approximated with limited number of terms. The round off error is introduced when the computer rounds up or down the number to a predetermined significant figures. The propagation error is caused by the inherent nature of the numerical techniques, where insignificant errors of individual steps of calculation builds up to a large amount.

Ordinary differential equations may be written in canonical form as:

$$\frac{dy_1}{dx}=f_1(y_1, y_2, \ldots, y_n, x) \tag{2.97.a}$$

$$\frac{dy_2}{dx}=f_2(y_1, y_2, \ldots, y_n, x) \tag{2.97.b}$$

$$\frac{dy_n}{dx}=f_n(y_1, y_2, \ldots, y_n, x) \tag{2.97.n}$$

With the initial conditions at x_0:

$$y_1(x_0)=y_{1,0} \tag{2.98.a}$$

$$y_2(x_0)=y_{2,0} \tag{2.98.b}$$

$$y_n(x_0)=y_{n,0} \tag{2.98.n}$$

These equations may be solved by the techniques outlined in Table 2.22. Most ordinary differential equations may be converted into canonical forms with appropriate transformations.

Example 2.27. Canonical form of a third order non-linear heterogeneous ordinary differential equation Transform the following third order non-linear heterogeneous ordinary differential equation into canonical form

$$\frac{d^3z}{dt^3}+3z\frac{d^2z}{dt^2}-7\frac{dz}{dt}+3z=e^{-t}$$

Solution We may rewrite this equation as

$$\frac{d^3z}{dt^3}=e^{-t}-3z+7\frac{dz}{dt}-3z\frac{d^2z}{dt^2}.$$

then substitute

$$y_1=e^{-t}, \text{ therefore } \frac{dy_1}{dt}=-e^{-t}=-y_1$$

$$z=y_2$$

$$\frac{dz}{dt}=\frac{dy_2}{dt}=y_3$$

$$\frac{d^2z}{dt^2}=\frac{dy_3}{dt}=y_4$$

$$\frac{d^3z}{dt^3}=\frac{dy_4}{dt}$$

Table 2.21 Numerical differentiation and integration formulas and order of magnitude of the involved error

i) First order differentiation:

		error
forward difference	$\frac{df(x)}{dx}=\frac{f(x+h)-f(x)}{h}$	$O(h)$
central difference	$\frac{df(x)}{dx}=\frac{f(x+h)-f(x-\Delta h)}{2h}$	$O(h^2)$
backward difference	$\frac{df(x)}{dx}=\frac{f(x)-f(x-h)}{h}$	$O(h)$
four point	$\frac{df(x)}{dx}=\frac{1}{12h}\{f(x-2h)$ $-8f(x-h)+8f(x+h)-f(x+2h)\}$	$O(h^4)$

ii) Second order differentiation:

		error
forward difference	$\frac{d^2f(x)}{dx^2}=\frac{f(x+2h)-f(x+h)-f(x)}{h^2}$	$O(h)$
central difference	$\frac{d^2f(x)}{dx^2}=\frac{f(x+h)-2f(x)+f(x-h)}{h^2}$	$O(h^2)$
backward difference	$\frac{d^2f(x)}{dx^2}=\frac{f(x)-2f(x-h)+f(x-2h)}{h^2}$	$O(h)$

iii) Integration fromulas: $A=\int_{x_0}^{x_n}f(x)dx$

		error
Trapezoidal rule	$A=\frac{h}{2}[f_0+2f_1+2f_2+\cdots+2f_{n-1}+f_n]$ where $f_n=f_0+nh$	$O(h^3)$
Simpson's rule	$A=\frac{h}{3}[f_0+4f_1+2f_2+4f_3$ $+\cdots+4f_{n-3}+2f_{n-2}+4f_{n-1}+f_n]$	$O(h^4)$

The set of the canonical equations will be

$$\frac{dy_1}{dt}=-y_1$$

$$\frac{dy_2}{dt}=y_3$$

$$\frac{dy_3}{dt}=y_4$$

Table 2.22 Solutions to ordinary differential equations in canonical form and order of magnitude of the involved error

method	calculation procedure	error
Euler's method	$y_{i+1} = y_i + k_1$	$O(h^2)$
Modified Euler's method	$y_{i+1} = y_i + \frac{1}{2}(k_1 + k_2)$	$O(h^2)$
Second order Runge-Kutta method	$y_{i+1} = y_i + k_3$	$O(h^2)$
Fourth order Runge-Kutta method	$y_{i+1} = y_i + \frac{1}{6}(k_1 + 2k_3 + 2k_4 + k_5)$	$O(h^4)$
	where $dy/dt = f(x,y)$ $h = x_{i+1} - x_i$ $k_1 = h\, f(x_i, y_i)$ $k_2 = h f(x_i + h, y_i + k_1)$ $k_3 = h f\left(x_i + \frac{1}{2}h, y_i + 1/2 k_1\right)$ $k_4 = h f\left(x_i + \frac{1}{2}h, y_i + 1/2 k_3\right)$ $k_5 = h f(x_i + h, y_i + k_4)$	

$$\frac{dy_4}{dt} = y_1 - 3\,y_2 + 7\,y_3 - 3\,y_2\,y_4$$

These are autonomous set of ordinary differential equations. If we know the initial conditions $y_1(t_0) = y_{1,0}$, $y_2(t_0) = y_{2,0}$, $y_3(t_0) = y_{3,0}$ and $y_4(t_0) = y_{4,0}$ we can easily solve them by using the techniques outlined in Table 2.22.

Example 2.28. Drying behavior of honey-starch mixtures (*Yener et al.*, 1987) Two consecutive falling rate periods were observed during drying of thin films of honey-starch mixture at 70 °C. Variation of the moisture content of these films were simulated after using (2.43.f) as:

$$\frac{dc}{dt} = \mathcal{K}_c(c^* - c)$$

In the first falling rate period the drying rate constant, i.e., mass transfer coefficient, was $\mathcal{K}_c = k_0 c + k_1$, and in the second falling rate period $\mathcal{K}_c = k_2$. With 10 % honey-starch mixture the constants were: $k_0 = 17.33$ (h kg moisture/kg dry solids)$^{-1}$, $k_1 = -0.36\ h^{-1}$, $k_2 = 0.116\ h^{-1}$, $c_0 = 0.078$ kg moisture/kg dry solids, $c^* = 0.015$ kg moisture/kg dry solids, the first falling rate prevailed about 4 hours, constant weight was attained in

30 hours. Solve the given drying rate expression with fourth order Runge-Kutta Method and compare the results with the experimental data given in Table E.2.28.

Table E.2.28 Variation of the moisture content of the honey starch mixtures during the course of drying (c: kg moisture/kg dry solids, t: h)

t	c	t	c	t	c	t	c	t	c	t	c
0	0.078	1.17	0.048	4	0.030	11	0.021	18	0.017	25	0.016
0.17	0.077	1.33	0.045	5	0.028	12	0.020	19	0.017	26	0.016
0.33	0.068	1.50	0.044	6	0.027	13	0.020	20	0.017	27	0.016
0.50	0.062	1.67	0.043	7	0.025	14	0.019	21	0.017	28	0.016
0.67	0.056	1.83	0.040	8	0.024	15	0.018	22	0.017	29	0.016
0.83	0.053	2	0.039	9	0.023	16	0.018	23	0.017	30	0.015
1	0.050	3	0.028	10	0.022	17	0.018	24	0.016		

Solution Fourth order Runge-Kutta method was outlined in Table 2.22 as:

$$c_{i+1} = c_i + \frac{1}{6}(k_1 + 2k_3 + 2k_4 + k_5) \quad \text{where}$$

$$k_1 = h\,f(t_i, c_i)$$

$$k_3 = h\,f\left(t_i + \frac{1}{2}h, c_i + \frac{1}{2}k_1\right)$$

$$k_4 = h\,f\left(t_i + \frac{1}{2}h, c_i + \frac{1}{2}k_3\right)$$

$$k_5 = h\,f(t_i + h, c_i + k_4)$$

i) The first falling rate period: The drying rate expression is

$$\frac{dc}{dt} = -17.33c^2 + 0.62c - 5.4 \times 10^{-3}$$

The solution is given in the spread sheet on the next page

column	parameter	function
A	time	$t_n = t_0 + nh$
B	c_{n+1}	
C	k_1	$0.62B - 17.33B^2 - 0.0054$
D	k_3	$0.62(B + 0.5C) - 17.33(B + 0.5C)^2 - 0.0054$
E	k_4	$0.62(B + 0.5D) - 17.33(B + 0.5D)^2 - 0.0054$
F	k_5	$0.62(B + 0.5E) - 17.33(B + 0.5E)^2 - 0.0054$
G	increment	$\frac{1}{6}(C + 2D + 2E + F)$
H	c_n	$B + G$

The first falling rate period takes four hours, therefore this worksheet will be used for the calculations when $0 \leqslant t \leqslant 4$ h, with $h = 1$ h.

ii) The second falling rate period: The drying rate expression is

$$\frac{dc}{dt} = 0.0017 - 0.116c$$

column	parameter	function
A	time	$t_n = t_0 + nh$
B	c_{n+1}	
C	k_1	$0.00174 - 0.116B$
D	k_3	$0.00174 - 0.116(B + 0.5C)$
E	k_4	$0.00174 - 0.116(B + 0.5D)$
F	k_5	$0.00174 - 0.116(B + 0.5E)$
G	increment	$\frac{1}{6}(C + 2D + 2E + F)$
H	c_n	$B + G$

The second falling rate period prevails when $4 \leqslant t \leqslant 30$ h. The following computer output summarizes the calculations:

A Time	B c_n	C k_1	D k_3	E k_4	F k_5	G increment	H k_{n+1}
0	0.0780	−0.062480	−0.014303	−0.050473	−0.032250	−0.032250	0.0458
1	0.0458	−0.013310	−0.007651	−0.010573	−0.009132	−0.009132	0.0366
2	0.0366	−0.005936	−0.004162	−0.005137	−0.004599	−0.004599	0.0320
3	0.0320	−0.003317	−0.002552	−0.003098	−0.002764	−0.002764	0.0293
4	0.0293	−0.001654	−0.001558	−0.001563	−0.001561	−0.001561	0.0277
5	0.0277	−0.001473	−0.001387	−0.001392	−0.001390	−0.001390	0.0263
6	0.0263	−0.001311	−0.001235	−0.001240	−0.001238	−0.001238	0.0251
7	0.0251	−0.001168	−0.001100	−0.001104	−0.001102	−0.001102	0.0240
8	0.0240	−0.001040	−0.000979	−0.000983	−0.000982	−0.000982	0.0230
9	0.0230	−0.000926	−0.000872	−0.000875	−0.000874	−0.000874	0.0221
10	0.0221	−0.000824	−0.000777	−0.000779	−0.000778	−0.000778	0.0213
11	0.0213	−0.000734	−0.000692	−0.000694	−0.000693	−0.000693	0.0206
12	0.0206	−0.000654	−0.000616	−0.000618	−0.000617	−0.000617	0.0200
13	0.0200	−0.000582	−0.000548	−0.000550	−0.000550	−0.000550	0.0195
14	0.0195	−0.000518	−0.000488	−0.000490	−0.000489	−0.000489	0.0190
15	0.0190	−0.000462	−0.000435	−0.000436	−0.000436	−0.000436	0.0185
16	0.0185	−0.000411	−0.000387	−0.000389	−0.000388	−0.000388	0.0182
17	0.0182	−0.000366	−0.000345	−0.000346	−0.000346	−0.000346	0.0178
18	0.0178	−0.000326	−0.000307	−0.000308	−0.000308	−0.000308	0.0175
19	0.0175	−0.000290	−0.000273	−0.000274	−0.000274	−0.000274	0.0172

A Time	B c_n	C k_1	D k_3	E k_4	F k_5	G increment	H k_{n+1}
20	0.0172	−0.000258	−0.000243	−0.000244	−0.000244	−0.000244	0.0170
21	0.0170	−0.000230	−0.000217	−0.000218	−0.000217	−0.000217	0.0168
22	0.0168	−0.000205	−0.000193	−0.000194	−0.000194	−0.000194	0.0166
23	0.0166	−0.000182	−0.000172	−0.000173	−0.000172	−0.000172	0.0164
24	0.0164	−0.000163	−0.000153	−0.000154	−0.000153	−0.000153	0.0162
25	0.0162	−0.000145	−0.000136	−0.000137	−0.000137	−0.000137	0.0161
26	0.0161	−0.000129	−0.000121	−0.000122	−0.000122	−0.000122	0.0160
27	0.0160	−0.000115	−0.000108	−0.000108	−0.000108	−0.000108	0.0159
28	0.0159	−0.000102	−0.000096	−0.000097	−0.000096	−0.000096	0.0158
29	0.0158	−0.000091	−0.000086	−0.000086	−0.000086	−0.000086	0.0157
30	0.0157	−0.000081	−0.000076	−0.000077	−0.000076	−0.000076	0.0156

Comparison of the model with the experimental data is shown in Figure E.2.28.

The problem solved in Example 2.28 is called an *initial value problem*, since the solution was propagated staring from the readily available IC. There are also cases where BCs, i.e., the final values of the dependent variable, are available. Such problems are called the *boundary value problems*. A boundary value problem may be converted into an initial value problem by using a *shooting method* (Costantinides, 1987). Newton's method is among the major shooting method and may be exemplified with the following equations (Costantinides, 1987):

$$\frac{dy_1}{dt} = f_1(t, y_1\, y_2) \tag{2.99.a}$$

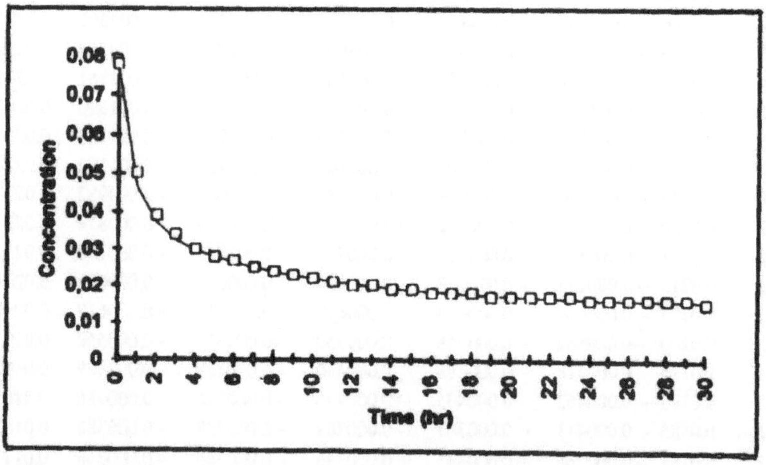

FIGURE E.2.28 Variation of the moisture content of the honey-starch mixture with time. —: model, □: experimental data.

$$\frac{dy_2}{dt} = f_2(t, y_1\, y_2) \tag{2.99.b}$$

with initial and boundary conditions

$$\text{IC. } y_1(t_0) = y_{1,0} \tag{2.100.a}$$

$$\text{BC. } y_2(t_f) = y_{2,f} \tag{2.100.b}$$

we guess the initial condition for y_2

$$y_2(t_0) = \gamma \tag{2.101}$$

Since the value of $y_2(t_0)$ was a guess the trajectory of y_2 may not satisfy BC2 and cause an error $\phi(\gamma)$:

$$\phi(\gamma) = y_2(t_f, \gamma) - y_{2,f} \tag{2.102}$$

where $y_2(t_f, \gamma)$ = the calculated value of y_2 at t_f. The desirable objective at BC was $y_2(t_f, \gamma) = y_{2,f}$. If the error $\phi(\gamma)$ is unacceptable, a new value may be assigned for γ as:

$$\gamma_{new} = \gamma_{old} + \alpha \Delta\gamma \tag{2.103}$$

where $\Delta\gamma = -\dfrac{\phi(\gamma)}{\partial y_2(t_f, \gamma)/d\gamma}$

$\partial y_2(t_f, \gamma)/\partial\gamma$ is the sensitivity of the endpoint trajectory of y_2 to the initial guess of the initial condition γ. Parameter α is a constant to prevent divergence ($0 < \alpha \leqslant 1$). The sensitivity may be calculated from the variational equations:

$$\frac{\partial}{\partial\gamma}\left(\frac{dy_1}{dt}\right) = \frac{\partial f_1}{\partial y_1}\frac{\partial y_1}{\partial\gamma} + \frac{\partial f_1}{\partial y_2}\frac{\partial y_2}{\partial\gamma} \tag{2.104.a}$$

$$\frac{\partial}{\partial\gamma}\left(\frac{dy_2}{dt}\right) = \frac{\partial f_2}{\partial y_1}\frac{\partial y_1}{\partial\gamma} + \frac{\partial f_2}{\partial y_2}\frac{\partial y_2}{\partial\gamma} \tag{2.104.b}$$

the variational equations may be rearranged (Costantinides, 1987):

$$\frac{dp_1}{dt} = \frac{\partial f_1}{\partial y_1} p_1 + \frac{\partial f_1}{\partial y_2} p_2 \tag{2.105.a}$$

$$\frac{dp_2}{dt} = \frac{\partial f_2}{\partial y_1} p_1 + \frac{\partial f_2}{\partial y_2} p_2 \tag{2.105.b}$$

where $p_1 = \dfrac{\partial y_1}{\partial\gamma}$ and $p_2 = \dfrac{\partial y_2}{\partial\gamma}$. The initial conditions for p_1 and are obtained from definitions of these variables:

$$p_1(t_0) = \left.\frac{\partial y_1}{\partial\gamma}\right|_{t=t_0} = 0 \tag{2.106.a}$$

$$p_2(t_0) = \frac{\partial y_2}{\partial \gamma}\bigg|_{t=t_0} = \frac{\partial(\gamma)}{\partial \gamma} = 1 \tag{2.106.b}$$

The Newton method may be outlined as:

(i) The ordinary differential equations (2.99. a) and (2.99.b) are given in canonical form with IC. (2.104.a) and BC (2.104.b).
(ii) The missing initial condition of the system equations was guessed by (2.101)
(iii) The system equations (2.99. a), (2.99.b) and the variational equations (2.105.a), (2.105.b) are solved simultaneously using the ICs (2.100.a), (2.101), (2.106.a) and (2.106.b)
(iv) The correction $\Delta\gamma$ to be applied to γ using (2.103). Parameter α is chosen intuitively.
(v) Steps iii and iv are repeated until $|\phi(\gamma)| \leqslant \varepsilon$, where ε = predetermined error.

Even when the truncation and round off errors are negligible, numerical methods are subject to instabilities arising from the differences between the analytical and the numerical solutions. Any numerical method which has a finite general stability bound is said to be conditionally stable. The solution to the differential equation

$$\frac{dy}{dt} = \lambda y \tag{2.107}$$

with Euler's method is

$$y_{n+1} = y_n + \lambda h\, y_n \tag{2.108}$$

Equation (2.108) may be rearranged as:

$$y_{n+1} - (1 + \lambda h)\, y_n = 0 \tag{2.109}$$

We may assume a trial solution as $y_n = C\mu^n$ and $y_{n+1} = C\mu^{n+1}$, then substitute in (2.109):

$$C\mu^{n+1} - (1 + \lambda h)\, C\mu^n = 0 \tag{2.110}$$

Equation (2.110) yields that

$$\mu = 1 + \lambda h \tag{2.111}$$

and the solution is

$$y_n = C(1 + \lambda h)^n \tag{2.112}$$

As n increases without bound, the solution remains stable if

$$\lim_{n\to\infty} = y_{n=0} \tag{2.113}$$

Equation (2.113) may be achieved only when $-2 \leqslant h\lambda \leqslant 0$ and $\lambda < 0$. After following a similar approach, it may be shown that $\mu = (1 - h\lambda/2)/(1 + h\lambda/2)$ with the modified Euler's Method the solution is stable when $\lambda < 0$. We may also show that $\mu = 1 + h\lambda + (1/2)h^2\lambda^2$ with the second and $\mu = 1 + h\lambda + (1/2)h^2\lambda^2 + (1/6)h^3\lambda^3 + (1/24)h^4\lambda^4$ with the fourth order Runge-Kutta Methods. Therefore the stability boundaries are $-2 \leqslant h\lambda \leqslant 0$ with the second and $-2.785 \leqslant h\lambda \leqslant 0$ with the fourth order Runge-Kutta Methods. An interested reader may refer to Costantinides (1987) for details.

The *finite difference method* replaces the derivatives in the differential equations with finite difference approximations (Tab. 2.21) and converts the differential equation into a large set of simultaneous non-linear algebraic equations. This method may be demonstrated with the following differential equations:

$$\frac{dy_1}{dt} = f_1(t, y_1, y_2) \tag{2.114}$$

$$\frac{dy_2}{dt} = f_2(t, y_1, y_2) \tag{2.115}$$

$$\text{BC1: } y_1(t_0) = y_{1,0} \tag{2.116}$$

$$\text{BC2: } y_2(t_f) = y_{2,f} \tag{2.117}$$

After replacing the derivatives with the forward difference numerical differentiation formula (Tab. 2.21) we will have:

$$y_{1,i+1} - y_{1,i} = h f_1(t, y_{1,i}\ y_{2,i}) \tag{2.119}$$

$$y_{2,i+1} - y_{2,i} = h f_1(t, y_{1,i}\ y_{2,i}) \tag{2.120}$$

$$\text{BC1: } y_1(t_0) = y_{1,0} \tag{2.121}$$

$$\text{BC2: } y_2(t_f) = y_{2,f} \tag{2.122}$$

The solution interval will be divided into n segments of equal length and resulting equations will be written for $i = 1, 2, \ldots, n-1$. They will form a set of $2n$ simultaneous non-linear algebraic equations in $2n + 2$ variables (there are $n + 1$ variables in each equation, including $y_{1,0}$; $y_{1,n}$; $y_{2,0}$ and $y_{2,n}$). The boundary conditions will provide values for two of these variables $y_1(t_0) = y_{1,0}$ and $y_2(t_f) = y_{2,n}$, then system of $2n$ equations in $2n$ unknowns will be solved.

In food and bioprocess modeling the finite difference method is very popular especially to solve partial differential equations. Table 2.23 may be used to convert the partial differential equations into the finite difference equations.

Table 2.23 Numerical partial differentiation formulas

i) First order partial derivative.

		error
forward difference	$\left.\frac{\partial u}{\partial x}\right\|_{i,j,k} = \frac{1}{h}(u_{i+1,j,k} - u_{i,j,k})$	$O(h)$
central difference	$\left.\frac{\partial u}{\partial x}\right\|_{i,j,k} = \frac{1}{2h}(u_{i+1,j,k} - u_{i-1,j,k})$	$O(h^2)$
backward difference	$\left.\frac{\partial u}{\partial x}\right\|_{i,j,k} = \frac{1}{h}(u_{i,j,k} - u_{i-1,j,k})$	$O(h)$

ii) Second order partial derivative:

		error
forward difference	$\left.\frac{\partial^2 u}{\partial x^2}\right\|_{i,j,k} = \frac{1}{h^2}(u_{i+2,j,k} - 2u_{i+1,j,k} + u_{i,j,k})$	$O(h)$
central difference	$\left.\frac{\partial^2 u}{\partial x^2}\right\|_{i,j,k} = \frac{1}{h^2}(u_{i+1,j,k} - 2u_{i,j,k} + u_{i-1,j,k})$	$O(h^2)$
backward difference	$\left.\frac{\partial^2 u}{\partial x^2}\right\|_{i,j,k} = \frac{1}{h^2}(u_{i,j,k} - 2u_{i-1,j,k} + u_{i-2,j,k})$	$O(h)$

iii) Second order mixed partial derivative:

		error
forward difference	$\left.\frac{\partial^2 u}{\partial x \partial y}\right\|_{i,j,k} = \frac{1}{hk}(u_{i+1,j+1,k} - u_{i,j+1,k} - u_{i+1,j,k} + u_{i,j,k})$	$O(h+k)$
central difference	$\left.\frac{\partial^2 u}{\partial x \partial y}\right\|_{i,j,k} = \frac{1}{4hk}(u_{i+1,j+1,k} - u_{i-1,j+1,k} - u_{i+1,j-1,k} + u_{i-1,j-1,k})$	$O(h^2+k^2)$
backward difference	$\left.\frac{\partial^2 u}{\partial x \partial y}\right\|_{i,j,k} = \frac{1}{hk}(u_{i,j,k} - u_{i,j-1,k} - u_{i-1,j,k} + u_{i-1,j-1,k})$	$O(h)$

Example 2.29. Drying behavior of an apple slice (*McCarthy et al.*, 1991) A 5 cm × 4.9 cm × 2.0 cm apple slice was placed in a sample holder and dried in a tunnel drier. Only the top and the bottom faces were exposed to air flow. Variation of the moisture content along the slice may be described

with the equation in rectangular coordinates:

$$\frac{\partial c}{\partial t}+\left(\frac{\partial N_x}{\partial x}+\frac{\partial N_y}{\partial y}+\frac{\partial N_z}{\partial z}\right)=R \tag{2.12}$$

Since there is no flux in y and z directions and water does not involve into any chemical reactions (2.12) will be simplified:

$$\frac{\partial c}{\partial t}+\frac{\partial N_x}{\partial x}=0 \tag{E.2.29.1}$$

parameter N_x, water flux in x direction, was defined as:

$$N_x = x_w(N_x + N_s) - D\frac{dc}{dz} \tag{2.10}$$

where $N_s = 0$ (flux of the solids) and $x_w = c/(c + c_s)$ (c_s = solid content of the apple = 0.3284 g solids/cm^3)
Equation (E.2.29.1) is rearranged after substituting (2.10) in (E.2.29.1):

$$\frac{\partial c}{\partial t}=\frac{D}{c_s}\left[\frac{\partial c}{\partial x}\right]^2+D\left(\frac{c}{c_s}+1\right)\frac{\partial c^2}{\partial x^2} \tag{E.2.29.2}$$

The origin of the coordinate system, i.e., $x = 0$, was located on the apple-air interface. The IC and the BCs are:

$$\text{IC: } c = c_0 = 0.627\ \text{g/cm}^3 \text{ at } t = 0 \text{ for all } x \tag{E.2.29.3}$$

$$\text{BC1: } c = c_e \text{ at } t \geqslant 0 \text{ for } x = 0 \tag{E.2.29.4}$$

(c_e = equilibrium moisture content = 0.054 g/cm^3)

$$\text{BC2: } \frac{dc}{dx} = 0 \text{ at } t \geqslant 0 \text{ for } x = L \tag{E.2.29.5}$$

($2L$ = 5 cm = thickness of the apple slab)

Equation (E.2.29.2) was converted into a finite difference equation by using the forward difference formulas:

partial derivative	finite difference equivalent
$\frac{\partial c}{\partial t}$	$\frac{1}{\Delta t}(c_{i,j+1} - c_{i,j})$
$\frac{\partial c}{\partial x}$	$\frac{1}{\Delta x}(c_{i+1,j} - c_{i,j})$
$\frac{\partial^2 c}{\partial x^2}$	$\frac{1}{\Delta x^2}(c_{i+1,j} - 2c_{i,j} + c_{i-1,j})$

After substituting the finite difference equivalents (E.2.29.2) becomes

$$\frac{c_{i,j+1}-c_{i,j}}{\Delta t}=\left(\frac{D}{c_s}\right)\frac{c_{i+1,j}^2-2c_{i+1,j}c_{i,j}+c_{i,j}^2}{(\Delta x)^2}+D\left(\frac{c_{i,j}}{c_s}+1\right)\frac{c_{i+1,j}-2c_{i,j}+c_{i-1,j}}{(\Delta x)^2} \quad \text{(E.2.29.6)}$$

Equation (E.2.29.6) may be rearranged as:

$$c_{i,j+1}=c_{i,j}+\Delta t\left[\left(\frac{D}{c_s}\right)\frac{c_{i+1,j}^2-2c_{i+1,j}c_{i,j}+c_{i,j}^2}{(\Delta x)^2}+D\left(\frac{c_{i,j}}{c_s}+1\right)\frac{c_{i+1,j}-2c_{i,j}+c_{i-1,j}}{(\Delta x)^2}\right] \quad \text{(E.2.29.7)}$$

Numerical values of the time and the location increments were assigned as $\Delta t = 2\,s$ and $\Delta x = 0.10\,\text{cm}$, respectively. All the parameters on the right hand side (with the exception of diffusivity D) are known when $t = 0$. After assigning a numerical value for D, it will be possible to determine $c_{i,j+1}$. for all i's. The same procedure may be repeated for all j's to simulate the drying process over the entire drying period. The correct estimate of D was $8.3 \times 10^{-9}\,\text{m}^2/\text{s}$, which simulated the experimental data very closely as shown in Figure E.2.29.

Example 2.30. Temperature profiles along a spherical potato tuber during blanching Calculate the temperature profile along a spherical potato tuber ($\alpha = 0.19\ 10^{-6}\,\text{m}^2/\text{s}$, $k = 0.554\,\text{W/mK}$, radius $= 4.1\,\text{cm}$, $T_0 = 18°\text{C}$) 5 and 15 minutes after dipping in blanching water ($T_w = 100°\text{C}$) with the convective heat transfer coefficient $h = 70\,\text{W/m}^2\text{K}$.

Solution Equation of energy in the spherical coordinates is:

$$\frac{\partial T}{\partial t}+v_r\frac{\partial T}{\partial r}+\frac{v_\theta}{r}\frac{\partial T}{\partial \theta}+\frac{v_\phi}{r\sin\theta}\frac{\partial T}{\partial \phi}=\frac{\psi_G}{\rho c_p}+\frac{1}{r^2}\frac{\partial}{\partial r}\left(r^2\alpha\frac{\partial T}{\partial r}\right)+\frac{1}{r^2\sin\theta}\frac{\partial}{\partial \theta}\left(\alpha\sin\theta\frac{\partial T}{\partial \theta}\right)+\frac{1}{r^2\sin^2\theta}\frac{\partial}{\partial \phi}\left(\alpha\frac{\partial T}{\partial \phi}\right) \quad (2.18)$$

When $v_r = v_\theta = v_\phi = \psi_G = \dfrac{\partial T}{d\theta} = \dfrac{\partial T}{d\phi} = 0$, equation (2.18) becomes

$$\frac{\partial T}{\partial t}=\frac{1}{r^2}\frac{\partial}{\partial r}\left(r^2\alpha\frac{\partial T}{\partial r}\right) \quad \text{(E.2.30.1)}$$

When α = constant (E.2.30.1) may be rearranged as:

$$\frac{\partial^2 T}{\partial r^2}+\frac{2}{r}\frac{\partial T}{\partial r}=\frac{1}{\alpha}\frac{\partial T}{\partial t} \quad \text{(E.2.30.2)}$$

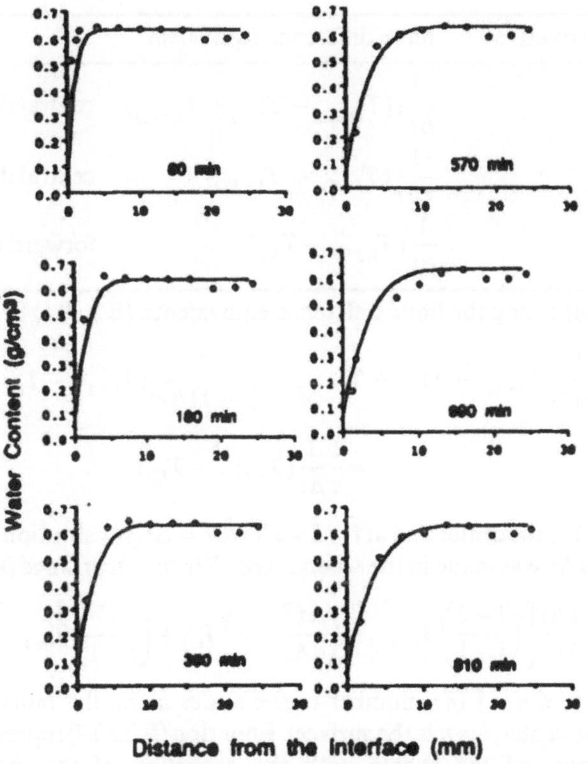

FIGURE E.2.29 Transient moisture profiles of the drying apple slab. Solid lines are the simulations; experimental data are shown in symbols (Reprinted with permission, McCarthy *et al.*, © 1991 ACS).

with

$$\text{IC: } T(r,0) = T_0 \tag{E.2.30.3}$$

$$\text{BC1: } \frac{dT}{dr} = 0 \text{ at } r = 0 \tag{E.2.30.4}$$

$$\text{BC2: } -k\frac{dT}{dr} = h\,(T_R - T_w) \text{ at } r = R \tag{E.2.30.5}$$

where R = radius of the sphere, k = thermal conductivity of the food, h = convective heat transfer coefficient and T_w = blanching water temperature. Each term in (E.2.30.1) is converted into finite difference notation:

partial derivative	finite difference equivalent	
$\frac{\partial^2 T}{\partial r^2}$	$\frac{1}{\Delta r^2}(T_{i+1,j}-2T_{i,j}+T_{i-1,j})$	central difference
$\frac{\partial T}{\partial r}$	$\frac{1}{2\Delta r}(T_{i+1,j}-T_{i-1,j})$	central difference
$\frac{\partial T}{\partial t}$	$\frac{1}{\Delta t}(T_{i,j+1}-T_{i,j})$	forward difference

After substituting the finite difference equivalents, (E.2.30.1) becomes:

$$\frac{1}{\Delta r^2}(T_{i+1,j}-2T_{i,j}+T_{i-1,j})+\frac{1}{(i-1)\Delta r^2}(T_{i+1,j}-T_{i-1,j})$$
$$=\frac{1}{\alpha}\frac{1}{\Delta t}(T_{i,j+1}-T_{i,j}) \tag{E.2.30.6}$$

Since $i=1$ is the center and at $i=2$ we have $r=\Delta r$, substitution of the form $r=(i-1)\Delta r$ was made in the second term. We may rearrange (E.2.30.6) as:

$$T_{i,j+1}=\frac{\alpha\Delta t}{\Delta r^2}\left[\left(\frac{i-2}{i-1}\right)T_{i-1,j}+\left(\frac{\Delta r^2}{\alpha\Delta t}-2\right)T_{i,j}+\left(\frac{i}{i-1}\right)T_{i+1,j}\right] \tag{E.2.30.7}$$

where $2 \leqslant i \leqslant n-1$ (n = number of the nodes along the radial direction; $i=1$ is the center; $i=n$ is the surface). Equation (E.2.30.7) represents all the inner points of the sphere with the exception of the center where $(2/r)(\partial T/\partial r)$ is not defined. We may use the L'Hospital's rule on the second term:

$$\lim_{r\to\infty}\left(\frac{2}{r}\frac{\partial T}{\partial r}\right)=\lim_{r\to\infty}\left\{2\frac{\partial}{\partial r}\left(\frac{\partial T}{\partial r}\right)\Big/\frac{\partial}{\partial r}(r)\right\}=\left.\frac{\partial^2 T}{\partial r^2}\right|_{r=0} \tag{E.2.30.8}$$

Therefore at $r=0$ (E.2.30.2) should be expressed as:

$$3\frac{\partial^2 T}{\partial r^2}=\frac{1}{\alpha}\frac{\partial T}{\partial t} \tag{E.2.30.9}$$

Finite difference equivalent of (E.2.30.9) is:

$$\frac{3}{\Delta r^2}(T_{2,j}-2T_{1,j}+T_{0,j})=\frac{1}{\alpha\Delta t}(T_{1,j+1}-T_{1,j}) \tag{E.2.30.10}$$

which may be rearranged as:

$$T_{1,j+1}=T_{1,j}+\frac{3\alpha\Delta t}{\Delta r^2}(T_{2,j}-2T_{1,j}+T_{0,j}) \tag{E.2.30.11}$$

$T_{1,j}$ is the center temperature, but $i=0$ is not defined, therefore is $T_{0,j}$ of (E.2.30.11) is fictitious temperature. We may express (E.2.30.4) in finite difference form as:

$$\frac{1}{2\Delta r}(T_{2,j} - T_{0,j}) = 0 \tag{E.2.30.12}$$

which indicate $T_{0,j} = T_{2,j}$; then (E.2.30.11) becomes:

$$T_{1,j+1} = \left\{1 - \frac{6\alpha\Delta t}{\Delta r^2}\right\} T_{1,j} + \frac{6\alpha\Delta t}{\Delta r^2} T_{2,j} \tag{E.2.30.13}$$

The surface boundary condition may be approximated with backwards difference:

$$-\frac{k}{\Delta r}(T_{n,j+1} - T_{n-1,j+1}) = h(T_{n,j+1} - T_w) \tag{E.2.30.14}$$

which may be rearranged as:

$$T_{n,j+1} = -\frac{1}{1+(h/k)\Delta r}\left\{T_{n-1,j} + \frac{h\Delta r}{k} T_w\right\} \tag{E.2.30.15}$$

Equations (E.2.30.7), (E.2.30.13) and (E.2.30.15) represent the temperature variation with time at the selected nodes in radial direction. We will choose eight nodes ($n = 8$) along the radial direction to calculate the temperature, therefore:

$$2 \leqslant i \leqslant 7$$

and

$$\Delta r = \frac{R}{8} = \frac{0.041}{7} = 0.005857\,m$$

Since the most restrictive stability criterion is associated with the surface equation (Incropera and De Witt, 1985):

$$Fo(1 + Bi) \leqslant \frac{1}{2} \tag{E.2.30.16}$$

After substituting $Bi = (h\Delta r/k) = 0.74$ in (E.2.30.16) we will obtain $Fo \leqslant 0.287$, since $Fo = (\alpha\Delta t/\Delta r^2)$ we should have $\Delta t \leqslant 51.88$ s and choose $\Delta t = 50$ s.

After substituting numbers in (E.2.30.7), (E.2.30.13) and (E.2.30.15):

$$T_1^{j+1} = -0.6616\, T_1^j + 1.6616\, T_2^j \tag{E.2.30.17.a}$$

$$T_2^{j+1} = 0.4461\, T_2^j + 0.5539\, T_3^j \tag{E.2.30.17.b}$$

$$T_3^{j+1} = 0.1385\, T_2^j + 0.4461\, T_3^j + 0.4154\, T_4^j \tag{E.2.30.17.c}$$

$$T_4^{j+1} = 0.1846\, T_3^j + 0.4461\, T_4^j + 0.3692\, T_5^j \tag{E.2.30.17.d}$$

$$T_5^{j+1} = 0.2077\,T_4^j + 0.4461\,T_5^j + 0.3462\,T_6^j \qquad \text{(E.2.30.17.e)}$$

$$T_6^{j+1} = 0.2215\,T_5^j + 0.4461\,T_6^j + 0.3323\,T_7^j \qquad \text{(E.2.30.17.f)}$$

$$T_7^{j+1} = 0.2308\,T_6^j + 0.4461\,T_7^j + 0.3221\,T_8^j \qquad \text{(E.2.30.17.g)}$$

$$T_8^{j+1} = 0.5744\,T_7^j + 158.64 \qquad \text{(E.2.30.17.h)}$$

At $t = 0$ $T_1^1 = T_2^1 = T_3^1 = T_4^1 = T_5^1 = T_6^1 = T_7^1 = T_8^1 = 291$ K. After substituting them in (E.2.30.17) we can easily calculate $T_1^2, T_2^2, T_3^2, T_4^2, T_5^2, T_6^2, T_7^2$ and T_8^2, the same procedure may be repeated to calculate the temperature at the same nodes at different times. A spread sheet program may be used to do the calculations. The first column of the program may contain the initial temperatures $T_1^1, T_2^1, T_3^1, T_4^1, T_5^1, T_6^1, T_7^1$ and T_8^1, the second column may be calculated using (E.2.30.17) and the first column. Calculations may be propogated column wise until covering the required heating period:

	A	B	C	D	E	F	G	H
1	291	291	291					
2	291	291	291					
3	291	291	291					
4	291	291	291					
5	291	291	291					
6	291	291	291					
7	291	291	302.3					
8	291	325.9	325.9					

Cells from A_1 to A_8 should contain the initial temperatures at each node, i.e., $A_1 = A_2 = \cdots = A_8 = 291$ K, thus

$$B_1 = -0.6616\,A_1 + 1.6616\,A_2 = 291\ \text{K}$$

$$B_2 = 0.4461\,A_2 + 0.5539\,A_3 = 291\ \text{K}$$

$$B_3 = 0.1385\,A_2 + 0.4461\,A_3 + 0.4154\,A_4 = 291\ \text{K}$$

$$B_4 = 0.1846\,A_3 + 0.4461\,A_4 + 0.3692\,A_5 = 291\ \text{K}$$

$$B_5 = 0.2077\,A_4 + 0.4461\,A_5 + 0.3462\,A_6 = 291\ \text{K}$$

$$B_6 = 0.2215\,A_5 + 0.4461\,A_6 + 0.3323\,A_7 = 291\ \text{K}$$

$$B_7 = 0.2308\,A_6 + 0.4461\,A_7 + 0.3221\,A_8 = 291\ \text{K}$$

$$B_8 = 0.5347\,A_7 + 158.64 = 325.9\ \text{K}$$

$C_1 = -0.6616\,B_1 + 1.6616\,B_2 = 291\ \text{K}$

$C_2 = 0.4461\,B_2 + 0.5539\,B_3 = 291\ \text{K}$

$C_3 = 0.1385\,B_2 + 0.4461\,B_3 + 0.4154\,B_4 = 291\ \text{K}$

$C_4 = 0.1846\,B_3 + 0.4461\,B_4 + 0.3692\,B_5 = 291\ \text{K}$

$C_5 = 0.2077\,B_4 + 0.4461\,B_5 + 0.3462\,B_6 = 291\ \text{K}$

$C_6 = 0.2215\,B_5 + 0.4461\,B_6 + 0.3323\,B_7 = 291\ \text{K}$

$C_7 = 0.2308\,B_6 + 0.4461\,B_7 + 0.3221\,B_8 = 302.3\ \text{K}$

$C_8 = 0.5347\,B_7 + 158.64 = 325.9\ \text{K}$

The remaining cells will be filled with the same procedure:

t	0	50	100	150	200	250	300	350	400	450
T_1	291	291	291	291	291	291	291	291	291.1	291.5
T_2	291	291	291	291	291	291	291	291.1	291.3	291.9
T_3	291	291	291	291	291	291	291.1	291.6	292.3	293.2
T_4	291	291	291	291	291	291.4	292.2	293.4	294.8	296.5
T_5	291	291	291	291	292.3	294.0	296.1	298.3	300.6	302.9
T_6	291	291	291	294.7	298.0	301.5	304.8	307.9	310.8	313.5
T_7	291	291	302.3	307.3	312.5	316.5	320.1	323.1	325.9	328.4
T_8	291	325.9	325.9	332.4	335.2	338.2	340.5	342.6	344.4	345.9

t	500	550	600	650	700	750	800	850	900
T_1	292.1	292.9	294.0	295.3	296.7	298.3	300.0	301.7	303.6
T_2	292.6	293.6	294.8	296.1	296.7	299.3	301.0	302.8	304.7
T_3	294.4	295.7	297.2	298.9	300.6	302.4	304.3	306.2	308.1
T_4	298.2	300.1	302.0	304.0	305.9	307.9	309.9	311.9	313.8
T_5	305.3	307.5	309.8	312.0	314.1	316.1	318.1	320.0	321.9
T_6	316.1	318.5	320.7	322.9	324.9	326.8	328.6	330.3	332.0
T_7	330.6	332.7	334.6	336.3	337.9	339.4	340.9	342.2	343.4
T_8	347.4	348.7	349.8	350.9	351.9	352.8	353.7	354.5	355.3

Example 2.31. Temperature profiles along a spherical potato tuber during thermal processing in a can A cylindrical can ($D = 10$ cm, $L = 15$ cm) is filled with $N = 150$ spherical potatoes ($r = 1$ cm, $k = 0.554$ W/mK, $\alpha = 0.19\ 10^{-6}$ m^2/s) and enough water to fill up the rest ($c_{pw} = 4.2$ kJ/kg K). The can is placed in a retort operating of constant temperature of 121 °C. Convective heat transfer coefficient between potatoes and water is 50 W/m^{2}K and the overall heat transfer coefficient from retort to can is 1800 W/m^{2}K. The contents of the can are mixed by natural convection, therefore

water may be assumed to have a uniform temperature. The governing equation for water temperature is:

$$m_w c_{pw} \frac{dT_w}{dt} = U_0 A_c(T_s - T_w) + h_p A_p(T_w - T_p(t, R)) \quad \text{(E.2.31.1)}$$

where m_w: mass of water, U_0: overall heat transfer coefficient from retort to can, A_c: heat transfer area between retort and can, h_p: convective heat transfer coefficient between potatoes and water, A_p: heat transfer area between potatoes and water, $T_p(t, R)$: surface temperature of the potatoes at time t. Equation (E.2.31.1) may be converted into a finite difference equation after substituting a forward difference formula for the derivative as:

$$m_w c_{pw} \frac{T_{w\,j+1} - T_{wj}}{\Delta t} = U_0 A_c(T_s - T_w) + h_p A_p(T_w - T_p(t, R)) \quad \text{(E.2.31.2)}$$

which may be solved for the water temperature at the new time:

$$T_{w\,j+1} = T_{wj} + \frac{\Delta t}{m_w c_{pw}}(U_0 A_c(T_s - T_w) + h_p A_p(T_w - T_p(t, R))) \quad \text{(E.2.31.3)}$$

Equations (E.2.30.7), (E.2.30.13) and (E.2.30.15) are employed to describe the governing equation of heat transfer within the particles with variable T_w. After solving the necessary difference equations, plot the variation of the water and can center temperatures with time. We may calculate the followings with the given data:

total volume of the potatoes $= V_{\text{potatoes}} = N\frac{4}{3}\pi r^3 = 0.000628\ \text{m}^3$

volume of the can $= V_{\text{can}} = \pi r^2 L = 0.001178\ \text{m}^3$

volume of water in the can $= V_{\text{water}} = V_{\text{can}} - V_{\text{potatoes}} = 0.00055\ \text{m}^3$

mass of water in the can $= m_{\text{water}} = \rho V_{\text{water}} = 0.55$ kg

heat transfer area between retort and the can $= A_c = 2\pi R^2 + 2\pi RL = 0.063$ m^3

heat transfer area between potatoes and water $= A_p = N 4\pi R^2 = 0.1885\ \text{m}^2$

We will choose $n = 11$ (number of nodes along the potato), therefore the internal nodes of the potato will be denoted by subscript i when $2 \leqslant i \leqslant 10$. With $\Delta r = 0.001$ cm the Biot number is:

$$Bi = \frac{h\Delta r}{k} = 0.090 \quad \text{(E.2.31.4)}$$

The stability criterion is (Incropera and De Witt, 1985):

$$Fo(1 + Bi) \leqslant \frac{1}{2} \quad \text{(E.2.31.5)}$$

After substituting $Bi = 0.090$ in (E.2.31.5) we will obtain $Fo \leqslant 0.459$, since $Fo = (\alpha \Delta t / \Delta r^2)$ we should have $\Delta t \leqslant 2.4$ s and choose $\Delta t = 2$ s.

After substituting numbers in (E.2.30.7), (E.2.30.13) and (E.2.30.15):

$$T_1^{j+1} = -0.28\,T_1^j + 1.28\,T_2^j \qquad \text{(E.2.31.6.a)}$$

$$T_2^{j+1} = 0.24\,T_2^j + 0.76\,T_3^j \qquad \text{(E.2.31.6.b)}$$

$$T_3^{j+1} = 0.19\,T_2^j + 0.24\,T_3^j + 0.57\,T_4^j \qquad \text{(E.2.31.6.c)}$$

$$T_4^{j+1} = 0.2533\,T_3^j + 0.24\,T_4^j + 0.5067\,T_5^j \qquad \text{(E.2.31.6.d)}$$

$$T_5^{j+1} = 0.285\,T_4^j + 0.24\,T_5^j + 0.475\,T_6^j \qquad \text{(E.2.31.6.e)}$$

$$T_6^{j+1} = 0.304\,T_5^j + 0.24\,T_6^j + 0.456\,T_7^j \qquad \text{(E.2.31.6.f)}$$

$$T_7^{j+1} = 0.3167\,T_6^j + 0.24\,T_7^j + 0.4433\,T_8^j \qquad \text{(E.2.31.6.g)}$$

$$T_8^{j+1} = 0.3257\,T_7^j + 0.24\,T_8^j + 0.4343\,T_9^j \qquad \text{(E.2.31.6.i)}$$

$$T_9^{j+1} = 0.3325\,T_8^j + 0.24\,T_9^j + 0.4275\,T_{10}^j \qquad \text{(E.2.31.6.j)}$$

$$T_{10}^{j+1} = 0.3378\,T_9^j + 0.24\,T_{10}^j + 0.4222\,T_{11}^j \qquad \text{(E.2.31.6.k)}$$

$$T_{11}^{j+1} = 0.9172\,T_{10}^j + 0.0828\,T_w^j \qquad \text{(E.2.31.6.l)}$$

After substituting numbers in (E.2.32.3):

$$T_w^{j+1} = 0.91\,T_w^j - 0.0082\,T_{11}^j + 38.58 \qquad \text{(E.2.31.6.m)}$$

where $T_{11} = T(t, R)$ = surface temperature of the potatoes. Equations (2.31.6.a)–(2.31.6.m) are solved with the same procedure as explained in the previous example:

t	0	2	4	6	8	10	12	14	16	18	20	22	24
T_1	291	291	291	291	291	291	291	291	291	291	291	291	291
T_2	291	291	291	291	291	291	291	291	291	291	291	291	291
T_3	291	291	291	291	291	291	291	291	291	291	291	291	291
T_4	291	291	291	291	291	291	291	291	291	291	291	291	291.1
T_5	291	291	291	291	291	291	291	291	291	291	291	291.1	291.2
T_6	291	291	291	291	291	291	291	291	291	291.1	291.2	291.3	291.4
T_7	291	291	291	291	291	291	291	291.1	291.2	291.3	291.5	291.7	291.9
T_8	291	291	291	291	291	291.1	291.2	291.3	291.5	291.8	292.1	292.5	292.9
T_9	291	291	291	291	291.1	291.3	291.6	292.0	292.4	292.9	293.5	294.1	294.7
T_{10}	291	291	291	291.3	291.7	292.3	292.9	293.6	294.3	295.1	295.9	296.7	297.6
T_{11}	291	291	291.8	292.5	293.5	294.4	295.5	296.5	297.6	298.7	299.8	300.9	301.9
T_w	291	300.6	309.4	317.3	324.6	331.2	337.1	342.6	347.5	352.0	356.1	359.8	363.1

The rest of the calculations can be done with the same procedure. Variation of the water and the center temperature of the potatoes with time are shown in Figure E.2.31.

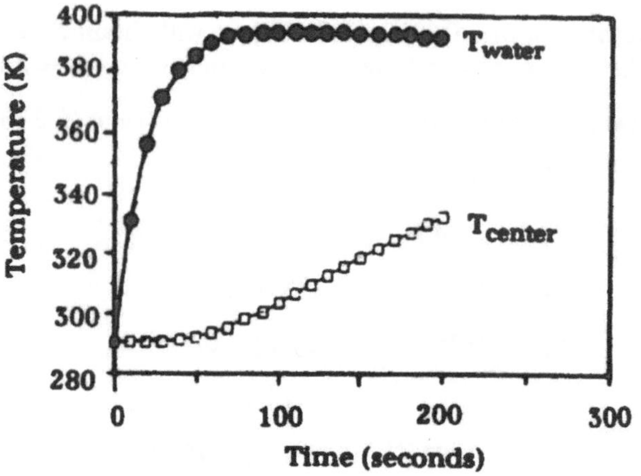

FIGURE E.2.31 Variation of the water and the center temperature of the potatoes with time.

REFERENCES

Altomare, R. (1988) Scale-up in the cooking of a meat analog, *Chemical Engineering Progress*, 84(5) 52–57.

American Society for Heating, Refrigerating and Air Conditioning Engineers. *Guide and Data Book, Applications for 1966 and 1967*. ASHRAE, New York, 1965.

Atkinson, B. and Mavituna, F. (1991) *Biochemical Engineering and Biotechnology Handbook*, 2nd ed. McMillan Pub. Ltd., UK.

Bailey, E. J. and Ollis, D. F. (1986) *Biochemical Engineering Fundamentals*, 2nd ed. McGraw-Hill Book Co. USA.

Berkman-Dik, T., Özilgen, M. and Bozoğlu, T. F. (1992) Salt, EDTA, and pH effects on rheological behavior of mold suspensions. *Enzyme and Microbial Technology* 14, 944–948.

Bird, R. B., Stewart, W. E. and Lightfoot, E. N. (1960) *Transport Phenomena*, Willey and Sons Inc., USA.

Brodkey, R. S. and Hershey, H. C. (1988) Transport Phenomena, McGraw-Hill, Singapore.

Carsten, H. R. F and Mc Kerrow, N. W. (1944) The tabulation of some Bessel functions $K_0(x)$ and $K_0^*(x)$ of fraction order, *Philosophy Magazine*, 35, 812–818.

Choi, Y. and Okos, M. R. Effects of temperature and composition on thermal properties of foods. In Food Engineering and Process Applications, Le Maguer, M. and Jelen, P. (eds) Vol. 1. Elsevier Applied Science Publishers, New York, pp. 93–102, USA.

Costantinides, A. (1987) *Applied Numerical Methods with Personal Computers*. McGraw-Hill.

Dik, T. and Özilgen, M. (1994) Rheological behavior of bentonite-apple juice dispersions. *Lebensmittel - Wissenschaft und Technologie*, **27**, 55–58 and 296.

Dik, T., Katnas, S. and Özilgen, M. (1996) Effects of bentonite combinations and gelatin on the rheologcal behavior of bentonite-apple juice dispersions, *Lebensmittel-Wissenschaft und Technologie*, **29**, 673–676.

Ece, M. C. and Cihan, A. (1993) A liquid diffusion model for drying rough rice. *Transactions of ASAE*, **36**, 837–840.

Gamero, M., Fiszman, S. M. and Duran, L. (1993) Stress relaxation of fruit gels. Evaluation of models and effects of composition. *Journal of Food Science*, **58**, 1125–1128, 1134.

Glasstone, S., Laidler, K. J. and Eyring, H. (1941). *Theory of Rate Processes*, McGraw-Hill, Newyork.

Hanks, R. W. and Ricks, B. L. (1974). Laminar-turbulent transition in flow of pseudoplastic fluids with yield stress. *Journal of Hydronautics*, **8**(4), 163–166.

Harper, J. C. and El-Sahrigi, A. F. (1965) Viscometric behavior of tomato concentrates. *Journal of Food Science*, **30**, 470–476.

Heisler, M. P. (1947) Temperature charts for induction and constant temperature heating. *Transactions of ASME*, **69**, 227–236.

Heldman, D. R. and Singh, R. P. (1981) *Food Process Engineering*, 2nd ed. AVI Pub. Co. Westport, CT.

Hunter, R. J. and Nicol, S. K. (1968) The dependence of plastic flow behavior of clay suspensions on surface properties. *Journal of Colloid and Interface Science*, **28**, 250–259.

Ilter, M., Özilgen, M. and Orbey, N. (1991) Modelling permeation of modified atmosphere gas mixtures through a low density polyethylene package film. *Polymer International*, **25**, 211–217.

Incropera, F. P. and De Witt D. P. (1985) *Fundamentals of Heat and Mass Transfer*, 2nd ed. John Wiley and Sons, Inc. Singapore.

Levine, L., (1992). Extrusion processes, in *Handbook of Food Engineering*, Heldman, D. R. and Lund, D. B. (eds.), Marcel Dekker, Inc. New York.

Lozano, J. E., Rotstein, E. and Urbicain, M. J. (1983). Shrinkage, Porosity and bulk density of foodstuffs at chainging moisture contents. *Journal of Food Science*, **48**, 1497–1502, 1553.

McCarthy, M. J., Perez, E. and Özilgen, M. (1991) Model for transient moisture profiles of a drying apple slab using the data obtained with magnetic resonance imaging. *Biotechnology Progress*, **7**, 540–543.

Metz, B., Kossen, N. W. F. and van Suijdam, J. C. (1979) The rheology of mould suspensions. *Advances in Biochemical Engineering*, **11**, 103–156.

Michaels, A. S. and Bolger, J. C. (1962) The plastic flow behavior of flocculated kaolin suspensions. *Industrial and Engineering Chemistry Fundamentals*, **1**, 153 – 162.

Miles, C. A., van Beek, G. and Veerkamp, C. H. (1983) Calculation of thermophysical properties of foods, in *Physical Properties of Foods*, Jowitt, R., Escher, F., Hallstrom, B., Meffert, H. F. Th., Spiess, W. E. L. and Vos, G (eds.), Appl. Sci. Pub., London.

Osorio, F. A. and Steffe, J. F. (1984) Kinetic energy calculations for non-Newtonian fluids in circular tubes, *Journal of Food Science*, **49**, 1295–1296, 1315.

Özilgen, M. and Bayindirli, L. (1992) Frequency factor – activation energy compensation for viscosity of the fruit juices. *Journal of Food Engineering*, **17**, 143–151.

Özilgen, M. (1993) Enthalpy–entropy and frequency factor – activation energy compensation relations for diffusion in starch and potato tissue. *Starch*, **45**, 48–51.

Özilgen, M. and Kauten R. J. (1994) NMR analysis and modeling of shrinkage and whey expulsion in rennet curd. *Process Biochemistry*, **29**, 373–397.

Özilgen, S. and Özilgen, M. (1996) Kinetic compensation relations: Tools for design in desperation. *Journal of Food Engineering*, **29**, 387–397.

Parnell-Clunies, E. M., Kakuda, Y. and Deman, J. M. (1986) Influence of heat treatment of milk on the flow properties of yoghurt, *Journal of Food Science*, **51**, 1459–1462.

Pekyardimci, Ş. and Özilgen, M. (1994) Solubilization and rheological behavior of raisins. *Process Biochemistry*, **29**, 465–473.

Peleg, M. (1983) Application of computers in food rheology studies, in *Computer aided techniques in food technology*, Saguy, I. (ed.) Marcel Dekker, Inc. New York.

Pham, Q. T. and Willix, J. (1989) Thermal conductivity of fresh lamb meat, offals and fat in the range −40 to +30°C: Measurements and correlations. *Journal of Food Science*, **54**, 508–515.

Purkayastha, S., Peleg, M., Johnson, E. A. and Normand, M. D. (1985) A computer aided characterization of the compressive creep behavior of potato and chaddar cheese. *Journal of Food Science*, **50**, 45–50, 55.

Qui, C. G. and Rao, M. A. (1988) Role of pulp content and particle size in yield stress of apple sauce. *Journal of Food Science*, **53**, 1165–1170.

Race, S. W. (1991) Improved product quality through viscosity measurement. *Food Technology*, **45**(7), 86–88.

Rao, M. A., Cooley, H. J., Anantheswaran, R. C. and Ennis, R. W. (1985) Convective heat transfer to canned liquid foods in steritort. *Journal of Food Science*, **50**, 150–154.

Rao, M. A., Cooley, H. J., Nogueria, J. N. and McLellan, M. R. (1986) Rheology of Apple sauce: Effect of apple cultivar, firmness, and processing parameters. *Journal of Food Science*, **51**, 176–179

Ross, S. L. (1989) Introduction to Ordinary Differential Equations, 4th ed., Willey, New York.

Sarikaya, A. and Özilgen, M. (1991) Kinetics of peroxidase inactivation during thermal processing of whole potatoes. *Lebensmittel-Wissenschaft und Technologie*, **24**, 159–163.

Steffe, J. F. (1992) *Rheological Methods in Food Process Engineering*, Freeman Press, East Lansing, Michigan, USA.

Steffe, J. F. and Morgan, R. G. (1986) Pipeline design and pump selection for non-Newtonian fluid foods. *Food Technology*, **40**(12), 78–85.

Tanglertpaibul, T. and Rao, M. A. (1987) Rheological properties of tomato concentrates as affected by particle size and methods of concentration. *Journal of Food Science*, **52**, 141–145.

Toledo, R. T. (1991) Fundamentals of Food Process Engineering, 2nd ed. Van Nostrand Reinhold, New York.

Turhan, M. and Kaletunc, G. (1992) Modeling of salt diffusion in white cheese during long-term brining. *Journal of Food Science*, **57**, 1082–1085.

Tütüncü, M. A., Özilgen, M. and Ungan, S. (1990) Weight loss behavior of refrigerated and frozed beef and ground beef. *Canadian Institute of Food Science and Technology Journal*, **23**, 76–78.

Vagenas, G. K. and Karathanos, V. T. (1991) Prediction of Moisture diffusivity in granular materials, with special applications to foods, *Biotechnology Progress*, **7**, 419–426.

Vitali, A. A. and Rao, M. A. (1984a) Flow properties of low-pulp concentrated orange juice: Serum viscosity and effect of pulp content. *Journal of Food Science*, **49**, 876–881.

Vitali, A. A. and Rao, M. A. (1984b) Flow properties of low-pulp concentrated orange juice: Effect of temperature and concentration. *Journal of Food Science*, **49**, 882–888.

Yener, E., Ungan, S. and Özilgen, M. (1987) Drying behavior of honey-starch mixtures. *Journal of Food Science*, **52**, 1054–1058.

Whitaker, S. (1977). *Fundamental Principles of Heat Transfer*, Pergamon Press, New York.

CHAPTER 3

Kinetic Modeling

3.1. KINETICS AND FOOD PROCESSING

Chemical reactions and microbial growth or product formation are among the most common causes of food spoilage. Typical examples of food deterioration reactions may include lipid oxidation, thermal or photo catalytic degradation processes. Toxins, enzymes or chemicals of microbial origin are the *microbial products* causing food deterioration. Microbial growth and product formation are achieved via metabolic activity involving sets of chemical reactions occurring mostly in the microbial cell. Some form of metabolic activity also continues in the plant tissues after harvesting, or inside the animal cells after slaughtering. Slowing down the deteriorative chemical reactions or microbial, post mortem or post harvest metabolic activity is indeed the goal of most food processing and preservation methods. The initial objective of the experimental kinetics studies is the development of a mathematical model to describe the reaction rate as a function of the experimental variables.

A chemical reaction may actually involve many elementary steps as explained in the following example:

$A_2 \rightleftarrows 2A^*$	elementary reaction 1
$A^* + B_2 \rightleftarrows AB + B^*$	elementary reaction 2
$A^* + B^* \rightarrow AB$	elementary reaction 3
$A_2 + B_2 \rightarrow 2\,AB$	overall reaction

In this example the sum of the *elementary reactions* gives the *overall reaction*, A^* and B^* are the intermediates, which have very short lives and are converted into other components immediately after being formed. The intermediates usually cannot be detected.therefore their concentrations may not be monitored with the conventional direct measurement techniques. Their presence may be understood by getting them to react with externally added chemicals. The overall reaction may also be called the *observed reaction.*

The order of an elementary reaction equals the number of molecules entering the reaction. The forward reaction $A_2 \rightarrow 2A^*$ in elementary step

1 is first order with respect to A_2, because only one molecule is converted into $2A^*$, but the reverse reaction $2A^* \rightarrow A_2$ is second order with respect to A^*, since two molecules of A^* are involved. The overall order of an elementary step is the sum of the orders with respect to each species. With the same reasoning we can easily find out that the overall orders of both the forward and reverse reactions of the elementary step 2 and that of step 3 are 2.

We have both forward and reverse reactions associated with the elementary steps 1 and 2, therefore they are called *reversible reactions*. There is only a forward reaction associated with the elementary step 3, and the reaction may not be reversed under the pertinent experimental conditions. The elementary step 3 is called an *irreversible reaction*.

A chemical reaction may be prevented via eliminating one of the reactants. Dipping the potatoes in water, or waxing the apples after cutting into half may reduce the availability of oxygen for the color change reactions. Fermenting eggs eliminate glucose, thus prevent it from entering into deteriorative reactions during dried eggs production (Hill and Sebring, 1990). Eliminating intermediates from the step-wise propagating reactions may slow down spoilage. Anti-oxidants react with free radicals and slow down the lipid oxidation reactions.

Increasing product concentration may establish equilibrium between the reactants and products of a reaction, thus slow down the metabolic activity. Carbon dioxide in controlled atmosphere gas mixtures may slow down the carbon dioxide producing reactions of the Krebs cycle, and consequently the whole metabolic activity.

A chemical reaction may be slowed down by eliminating the enzyme catalyzing the reaction. In thermal processing heat is applied to convert the enzymes into inactive proteins (Chapter 4.1). Off-odors in unblanched or under blanched frozen vegetables may be caused by enzymatic oxidation of lipids. Lipoxigenase is the catalyst involved into these oxidation reactions. It is inactivated upon blanching and development of off-odors during the subsequent storage is prevented (Fennema *et al.*, 1973). Ionizing radiation radiolizes water and may cause the formation of reactive intermediates including excited water $(H_2O)^*$, free radicals ($OH\cdot$ and $H\cdot$) ionized water molecules $(H_2O)^+$ and hydrated electron (e^-_{aq}), which may consequently react with food components including the proteins. Ionizing radiation may cause cross linking or scission of the proteins (Karel, 1975). When the enzymes undergo such changes they lose their activity and may not be able to catalyze the deteriorative reactions.

Changing the pH of food with acid addition or fermentation may denature the enzymes. Drying the foods may eliminate water needed for the enzyme activity. Some chemical reactions occur with the effect of light. Opaque packaging prevents light reaching light sensitive foods or beverages.

Refrigeration is among the most common food preservation method and slows down spoilage by reducing the rate constants of the deterioration reactions.

Food deterioration usually occurs via a large number of chemical reactions. The slowest reaction involving into a process is referred to as the *rate determining step* and its rate may be equal the spoilage rate. Preventing a single reaction in a set of deterioration reactions may create a new rate determining step and slow down the process and increase the shelf life of a food. Since chemical reactions are among the major causes of food spoilage, their kinetics deserve special attention.

3.2. THE RATE EXPRESSION

The consumption rate R_A of the species A, with reaction $A \xrightarrow{k_1} 2B$ may be expressed as $R_A = -dc_A/dt = k_1 c_A$. The minus signs imply that A is a reactant, therefore its concentration c_A is decreasing with time. The parameter k_1 is called the reaction rate constant. The term - $k_1 c_A$ implies that the higher the concentration of reactant A, the higher is the rate. Reaction rate also increases with k_1. If this reaction should describe loss of a nutrient in storage, the preservation methods would try to minimize the value of k_1 to slow down the nutrient loss. Production rate of B is $R_B = dc_B/dt = -2\,dc_A/dt = 2k_1 c_A$ implying that two molecules of B are produced for each molecule of A consumed.

With the reaction $2A + B \underset{k_2}{\overset{k_1}{\rightleftharpoons}} C$ the consumption rates of A and B and production rate of C are $-dc_A/dt = -2\,dc_B/dt = 2\,dc_C/dt = k_1 c_A^2 c_B - k_2 c_C$. There are actually two reactions represented here. The forward reaction implies that two molecules of A and one molecule of B react together with the rate constant k_1 to produce one molecule of C. The reverse reaction implies that one molecule of C disintegrates with the reaction rate constant k_2 to produce two molecules of A and one molecule of B. When the rate of the forward reaction, $k_1 c_A^2 c_B$, is larger than that of the reverse reaction, $k_2 c_C$, components A and B are depleted and C is produced with the net rate predicted by the rate expression. When the rate of the forward reaction becomes the same as that of the reverse reaction production rate of each component becomes the same as its depletion rate, and their concentrations do not change with time. This is called *chemical equilibrium*. Since two molecules of A are involved into reaction the power of c_A is 2 on the right hand side of the rate expression. When a reaction represents the actual mechanism at the molecular level, the power of the concentration term on the right hand side of the rate expression is the number of molecules involved the reaction. It is also possible that a chemical conversion may involve many reactions and the net change may be expressed with an apparent expression summing up the involving steps. In the rate expression of such reactions, powers of the species concentration may not be equal to the net number of molecules involved in the apparent expression. The forward rate expression is second order in A and first order in B. The order of the overall forward rate expression is three, i.e., the sum of the orders with respect to each components. There is only one mole of chemical species involved in the reverse reaction and it is first order.

Deterioration of foods may be simulated in analogy with first or zero order irreversible monatomic reactions (Labuza, 1980):

$$\frac{dc_A}{dt} = -kc_A \tag{3.1}$$

or

$$\frac{dc_A}{dt} = -k \tag{3.2}$$

This does not mean that the reactions taking place are very simple reactions, but rather shows that complex systems can be simulated with simple apparent mathematical models. Other simple chemical reactions involved in food processing and preservation, with their differential and integrated rate expressions are given in Table 3.1.

The unit of the rate constant is determined by the reaction. With any elementary reaction units of k are expressed in $(\text{concentration})^{1-n}/\text{time}$, where n = overall reaction rate order. With the zero order reaction, i.e., (3.2), k has the units of concentration/time, with a first order reaction, i.e., (3.1), k has the units of 1/time.

The rate expressions are arranged such that the rate constants are evaluated from the slopes of the differential or integrated rate expressions. The *differential methods* are based on the rate expressions evaluated via differentiation of the experimentally determined concentration versus time data. The integral methods (Examples 3.1–3.3) are based on integration of the reaction rate expression. The numerical differentiation techniques are generally usually unstable, therefore the integral methods are generally preferred over the differential methods.

Example 3.1. Ascorbic acid loss in packaged and non-packaged broccoli Reduced ascorbic acid retention in packaged and non-packaged broccoli were measured by (Barth *et al.*, 1993) as:

	packaged	non-packaged
t (h)	c_A (mg/g)	c_A (mg/g)
0	5.6	5.6
24	5.38	4.82
48	4.76	3.86
72	4.59	3.75
96	4.42	3.25

Table 3.1 Elementary reactions and their rate expressions

reaction	differential rate expression	integrated rate expression
$A \xrightarrow{k} \text{products}$	$\frac{dc_A}{dt} = -k$	$c_A = c_{A0} - kt$
$A \xrightarrow{k} \text{products}$	$\frac{dc_A}{dt} = -kc_A$	$\ln c_A = \ln c_{A0} - kt$
$A + B \xrightarrow{k} \text{products}$	$\frac{dc_A}{dt} = \frac{dc_B}{dt} = -k c_A c_B$	$\int_0^{x_A} \frac{dx_A}{(1-x_A)(\gamma - x_A)} = kc_{A0} \int_0^t dt$ where $\gamma = \frac{c_{B0}}{c_{A0}}$, $x_A = \frac{c_{A0} - c_A}{c_{A0}}$
without catalyst $A \xrightarrow{k1} R$	$\frac{dc_A}{dt} = -kc_A$	
with catalyst $A + \text{cat} \xrightarrow{k2} R + \text{cat}$	$\frac{dc_A}{dt} = -k_c c_A$	$\ln c_A = \ln c_{A0} - k_c t$ where $k_c = k_1 + k_2 c_{cat}$
$A \xrightarrow{k1} R$ $A \xrightarrow{k2} S$	$\frac{dc_A}{dt} = -(k_1 + k_2)c_A$ $\frac{dc_R}{dt} = k_1 c_A$ $\frac{dc_S}{dt} = k_2 c_A$	$\ln c_A = \ln c_{A0} - (k_1 + k_2)t$ $c_R = c_{R0} + \frac{k_1}{k_2}(c_S - c_{S0})$
$A \underset{k2}{\overset{k1}{\rightleftarrows}}$	$\frac{dc_A}{dt} = -\frac{dc_R}{dt} = -k_1 c_A + k_2 c_B$	$-\ln\left(1 - \frac{x_A}{x_{Ae}}\right) = \frac{k_1(\gamma + 1)}{(\gamma + x_{Ae})} t$ where x_{Ae} = value of x_A at equilibrium

a) Find out if the data may be represented by (3.1) or (3.2).

Solution Integrating (3.1) gives: $\ln(c_A) = \ln(c_{A0}) - kt$ where c_{A0} is the initial concentration of the reduced ascorbic acid in broccoli. Substituting the data into the integrated equation gives: $\ln(c_A) = 1.69 - 5.54\ 10^{-3}t$ with $r = -0.98$ and $s_e = 0.037$ with non-packaged broccoli and $\ln(c_A) = 1.72 - 2.58\ 10^{-3}t$ with $r = -0.97$ and $s_e = 0.006$ with packaged broccoli.

Integrating (3.2) gives: $c_A = c_{A0} - kt$. Substituting the data into the integrated equation gives: $c_A = 5.41 - 0.024t$ with $r = -0.97$ and $s_e = 0.28$ with non-packaged broccoli and $c_A = 5.58 - 0.013t$ with $r = -0.97$ and $s_e = 0.011$ with packaged broccoli. Since the correlation coefficient and the standard error of both models are very good we may conclude that both models may represent the data very well. It should be reminded that these models are simple empirical equations and do not describe the actual mechanism of the reaction.

When we use (3.1) $k = 5.54\ 10^{-3}\,h^{-1}$ with the non packaged and $k = 2.58\ 10^{-3}\,h^{-1}$ with the packaged broccoli. When we use (3.2) $k = 0.024$ mg/g h with the non packaged and $k = 0.013$ mg/g h with the packaged broccoli. With both models the rate constants were greater with the non packaged broccoli, implying that the reduced ascorbic acid loss was faster when no packaging was used.

b) Use (3.1) and (3.2) to find out what percent of the reduced ascorbic acid will remain after 100 hours of storage of the packaged broccoli?

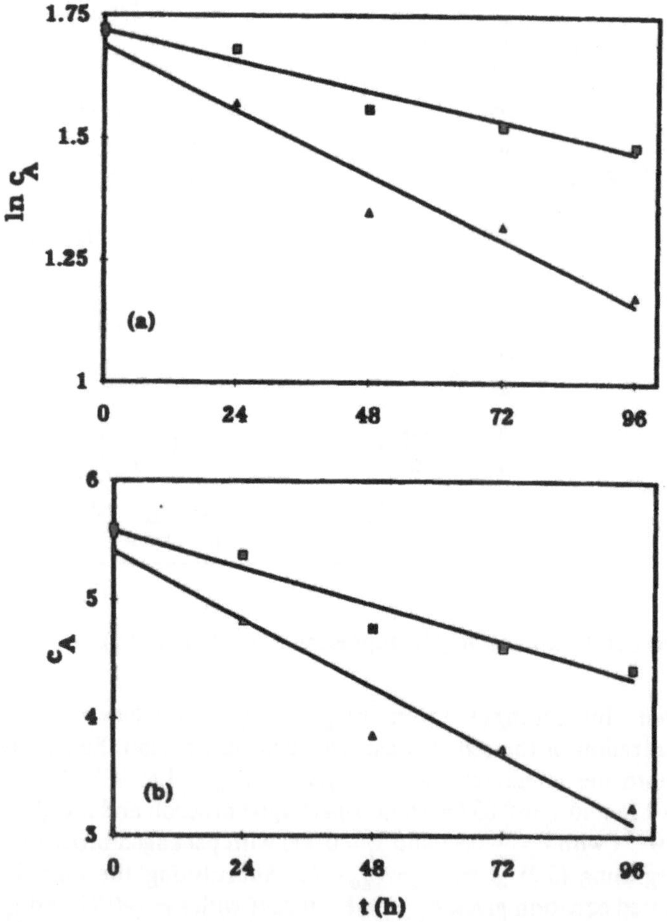

FIGURE E.3.1 Comparison of (3.1) and (3.2) with the experimental data (–■– : packaged; –▲– : non-packaged).

Solution Equations (3.1) and (3.2) are empirical equations and expected to simulate the data within the range of the experiments only. Although 100 h of storage time is not in this range, it is close enough not to expect a drastic change in trend with packaged broccoli. The integrated form of (3.1) is $\ln(c_A) = \ln(c_{A0}) - kt$, after substituting $c_{A0} = 5.6$ mg/g and $k = 2.58$ 10^{-3} h^{-1} and $t = 100$ h we obtain $c_A = 4.32$ mg/g.

Integrated form of (2.2) is $c_A = c_{A0} - kt$, after substituting $c_{A0} = 5.6$ mg/g and $k = 0.013$ mg/g h and $t = 100$ h we obtain $c_A = 4.30$ mg/g.

Generally it is not expected to have two models to simulate the same model equally well especially when we have large changes during the experiments. The small range of the reduced ascorbic acid loss during the storage period may contribute to our observation.

Example 3.2. Simultaneous nutrient and toxin degradation during thermal processing The half life of a bacterial toxin and a nutrient are 3 and 180 minutes, respectively in a food at 121°C. Degradation processes may be described by (3.1). Four log cycles of reduction are required in toxin for safe food production. How much of the nutrient survives the heat treatment?

Solution We have two reactions occurring in the same medium. Although these reactions do not interfere with each other, they inevitably take the same time. Half life ($t_{1/2}$) of the chemical species is the time required to lose half of the initial value. Integrated form of (2.1) is $\ln(c_A/c_{A0}) = -kt$ after substituting $c_A/c_{A0} = 0.5$, and $t_{1/2} = 3$ min we obtain $k = 0.231$ min^{-1} for the toxin. Similarly with $c_A/c_{A0} = 0.5$, and $t_{1/2} = 180$ we find $k = 0.004$ min^{-1} for the nutrient. After substituting $k = 0.231\ \text{min}^{-1}$ and $c_A/c_{A0} = 10^{-4}$ in the integrated model we will get $t = 40$ min = time required to reduce the toxin content by four log cycles. After substituting $k =$ 0.004 min^{-1} and $t = 40$ min in the same integrated model, we will calculate $c_A/c_{A0} = 0.85$, implying that 85% of the nutrient will survive the heat treatment.

Example 3.3. Shelf life calculation based on nutrient loss A micro-nutrient A undergoes a reaction $A + B \xrightarrow{k} C$ during storage of a food. Initial concentration of A is 2 g/kg and that of B is 75 g/kg. When the micro-nutrient concentration falls to 75% of its initial level the food becomes inedible. Calculate the shelf life of the food with $k_1 = 0.0001$ week^{-1} $(\text{g/kg})^{-1}$.

Solution The rate expression is $dc_A/dt = -kc_Ac_B$. Variation in B is negligible even after all of A is consumed with the reaction, since c_B is much larger than c_A. We may consider $k' = kc_B =$ constant, and the rate expression becomes $-dc_A/dt = -k'c_A$. This is called a *pseudo first order rate expression.* Integrated pseudo first order rate expression is $\ln(c_A/c_{A0}) =$ $-k'\,t$. After substituting $k' = 0.0075$ week^{-1} and $c_A/c_{A0} = 0.75$ we may calculate the shelf life of the food $t = 38$ weeks.

Example 3.4. Kinetics of nutrient loss with sequential chemical reactions

Nutrient A undergoes a degradation reaction $A \rightleftarrows B \rightarrow C$. The rate expressions for this reaction are:

$$\frac{dc_A}{dt} = -k_1 c_A + k_2 c_B \tag{E.3.4.1}$$

$$\frac{dc_B}{dt} = k_1 c_A - (k_2 + k_3) c_B \tag{E.3.4.2}$$

$$\frac{dc_C}{dt} = k_3 c_B \tag{E.3.4.3}$$

where the rate constants are $k_1 = 0.6$ weeks^{-1}, $k_2 = 0.2$ weeks^{-1} and $k_3 = 0.1$ weeks^{-1}. Initial substrate concentrations were $c_{A0} = 35$ g/L, $c_{B0} = c_{C0} = 0$ g/L. Solve these differential equations simultaneously and plot variations of c_A, c_B and c_C with time. When $c_C = 20$ g/L the food is considered inedible. Determine the shelf life of the food from the plot.

Solution We may rearrange the equations and solve as described in Table 2.6:

$$(D + k_1)c_A - k_2 c_B = 0$$

$$-k_1 c_A + (D + k_2 + k_3)c_B = 0$$

$$-k_3 c_B + D c_C = 0$$

where $D = \dfrac{d}{dt}$

$$\Delta = \begin{vmatrix} D + k_1 & -k_2 & 0 \\ -k_1 & D + k_2 + k_3 & 0 \\ 0 & -k_3 & D \end{vmatrix}$$

$$= (D + k_1)\begin{vmatrix} D + k_2 + k_3 & 0 \\ -k_3 & D \end{vmatrix} - (-k_1)\begin{vmatrix} -k_2 & 0 \\ -k_3 & D \end{vmatrix} + (0)\begin{vmatrix} -k_2 & 0 \\ D + k_2 + k_3 & 0 \end{vmatrix}$$

$$= (D + k_1)[(D + k_2 + k_3)(D) - (-k_3)(0)] + (k_1)[(-k_2)(D) - (-k_3)(0)]$$

$$= D^3 + (k_1 + k_2 + k_3)D^2 + k_1 k_3 D$$

$$\mathscr{D}c_A = \begin{vmatrix} 0 & -k_2 & 0 \\ 0 & D + k_2 + k_3 & 0 \\ 0 & -k_3 & D \end{vmatrix} = 0, \quad \mathscr{D}c_B \begin{vmatrix} D + k_1 & 0 & 0 \\ -k_1 & 0 & 0 \\ 0 & 0 & D \end{vmatrix} = 0,$$

$$\mathscr{D}c_C = \begin{vmatrix} D + k_1 & -k_2 & 0 \\ -k_1 & D + k_2 + k_3 & 0 \\ 0 & -k_3 & 0 \end{vmatrix} = 0$$

$$\Delta c_A = \mathscr{D} c_A$$

$$\frac{d^3c_A}{dt^3} + (k_1+k_2+k_3)\frac{d^2c_A}{dt^2} + k_1k_3\frac{dc_A}{dt} = 0 \quad \text{(E.3.4.4)}$$

$$\Delta c_B = \mathscr{D} c_B$$

$$\frac{d^3c_B}{dt^3} + (k_1+k_2+k_3)\frac{d^2c_B}{dt^2} + k_1k_3\frac{dc_B}{dt} = 0 \quad \text{(E.3.4.5)}$$

and

$$\Delta c_C = \mathscr{D} c_C$$

$$\frac{d^3c_C}{dt^3} + (k_1+k_2+k_3)\frac{d^2c_C}{dt^2} + k_1k_3\frac{dc_C}{dt} = 0 \quad \text{(E.3.4.6)}$$

Assume a solution $y = Ce^{\lambda t}$, therefore $dy/dt = \lambda e^{\lambda t}$, $d^2y/dt^2 = \lambda^2 e^{\lambda t}$ and $d^3y/dt^3 = \lambda^3 e^{\lambda t}$ substitute all in the differential equation and obtain the characteristic equation:

$$\lambda^3 + (k_1+k_2+k_3)\lambda^2 + k_1k_3\lambda = 0 \quad \text{(E.3.4.7)}$$

Solutions of the characteristic equation (after substituting values of the rate constants) $\lambda_1 = 0$, $\lambda_2 = -0.072$ and $\lambda_3 = -0.827$, therefore:

$$c_A = K_1 + K_2e^{-0.072t} + K_3e^{-0.827t} \quad \text{(E.3.4.8)}$$

$$c_B = K_4 + K_5e^{-0.072t} + K_6e^{-0.827t} \quad \text{(E.3.4.9)}$$

$$c_C = K_7 + K_8e^{-0.072t} + K_9e^{-0.827t} \quad \text{(E.3.4.10)}$$

We have three equations with nine unknown constants. We may determine the constants if we can reduce their number to three. After substituting the solutions for c_A and c_B in $dc_A/dt = -k_1c_A + k_2c_B$ we will get

$$-0.072\,K_2e^{-0.072t} - 0.827\,K_3e^{-0.827t} = -0.6(K_1 + K_2e^{-0.072t} + K_3e^{-0.827t}) + 0.2(K_4 + K_5e^{-0.072t} + K_6e^{-0.827t})$$

same exponential terms must have the same coefficients on both sides of the equation, therefore

term	equal coefficients	relation between constants
e^0	$0 = 0.6\,K_1 + 0.2\,K_4$	$K_4 = -0.3\,K_1$
$e^{-0.072t}$	$-0.072\,K_2 = -0.6\,K_2 + 0.2\,K_5$	$K_5 = 2.64\,K_2$
$e^{-0.827t}$	$-0.827\,K_3 = -0.6\,K_3 + 0.2\,K_6$	$K_6 = -1.135\,K_3$

After substituting the solutions for c_B and c_C in $dc_C/dt = k_3\,c_B$ we will get

$$-0.072\,K_8e^{-0.072t} - 0.827\,K_9e^{-0.827t} = 0.1(-0.3\,K_1 + 2.64\,K_2e^{-0.072t} - 1.135\,K_3e^{-0.827t})$$

same exponential terms must have the same coefficients on both sides of the equation, therefore

term	equal coefficients	relation between constants
e^{0}	$0 = -0.03\,K_1$	$K_1 = 0$
$e^{-0.072t}$	$-0.072\,K_8 = 0.264\,K_2$	$K_8 = -3.67\,K_2$
$e^{-0.827t}$	$-0.827\,K_9 = -0.114\,K$	$K_9 = 0.137\,K_3$

After substituting the constants we will have

$$c_A = K_2 e^{-0.072t} + K_3 e^{-0.827t} \tag{E.3.4.11}$$

$$c_B = -3.67\,K_2 e^{-0.072t} + 0.137\,K_3 e^{-827t} \tag{E.3.4.12}$$

$$c_C = K_7 + 5.5\,K_2 e^{-0.072t} + 0.19\,K_3 e^{-0.827t} \tag{E.3.4.13}$$

Equations (E.3.4.11)–(E.3.4.13) have three unknown constants (K_2, K_3, K_7) we may solve these constants by using the initial conditions:

$$35 = K_2 + K_3$$

$$0 = 2.64\,K_2 - 1.135\,K_3$$

$$0 = K_7 - 3.667\,K_2 + 0.137\,K_3$$

Therefore $K_2 = 10.5$, $K_3 = 24.48$, $K_7 = 35.23$ and

$$c_A = 10.5\,e^{-0.072t} - 24.47\,e^{-0.827t} \tag{E.3.4.14}$$

$$c_B = 27.72\,e^{-0.072t} - 27.78\,e^{-0.827t} \tag{E.3.4.15}$$

$$c_C = 35.23 - 38.5\,e^{-0.072t} + 3.35\,e^{-0.827t} \tag{E.3.4.16}$$

Variations of c_A, c_B and c_C during the storage period are shown in Figure E.3.4. Shelf life of the food ($c_C = 20$ g/L) is about 13 weeks.

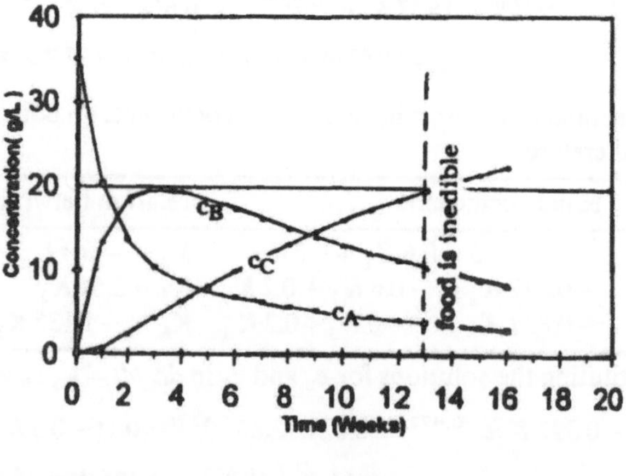

FIGURE E.3.4 Variation of c_A, c_B and c_C in storage.

3.3. WHY DO THE CHEMICALS REACT?

A chemical reaction occurs spontaneously if the total Gibbs free energy of formation of the products is smaller than that of the reactants. Due to the structural complexity of food and biological systems, the rates of the reactions associated with them are determined experimentally, but theoretical explanations are also available for some simple systems:

i) *Collision Theory*: The reaction rate for a bimolecular reaction (A + B → products) is:

$$R = k\, c_A c_B \tag{3.3.a}$$

The collision rate of gas molecules (Z_{AB} = number of collisions of A with B in unit volume in unit time) is expressed with the basic gas laws as:

$$Z_{AB} = \left(\frac{\sigma_A + \sigma_B}{2}\right)^2 \frac{N_{Av}^2}{10^6} \sqrt{8\pi\kappa T\left(\frac{1}{M_A} + \frac{1}{M_B}\right)} \tag{3.3.b}$$

where σ = diameter of a molecule; M_A, M_B = molecular weights, κ = Boltzman constant. Only the collisions involving more than a given minimum energy E_a may lead to a reaction. Fraction of all bimolecular collisions involving more energy than E_a is $(-E_a/RT)$, therefore the rate of the reaction and the rate constant are

$$R = Z_{AB} \exp\left(-\frac{E_a}{RT}\right) c_A c_B \tag{3.3.c}$$

$$k = \left(\frac{\sigma_A + \sigma_B}{2}\right)^2 \frac{N_{Av}^2}{10^6} \sqrt{8\pi\kappa T\left(\frac{1}{M_A} + \frac{1}{M_B}\right)} \exp\left(-\frac{E_a}{RT}\right) \tag{3.3.d}$$

ii) *Transition State Theory:* This model assumes that the reactants A and B combines to form an activated complex AB*

$$A + B \underset{k_2}{\overset{k_1}{\rightleftharpoons}} AB^* \tag{3.4.a}$$

There is an equilibrium between the reactants and the activated complex at all times:

$$K_{eq} = \frac{k_1}{k_2} = \frac{c_{AB}^*}{c_A c_B} \tag{3.4.b}$$

The activated complex undergoes decomposition as

$$AB^* \xrightarrow{k_3} \text{products} \tag{3.4.c}$$

The rate constant of decomposition is the same for all reactions:

$$k_3 = \frac{\kappa T}{h} \tag{3.4.d}$$

where h = Planck constant.

The product formation rate and the rate constant are

$$R = k_3 c^*_{AB} = \frac{\kappa T}{h} K_{eq} c_A c_B \tag{3.4.e}$$

$$k = \frac{\kappa T}{h} K_{eq} \tag{3.4.f}$$

Variation of the Gibbs free energy with the reaction coordinates is described in Figure 3.1. The higher the Gibbs free energy of the activated complex, the smaller is the fraction of molecules which can gain sufficient energy to form the activated complex and pass the energy barrier. A catalyst changes the structure of the activated complex and lowers the activation energy barrier, therefore a higher fraction of the molecules may pass through it and the reaction rate increases.

3.4. TEMPERATURE EFFECTS ON THE REACTION RATES

Temperature effects on the rate constants may be described with the *Arrhenius expression*:

$$k = k_0 \exp\left\{-\frac{E_a}{RT}\right\} \tag{3.5}$$

where k = rate constant, k_0 = pre-exponential constant, E_a = activation energy, R = gas constant, and T = absolute temperature.

Example 3.5. Vitamin loss in a snack food Loss of a vitamin in a snack food is described by (3.1). Estimate the time required to lose 15% of the initial vitamin content at 22°C if half lives of the vitamin at different storage temperatures are:

T(°C)	10	15	20	25
$t_{1/2}$ (days)	2900	1600	925	530

Solution Half life ($t_{1/2}$) is the time required to lose half of the initial vitamin content. Integrated form of (3.1) is $\ln(c_A/c_{A0}) = -kt$ after substituting $c_A/c_{A0} = 0.5$ and $t = t_{1/2}$ we obtain

T(°C)	10.0	15.0	20.0	25.0
k (day^{-1})	2.4 10^{-4}	4.3 10^{-4}	7.5 10^{-4}	1.3 10^{-3}

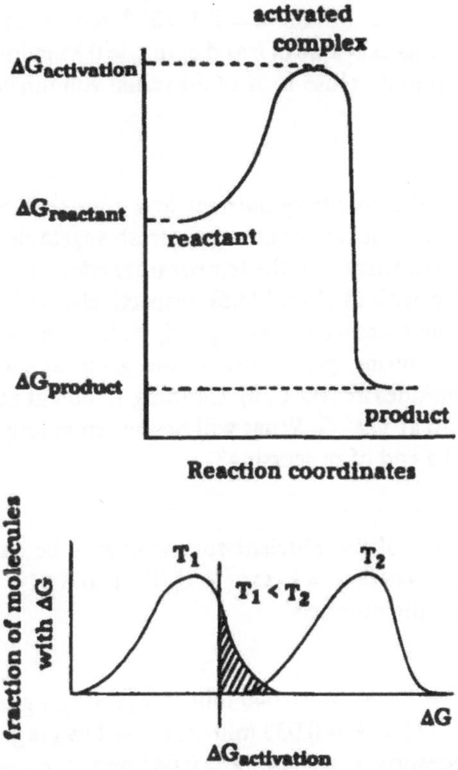

FIGURE 3.1 Idealized plots describing the transition state theory (upper figure) and the effect of the temperature variations on the reaction rates (lower figure). A different activated complex with smaller activation free energy forms when a catalyst is used. The shaded area in the lower figure is proportional with the fraction molecules which have sufficient energy to pass through the activation energy barrier at temperature T_1. Almost all the molecules have Gibbs free energy larger than $\Delta G_{activation}$ at T_2, therefore the reaction proceeds at a higher rate at this temperature. An interested reader may refer to Lienhard (1973) for discussion of the transition state theory as applied to enzymatic catalysis.

Equation (3.5) may be rewritten as: $\ln k = \ln k_0 - E/RT$, then data are:

$1/T(K^{-1})$	$3.53\ 10^{-3}$	$3.47\ 10^{-3}$	$3.41\ 10^{-3}$	$3.36\ 10^{-3}$
$\ln k$	-8.33	-7.75	-7.20	-6.65

After plotting $\ln k$ versus $1/T$(Figure E.3.5) we may obtain the best line ($\ln k = 26.23 - 9.8\ 10^3/T$($r = -1.0$) with the intercept $\ln k_0 = 26.23$ and the slope $E_a/R = 9.8\ 10^3 K^{-1}$. After substituting these parameters and

$T = 295$ K in (3.5) we obtain $k = 9.2\ 10^{-4}$ day^{-1} at 22°C. We may substitute the calculated value of k and $c_A/c_{A0} = 0.85$ in $\ln(c_A/c_{A0}) = -kt$ to estimate time required to lose 15% of the initial vitamin content at 22°C as 177 days.

Example 3.6. Total amounts of nutrient loss after sequences of a canning process The initial nutrient content of a fresh vegetable is 5 g/kg. Degradation rate of the nutrient and the temperature effects on the rate constant may be described with (3.1) and (3.5), respectively, with frequency factor $k_0 = 0.2$ min^{-1} and activation energy $E_a = 5030$ J/mole. The following operations occur during processing at the given average temperatures: i) Blanching 5 minutes at 100°C, ii) Canning 15 min at 60°C, iii) Thermal processing 20 min at 121° C. What will be the remaining concentration of the nutrient at the end of processing?

Solution Amounts of the nutrient surviving may be calculated from $c_A = c_{A0} \exp\{-kt\}$, where $k = k_0 \exp\{-E_a/RT\}$ and $R = 8.314$ J/mole K. After substituting the numbers:

i) Blanching $T = 373$ K, $k = 0.040$ min^{-1}, $c_{A0} = 5$ g/kg, $c_A = 4.09$ g/kg.
ii) Canning $T = 333$ K, $k = 0.033$ min^{-1}, $c_{A0} = 4.09$ g/kg, $c_A = 2.60$ g/kg.
iii) Thermal processing $T = 394$ K, $k = 0.043$ min^{-1}, $c_{A0} = 2.60$ g/kg, $c_A = 1.10$ g/kg = remaining concentration of the nutrient at the end of processing.

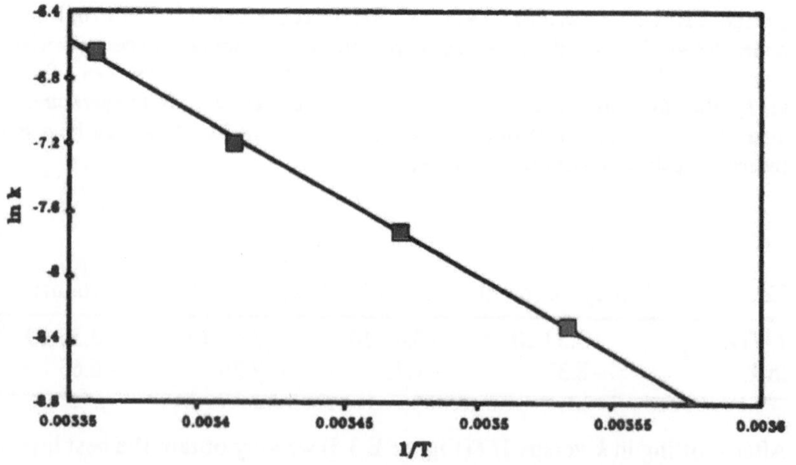

FIGURE E.3.5 Comparison of the best fitting line with the data.

3.5. PRECISION OF REACTION RATE CONSTANTS AND ACTIVATION ENERGY MEASUREMENTS

A general n^{th} order rate expression ($n \neq 1$) is

$$\frac{dc_A}{dt} = k\,c_A^n \tag{3.6.a}$$

after rearrangement and integration, the rate constant will be

$$k = \frac{c_{A1}^{n-1} - c_{A2}^{n-1}}{(n-1)(t_2 - t_1)c_{A2}^{n-1}c_{A1}^{n-1}} \tag{3.6.b}$$

where subscripts 1 and 2 denote the beginning and the end of an interval, respectively. In the completely general case of dependent variable $y = f(x_1, x_2, \ldots, x_n)$ the relative error in y due to the relative errors of $x_1, x_2, \ldots, x_n$ is given by

$$\left(\frac{\Delta y}{y}\right)^2 = \sum_{i=1}^{n}\left(\frac{\partial f}{\partial x_i}\right)^2\left(\frac{\Delta x_i}{x_i}\right)^2 \tag{3.6.c}$$

If we can assume that errors in c_{A1}, c_{A2}, t_1 and t_2 are independent we may calculate the relative error in k after using (3.6.c) as:

$$\left(\frac{\Delta k}{k}\right)^2 = \left(\frac{\Delta t_1}{t_2 - t_1}\right)^2 + \left(\frac{\Delta t_2}{t_2 - t_1}\right)^2 + \left\{\frac{(n-1)c_{A2}^{n-1}}{c_{A1}^{n-1} - c_{A2}^{n-1}}\right\}^2\left(\frac{\Delta c_{A1}}{c_{A1}}\right)^2$$
$$+\left\{\frac{(n-1)c_{A1}^{n-1}}{c_{A1}^{n-1} - c_{A2}^{n-1}}\right\}^2\left(\frac{\Delta c_{A2}}{c_{A2}}\right)^2 \tag{3.6.d}$$

In the case of $n = 1$ the corresponding equation is:

$$\left(\frac{\Delta k}{k}\right)^2 = \left(\frac{\Delta t_1}{t_2 - t_1}\right)^2 + \left(\frac{\Delta t_2}{t_2 - t_1}\right)^2 + \left\{\frac{1}{\ln(c_{A1}/c_{A2})}\right\}^2\left(\frac{\Delta c_{A1}}{c_{A1}}\right)^2$$
$$+\left\{\frac{1}{\ln(c_{A1}/c_{A2})}\right\}^2\left(\frac{\Delta c_{A2}}{c_{A2}}\right)^2 \tag{3.6e}$$

Example 3.7. Precision of the rate constant and the Arrhenius expression parameters of a second order reaction (Hill and Grieger-Block, 1980) When $n = 2$, $t_2 - t_1 = 100$ s, the uncertainty in each time measurements (Δt_1 and Δt_2) is 1 s, $c_{A2} = 0.9\,c_{A1}$ (10% reaction), and relative uncertainty ($\Delta c/c$) in each concentration measurement is 1%. After substituting these values in (3.6.d) we will have

$$\left(\frac{\Delta k}{k}\right)^2 = \left(\frac{1}{100}\right)^2 + \left(\frac{1}{100}\right)^2 + \left\{\frac{(2-1)0.9c_{A1}}{(1-0.9)c_{A1}}\right\}^2(0.01)^2$$

$$+\left\{\frac{(2-1)0.9c_{A1}}{(1-0.9)c_{A1}}\right\}^2(0.01)^2=0.0164$$

therefore $(\Delta k/k) = \pm 0.128$

Activation energy E_a may be calculated from the Arrhenius expression (3.5) as:

$$E_a=\frac{RT_1T_2}{T_2-T_1}\ln(k_2/k_1) \tag{E.3.7.1}$$

where subscripts 1 and 2 denote the beginning and the end of an interval, if the errors in each of the quantities k_1, k_2, T_1 and T_2 are random, the relative error in the Arrhenius activation energy is given after using (3.6.c) as:

$$\left(\frac{\Delta E_a}{E_a}\right)^2=\left(\frac{T_2}{T_2-T_1}\right)^2\left(\frac{\Delta T_1}{T_1}\right)^2+\left(\frac{T_1}{T_2-T_1}\right)^2\left(\frac{\Delta T_2}{T_2}\right)^2$$
$$+\left\{\frac{1}{\ln(k_2/k_1)}\right\}^2\left\{\left(\frac{\Delta k_1}{k_1}\right)^2+\left(\frac{\Delta k_2}{k_2}\right)^2\right\} \tag{E.3.7.2}$$

Equation (E.3.7.2) shows that the relative error in the activation energy is strongly dependent on the size of the temperature interval chosen.

3.6. IDEAL REACTOR DESIGN

Vessels used to conduct chemical reactions are called reactors. There are three basic types of ideal reactors: Batch, Continuously Stirred Tank (CSTR) and Plug Flow (PF). A *batch reactor* is a closed vessel with no input and output streams. Reactants are charged into the reactor, left for a certain period with well mixing and removed when the required conversion is achieved. A *CSTR reactor* is equipped with input and output streams. Due to the well mixing every point in the reactor and at the exit are expected to have the same concentration. *Plug flow reactors* are tubular flow reactors, there is no velocity gradient in the radial direction and composition of the fluid varies in the flow direction only.

Material for a specific chemical balance around batch or CSTR reactor requires:

$$\begin{bmatrix}\text{input rate}\\ \text{into the}\\ \text{reactor}\end{bmatrix}-\begin{bmatrix}\text{ourput}\\ \text{rate from the}\\ \text{reactor}\end{bmatrix}+\begin{bmatrix}\text{generation}\\ \text{rate in the}\\ \text{reactor}\end{bmatrix}=\begin{bmatrix}\text{accumulation}\\ \text{rate in the}\\ \text{reactor}\end{bmatrix} \tag{3.7}$$

In a batch reactor where reaction $A \xrightarrow{k}$ is occurring, with the rate expression $R_A = -kc_A$ we have (input rate of A into the reactor) = (output rate from the reactor) = 0 (3.8.a)

(generation of A rate in the reactor) = $R_A V$ (3.8.b)

$$\text{(accumulation rate of A in the reactor)} = \frac{d(c_a V)}{dt} \quad (3.8.c)$$

With a constant reactor volume V, after substituting (3.8.a) – (3.8.c) into (3.7) we will obtain:

$$\frac{dc_A}{dt} = R_A \quad (3.9)$$

After rearranging (3.9) we may calculate the time required to achieve a certain final reactant concentration c_{Af} after starting with the initial concentration c_{A0} as:

$$t = \int_{C_{A0}}^{C_{Af}} \frac{dc_A}{R_A} \quad (3.10)$$

Example 3.8. Acid hydrolysis of lactose in a batch reactor Large amounts of lactose is produced in the cheese industry. It is not a good food ingredient because of its low solubility, and limited sweetness. It is a pollutant when disposed into the environment. Lactose hydrolysis is generally achieved with an enzymatic process. Resin catalyzed hydrolysis may be preferred over enzymatic process in some applications because of the higher temperature and conversion rates, and lower pH which prevents microbial contamination. Chen and Zall (1983) have shown that resin catalyzed lactose hydrolysis may be expressed with an irreversible first order apparent rate expression as:

$$\text{Lactose} \xrightarrow{k} \text{products} \quad (E.3.8.1)$$

The major products glucose and galactose have higher solubility and sweetness than lactase. An Arrhenius type of rate expression was suggested to express the temperature effects on the apparent rate constant k with $k_0 = 2.54 \times 10^{21}\ h^{-1}$ and $E_a = 1.543 \times 105$ J/mole. A batch reactor will be used to convert a 10% (w/w) lactose solution to obtain $c_{Af}/c_{A0} = 0.60$. Initial temperature of the reactor will be 95°C. Chen and Zall (1983) observed formation of browning products during their studies, therefore a linearly decreasing temperature profile

$$T = T_0 - \alpha t \quad (E.3.8.9)$$

will be used during the hydrolysis experiments to reduce the formation of the browning reactions ($T_0 = 95$°C, $\alpha =$ constant). Parameter α will be chosen such that the temperature of the reactor will be 75°C at the end of the process. What should the process time and parameter α be?

Solution Equation (3.9) requires

$$\frac{dc_A}{dt} = R_A \tag{3.9}$$

where

$$R_A = kc_A = k_0\, c_A \exp\{-E_a/RT\} \tag{E.3.8.10}$$

After combining (3.9), (E.3.8.10) and substituting values of constants we obtain

$$\frac{dc_A}{dt} = 2.54 \times 10^{21}\, c_A \exp\left\{-\frac{18559}{T}\right\} \tag{E.3.8.11}$$

we may use (E.3.8.9) to calculate the derivative

$$\frac{dT}{dt} = -\alpha \tag{E.3.8.12}$$

after combining (E.3.8.11) and (E.3.8.12), and rearranging we obtain

$$\int_{C_{A0}}^{0.6C_{A0}} \frac{dc_A}{c_A} = -\int_{368}^{348} \frac{2.54 \times 10^{21}}{\alpha} \exp\left\{\frac{18559}{T}\right\} dT \tag{E.3.8.13}$$

We may define the integral as:

$$I = \int_{368}^{348} \exp\left\{-\frac{18559}{T}\right\} dT$$

and use Simpson's method to calculate it with $\Delta T = 2$ K, $N = (368 - 348)/\Delta T = 10$:

$$I = \tfrac{2}{3}\{f_0 + 4(f_1 + f_3 + f_5 + f_7 + f_9) + 2(f_2 + f_4 + f_6 + f_8) + f_{10}\}$$

where

$$f_0 = \exp\left\{-\frac{18559}{368}\right\} = 1.25 \times 10^{-22}, f_1 = \exp\left\{-\frac{18559}{366}\right\} = 9.5 \times 10^{-23},$$

$$f_2 = \exp\left\{-\frac{18559}{364}\right\} = 7.19 \times 10^{-23},$$

$$f_3 = \exp\left\{-\frac{18559}{362}\right\} = 5.42 \times 10^{-23}\ldots\ldots$$

After substituting the numbers in (E.3.8.13) we will calculate that $\alpha = 4.15$, therefore the temperature profile should be $T = 368 - 4.15t$. Since the final temperature is required to be 348 K the reaction time should be 4.8 h.

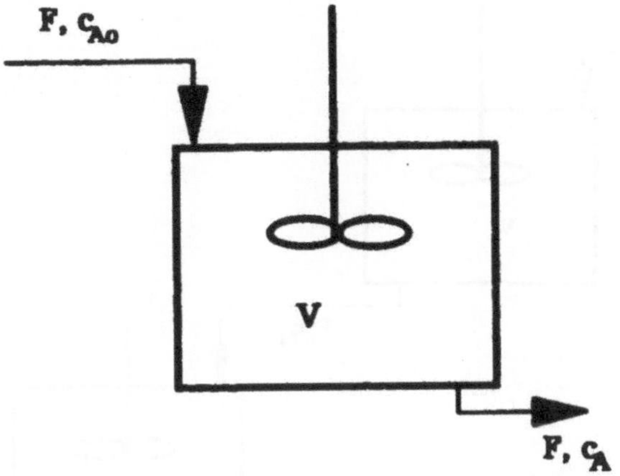

FIGURE 3.2 A continuously stirred tank reactor (CSTR).

In a CSTR (Fig. 3.2) with input and output flow rates of F, where reaction $A \xrightarrow{k}$ products is occurring (rate expression $R_A = -kc_A$) we have

(input rate of A into the reactor) $= Fc_{A0}$ (3.11.a)

(output rate from the reactor) $= Fc_A$ (3.11.a)

(generation rate of A in the reactor) $= R_A V$ (3.11.c)

(accumulation rate of A in the reactor) $= \frac{d(c_A V)}{dt}$ (311.d)

after substituting (31.a) – (3.11.d) into (3.7) we will obtain:

$$Fc_{A0} - Fc_A + R_A V = \frac{d(c_A V)}{dt} \tag{3. 12}$$

With a constant reactor volume V, under steady state conditions, i.e., $d(c_A V)/dt = 0$, (3.12) may be rearranged as:

$$\frac{V}{F} = -\frac{c_{A0} - c_A}{R_A} \tag{3.13}$$

The ratio $V/F = \tau$ is called the residence time. It is the average time spent in the reactor by the entering liquid.

Example 3.9. Acid hydrolysis of lactose in a CSTR The lactose hydrolysis process described in Example 3.8 will be conducted in two series CSTRs (Fig. E.3.9). The first reactor has a volume of 1 m^3 and operated at 95°C. The second reactor has a volume of 3 m^3 and operated at 50°C. What should the volumetric flow rate F(m^3/h) be to obtain $c_{A2}/c_{A0} = 0.60$?

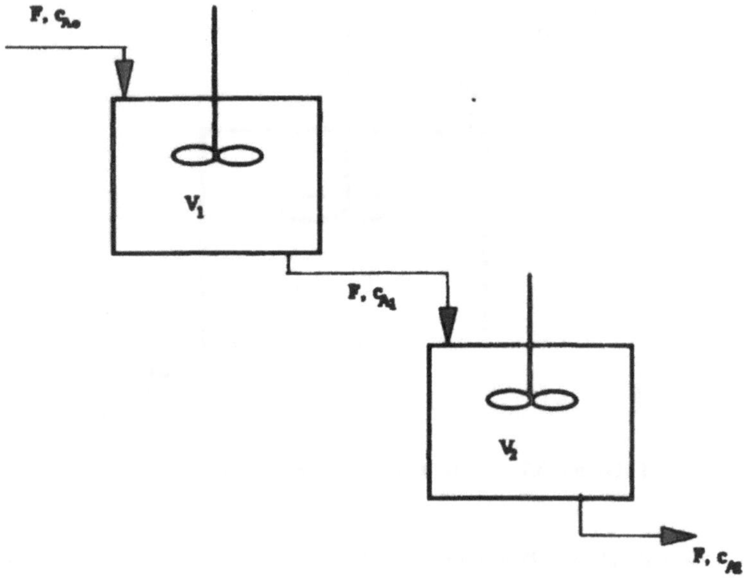

FIGURE E.3.9 Series CSTRs for lactose hydrolysis process.

Solution Under steady state conditions, lactose balance around the first reactor is performed similary as (3.12) to obtain:

$$Fc_{A0} - Fc_{A1} - k_1 c_{A1} V_1 = 0 \tag{E.3.9.1}$$

a similar balance is performed for the second reactor:

$$Fc_{A1} - Fc_{A2} - k_2 c_{A2} V_2 = 0$$

After rearanging (E.3.9.1) we will obtain

$$c_{A1} = \frac{Fc_{A0}}{F + k_1 V_1} \tag{E.3.9.3}$$

Equation (E.3.9.3) is substituted in (E.3.9.2) to obtain

$$F^2 \frac{c_{A0}}{F + k_1 V_1} - Fc_{A2} - k_2 c_{A2} V_2 = 0 \tag{E.3.9.4}$$

All the terms of (E.3.9.4) will be devided in c_{A0} to obtain

$$F^2 \frac{1}{F + k_1 V_1} - \frac{c_{A2}}{c_{A0}} F - k_2 V_2 \frac{c_{A2}}{c_{A0}} = 0 \tag{E.3.9.5}$$

Arrhennius expression is:

$$k = k_0 \exp\left\{ -\frac{E_a}{RT} \right\} \tag{3.5}$$

After substituting the numbers in (3.5) we will calculate the reaction rate constant as 0.317 h^{-1} at 95°C and at 2.82×10^{-4} h^{-1} 50°C. After substituting $c_{A2}/c_{A0} = 0.6$ and values of the rate constants and the reactor volumes in (E.3.9.5) we will calculate $F = 0.44$ m^3/h.

In a PF reactor where reaction $A \xrightarrow{k}$ products is occurring, with the rate expression $R_A = -k\,c_A$, material balance described in (3.7) may be worked around a volume element of a PF reactor as described in Figure 3.3.

Each term appearing in the material balance may be written as follows:

$$\text{(input rate of A into the volume element)} = Fc_{A|z} \tag{3.14.a}$$

$$\text{(output rate of A from the volume element)} = Fc_{A|z+\Delta z} \tag{3.14.b}$$

$$\text{(generation of A rate in the reactor)} = R_A \Delta z S \tag{3.14.c}$$

where $S =$ cross sectional area of the PF reactor

$$\text{(accumulation rate of A in the volume element)} = \frac{d(c_A \Delta z S)}{dt} \tag{3.14.d}$$

Under steady state conditions, i.e., $d(c_A\Delta zS)/dt$, after substituting (3.14.a) – (3.14.d) into (3.7) we will obtain:

$$Fc_{A|z} - Fc_{A|z+\Delta z} + R_A \Delta z S = 0 \tag{3.15}$$

Equation (3.15) may be rearranged as:

$$F\frac{c_{A|z+\Delta z} - c_{A|z}}{\Delta z} = R_A S \tag{3.16}$$

After taking the limit of the left hand side of (3.16) as $\Delta z \to 0$ we will obtain:

$$F\frac{dc_A}{dz} = R_A S \tag{3.17}$$

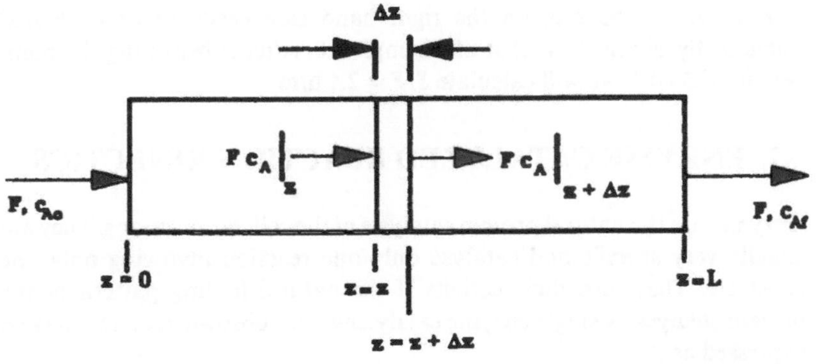

FIGURE 3.3 The volume element in a PF reactor used for material balance.

We may calculate the reactor length (L) for required conversion from after rearranging and integrating (3.17):

$$L = \frac{F}{S}\int_{c_{A0}}^{c_{A1}} \frac{dc_A}{R_A} \tag{3.18}$$

Example 3.10. Acid hydrolysis of lactose in a plug flow reactor The lactose hydrolysis process described in Example 3.8 will be conducted in a PF reactor. Temperature is 95°C at the inlet of the reactor and will decrease linearly as $T = T_0 - \beta z$ along the reactor to 75°C at the exit. What should the S/F ratio (in (h/m) be to obtain $c_A/c_{A0} = 0.60$ in a 2 m long reactor?

Solution Under steady state conditions, lactose balance around a shell element of the reactor is performed similary to (3.17) to obtain:

$$F\frac{dc_A}{dz} = -kc_A S \tag{E.3.10.1}$$

The data implies that $\beta = 10$°C/m and the temperature profile along the reactor is

$$T = 95 - 10z \tag{E.3.10.2}$$

The reaction rate constant is expressed in terms of the Arrhenius expression as:

$$k = 2.54 \times 10^{21} \exp\left\{-\frac{18\,559}{273 + 95 - 10z}\right\} \tag{E.3.10.3}$$

After combining (E.3.10.1) and (E.3.10.3) and rearranging we will obtain

$$-\frac{1}{2.54 \times 10^{21}}\int_{c_{A0}}^{c_{A1}} \frac{dc_A}{c_A} = \frac{S}{F}\int_0^2 \exp\left\{-\frac{18\,559}{368 - 10z}\right\} dz \tag{E.3.10.4}$$

The integral appearing on the right hand side needs to be evaluated numerically, similarly as that of Example 3.8. After substituting the numbers in (E.3.10.4) we will calculate $S/F = 2.4$ h/m.

3.7. ENZYME CATALYZED REACTION KINECTICS

Enzymes are the natural protein catalysts of the cellular reactions. They are usually very specific and catalyze only one reaction involving only one substrate. They lose their activity if the natural folding pattern of the protein changes. A single enzyme catalyzing one substrate reaction may be expressed as:

$$S \xrightarrow{E} P \tag{3.19.a}$$

where the reaction rate is

$$v = -\frac{dc_s}{dt} = \frac{dc_p}{dt} \tag{3.19.b}$$

The term $-dc_s/dt$ is the substrate consumption rate and dc_p/dt is the product formation rate. The rates of the enzyme catalyzed reactions were referred to as *velocity* in the pioneering biology literature, therefore they are conventionally denoted with the letter v. The mechanism for a single enzyme catalyzed one substrate reaction was first suggested by Michaelis and Menten in 1913 as:

$$S + E \underset{k_2}{\overset{k_1}{\rightleftarrows}} SE \tag{3.19.c}$$

$$SE \overset{k_3}{\rightarrow} P + E \tag{3.19.d}$$

The first elementary reaction of this mechanism is considered as an equilibrium step with the dissociation constant

$$K_m = \frac{c_E c_S}{c_{ES}} \tag{3.19.e}$$

where K_m = Michaelis constant, c_E = enzyme concentration, c_S = substrate concentration and c_{ES} = concentration of the ES complex. The total enzyme concentration was initially c_{EO}, after making the ES complex, and concentration of the free enzyme c_E may be calculated as

$$c_E = c_{EO} - c_{ES} \tag{3.19.f}$$

The second reaction is slow, therefore its rate is the same as the rate of the overall apparent reaction. The slowest reaction in such a mechanism is called the rate determining step. The rate of the second elementary reaction is

$$v = \frac{dc_p}{dt} = k_3 c_{ES} \tag{3.19.g}$$

It is not usually possible to measure c_{ES}, therefore we may use (3.19.a)–(3.19.f) to rearrange (3.19.g) as:

$$v = \frac{v_{max} c_S}{K_m + c_S} \tag{3.20}$$

This is called the Michaelis-Menten equation, where

$$v_{max} = k_3 c_{EO} \tag{3.21}$$

It was later claimed by Briggs and Haldane that (3.19.c) may not be an equilibrium step, and the material balances for the substrate and the intermediary complex ES were expressed as:

$$v = -\frac{dc_s}{dt} = k_1 c_S c_E - k_2 c_{ES} \tag{3.22.a}$$

$$\frac{dc_{ES}}{dt} = k_1 c_S c_E - k_2 c_{ES} - k_3 c_{ES} \tag{3.22.b}$$

The complex do not accumulate, i.e.,

$$\frac{dc_{ES}}{dt} = 0 \tag{3.22.c}$$

After using (3.22.a)–(3.22.c), (3.20) is obtained with the Michaelis constant

$$K_m = \frac{k_2 + k_3}{k_1} \tag{3.22.d}$$

Enzymes belonging to the classification of hydrolases are typical examples of single enzyme catalyzed one substrate reactions.

The Michaelis-Menten Equation has a variable apparent order. When $K_m \gg c_S$ the apparent rate is $v = k_{app} c_S$ (first order in c_S), where $k_{app} = v_{max}/K_m$; when $K_m \ll c_S$ the apparent rate is $v = v_{max}$ (zero order in c_S).

The Michaelis-Menten equation is generally arranged in three different linear forms to evaluate the apparent constants v_{max} and K_m from the slopes and the intercepts of the plots of the experimental data (Tab. 3.2). A number of advantages and disadvantages are associated with each type of plot. Even spacing of the data points along the line and best fit of the data points to a straight line are among the factors to be considered while making such a decision. More workers use the Lineweaver-Burk method than the other two combined.

Example 3.11. Kinetics of linolenic acid peroxidation by sunflower lipoxygenase Strong lipoxygenase activity is observed during the first days of sunflower seed germination, which may cause lipid peroxidation

Table 3.2 Linear arrangements of the Michaelis-Menten equation

Lineweaver-Burk arrangement	$\frac{1}{v} = \frac{1}{v_{max}} + \frac{K_m}{v_{max}} \frac{1}{c_S}$
Eadie-Hofstee arrangement	$v = v_{max} - K_m \frac{v}{c_S}$
Hanes-Woolf arrangement (also called the Augustinsson arrangement)	$\frac{c_S}{v} = \frac{k_m}{v_{max}} + \frac{1}{v_{max}} c_S$

under unfavorable storage conditions. Linolenic acid is a substrate for sunflower lipoxygenase. The following data was evaluated from a publication by Leoni, *et al.* (1985):

c(mM)	0.0025	0.0033	0.0052	0.0080	0.012	0.05	0.015	0.25
v(U/mg protein)	21	27	33	43	50	60	53	54

If the apparent reaction agrees with the Michaelis-Menten scheme, determine the constants of the rate expression.

i) Double reciprocal (Lineweaver-Burk) plot:

Equation (3.20) and the data may be rearranged as:

$$\frac{1}{v} = \frac{1}{v_{max}} + \frac{K_m}{v_{max}} \frac{1}{c}$$

$1/c(\text{mM})^{-1}$	400	303	192	125	83	20	6.6	4
$1/v(\text{U/mg protein})^{-1}$	0.0476	0.0370	0.0303	0.0232	0.0200	0.0167	0.0189	0.0185

The best fitting line to the data is

$$\frac{1}{v} = 0.0161 + 7.35 \times 10^{-5} \frac{1}{c} \quad (r = 0.98)$$

The intercept is $1/v_{max}$ therefore $v_{max} = 62$ U/mg protein and the slope is K_m/v_{max}, therefore $K_m = 5 \times 10^{-6}$ M. The Lineweaver-Burk is shown in Figure E.3.11.1:

ii) Eadie-Hofstee plot:
Equation (3.20) and the data may be rearranged as:

$$v = v_{max} - K_m \frac{v}{c}$$

v/c (U/mg protein mM)	8400	8181.8	6346.1	5375	4166.7	1200	353.3	216
v(U/mg protein)	21	27	33	43	50	60	53	54

The best fitting line to the data is

$$v = 63 - 0.0044 \frac{v}{c} \quad (r = -0.94)$$

The intercept is v_{max} therefore $v_{max} = 63$ U/mg protein and the slope is K_m, therefore $K_m = 4.4 \times 10^{-6}$ M. The Eadie-Hofstee plot is shown in Figure E.3.11.2.

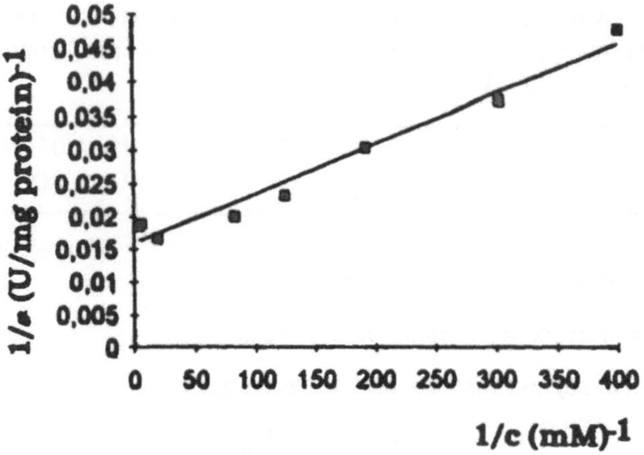

FIGURE E.3.11.1 The Lineweaver-Burk plot to determine kinetic constants K_m and v_{max}.

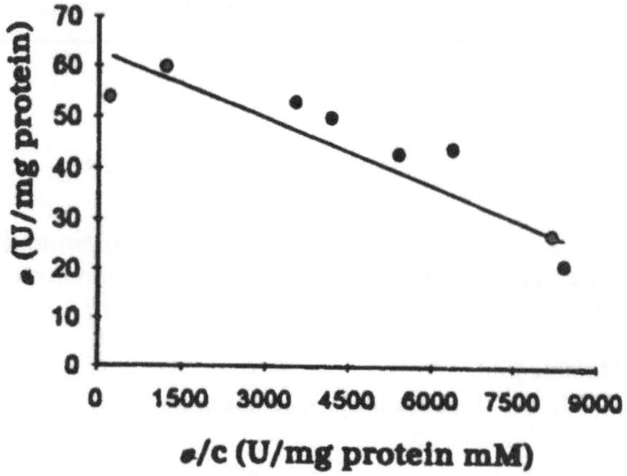

FIGURE E.3.11.2 The Eadie-Hofstee plot to determine kinetic constants K_m and v_{max}.

Enzymatic reactions involved in food processing and preservation may also involve two substrates:

$$A + B \overset{E}{\rightleftharpoons} P + Q \tag{3.23}$$

Three different mechanisms are suggested for these reactions (Whitaker, 1972). The ordered and the random mechanisms suggest that the products may be released only after both of the substrates are bound to the enzyme. In the *ordered mechanism* always the same substrate is bound to the enzyme first, and the same product is released first. In the *random mechanism* there is no priority in binding of the substrates or removal of the products. The ordered and random mechanisms result in the same rate expression:

$$v = \frac{v_{max} c_A c_B}{(c_A + K_A)(c_B + K_B)} \tag{3.24}$$

In *ping pong mechanism* the first product is released after binding the first substrate; then the second substrate is bound and subsequently the second product is formed, leading the rate expression:

$$v = \frac{v_{max} c_A c_B}{c_A c_B + K_A c_B + K_B c_A} \tag{3.25}$$

$K_A K_B$ term is missing in the denominator of (3.25) since there is no ternary complex in the mechanism. It should be noticed (3.24) and (3.25) are valid when c_A and c_B are maintained at constant levels and there is no product accumulation in the reaction medium. Kinetic constants of (3.24) and (3.25) may be evaluated by performing two sets of experiments. In the first set of experiments where c_A is constant and c_B is variable the first set of apparent kinetic constants may be obtained treating the data as explained for the single substrate reactions. The remaining kinetic constants may be obtained after making experiments with variable c_A and constant c_B (Whitaker, 1972).

Certain enzymes have ionic groups on their active sites and these enzymes must be in a suitable form (acid or base) to function. Variations in the pH of the medium result in changes in the ionic form of the active site and changes in the activity of the enzyme hence the reaction rate. The following scheme may be used to describe the pH dependence of the enzymatic reaction rate for ionizing enzymes (Shuler and Kargi, 1992):

$$\begin{array}{l} E^- + H^+ \\ \downarrow\uparrow K_2 \\ EH + S \overset{K_s}{\rightleftharpoons} EHS \xrightarrow{k_2} EH + P \\ \qquad\quad + \\ \qquad\quad H^+ \\ \qquad\quad \downarrow\uparrow K_1 \\ \qquad\quad EH_2^+ \end{array} \tag{3.26a}$$

The mechanism given in (3.26.a) may be valid within a reasonable narrow pH range. The enzyme may start denaturation due to the changes in the overall structure when subjected to severe pH effects.

where

$$K_1 = \frac{c_{EH} c_{H^+}}{c_{EH_2^+}} \tag{3.26.b}$$

$$K_2 = \frac{c_{E^-} c_{H^+}}{c_{EH}} \tag{3.26.c}$$

and

$$K_3 = \frac{c_{EH} c_S}{c_{EHS}} \tag{3.26.d}$$

The product formation rate is:

$$v = \frac{dc_P}{dt} = k_p c_{EHS} \tag{3.26.e}$$

Equation (3.22.b) may be rearranged as:

$$v = k_p \frac{c_{E0}}{c_{E^-} + c_{EH} + c_{EH_2^+} + c_{EHS}} c_{EHS} \tag{3.26.f}$$

where $c_{E0} = c_{E^-} + c_{EH} + c_{EH_2^+} + c_{EHS}$. After substituting $v_{max} = k_p c_{E0}$ and using (3.26.b)–(3.26.f) we will obtain:

$$v = \frac{v_{max} c_S}{K_3\left(1 + \frac{K_2}{c_{H^+}} + \frac{c_{H^+}}{K_1}\right) + c_S} \tag{3.27}$$

Theoretical prediction of the pH optimum of enzymes requires knowledge of the active site characteristics, which is very difficult to obtain. The pH optimum for an enzyme is usually determined experimentally as depicted with a typical example in Figure 3.4.

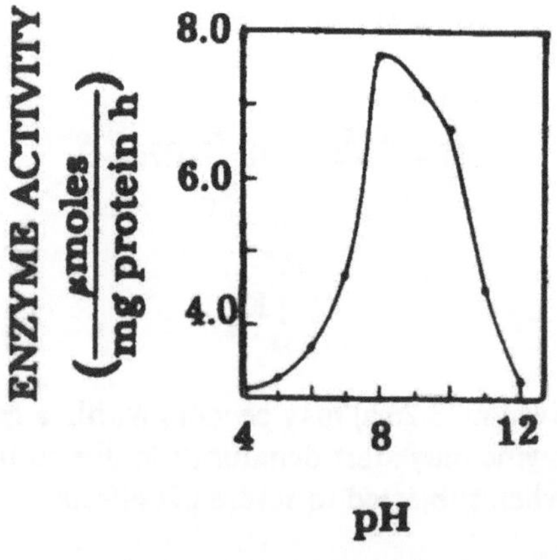

FIGURE 3.4 Effect of pH on the activity profile of alkaline proteinase from shrimp muscle at 60°C (Doke and Ninjoor, 1987; © IFT, reproduced by permission).

Any substance which reduces the rate of an enzyme catalyzed reaction is called an *inhibitor*. A competitive inhibitor competes with the substrate for the active site of the enzyme. If the Michaelis-Menten mechanism prevails, after including the inhibitor we will have:

$$\begin{array}{l} E + S \overset{K_s}{\rightleftharpoons} SE \xrightarrow{k_p} P + E \\ + \\ I \\ \uparrow\downarrow K_i \\ EI \end{array} \tag{3.28.a}$$

where

$$K_s = \frac{c_E c_S}{c_{ES}} \tag{3.28.b}$$

and

$$K_i = \frac{c_E c_I}{c_{EI}} \tag{3.28.c}$$

The product formation rate is:

$$v = \frac{dc_P}{dt} = k_p c_{ES} \tag{3.28.d}$$

Equation (3.28.b) may be rearranged as:

$$v = k_p \frac{c_{EO}}{c_E + c_{ES} + c_{EI}} c_{ES} \tag{3.28.e}$$

where $c_{EO} = c_E + c_{ES} + c_{EI}$. After substituting $v_{max} = k_p c_{EO}$ and using (3.28.b) and (3.28.c) to eliminate c_{ES} and c_{EI}, (3.28.e) becomes:

$$v = \frac{v_{max} c_S}{c_S + K_S\left(1 + \frac{c_I}{K_i}\right)} \tag{3.28.f}$$

After comparing (3.28.f) with (3.20) we see that the competitive inhibitor increased the Michaelis constant K_m by a factor of $(1 + c_I/K_i)$. Equation (3.28.f) implies that v decreases with c_I and increases with K_i. When $c_s \gg K_s(1 + c_I/K_i)$ inhibitor may not have considerable effect on the reaction rate. After substituting $K_I = K_s(1 + c_I/K_i)$ (3.28.f) may be rearranged as:

$$\frac{1}{v} = \frac{1}{v_{max}} + \frac{K_I}{v_{max}} \frac{1}{c_S} \tag{3.29}$$

Parameters v_{max} and K_1 may be evaluated from the intercept and slope as explained with a typical example in Figure 3.5; where it should be noticed that $1/v_{max}$ is (approximately) the same under all the experimental conditions implying that v_{max} is not affected by the inhibitor.

A non-competitive inhibitor binds to a site different than that of the substrate. If the Michaelis-Menten mechanism prevails, after including the inhibitor we will have:

$$\begin{array}{ccc} E+S & \overset{K_s}{\rightleftarrows} SE \overset{k_2}{\rightarrow} & P+E \\ + & + & \\ I & I & \\ \downarrow\uparrow K_i & \downarrow\uparrow K_i & \\ EI+S & \overset{K_s}{\rightleftarrows} ESI & \end{array} \quad (3.30.a)$$

where the dissociation constants are

$$K_s = \frac{c_E c_S}{c_{ES}} = \frac{c_{EI} c_S}{c_{ESI}} \quad (3.30.b)$$

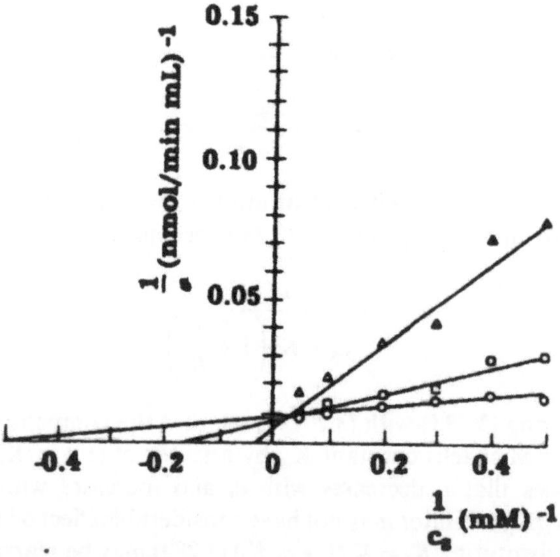

FIGURE 3.5 Double reciprocal plot to evaluate kinetic constants under competitive inhibition of β-glucosidase by glucose. (Δ) 5 mM glucose, (□) 2 mM glucose, (o) 0 mM glucose (Data from Woodward and Arnold, 1981).

$$K_i = \frac{c_E c_I}{c_{EI}} = \frac{c_{ES} c_I}{c_{ESI}} \tag{3.30.c}$$

The product formation rate is:

$$v = \frac{dc_P}{dt} = k_p c_{ES} \tag{3.30.d}$$

Equation (3.30.d) may be rearranged as:

$$v = k_p \frac{c_{EO}}{c_E + c_{ES} + c_{EI} + c_{ESI}} c_{ES} \tag{3.30.e}$$

where $c_{EO} = c_E + c_{ES} + c_{EI} + c_{ESI}$. After substituting $v_{max} = k_p c_{EO}$ and using (3.30.b) and (3.30.c) to eliminate c_{ES} and c_{EI}, (3.30.e) becomes:

$$v = \frac{v_{app} c_S}{c_S + K_s} \tag{3.30.f}$$

where $v_{app} = v_{max}/(1 + c_I/K_i)$. After comparing (3.30.f) with (3.20) we see that the non-competitive inhibitor decreased the apparent maximum rate v_{max} by a factor of $(1 + c_I/K_i)$. Equation (3.30.f) implies that v decreases with c_I. Increasing c_S does not affect v_{app}. Equation (3.30.f) may be rearranged as:

$$\frac{1}{v} = \frac{1}{v_{app}} + \frac{K_S}{v_{app}} \frac{1}{c_S} \tag{3.31}$$

Parameters v_{app} and K_S may be evaluated from intercept and slope as explained with a typical example in Figure 3.6; where it should be noticed that $1/v_{app}$ changes with the experimental conditions.

Temperature effects on enzyme activity is shown with a typical example in Figure 3.7.

The maximum attainable rate v_{max} in (3.20) was defined when the initial enzyme activity c_{EO} was constant as:

$$v_{max} = k_3 c_{EO} \tag{3.21}$$

Temperature effects on the rate constant k_3 may be simulated with the Arrhenius expression:

$$k_3 = k_0 \exp\left\{-\frac{E_a}{RT}\right\} \tag{3.5}$$

Denaturation of the enzyme may be expressed with a first order rate expression:

$$\frac{dc_E}{dt} = -k_d c_E \tag{3.32}$$

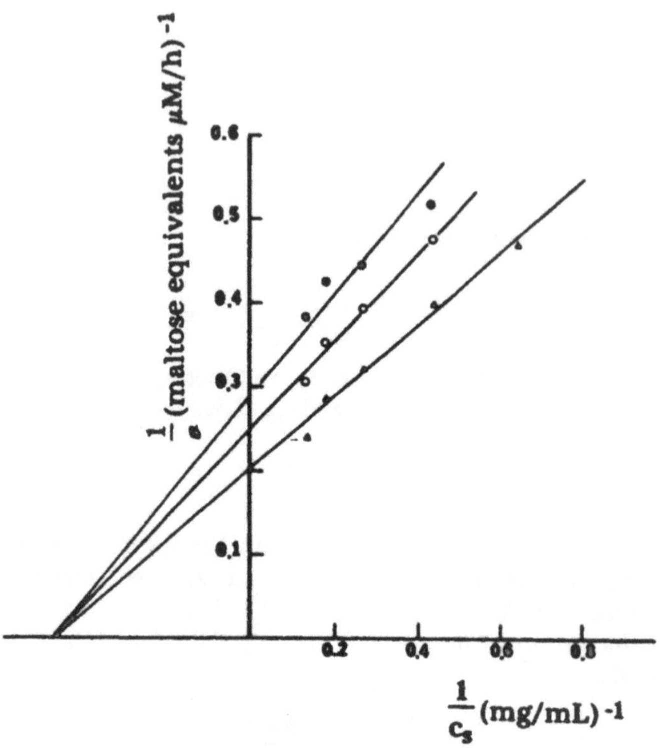

FIGURE 3.6 Double reciprocal plot to evaluate kinetic constants under non-competitive inhibition of porcine pancreatic α-amylase in the absence (▲) and in the presence of two different inhibitors (o: I − 1 and : ● I − 2) from black bean. Substrate was starch and υ was evaluated after expressing the product in maltose equivalents (Frels and Rupnow, 1985; © IFT, reproduced by permission).

At a constant temperature T when enzyme undergoes denaturation, after integrating (3.32) and combining with (3.5) and (3.21) we may express the temperature dependence of the enzyme activity as:

$$\upsilon = \frac{c_S c_{E0} k_0 e^{-k_d t} e^{-Ea/RT}}{K_m + c_S} \tag{3.33}$$

Example 3.12. Kinetic compensation relations for pectinesterase inactivation during pasteurization of orange juice (Ülgen and Özilgen, 1991) Enzyme inactivation during thermal processing of the foods may be described in analogy with uni-molecular, irreversible, first order chemical reaction:

$$\underset{\text{active enzyme}}{E} \rightarrow \underset{\text{inactive enzyme}}{E_i}$$

$$\frac{dc_E}{dt} = -k_d c_E \tag{3.32}$$

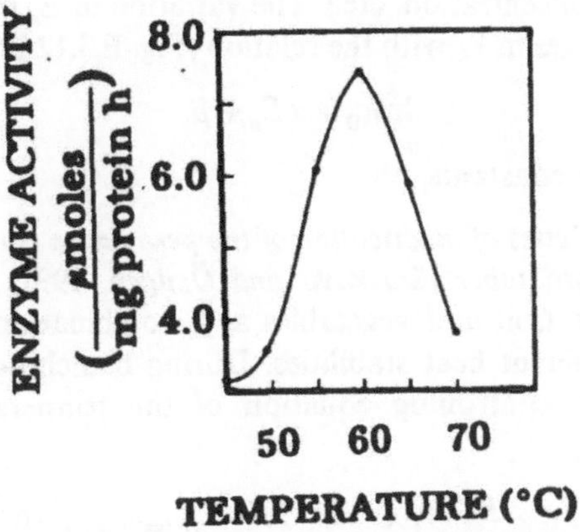

FIGURE 3.7 Effect of temperature on the activity profile of alkaline proteinase from shrimp muscle at pH = 8.0 (Doke and Ninjoor, 1987; © IFT, reproduced by permission).

After integrating (3.32) we will have:

$$\ln c_E = \ln c_{E0} - k_d t \tag{E.3.12.1}$$

When we plot $\ln c_E$ versus time during inactivation at a constant temperature, the slope of the line gives k_d as exemplified in Figure E.3.12.1.

Kinetic constants k_0 and E_a of the Arrhenius expression are not usually independent of each other in a family of related systems where parameters k_0 and E_a change due to slight variations in the experimental conditions

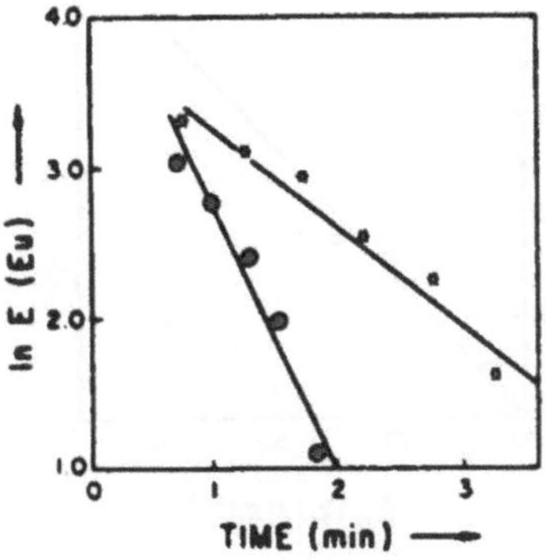

FIGURE E.3.12.1 Typical plots for pectinesterase inactivation during pasteurization of orange juice (pH = 3.5) at (*) 60°C and (●) 70°C (Ülgen and Özilgen, 1991; © SCI, reproduced by permission).

(like pH, sugar concentration, etc.). The variation in E_a may be compensated by the changes in k_0 with the relation (Fig. E.3.12.2):

$$\ln k_0 = \alpha E_a + \beta \tag{E.3.12.2}$$

where α and β are constants.

Example 3.13. Kinetics of inactivation of the peroxidase iso enzymes during blanching of potato tuber (Sarikaya and Özilgen, 1991) Peroxidase is usually present in fruit and vegetables as a combination of various iso enzymes with different heat stabilities. During blanching of a spherical potato tuber the controlling equation of the temperature profile is (Example 2.9):

$$\frac{T - T_1}{T_0 - T_1} = \frac{R}{r}\left(\frac{2}{\pi}\right)\sum_{n=0}^{\infty}\left\{\frac{(-1)^{n+1}}{n} e^{-(\pi n)^2\tau} \sin\left(\frac{\pi n r}{R}\right)\right\} \tag{E.3.13.1}$$

Inactivation kinetics of the enzyme may be described with separate first order reactions for heat stable and heat labile fractions:

$$\frac{dc_{E1}}{dt} = -k_1 c_{E1} \tag{E.3.13.2}$$

and

$$\frac{dc_{E2}}{dt} = -k_2 c_{E2} \tag{E.3.13.3}$$

Total enzyme activity is:

$$c_E = c_{E1} + c_{E2} \tag{E.3.13.4}$$

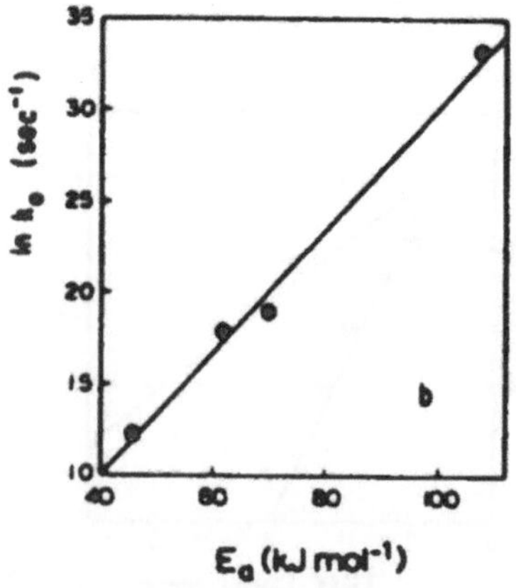

FIGURE E.3.12.2 Kinetic compensation relation for pectinesterase inactivation during pasteurization of orange juice. Equation of the line: ln $k_0 = 3.37 \times 10^{-4}$ $E_a - 3.50$ (Ülgen and Özilgen, 1991; © SCI, reproduced by permission).

Temperature effects on the inactivation rate constants k_1 and k_2 are described with the Arrhenius expression:

$$k_1 = k_{10}\exp\left\{-\frac{E_{a1}}{R_g T}\right\} \tag{E.3.13.5}$$

and

$$k_2 = k_{20}\exp\left\{-\frac{E_{a2}}{R_g T}\right\} \tag{E.3.13.6}$$

where R_g = gas constant. Model constants were assumed with a trial and error procedure:

i. The measured values of r, R, T_0, T_1 and an estimate of apparent thermal diffusivity α were used to solve (E.3.13.1) with 0.5 minutes of time intervals to predict T as a function of r and t.
ii. The trial and error procedure was continued by introducing a new value for α until all the individual model temperatures agreed with the measured temperatures with less than $\pm 1°C$ difference. Apparent thermal diffusivity was found to be 1.93×10^{-7} m^2/s and it was the only parameter determined by using (E.3.13.1) and the temperature data. Comparison of the model with the experimental data was exemplified with typical plots in Figure E.2.9.a.
iii. Values of the constants k_{10}, k_{20}, E_{a1} and E_{a2} and the initial enzyme activities $c_{E1,0}$ and $c_{E2,0}$ were determined by trial and error computation procedure. These constants were used in (E.3.13.5) and (E.3.13.6) to predict k_1 and k_2, these parameters were then used in (E.3.13.2) and (E.3.13.3) to predict the remaining activity of each peroxidase fraction.
iv. New estimates were introduced into the trial and error calculations until the standard error of estimate of the remaining enzyme activity model and the experimental data were reduced to about 0.001 EU (enzyme units) in each run. A typical plot for the comparison of the model with the data is shown in Figure E.3.13.

Enzyme activity may re-appear some time after thermal processing, if heat treatment is not sufficient. The total Gibbs free energy of an enzyme suspension is (Shulz and Schirmer, 1979):

$$\Delta G_{total} = \Delta H_{chain} - T\Delta S_{chain} + \Delta G_{solvent} \tag{3.34}$$

where ΔG_{total} is the total Gibbs free energy of enzyme plus the solvent, ΔH_{chain} is the binding enthalpy of the chain in vacuum provided mostly by hydrogen bonding and van der Waals interactions in the chain, ΔS_{chain} is the chain entropy, T is the absolute temperature and $\Delta G_{solvent}$ is the Gibbs free energy of solvent. The chain plus solvent system attain the minimum ΔG_{total} (and maximum ΔS_{chain}) corresponding to the active folding pattern of the enzyme in its native environment, i.e., the vegetable or animal tissue. The

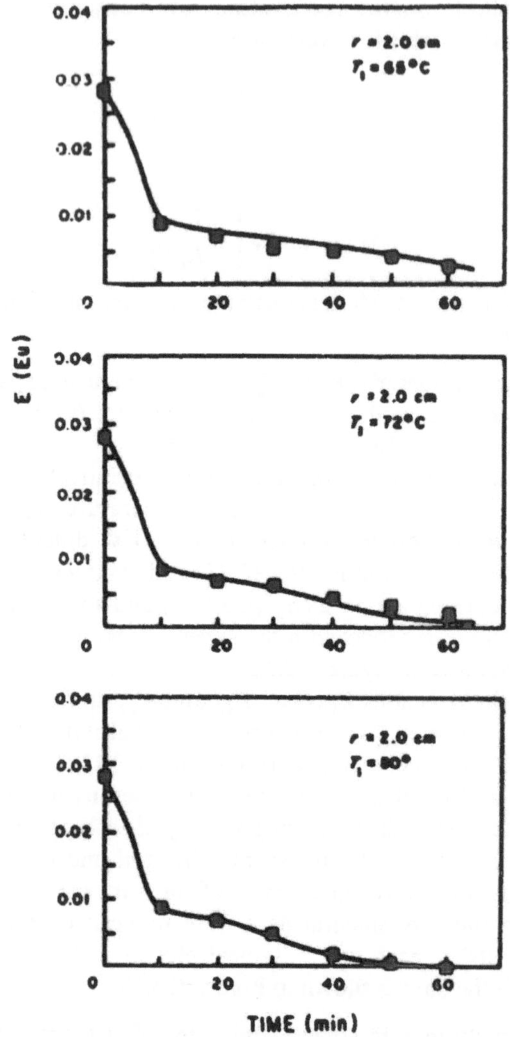

FIGURE E.3.13 Comparison of the enzyme inactivation model (—) with experimental peroxide activities (■) during thermal processing of whole potatoes. $R = 4.0$ cm, $T_a = 17.5 \pm 1$°C (Sarikaya and Özilgen, 1991; © Academic Press, reproduced by permission).

secondary, tertiary and the quaternary structure of the proteins are dictated by their primary structure. Destroying the protein structure at any level eliminates the enzyme activity, but the enzyme may fold back to the original structure and re-gain its activity due to the thermodynamic reasons, if the primary structure should not be destroyed. The primary structure of the proteins or their post production modifications, including the —S—S—

bonds, are required to be destroyed irreversibly to prevent re-gaining of the enzyme activity. The denaturation mechanism of a protein, α-lactalbumin, and the associated Gibbs free energy changes are given in Figure 3.8.

In biological reactors or sensors enzymes are frequently immobilized by using a carrier. Immobilization may be achieved via entrapment in the network or binding on the surface of a carrier. Either pure or crude enzyme preparations or the whole cell may be immobilized. When we work with immobilized enzymes the reaction rate expression R_A is replaced with the apparent reaction rate $R_{A,app}$ in (3.7) and in the subsequent equations including (3.10), (3.13) and (3.18). Apparent rate of a reaction catalyzed by the immobilized enzymes may be related to that of the free enzymes in terms of an effectiveness factor η:

$$\eta = \frac{R_{A,app}}{R_A} \tag{3.35}$$

When the enzymes are immobilized into spherical particles (2.14) (equation of continuity in spherical coordinate system) may be simplified under steady state conditions as:

$$\frac{d^2c}{dr^2} + \frac{2}{r}\frac{dc}{dr} = \frac{1}{D_e}\frac{v_{app}c}{c + K_m} \tag{3.36.a}$$

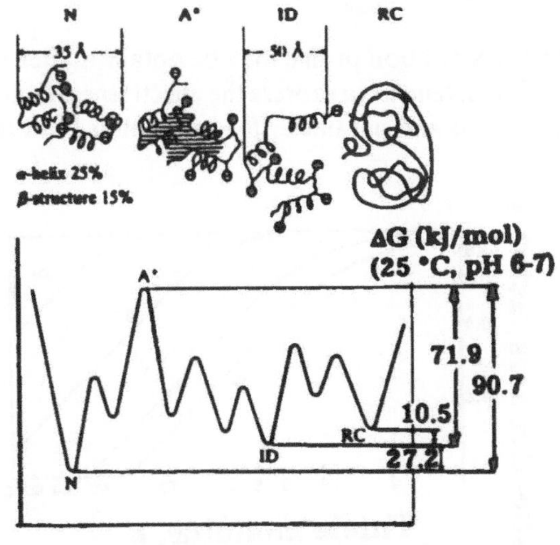

FIGURE 3.8 Schematic illustration of the stages of unfolding of the protein α-lactalbumin with corresponding data. N = native folding pattern, A* = critically activated state, ID = incompletely disordered folding pattern, RC = fully denatured random coil (Kuwajima, 1977; Academic Press, reproduced by permission).

where $c = c(r)$ = substrate concentration in the particle and D_e = effective diffusivity of the substrate within the particle, v_{app} is the maximum attainable apparent rate of the immobilized enzyme and K_m is the Michaelis constant of the free enzyme. The boundary conditions are:

$$c = c_b \quad \text{at } r = R \tag{3.36.b}$$

$$\frac{dc}{dr} = 0 \quad \text{at } r = 0 \tag{3.36.c}$$

It should be noticed that in (3.36.b) the concentration of the substrate on the surface of the particle was assigned the bulk concentration c_b after assuming that the mass transfer resistance to the substrate was negligible outside the surface. We may define dimensionless variables as:

$$c^* = \frac{c}{c_b}, \quad r^* = \frac{r}{R}, \quad \beta = \frac{K_m}{c_b} \quad \text{and} \quad \phi = R\sqrt{\frac{v_{app}}{D_e K_m}}$$

where parameter ϕ is called the *Thiele modulus*, then (3.36.a)–(83.36.c) will be rewritten as:

$$\frac{d^2c^*}{dr^{*2}} + \frac{2}{r^*}\frac{dc^*}{dr^*} = \phi^2 \frac{c^*}{c^* + \beta} \tag{3.37.a}$$

$$c^* = 1 \quad \text{at } r^* = 1 \tag{3.37.b}$$

$$\frac{dc^*}{dr} = 0 \quad \text{at } r^* = 0 \tag{3.37.c}$$

The substrate concentration profile may be obtained after solving (3.37.a) numerically. It is convenient to express the effectiveness factor η in terms of the Thiele modulus ϕ and parameter β as exemplified in Figure 3.9.

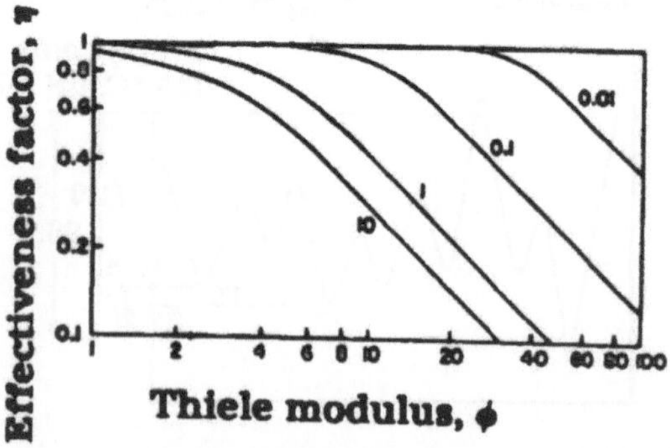

FIGURE 3.9 The theoretical relationship between the effectiveness factor η and the modulus ϕ of a spherical porous immobilized particle for various values of β (shown beside curves) (Wang *et al.*, 1979; © Wiley & Sons Inc., reproduced by permission).

When designing an immobilized enzyme system using a particular support the main variables will be v_{app} and R since the substrate concentration will often be predetermined and K_m and D_e will be fixed, by the support, solute, substrate and the enzyme. Once the particle size is selected, it is necessary to choose an optimum content to be immobilized (Wang *et al.*, 1979). The influence of the amount of enzyme on the effectiveness factor is exemplified in Figure 3.10.

In (3.36.a) it was assumed that in an immobilized enzyme system v_{app} was different and K_m was the same as that of the free enzyme. This is actually a special case and the apparent value of Michaelis constant K_m may also change upon immobilization. The support may reduce the activity of an immobilized enzyme towards large molecules via steric hindrance of the active site. An immobilized enzyme is also in a different medium than its native environment. When the support has a net charge the pH of the microenvironment of the enzyme may be different than the pH of the bulk medium due to the electrostatic interactions (Goldstein *et al.*, 1964):

$$\Delta pH = pH^i - pH^b = 0.43\frac{zF\psi}{RT} \tag{3.38}$$

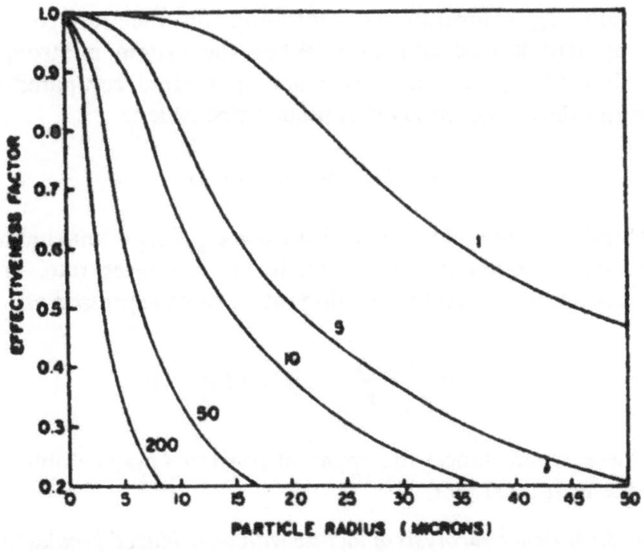

FIGURE 3.10 The influence of radius on the effectiveness factor of a spherical porous immobilized enzyme particle with various enzyme contents (mg/cm^3, values shown beside curves). Enzyme specific activity was 100 μ moles/(min mg enzyme): $c_b/K_m = 10$, $D_e = 4 \times 10^{-4}$ cm^2/min (Regan *et al.*, 1974; © Wiley & Sons Inc., reproduced by permission).

where pH^i = pH of the microenvironment of the enzyme, pH^b = pH of the bulk medium, z = net charge on the diffusing substrate, F = Faraday constant, ψ = electrostatic potential, R = gas constant. The intrinsic activity of the enzyme is altered by the local changes in pH and ionic constituents. Further alterations in the apparent kinetics are due to the repulsion or attraction of substrates or inhibitors. A similar expression may also be suggested for the variation of K_m upon immobilization (Wang *et al.*, 1979):

$$\Delta pK_m = pK_m^i - pK_m^b = \log\left[\frac{K_m^b}{K_m^i}\right] = 0.43\frac{zF\psi}{RT} \tag{3.39}$$

where K_m^b = Michaelis constant in the bulk medium and K_m^i = Michaelis constant in the microenvironment.

When the enzymes are bound and evenly distributed on the surface of a non-porous support material, the substrate diffuses through a thin liquid film surrounding the support to reach the active sites of the enzymes. Under steady state conditions the reaction rate equals the mass transfer rate:

$$k(c_{Ab} - c_{As}) = \frac{v_{app} c_{As}}{c_{As} + K_{app}} \tag{3.40}$$

where k = mass transfer coefficient, c_{Ab} = substrate concentration in the bulk liquid, c_{As} = substrate concentration on the surface, v_{app} and K_{app} = apparent kinetic constants. When the system is strongly mass transfer limited $c_{As} \cong 0$, since the reaction is rapid compared to mass transfer and the system behaves as pseudo first order:

$$v = kc_{Ab} \text{ (when } Da \gg 1) \tag{3.41}$$

where Damköhler number is defined as $Da = v_{app}/kc_{Ab}$ = maximum attainable reaction rate/maximum attainable mass transfer rate. When the system is reaction limited the reaction rate is often expressed as:

$$v = \frac{v_{app} c_{As}}{c_{As} + K_{app}} \text{ (when } Da \ll 1) \tag{3.42}$$

Under these circumstances the apparent constants may be obtained from the double-reciprocal plot.

Example 3.14. Acid hydrolysis of lactose with immobilized β-galactosidase in CSTR and plug flow reactors a) Kinetic constants for lactose hydrolysis by β-galactosidase from *Escherichia coli* were reported as $K_m = 1.9$ moles/m^3 and $v_{max} = 6.55 \times 10^{-6}$ moles/(min mg enzyme) at 20°C and pH = 7.6 (Whitaker, 1972). If this enzyme should be immobilized on non-porous support to obtain 10 mg enzyme/cm^3 of reactor with $\eta = 0.8$, what should the volumetric input flow rate of 50 moles/m^3 lactose solution be to a 0.7 m^3 CSTR to achieve $c_{exit}/c_0 = 0.30$?

Solution It is required that $c_{A0} = 50$ moles/m^3 and $c_A = 15$ moles/m^3. The apparent reaction rate is:

$$R_{App} = -\eta \frac{c_A v_{max}}{K_m + c_A} \tag{E.3.14.1}$$

After substituting the numbers we will calculate that $R_{App} = -4.65 \times 10^{-6}$ moles/min mg enzyme. Since we have 10 mg enzyme/cm^3 of reactor the apparent rate may be expressed in terms of the reactor volume as $R_{App} = -46.5$ moles/min m^3. Lactose balance around the reactor under steady state conditions requires:

$$Fc_{A0} - Fc_A + R_{App}V = 0 \tag{E.3.14.2}$$

Equation (E.3.14.2) will be rearranged as:

$$F = -V\frac{R_{App}}{c_{A0} - c_A} \tag{E.3.14.3}$$

After substituting the numbers in (E.3.14.3) we will obtain $F = 0.93$ m^3/min.

b) The same enzyme immobilized support is filled into a packed bed plug flow reactor (reactor diameter = 10 cm). What should the reactor length be to achieve the same conversion as in section (a)?

Solution Lactose balance around the reactor under steady state conditions requires:

$$L = \frac{F}{S}\int_{c_{A0}}^{c_{Af}} \frac{dc_A}{R_{App}} \tag{3.18}$$

We have already calculated that $F = 0.93$ m^3/min, it is also known that $c_{A0} = 50$ moles/m^3 and $c_A = 15$ moles/m^3 and the cross sectional area of the reactor is $S = \pi(\text{reactor radius})^2 = 7.85 \times 10^{-3}$ m^2. The apparent reaction rate is $R_{App} = -52.4c_A/1.9 + c_A$ moles/min m^3. After substituting the numbers in (3.18) we will obtain

$$L = -\frac{0.93\,\text{m}^3/\text{min}}{7.85 \times 10^{-3}\,\text{m}^2}\int_{50}^{15} \frac{(1/52.4\,c_A)\,dc_A}{1.9 + c_A} = 84.3\,\text{m}$$

The volume of the reactor $= L_x S = 0.66$ m^3 and smaller than the reactor volume required for the same conversion in a CSTR.

c) If the immobilized enzyme loses 10% of its activity after 24 h of operation with reaction

$$\underset{\text{active enzyme}}{E} \rightarrow \underset{\text{inactive enzyme}}{E_i}$$

what will be c_{Af}/c_0 ratio after 3 days of operation with both CSTR and PF reactors?

Solution We have already described enzyme inactivation rate as

$$\frac{dc_E}{dt} = -k_d c_E \tag{3.32}$$

After integrating and rearranging (3.32) we will have:

$$k_d = -\frac{1}{t}\ln\left(\frac{c_E}{c_{E0}}\right) \tag{E.3.14.4}$$

Since $c_E/c_{E0} = 0.90$ when $t = 24$ h we may use (E.3.14.4) to calculate $k_d = 4.4 \times 10^{-3}\,h^{-1}$. Equation (E.3.14.4) may be rearranged to calculate the fraction of the remaining enzyme activity after three days (72 h) of operation as:

$$\frac{c_E}{c_{E0}} = (-k_d t) = 0.72$$

The initial maximum enzyme activity v_{max} was defined as

$$v_{max} = k_3 c_{E0} \tag{3.21}$$

The remaining maximum enzyme activity after three days is:

$v_{max}^{3\ days} = (k_3 c_{E0})(c_E/c_{E0}) = (6.55 \times 10^{-6})(0.72) = 4.72 \times 10^{-6}$ moles/(min mg enzyme) since we have initially 10 mg enzyme immobilized/cm^3 of reactor $v_{max}^{3\ days} = 47.2$ moles/min m^3.

i. The c_{Af}/c_0 ratio after 3 days of operation with CSTR

We will substitute in $v_{max}^{3\ days}$ in (E.3.14.1) for v_{max} to calculate R_{App}. Under these conditions (E.3.14.2) will be rewritten as

$$Fc_{A0} - Fc_A - \eta v_{max}^{3\ days} \frac{c_A}{K_m + c_A} V = 0 \tag{E.3.14.5}$$

After substituting the numbers we will calculate $c_A = 23.7$ moles/m^3 from (E.3.14.5), implying that $c_{Af}/c_{A0} = 0.474$.

ii. The c_{Af}/c_0 ratio after 3 days of operation with plug flow reactor

We will substitute $v_{max}^{3\ days}$ in (E.3.14.1) for v_{max} to calculate R_{App}. Under these conditions (3.18) will be rewritten as

$$84.3\,\mathrm{m} = -\frac{0.93\,\mathrm{m^3/min}}{7.85\times10^{-3}\,\mathrm{m^2}}\int_{50}^{c_{Af}}\frac{(1/37.7\,c_A)\,dc_A}{1.9+c_A} \qquad \text{(E.3.14.6)}$$

Equation (E.3.14.6) will be simplified as

$$16.7 = \ln c_{Af} + 0.54\,c_{Af} \qquad \text{(E.3.14.7)}$$

After solving (E.3.14.7) we will obtain $c_{Af} = 23.8\,\mathrm{moles/m^3}$, implying that $c_{Af}/c_{A0} = 0.474$.

3.8. ANALOGY KINETIC MODELS

Variation of the sensory properties of the foods may be studied in analogy with chemical kinetics. There are numerous examples to this approach in the literature.

Example 3.15. Kinetics of browning of milk (*Pagliarini et al.*, 1990) The difference of color ΔC between samples and raw milk reference were determined with the following equation:

$$\Delta C = \sqrt{\Delta a^2 + \Delta b^2 + \Delta L^2} \qquad \text{(E.3.15.1)}$$

where a, b and L are the hunter color scale parameters. The kinetics of browning are expressed in analogy with zero order kinetics:

$$\frac{d(\Delta C)}{dt} = k \qquad \text{(E.3.15.2)}$$

After integration we will have

$$\Delta C = \Delta C_0 + kt \qquad \text{(E.3.15.3)}$$

Where ΔC_0 is the intercept with $t = 0$ axis. Comparison of the integrated equation with the experimental data is shown in Figure E.3.15.
Further analysis of the data indicated that

$$\Delta C_0 = -13.8909 + 0.0385785\,T \qquad \text{(E.3.15.4)}$$

where T is the heating temperature in K. The rate constant was expressed with the Arrhenius equation:

$$k = 1.5\times10^{-3}\exp\left\{-\frac{101.8}{8.314\times10^{-3}\,T}\right\} \qquad \text{(E.3.15.5)}$$

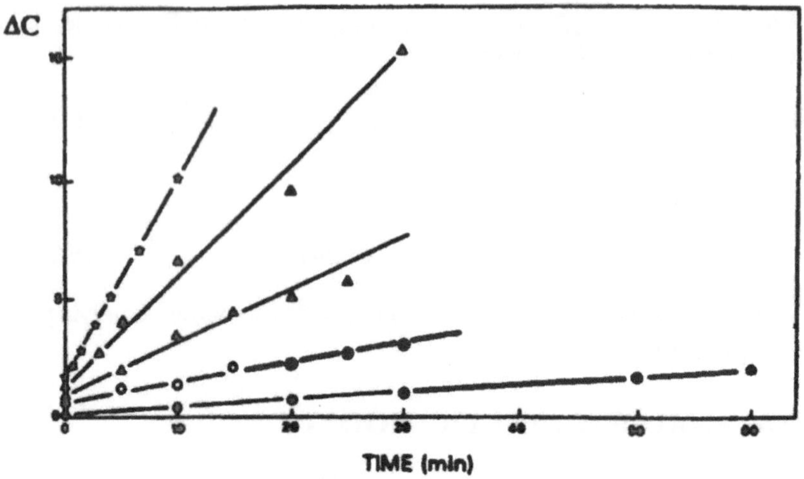

FIGURE E.3.15 Variation of ΔC with time at 90 (●), 102 (○), 110 (▲), 120 (Δ) and 130 (★) °C (Pagliarini *et al.*, 1990; © IFT, reproduced by permission).

Example 3.16. Kinetics of the color change and polyphenol oxidase inactivation during blanching of Sultana grapes (Aguilera et al., 1987) Sultana grapes were blanched in water then dried in a tray dryer. The effect of the blanching time on Hunterlab L values of raisins was expressed by analogy to the zero order rate expression

$$\frac{dL}{dt} = k_L \tag{E.3.16.1}$$

where k_L is the apparent rate constant. After integration

$$L = L_0 + k_L t \tag{E.3.16.2}$$

where L_0 is the initial Hunterlab L value. Inactivation of polyphenol oxidase (PPO) was modeled with zero order rate expression:

$$\frac{d(PPO)}{dt} = -k_{PPO} \tag{E.3.16.3}$$

where k_{PPO} is the apparent rate constant. After integration

$$PPO = PPO_0 - k_{PPO}t \tag{E.3.16.4}$$

where PPO_0 is the initial polyphenol oxidase activity. Percentage of inactivation of PPO may be expressed as:

$$\frac{PPO_0 - PPO}{PPO_0} \times 100 = k_{PPO} \frac{100}{PPO_0} t \tag{E.3.16.5}$$

Equations (E.3.16.2) and (E.3.16.5) were compared with the experimental data in Figures E.3.16.a and b, respectively. The apparent reaction rate constants were also found to agree with the Arrhenius expression (Fig. E.3.16.c).

Example 3.17. Kinetics of the change of texture and taste of the potatoes during cooking (*Harada et al.*, 1985) Changes in texture and taste of the potatoes during cooking were simulated with zero order models:

$$\frac{d(\text{texture judgement})}{dt} = -k_{\text{texture}} \qquad \text{(E.3.17.1)}$$

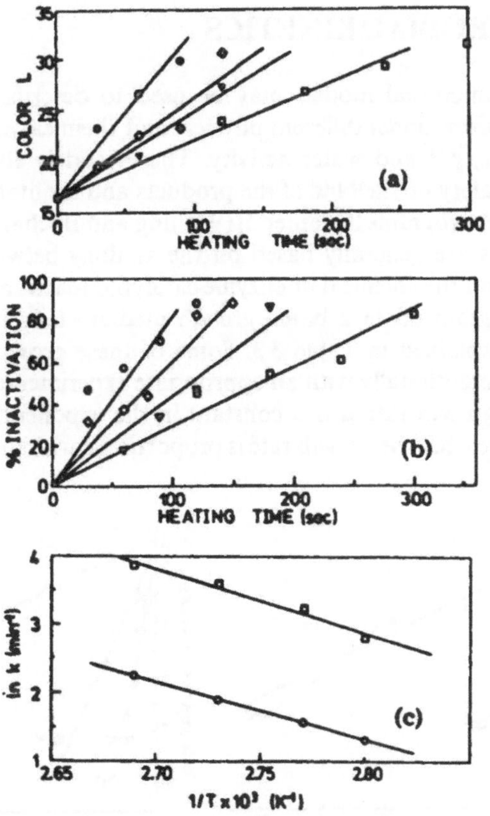

FIGURE E.3.16 (a) Comparison of (a) (E.3.16.2) and (b) (E.3.16.5) with the experimental data. Data are shown in symbols (□ 83 °C, ∇ 88 °C, ◇ 93 °C and ○ 98 °C), models are shown in solid lines. (c) Arrhenius plots for the inactivation rate constant of *PPO* (□) and Hunterlab *L* value (○). (Aguilera *et al.*, 1987; © IFT, reproduced by permission).

$$\frac{d(\text{taste judgement})}{dt} = -k_{\text{taste}} \qquad \text{(E.3.17.2)}$$

where k_{texture} and k_{taste} are the apparent rate constants. After integration we will have:

$$(\text{texture judgment}) = (\text{texture judgment})_0 - k_{\text{texture}}\, t \qquad \text{(E.3.17.3)}$$

$$(\text{taste judgment}) = (\text{taste judgment})_0 - k_{\text{taste}}\, t \qquad \text{(E.3.17.4)}$$

where $(\text{texture judgment})_0$ and $(\text{taste judgment})_0$ are the numerical values of the texture and taste judgments at $t = 0$ (both of them should be theoretically 1.0). Comparison of (E.3.17.3) and (E.3.17.4) with the experimental data is shown in Figures E.3.17.a and b.

3.9. MICROBIAL KINETICS

Predictive microbial models may be used to describe the behavior of microorganisms under different physical and chemical conditions, such as temperature, pH and water activity. These models allow prediction of microbial safety or shelf life of the products and facilitate development of the HACCP programs (Chapter 5) (Whiting and Buchanan, 1994). Microbial kinetics are generally based on the analogy between the microbial processes and the chemical or enzyme catalyzed reactions. Proliferation of the microorganisms in a batch growth medium may include four ideal stages as explained in Table 3.3. Some of these growth phases may be prevented intentionally with an appropriate experimental design.

Specific growth rate μ is a constant in the exponential growth phase, which implies that the growth rate is proportional with the viable microbial

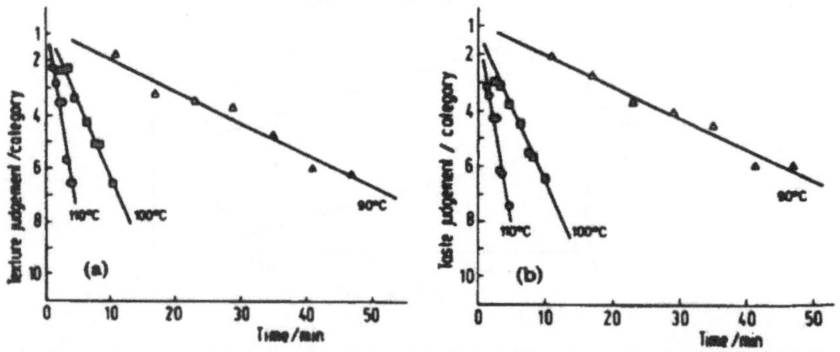

FIGURE E.3.17 Development of texture (a) and taste (b) during cooking of "Mentor" potatoes (texture category: 1 = very hard, 6 = optimal texture, 11 = pulpy; taste category: 1 = completely raw, 6 = optimal taste, 11 = completely over cooked) (Harada *et al.*, 1985; © IFT, reproduced by permission).

Table 3.3 Ideal growth phases of microorganisms in batch medium

growth phase	growth rate expression		comments
lag phase	$\frac{dx}{dt}=0$	(3.43)	time required for adaptation of the microorganisms to a new medium or recovery from injury
exponential phase	$\frac{dx}{dt}=\mu x$	(3.44)	variation of the population size is proportional with the number of microorganisms in the culture
stationary phase	$\frac{dx}{dt}=0$	(3.45)	microbial growth may stop because of lack of a limited nutrient or product inhibition
death phase	$\frac{dx}{dt}=-k_d x$	(3.46)	microbial death may start after depletion of the cellular reserves in the stationary phase or with the effect of the unfavorable factors

population, where all the members have equal potential for growth. The specific growth rate may be regarded as the frequency of producing new microorganisms by the ones already present. When microbial proliferation occurs in a substrate limited medium, specific growth rate may be related with the substrate concentration:

$$\mu=\frac{\mu_{max}c_S}{c_S+K} \tag{3.47}$$

where μ_{max} is the maximum attainable specific growth rate. Equation (3.47) is called the Monod equation. There are numerous variations of the Monod equation available in the literature (Mulchandani and Luong, 1989). The most common empirical modifications of (3.47) to include substrate and product inhibition are:

$$\mu=\frac{\mu_{max}c_S}{c_S+K+\frac{c_s^2}{K_s}} \tag{3.48}$$

and

$$\mu=\frac{\mu_{max}c_S}{c_S+K}\frac{K_p}{K_p+c_p} \tag{3.49}$$

where K_s and K_p are constants, c_p is the product concentration.

The logistic model is frequently used to simulate microbial growth when a microbial population inhibits its own growth via depletion of a limited nutrient, product accumulation, or unidentified reasons (Example 1.2):

$$\frac{dx}{dt} = \mu x\left(1 - \frac{x}{x_{max}}\right) \tag{3.50}$$

where μ is initial specific growth rate and x_{max} is the maximum attainable value of x. The logistic equation is an empirical model, it simulates the data when microbial growth curve follows a sigmoidal path to attain the stationary phase. It is based on experimental observations only. When $x \ll x_{max}$, the term in parenthesis is almost one and neglected, then the equation simulates the exponential growth, i.e., (3.44). When x is comparable with x_{max}, the term in parenthesis becomes important and simulates the inhibitory effect of over-crowding on microbial growth. When $x = x_{max}$ the term in parenthesis becomes zero, and the equation will predict no growth, i.e., stationary phase as described with (3.45). The logistic equation may be integrated as:

$$x = \frac{x_0 e^{\mu t}}{1 - \frac{x_0}{x_{max}}(1 - e^{\mu t})} \tag{3.51}$$

Equation (3.51) may be rearranged and linearized as:

$$\ln\left(\frac{\mathscr{x}}{1-\mathscr{x}}\right) = -\ln\left(\frac{x_{max}}{x_0} - 1\right) + \mu t \tag{3.52}$$

where $\mathscr{x} = x/x_{max}$. Usually x_{max} may be evaluated from experimental data. Equation (3.54) implies that, plot of $\ln \mathscr{x}/(1-\mathscr{x})$ versus time is a line (Fig. 3.11) with intercept $= -\ln(x_{max}/x_0 - 1)$ and slope $= \mu$. Intercept of the line may be used to evaluate x_0.

Exponential growth model simulated with (3.44) may be modified after substituting:

$$\mu = \mu_0 + \mu_1 x - \mu_2 x^2 \tag{3.53}$$

to simulate the *Allee effect*, which represents a population with maximum specific growth rate at intermediate microbial concentrations when μ_0, μ_1 and μ_2 are positive constants (Edelstein-Keshet, 1988).

When μ of equation (3.44) is a function of time such that

$$\frac{d\mu}{dt} = -\alpha\mu \tag{3.54}$$

we obtain the Gompertz model, which may also be used to simulate the sigmoidal behavior of the microbial growth curve (α = constant). The

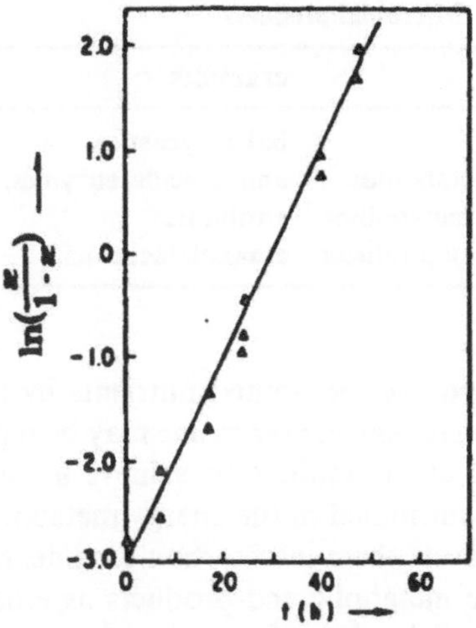

FIGURE 3.11 A typical linearized plot for the evaluation of the constants x_0 and μ based on the data obtained during cultivation of *Brevibacterium flavum* S-225 for lysine. HCl production. Maximum attainable biomass concentration was evaluated from the stationary phase data of the original growth curve as $x_{max} = 29$ ($OD \times 26$). Equation of the line: $\ln(x/(1-x)) = 3 + 0.1\,t$, implying that $\mu = 0.1\ \mathrm{h}^{-1}$ and $x_0 = 1.4$ ($OD \times 26$). Reprinted by permission of the publisher from Kinetics of Amino Acid Production by Over-Producer Mutant Microorganisms, Özilgen, *Enzyme and Microbial Technology*, 10(2), 110–114. © 1988 by Elsevier Science Ltd.

Gompertz model is usually expressed in three equivalent versions (Edelstein-Keshet, 1988):

$$\frac{dx}{dt} = \mu x, \quad \frac{d\mu}{dt} = -\alpha\mu \tag{3.55.a}$$

$$\frac{dx}{dt} = (\lambda e^{-\alpha t})x \tag{3.55.b}$$

and

$$\frac{dx}{dt} = (\kappa \ln x)x \tag{3.55.c}$$

where λ and κ are constants.

Microbial products may be roughly categorized in four groups as depicted in Table 3.4.

Primary metabolites are produced by the microorganism for its own metabolic activity. Secondary metabolites are usually produced against the external factors, i.e., production of antibiotics starts in the stationary phase

Table 3.4 Microbial products

product	examples
biomass	bakers yeast
primary metabolites	amino acids, enzymes, vitamins
secondary metabolites	antibiotics
metabolic by products	ethanol, lactic acid

to prevent consumption of the limited nutrients by the other microbial species. The yeast *Saccharomyces cerevisiae* may be regarded as a product itself when produced as an additive to achieve leavening in the bakery industry. Sugars are consumed in the energy metabolism, some microorganisms may not convert them into carbon dioxide, but follow a shorter path and excrete the metabolic end-products as ethanol or lactic acid. Product formation models relates the product formation rate to fermentation variables, i.e., growth rate, biomass or substrate concentration, etc. The Luedeking-Piret (1959) model is among the most popular product formation models of food processing interest:

$$\frac{dc_{Pr}}{dt} = \alpha x + \beta \frac{dx}{dt} \tag{3.56}$$

where c_{Pr} is product concentration, α and β are constants. The term αx represents the product formation rate by the microorganisms regardless of their growth; $\beta dx/dt$ represents the additional product formation rate during growth in proportion with the growth rate. This is an empirical equation, because it simply relates the experimental observations without considerable theoretical basis. When growth-associated product formation rates are much greater than the non-growth-associated product formation rates (3.56) may be written as:

$$\frac{dc_{Pr}}{dt} = \beta \frac{dx}{dt} \tag{3.57}$$

when non-growth associated product formation rates are much greater than the growth-associated product formation rates, (3.56) becomes:

$$\frac{dc_{Pr}}{dt} = \alpha x \tag{3.58}$$

Structured and age distribution models relate cellular structure or age distribution to growth and product formation rates, but they need more information for application, and are generally difficult to use and not widely employed in food research, therefore not considered here; an interested reader may refer to Bailey and Ollis (1986) for a detailed discussion.

In a fermentation process a substrate, i.e., nutrient, is allocated to three basic uses:

$$\begin{bmatrix}\text{total rate of}\\ \text{substrate}\\ \text{utilization}\end{bmatrix}=\begin{bmatrix}\text{rate of substrate}\\ \text{utilization for}\\ \text{biomass synthesis}\end{bmatrix}+\begin{bmatrix}\text{rate of substrate}\\ \text{utilization for}\\ \text{maintenance}\end{bmatrix}+\begin{bmatrix}\text{rate of substrate}\\ \text{utilization for}\\ \text{product formation}\end{bmatrix} \quad (3.59)$$

Substrate consumption to keep cells alive without growth or product formation is referred to as *maintenance*. Equation (3.59) may be expressed in mathematical terms as:

$$-\frac{dc_s}{dt}=\frac{1}{Y_{x/s}}\frac{dx}{dt}+k_m x+\frac{1}{Y_{p/s}}\frac{dc_{Pr}}{dt} \quad (3.60)$$

where $Y_{x/s}$ is the cell yield coefficient defined as grams of biomass produced per grams of substrate used, and $Y_{p/s}$ is the product yield coefficient, i.e., grams of product produced per grams of substrate used. The yield coefficients usually remains constant as long as the fermentation behavior remains unchanged, but there are also examples of variable yield coefficients in the literature, which may indicate a shift in substrate preference, availability of oxygen, etc. during the course of the process (Özilgen *et al.*, 1988).

Microorganisms produce heat as a by-product of the energy metabolism which may be summarized with typical examples as:

$$\text{Glucose}+36\,P_i+36\,\text{ADP}+6\,O_2 \xrightarrow{\text{respiration}} 6\,CO_2+42\,H_2O+36\,\text{ATP}+Q \quad (3.61)$$

$$\text{Glucose}+2\,P_i+2\,\text{ADP} \xrightarrow{\text{anaerobic fermentation}} 2\,\text{lactate}+2\,H_2O+2\,\text{ATP}+Q \quad (3.62)$$

where P_i and Q represent inorganic phosphate atoms and metabolic heat generation, respectively. ADP and ATP are abbreviations for adenosine diphosphate and adenosine triphosphate, respectively. ATP is the energy currency of the cell. ATP to ADP conversion is coupled with the energy consuming metabolic reactions, i.e., biosynthesis, cellular transport, etc. and make them thermodynamically feasible. ADP is converted back to

ATP through (3.61) or (3.62). Metabolic heat generation rate may be related with the growth rate as:

$$\begin{bmatrix}\text{metabolic heat}\\ \text{generation rate}\\ \text{per unit volume}\\ \text{of fermentor}\end{bmatrix} = \frac{1}{Y_\Delta}\frac{dx}{dt} \tag{3.63}$$

where Y_Δ is the heat generation coefficient defined as grams of biomass production coupled with one unit, i.e., kJ, of heat evaluation.

Most food and beverage fermentations involve a mixed culture of microorganisms. Typical examples may include yogurt cultures (*Streptococcus thermophilus and Lactobacillus bulgaricus*); wine (*Saccharomyces cerevisiae* and wild microbial species), and cheese production (mixed culture of various microorganisms). The general interaction methods of two microorganisms are summarized in Table 3.5.

Table 3.5 indicates that when both microbial species 1 and 2 benefit from an interaction it is called mutualism; when microbial species 1 benefits but microbial species 2 not affected from an interaction it is called commensalism, etc. Interactions shown in Table 3.4 depend on the culture conditions. An initially neutral relation may turn in competition during the course of fermentation with depletion of the limiting substrate, or two microbial populations may have a mutualistic relation in one medium, but a competitive relation in another medium.

Example 3.18. Mixed culture interactions between P. vulgaris *and* S. cerevisiae (*Tseng and Phillips,* 1981) *Proteus vulgaris* prefers to utilize sodium citrate when both sodium citrate and glucose are available, also nicotinic acid is needed for its growth. *Saccharomyces cerevisiae* can utilize only glucose and produces nicotinic acid. By varying the concentrations of the medium components, various mixed culture interactions may be created as depicted in Table E.3.18.

Chemical reactor and fermentor design are based on similar principles. A CSTR is called a chemostat when there is only one limited substrate.

Table 3.5 Genereal interaction ways of microbial populations (symbols +, − and 0 indicates positive, negative and no effect on the related microbial population)

(Effect on the second microorganism)→ (Effect on the first microorganism)↓	+	−	0
+	mutualism	predation	commensalism
−		competition	amensalism
0			neutralism

Table E.3.18 Mixed culture interactions between *P.vulgaris* and *S.cerevisiae*

sodium citrate	glucose	nicotinic acid	interaction
limiting concentration	limiting concentration	not added	commensalism
not added	limiting concentration	not added	commensalism plus competition
not added	limiting concentration	sufficient amount (not limiting)	competition
excess amount	limiting concentration	not added	mutualism
limiting concentration	limiting concentration	sufficient amount (not limiting)	neutralism

Equation (3.7) may be applied to a chemostat for population balance of the microorganisms to yield:

$$Fx_0 - Fx + VR_x = \frac{d(Vx)}{dt} \tag{3.64}$$

where x_0 is the concentration of the microorganisms in the input stream. Under steady state conditions $d(Vx)/dt = 0$, when the microorganisms are in exponential growth phase $R_x = \mu_x$ we may rearrange (3.64) to obtain:

$$x = \frac{Dx_0}{D - \mu} \tag{3.65}$$

where x = microbial concentration in or at the exit of the fermentor,

$$D = \frac{F}{V} = \text{dilution rate} = \frac{1}{\tau} = \frac{1}{\text{residence time in the fermentor}} \tag{3.66}$$

While working with sterile nutrients, i.e., $x_0 = 0$, (3.65) will become

$$(D - \mu)\,x = 0 \tag{3.67}$$

Equation (3.67) implies that a non-zero population may be maintained under steady state conditions only when

$$\mu = D \tag{3.68}$$

It should be noticed that even under these conditions the output stream will have the same biomass concentration as in the chemostat (Bailey and Ollis, 1986).

When nutrients are sterile and biomass growth agrees with Monod equation (3.64) will become

$$D = \mu_{max}\frac{c_s}{K_s + c_s} \tag{3.69}$$

where c_s is the substrate concentration in the fermentor. Equation (3.69) may be rearranged as

$$c_S = \frac{DK_s}{\mu_{max} - D} \tag{3.70}$$

Biomass yield coefficient $Y_{x/s}$ may be calculated as

$$Y_{x/s} = \frac{x - x_0}{c_{S0} - c_S} \tag{3.71}$$

where c_{s0} is the substrate concentration in the input stream. When we use sterile nutrients ($x_0 = 0$) (3.70) and (3.71) may be combined to calculate the biomass concentration in (also at the exit of) the fermentor:

$$x = Y_{x/s}\left(c_{S0} - \frac{DK_s}{\mu_{max} - D}\right) \tag{3.72}$$

Equation (3.67) implies that stable fermentation operation may be maintained only when $c_{S0} > DK_s/(\mu_{max} - D)$.

Example 3.19. Kinetics of microbial growth, gas production and increase in dough volume during leavening (*Akdoğan and Özilgen*, 1992) Commercial Baker's yeast contains strains of *Saccharomyces cerevisiae*, lactic acid bacteria and low numbers of contaminating microorganisms. The yeast utilizes the simple sugars derived from flour and produces carbon dioxide. Growth of the microorganisms in the dough was simulated with the logistic equation:

$$\frac{dx}{dt} = \mu x\left(1 - \frac{x}{x_{max}}\right) \tag{3.50}$$

The Luedeking-Piret equation was used to model gas production:

$$\frac{dG}{dt} = \alpha x + \beta\frac{dx}{dt} \tag{3.56}$$

A fraction of the gas produced by the Baker's yeast is retained in the dough and the remaining gas diffuses out. The retained gas increases the volume of the dough. The volume increase rates are expected to slow down as the volume of the dough gets larger and the walls of the gas cell get thinner due to stretching. The rate of the dough volume increase may be related to the gas production rate:

$$\frac{dV}{dt} = \phi\left(1 - \frac{V}{V_{max}}\right)\frac{dG}{dt} \tag{E.1.19.1}$$

where V and V_{max} are the volume and the maximum attainable volume of the dough, respectively. Parameter ϕ is the ratio of the initial dough volume increase rate to the initial gas production rate. Comparison of the model with a typical set of experimental data is shown in Figure E.3.19.1.

The dough was not kneaded during leavening process, and therefore the microorganisms proliferated at the original sites of the inoculum cells. The logistic equation actually simulated saturation of these fixed locations. Since the dough was not kneaded after beginning the leavening process, the inoculum size x_0 affected the gas production and holding kinetics substantially. This effect may arise especially due to the decrease of the thickness of the gas cell walls with large amounts of gas production resulting from higher inoculum size. The effect of the initial biomass concentration on the model constants α, β, and ϕ are shown in Figure E.3.19.2.

Example 3.20. Kinetics of microbial growth and lactic acid production by mixed cultures of Streptococcus thermophilus *and* Lactobacillus bulgaricus *in yogurt production (Özen and Özilgen,* 1992) Lactic acid production by mixed cultures of *Streptococcus thermophilus* and *Lactobacillus bulgaricus* is the most important process involved in yogurt production. Interaction of these microorganisms in milk is favorable for both of the species, but not

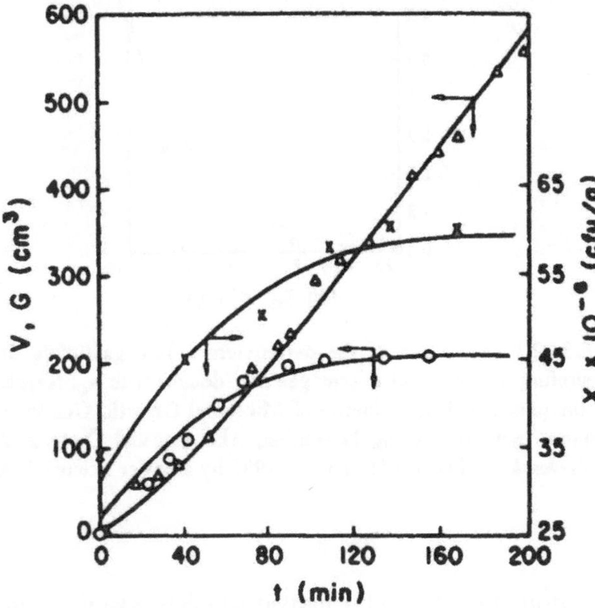

FIGURE E.3.19.1 Sample plot ($T = 25\,°C$) for comparison of the model (–) with the experimental data of microbial growth (x), gas production (Δ) and dough volume increase (○). Reprinted by permission of the publisher from Kinetics of Microbial Growth, Gas Production, and Dough Volume Increase During Leavening, Akdoğan and Özilgen, *Enzyme and Microbial Technology*, 14(2), 141–143. © 1992 by Elsevier Science Ltd.

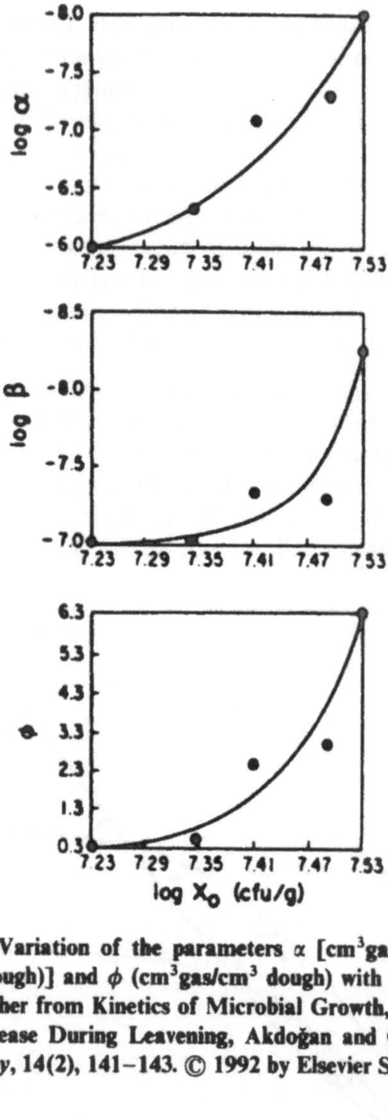

FIGURE E.3.19.2 Variation of the parameters α [cm^3gas/(cfu/g dough) min], β [cm^3 gas/(cfu/g dough)] and ϕ (cm^3gas/cm^3 dough) with x_0. Reprinted by permission of the publisher from Kinetics of Microbial Growth, Gas Production, and Dough Volume Increase During Leavening, Akdoğan and Özilgen, *Enzyme and Microbial Technology*, 14(2), 141–143. © 1992 by Elsevier Science Ltd.

obligatory. Both numbers of the individual microorganisms and the total amounts of lactic acid produced by mixed cultures of these microorganisms are considerably higher than those obtained with pure cultures of each microorganism. In mixed cultures, stimulation of growth of *S. thermophilus* was attributed to the production of certain amino acids, i.e., glycine, histidine, valine, leucine and isoleucine, by *L. bulgaricus*. *S. thermophilus* stimulates growth of *L. bulgaricus* by producing formic acid. The logistic equation was modified to simulate the mixed culture growth of the microorganisms as:

$$\frac{dx_S}{dt} = \mu_S x_S \left(1 - \frac{\tau_S x_S + \tau_L x_L}{x_{S\max} + x_{L\max}}\right) \tag{E.3.20.1}$$

and

$$\frac{dx_L}{dt} = \mu_L x_L \left(1 - \frac{\tau_S x_S + \tau_L x_L}{x_{S\max} + x_{L\max}}\right) \tag{E.3.20.2}$$

where subscript L and S refers to *L. bulgaricus* and *S. thermophilus*, respectively. Some of the microbial species died at the end of the fermentation process, the death rates were simulated as:

$$\frac{dx}{dt} = -kx \tag{E.3.20.3}$$

The Luedeking-Piret equation was modified to simulate lactic acid production by the mixed culture of the microorganisms as:

$$\frac{dc_{Pr}}{dt} = \alpha_S x_S + \alpha_L x_L + \beta_S \frac{dx_S}{dt} + \beta_L \frac{dx_L}{dt} \tag{E.3.20.4}$$

Comparison of the solutions of (E.3.20.1)–(E.3.20.4) with the experimental data is shown with a typical example in Figure E.3.20.1.

The kinetic analysis clearly showed that the contribution of each microbial species to the mixed culture growth process changed drastically when the substrate concentration was about 15% (Fig. E.3.18.2). This optimum initial substrate concentration was in agreement with the result of the previous studies and the optimum find by trial and error procedure in the commercial yogurt production.

Example 3.21. Kinetics of spontaneous wine production (*Özilgen et al.*, 1991) The winemaking process may be analyzed in two phases: alcoholic fermentation and malo-lactic fermentation. During the alcoholic fermentation process wine microorganisms consume the fermentable sugars and produce ethanol. A spontaneous wine production process is actually a mixed-culture and multi-product process, started by the natural microorganisms of the grapes. Natural grape microorganisms consist of various genera of molds, yeasts and lactic acid bacteria. Generally wine yeast, *Saccharomyces cerevisiae*, is extremely low in population on the grapes. In the wine making process it multiplies with a strong fermentative capacity, excludes most of the other microorganisms from the medium and eventually invades the raw grape juice.

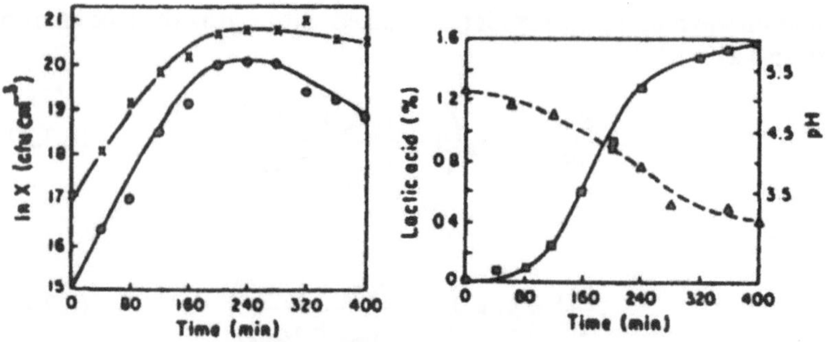

FIGURE E.3.20.1 Variation of the concentrations of *L. bulgaricus* (●), *S. thermophilus* (*x*), lactic acid (■) and pH (▲) in cultivations with 20% NFDM (non-fat dry milk) medium. Experimental data are shown in symbols (●, *x*, ■, ▲). Simulations are shown with solid lines. Variation of pH values were not simulated (Özen and Özilgen, 1992; © SCI, reproduced by permission). Constants: $\mu_L = 0.046\ \text{min}^{-1}$, $\mu_S = 0.032\ \text{min}^{-1}$, $\tau_L = 0.19$, $\tau_S = 1.34$, $\alpha_L = 5.1 \times 10^{-12}$ %cm^{3}cfu^{-1}min^{-1}, $\alpha_S = 6.1 \times 10^{-15}$ %cm^{3}cfu^{-1}min^{-1}, $\beta_L = 1.9 \times 10^{-10}$ %cm^{3}cfu^{-1}, $\beta_S = 4.0 \times 10^{-10}$ % cm^{3}cfu^{-1}, $x_{L\max} = 5.0 \times 10^8$ cfucm^{-3}, $x_{S\max} = 1.1 \times 109$ cfucm^{-3}, $k_L = 9.4 \times 10^{-3}\ \text{min}^{-1}$ and $k_S = 2.0 \times 10^{-3}\ \text{min}^{-1}$.

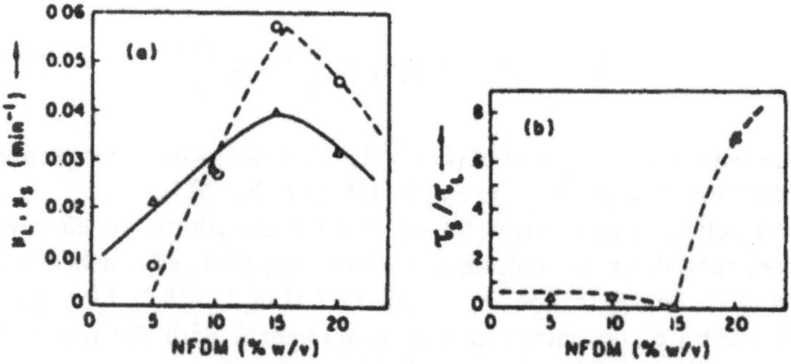

FIGURE E.3.20.2 (a) Variation of the specific growth rates, i.e., parameters μ_L and μ_S with the initial substrate concentration. (▲) *S. thermophilus*, (○) *L. bulgaricus*. (b) Variation of the ratio τ_S/τ_L with the initial NFDM concentration (Özen and Özilgen, 1992; © SCI, reproduced by permission).

In the alcoholic fermentation process growth of the biomass may be simulated in three phases:

$$\text{(i) Exponential growth: } \frac{dx}{dt} = \mu x \tag{3.44}$$

$$\text{(ii) Stationary phase: } \frac{dx}{dt} = 0 \tag{3.45}$$

(iii) Death phase: $\frac{dx}{dt} = -k_d x$ (3.46)

Total biomass concentration, denoted by x, is actually a mixed culture and the microbial species contributing to x actually change with time. In spontaneous wine fermentation, as the process proceeds, alcohol-sensitive microorganisms are inhibited and the alcohol tolerant microorganisms dominate the culture. Alcohol-tolerant microorganisms are generally better alcohol producers, therefore parameter β of the Luedeking-Piret equation may be related with the ethanol concentration in the medium:

$$\beta = \beta_0 + \beta_1 c_{pr} + \beta_2 c_{pr}^2 \tag{E.3.21.1}$$

where β_0, β_1 and β_2 are constants. After substituting (E.3.21.1) in (3.56):

$$\frac{dc_{pr}}{dt} = \alpha x + (\beta_0 + \beta_1 c_{pr} + \beta_2 c_{pr}^2)\frac{dx}{dt} \tag{E.3.21.2}$$

Substrate utilization in the exponential growth phase was:

$$-\frac{dc_S}{dt} = \frac{1}{Y_{x/s}}\frac{dx}{dt} + \frac{1}{Y_{p/s}}\frac{dc_{pr}}{dt} \tag{E.3.21.3}$$

Lactic acid bacteria are the major species contributing to the total biomass concentration x in the malo-lactic fermentation phase, where lactic acid bacteria uses malic acid as a substrate to produce lactic acid. The logistic equation was employed to simulate microbial growth in the malo-lactic fermentation phase:

$$\frac{dx}{dt_M} = \mu_M x\left(1 - \frac{x}{x_{max}}\right) \tag{E.3.21.4}$$

where t_M = time after commencement of the malo-lactic fermentation, and μ_M = specific growth rate of the malo-lactic fermentation.

Temperature changes in the fermentation vessel were modeled after making the thermal energy balance:

$$-\sum_{i=1}^{n} U_i A_i (T - T_{env}) + \frac{V}{Y_\Delta}\frac{dx}{dt} = \frac{d(\rho c V T)}{dt} \tag{E.3.21.5}$$

where the term $-\Sigma_{i=1}^{n} U_i A_i (T - T_{env})$ represents the energy loss from the fermentor surfaces, and the term $(V/Y_\Delta)(dx/dt)$ represents the thermal energy generation coupled with microbial growth. This term becomes zero after the end of the exponential growth in the alcoholic fermentation. The term $d(\rho c V T)/dt$ is thermal energy accumulation. Equation (E.3.21.5) was rearranged after substituting (3.44) for dx/dt:

$$\frac{dT}{dt} = K_1(T - T_{env}) + K_2 e^{\mu t} \tag{E.3.21.6}$$

where $K_1 = \Sigma_{i=1}^{n}(U_i A_i/\rho c V)$ and $K_2 = (\mu x_0/Y_\Delta \rho c V)$. Numerical value of parameter K_2 was zero after the end of the exponential growth phase of the alcoholic fermentation. The model is compared with a typical set of experimental data in Figure E.3.21.

Example 3.22. Kinetics of Aspergillus oryzae cultivations on starch (*Bayindirli et al.*, 1991) Cultivation of *Aspergillus oryzae* on starch is described as combination of two rate processes: starch hydrolysis, and uptake of some of the fermentable hydrolysis products for cellular activities, including growth, enzyme production and maintenance. These processes are interrelated in a cyclic way and neither of them can be accomplished without the other (Fig. E.3.21.1). The first link between these processes is starch hydrolyzing extracellular enzymes, i.e., glucoamylases and amylases. These are produced by the microorganism in the first process, i.e., with the cellular activities of the microorganism, and catalyze the second process, i.e., starch hydrolysis. The second link between these processes is the starch hydrolysis products. They are produced in the second process and consumed in the first process.

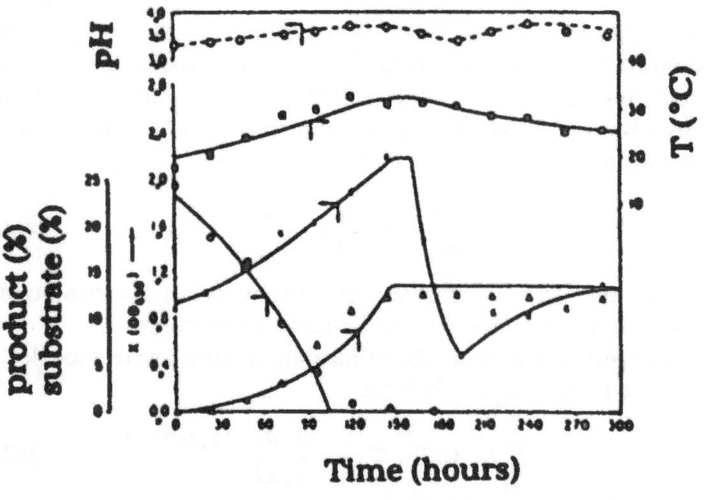

FIGURE E.3.21 Course of fermentation with slow fermented wine. Experimental data are shown in symbols: (○) pH, (■) temperature, (●) reducing sugars, (▲) ethanol, (x) biomass. Simulations are shown with solid lines (–). Numerical values of the model constants are: $\mu = 0.006\ h^{-1}$, $k_d = 0.042\ h^{-1}$, $\alpha = 0.0045$ % ethanol/OD_{650} h, $\beta = 3.6 + 1.25\,cP + 0.004\,c_p^2$ % ethanol/OD_{650}, $Y_{x/s} = 0.04\ OD_{650}$/ % reducing sugars, $Y_{p/s} = 10\%$ ethanol/% reducing sugars, $K_1 = 0.004\ h^{-1}$, $K_2 = 0.08\ ^\circ C\ h^{-1}$, $\mu_M = 0.032\ h^{-1}$, $x_{max} = 1.08\ OD = 650$. Reprinted by permission of the publisher from Kinetics of Spontaneous Wine Production, Özilgen *et al.*, *Enzyme and Microbial Technology*, 13(3), 252–256. © 1991 by Elsevier Science Ltd.

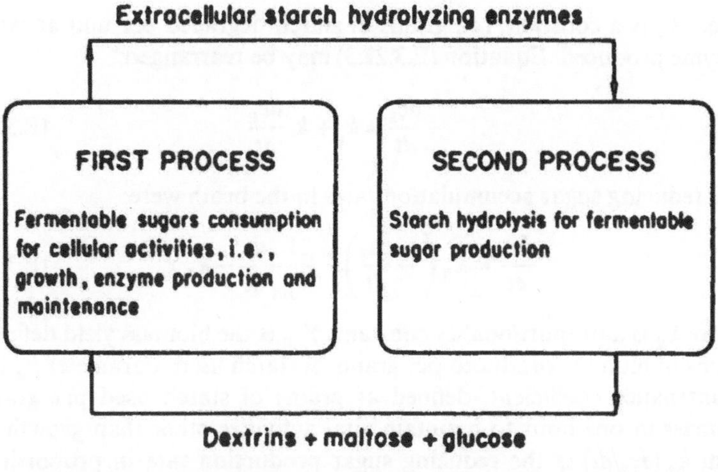

FIGURE E.3.22.1 Schematic description of the rate processes involved in *A. oryzae* cultivation on starch (Bayindirli *et al.*, 1991; © Harwood Academic Publishers GmbH, reproduced by permission).

Biomass production was simulated with the logistic equation:

$$\frac{dx}{dt} = \mu x\left(1 - \frac{x}{x_{max}}\right) \quad (3.50)$$

Total starch hydrolyzing enzyme production was:

$$\frac{dx_E}{dt} = k_M \frac{dx}{dt} - k_N x \quad (E.3.22.1)$$

where k_M and k_N are constants. The first term $k_M(dx/dt)$ indicates that the enzyme production rate is proportional to the growth rate of the microorganism. The second term $-k_N x$ shows that the microorganisms degrade starch hydrolyzing enzyme in proportion to their own concentration. *A. oryzae* produces extracellular proteases, which may be among the causes of the starch hydrolyzing enzyme degradation.

Starch hydrolyzing enzymes of *A. oryzae* are extracellular. If the microorganisms are separated from the broth during cultivation, the remaining enzymes will continue to degrade starch in the broth. When the total starch hydrolyzing enzyme activity is constant, the amount of starch degraded in time interval dt will be:

$$-dc_S = k_s\, dt \quad (E.3.22.2)$$

where k_s is the steady state starch degradation rate. If additional enzyme is produced within this time interval, additional amounts of enzyme will be degraded in proportion with enzyme production:

$$-dc_S = k_s\, dt + k_u\, dc_E \quad (E.3.22.3)$$

where k_u is a constant, i.e., grams of starch degraded per unit activity of enzyme produced. Equation (E.3.22.3) may be rearranged:

$$-\frac{dc_S}{dt} = k_s + k_u \frac{dc_E}{dt} \quad \text{(E.3.22.4)}$$

The reducing sugar accumulation rates in the broth were:

$$\frac{dc_R}{dt} = k_R\left(-\frac{dc_S}{dt}\right) - \frac{1}{Y_{x/R}}\frac{dx}{dt} - k_R x \quad \text{(E.3.22.5)}$$

where k_R is a proportionality constant; $Y_{x/R}$ is the biomass yield defined as grams of biomass produced per grams of starch used. Parameter k_R is the maintenance coefficient, defined as grams of starch used per gram of biomass in one hour to maintain vital activities other than growth. The term $k_R(dc_S/dt)$ is the reducing sugar production rate in proportion to starch degradation. The terms $(1/Y_{x/R})(dx/dt)$ and $k_R x$ are the reducing sugar consumption rates for growth and maintenance, respectively.

Glucose constitutes only a fraction of the reducing sugars. Glucose accumulation rates in the medium may be expressed similarly to the reducing sugar accumulation as:

$$\frac{dc_G}{dt} = k_G\left(-\frac{dc_S}{dt}\right) - \frac{1}{Y_{x/G}}\frac{dx}{dt} - k_G x \quad \text{(E.3.22.6)}$$

where the model constants have similar meanings to those of (E.3.22.5).

Comparison of the mathematical model with the experimental data is exemplified with a typical set of data in Figure E.3.22.2.

A value of $Y_{x/G} = 1.4$ g cell/g glucose indicates that reducing sugars, other than glucose has been utilized mostly to achieve the growth.

Example 3.23. Dynamic response to perturbations in a chemostat during ethanol production (Gupta and Chand, 1994) Dynamic response to perturbations of process parameters during conversion of sugars into ethanol was studied using a chemostat (Fig. E.3.23.1).

Substrate and biomass balances around the process vessel was stated in analogy with (3.12) as:

$$Fc_{S0} - Fc_S - R_S V = V\frac{d(c_S)}{dt} \quad \text{(E.3.23.1)}$$

$$Fx_0 - Fx + R_x V = V\frac{d(x)}{dt} \quad \text{(E.3.23.2)}$$

Both substrate and product inhibition involves in R_p and R_x:

$$R_p = xv_m\left(1 - \frac{c_p}{c^*_{pm}}\right)\left[\frac{c_S}{c_S + K^*_S + (c_s^2/K^*_S\omega^*)}\right] \quad \text{(E.3.23.3)}$$

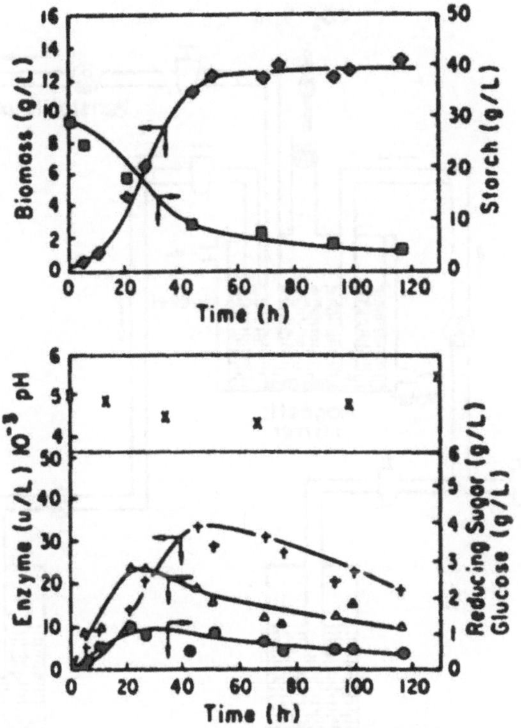

FIGURE E.3.22.2 Time course of *A. oryzae* growth with 3% initial starch concentration. Symbols: x pH, ■ starch, ◆ biomass, + total starch hydrolyzing enzyme activity, ▲ reducing sugars, ● glucose, – model (Bayindirli *et al.*, 1991; Harwood Academic Publishers GmbH, reproduced by permission). $\mu = 13.3 \times 10^{-2}\,h^{-1}$, $x_{max} = 12.5$ g/L, $k_M = 3.3 \times 103$ Eu/g cell, $k_N = 20$ Eu/g cell h, $k_s = 15.8 \times 10^{-2}$ g starch/L h, $k_u = 45.0 \times 10^{-5}$ g starch/Eu, $k_R = 1.09$ g reducing sugars/g starch, $Y_{x/R} = 0.55$ g cell/g reducing sugars, $k_R = 4.5 \times 10^{-3}$ g reducing sugars/g cell h, $k_G = 0.42$ g glucose/g starch, $Y_{x/G} = 1.4$ g cell/g glucose and $k_G = 1.8 \times 10^{-3}$ g glucose/g cell h.

$$R_x = x\mu_m\left(1 - \frac{c_P}{c_{pm}}\right)\left[\frac{c_S}{c_S + K_S + (c_s^2/K_S\omega)}\right] \qquad \text{(E.3.23.4)}$$

where v_m = maximum specific ethanol production rate, μ_m = maximum specific growth rate, c_{pm}^* = maximum ethanol concentration above which cells do not produce ethanol, c_{pm} = maximum ethanol concentration above which cells do not grow, K_S, K_S^*, ω and ω^* are constants. Ethanol production rate was related with the substrate utilization rate as:

$$\frac{dc_P}{dt} = -Y_{P/S}\frac{dc_S}{dt} \qquad \text{(E.3.23.4)}$$

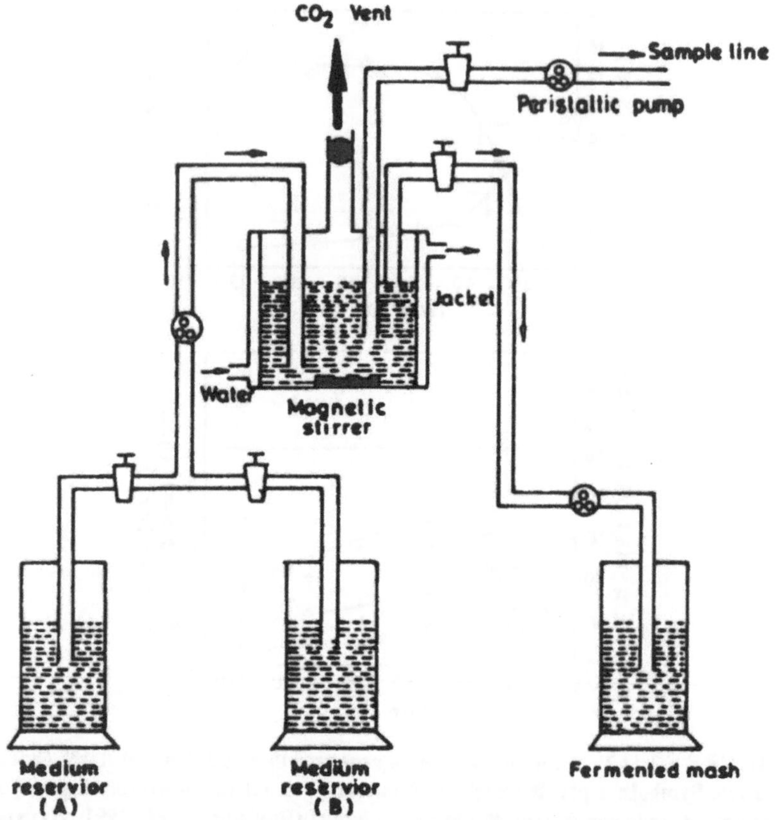

FIGURE E.3.23.1 Schematic diagram of the CSTR. Reprinted from *Process Biochemistry* 29(5), Gupta and Chand, Bioconversion of sugars to ethanol in a chemostat employing *S. cerevisiae*-Dynamic response to perturbations in process parameters, pages 343–354, © 1994 with kind permissions from Elsevier Science Ltd, The Boulevard, Langford Lane, Kidlington 0X5 1GB, UK.

After substituting $D = F/V$ the above equations may be combined to give:

$$\frac{dc_S}{dt} = Dc_{S0} - Dc_S - \frac{1}{Y_{P/S}} x v_m \left(1 - \frac{c_P}{c^*_{pm}}\right)\left[\frac{c_S}{c_S + K^*_S + (c_s^2/K^*_S \omega^*)}\right] \quad \text{(E.3.23.5)}$$

$$\frac{dx}{dt} = -Dx + x\mu_m \left(1 - \frac{c_P}{c_{pm}}\right)\left[\frac{c_S}{c_S + K_S + (c_s^2/K_S \omega)}\right] \quad \text{(E.3.23.6)}$$

$$\frac{dc_P}{dt} = -Dc_P + x v_m \left(1 - \frac{c_P}{c^*_{pm}}\right)\left[\frac{c_S}{c_S + K^*_S + (c_s^2/K^*_S \omega^*)}\right] \quad \text{(E.3.23.7)}$$

Equations (E.3.23.5)–(E.3.23.7) were solved with the fourth-order Runge-Kutta method subject to the following initial conditions:

$$c_s = c_{si},\ x = x_i,\ c_P = c_{pi} \quad \text{at} \quad t = 0 \tag{E.3.23.8}$$

where c_{si}, x_i and c_{pi} are the initial steady state values of sugar, cell and ethanol concentration, respectively. Transient values of specific growth rates μ and product formation rate v were computed from the following equations:

$$\mu = -D + \mu_m\left(1 - \frac{c_P}{c_{pm}}\right)\left[\frac{c_S}{c_S + K_S + (c_s^2/K_S\omega)}\right] \tag{E.3.23.9}$$

$$v = -\frac{D}{x}c_P + v_m\left(1 - \frac{c_P}{c_{pm}^*}\right)\left[\frac{c_S}{c_S + K_S^* + (c_s^2/K_S^*\omega^*)}\right] \tag{E.3.23.10}$$

Equations (E.3.23.5)–(E.3.23.10) were used to simulate dynamic behavior of the process under various conditions. Two examples are given in Figures E.3.23.2. and E.3.23.3. to show the agreement of the model with the data.

3.10. KINETICS OF MICROBIAL DEATH

Microbial death kinetics has a significant importance in food processing since it is among the major phenomena occurring during pasteurization and sterilization processes. Microbial death is generally described in analogy with uni-molecular, irreversible, first order rate expression:

$$x \rightarrow x_d$$

live microorganism dead microorganism

$$\frac{dx}{dt} = -k_d x \tag{3.67}$$

An alternative expression for (3.67) is:

$$\frac{d(\log\ x)}{dt} = -\frac{1}{D_T} \tag{3.68}$$

with

$$D_T = \frac{2.303}{k_d} \tag{3.69}$$

The D_T value is defined as the heating time at constant temperature T to reduce microbial population by one log cycle, or to 10% of its initial value.

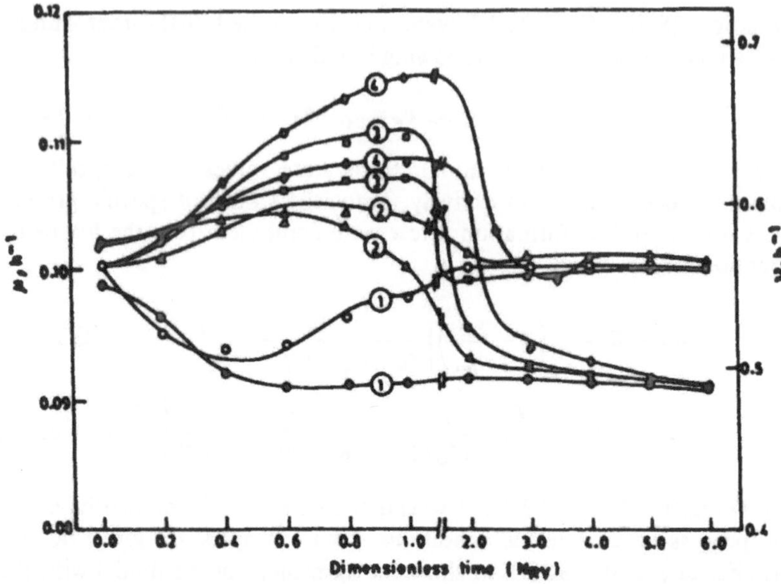

FIGURE E.3.23.2 Variation of the specific growth and product formation rates following shift down in feed sugar concentration at $D = 0.10\,\mathrm{h}^{11}$, (○,●) $c_{s0} = 120 \rightarrow 100$ g/L, (△,■) $c_{SO} = 140 \rightarrow 100$ g/L, (□,■) $c_{SO} = 160 \rightarrow 100$ g/L, (◇,◆) $c_{SO} = 180 \rightarrow 100$ g/L. Hollow and dark symbols indicate μ and v, respectively. Solutions of the model equations (E.3.25.9) and (E.3.25.10) are shown with solid lines. $c_{pm} = 91.0$ g/L, $c_{pm}^* = 117.5$ g/L, $K_S^* = 0.644$ g/L, $K_S = 0.455$ g/L, $\mu_m = 0.265\,\mathrm{h}^{-1}$, $v_m = 1.15\,\mathrm{h}^{-1}$, $\omega = 422.5$, $\omega^* = 456.0$, $Y_{P/S} = 0.463$ g substrate/g ethanol N_{Rv} = number of reactor volumes of fermentation broth passing through the reactor until reaching the indicated state of the operation. Reprinted from *Process Biochemistry* 29(5), Gupta and Chand, Bioconversion of sugars to ethanol in a chemostat employing *S. cerevisiae*-Dynamic response to perturbations in process parameters, pages 343–354, © 1994 with kind permissions from Elsevier Science Ltd, The Boulevard, Langford Lane, Kidlington 0X5 1GB, UK.

The z value is defined as the temperature difference required to change the D_T value by a factor of ten, or one log cycle:

$$\frac{d(\log D_T)}{k_d} = -\frac{1}{z} \tag{3.70}$$

Equation (3.70) may be rearranged and integrated as:

$$D_T = D_{T_{ref}}\, 10^{(T_{ref} - T)/z} \tag{3.71}$$

The z value is related with the activation energy of the Arrhenius expression:

$$z = \frac{2.303\, RTT_{ref}}{E_a} \tag{3.72}$$

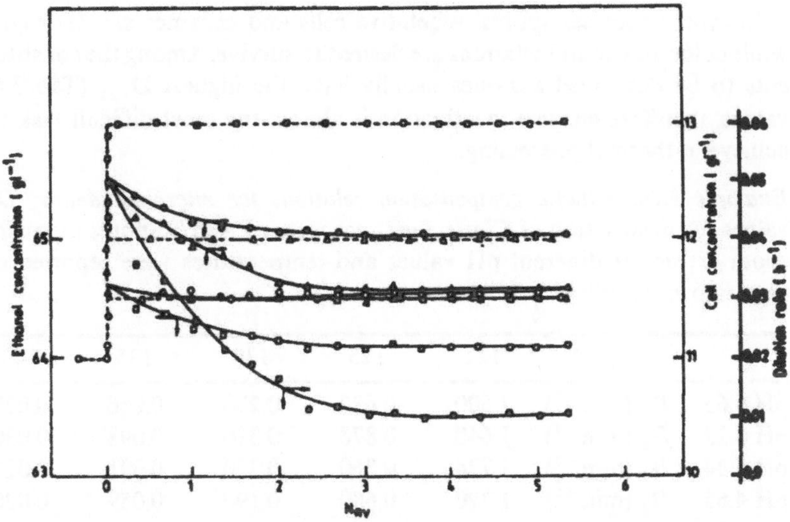

FIGURE E.3.23.3 Transient response to step-up in dilution rate while the system was operated at $c_{s0} = 140$ g/L: (○,●) $D = 0.02 \rightarrow 0.03\ h^{-1}$, (△,▲) $D = 0.02 \rightarrow 0.04\ h^{-1}$, (□,■) $D = 0.02 \rightarrow 0.06\ h^{-1}$. Broken lines represent corresponding dilution rate matching with legend. The filled and hollow symbols represent exit cell and ethanol concentrations, respectively. Reprinted from *Process Biochemistry* 29(5), Gupta and Chand, Bioconversion of sugars to ethanol in a chemostat employing *S. cerevisiae*-Dynamic response to perturbations in process parameters, pages 343–354, © 1994 with kind permissions from Elsevier Science Ltd, The Boulevard, Langford Lane, Kidlington 0X5 1GB, UK.

During thermal processing of foods, microbial death and inactivation of the enzymes are always accompanied with loss of nutrients due to thermal degradation, since they share the same medium. Although death of the microorganisms and destruction of the enzymes and toxins are desired, loss of the nutrients is not. The range of the kinetic constants describing the resistance of the food components to thermal processing are given in Table 3.6.

Table 3.6 Kinetic constants describing the resistance of the food constituents to thermal processing (Lund, 1977; Hallström *et al.*, 1988)

Constituent	z (°C)	E_a (Joules/mol)	D_{121} (min)
Vitamins	20–30	$8.4 \times 10^4 - 12.5 \times 10^4$	100–1000
Color, texture, flavor	25–45	$4.2 \times 10^4 - 12.5 \times 10^4$	5–500
Enzymes	12–56	$5.0 \times 10^4 - 41.8 \times 10^4$	1–10
Vegetative cells	4–7	$41.8 \times 10^4 - 50.2 \times 10^4$	0.002–0.02
Spores	7–12	$22.2 \times 10^4 - 34.7 \times 10^4$	0.1–5.0
Cooking value (over-all quality estimation)	15–45	$5.5 \times 10^4 - 17.0 \times 10^4$	1.2–12.5

In most processes spores, vegetative cells and enzymes are destroyed while color, flavor and vitamins are desired to survive. Among the constituents to be destroyed enzymes usually have the highest D_{121} (Tab. 3.6) values, therefore enzyme inactivation is almost the most difficult task to achieve in thermal processing.

Example 3.24. Kinetic compensation relations for microbial death D_T values for destruction of *Clostridium sporogenes* PA3679 spores in mushroom extract at different pH values and temperatures were reported as (Rodrigo *et al.*, 1993):

	T (°C)	121	125	130	135	140
pH 6.65	D_T (min^{-1})	1.500	0.630	0.270	0.086	0.029
pH 6.22	D_T (min^{-1})	1.640	0.870	0.310	0.098	0.030
pH 5.34	D_T (min^{-1})	1.774	0.860	0.320	0.091	0.027
pH 4.65	D_T (min^{-1})	1.720	0.680	0.190	0.059	0.020

Check the validity of the compensation relation with these data.

Solution Equation (3.69) may be used to convert the D values into the death rate constants:

$$D_T = \frac{2.303}{k_d} \tag{3.69}$$

	T (K)	394	398	403	408	413
pH 6.65	k_d (s^{-1})	0.026	0.061	0.142	0.446	1.324
pH 6.22	k_d (s^{-1})	0.023	0.044	0.124	0.392	1.279
pH 5.34	k_d (s^{-1})	0.022	0.045	0.120	0.422	1.422
pH 4.65	k_d (s^{-1})	0.022	0.056	0.202	0.651	1.919

This data set may be rewritten after calculating the inverse of the temperatures and taking the natural logarithm of the death rate constants:

	$1/T$ (K^{-1})	2.54×10^{-3}	2.51×10^{-3}	2.48×10^{-3}	2.45×10^{-3}	2.42×10^{-3}
pH 6.65	$\ln k_d$	−3.65	−2.80	−1.95	−0.81	0.28
pH 6.22	$\ln k_d$	−3.77	−3.12	−2.09	−0.94	0.25
pH 5.34	$\ln k_d$	−3.82	−3.10	−2.12	−0.86	0.35
pH 4.65	$\ln k_d$	−3.82	−2.88	−1.60	−0.43	0.65

Temperature effects on the death rate constants were given with the Arrhenius expression:

$$k = k_0 \exp\left\{-\frac{E_a}{RT}\right\} \qquad (3.5)$$

Equation (3.5) may be linearized:

$$\ln k_d = \ln k_{d0} - \frac{E_a}{RT}$$

parameters $\ln k_{d0}$ and E_a may be determined from the slope and the intercept of the above equation:

	$\ln k_{d0}$	E_a	r
pH 6.65	79.64	272974	−0.998
pH 6.22	82.55	283233	−0.995
pH 5.34	85.55	293210	−0.994
pH 4.65	92.54	315658	−0.999

The compensation relation may be determined with these data as:

$$\ln k_{d0} = -3.29 + 3.03 \times 10^{-4} \times E_a \qquad (E.3.24.1)$$

Equation (E.3.24.1) is compared with the experimental data in Figure E.3.24.

Example 3.25. Relation between the death rate and the Arrhenius expression constants and the D and z values. The surviving number of microorganisms

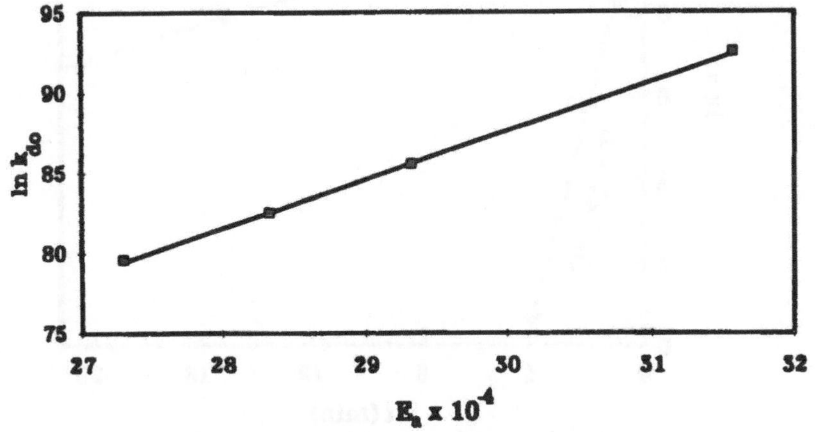

FIGURE E.3.24 Comparison of the experimental data with (E.3.24.1).

were counted after processing a certain food for given times at 110°C and 121°C:

t(min) at 110°C	0	5	10	15	20
x (cfu/ml)	60000	21000	7400	2650	930

t(min) at 121°C	0	1	2	3	4
x (cfu/ml)	60000	41000	290	19	2

a) Determine the death rate constant at 110°C and 121°C.

Solution Integrating (3.67) gives: $\ln x = \ln x_0 - k_d t$. Where x_0 is the initial cfu/ml of the microorganism. Substituting the data obtained at 110°C into the integrated equation gives: $\ln x = 10.99 - 0.208\,t$ with $r = -1.00$, implying that $k_d = 0.208\ \text{min}^{-1}$. With the data obtained at 121°C: $\ln x = 10.98 - 2.62\,t$ with $r = -1.00$ and $k_d = 2.62\ \text{min}^{-1}$. The best fitting lines are shown in Figure E.3.25.1.

b) Convert these death rate constants into D values.

Solution Equation (3.69) requires $D_T = 2.303/k_d$, after substituting values of k_d we will obtain $D_T = 11.1$ min at 110°C and $D_T = 0.88$ min at 121°C.

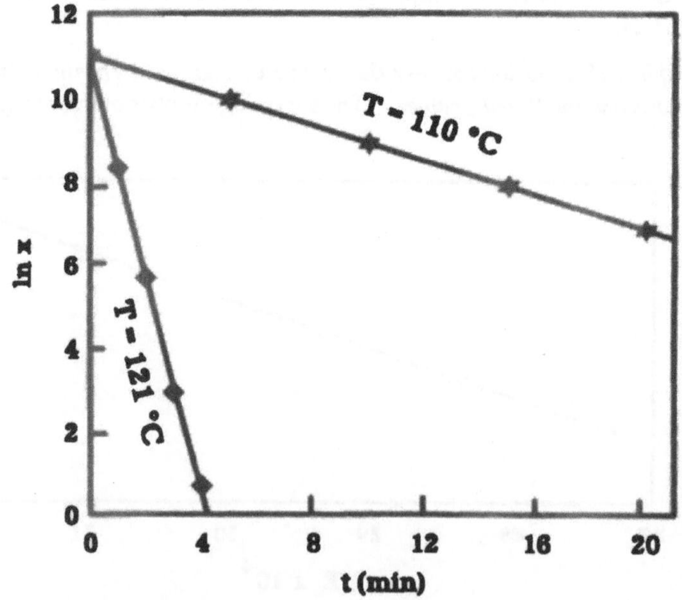

FIGURE E.3.25.1 Comparison of the integrated form of (3.67) with the data.

c) Death rate constant of the microorganism was calculated at different temperatures as:

T(°C)	110	113	115	118	121	124
k_d(min^{-1})	0.208	0.397	0.743	1.44	2.62	5.01

Determine the activation energy and frequency factor of the Arrhenius expression for the death rate constant.

Solution Equation (3.5) may be linearized as:

$$\ln k_d = \ln k_{d0} - \frac{E_a}{RT} \tag{E.3.25.1}$$

After converting temperatures in K and substituting the data we will obtain $\ln k_d = 91.30 - 35562/T$, ($r = -0.996$), implying that $k_{d0} = 4.48 \times 10^{39}$ min^{-1} and $E_a = 2.96 \times 10^5$ J/mol (Fig. E.3.25.2).

d) Calculate the z value from the activation energy at 110°C and 124°C ($T_{ref} = 121$°C). What is the average z value?

Solution Equation (3.72) requires $z = 2.303\, R\, TT_{ref}/E_a$, after substituting the numbers we will obtain, $z = 9.76$°C at 110°C, and $z = 10.12$°C at 124°C. The average z value is 9.94°C.

Example 3.26. Mechanisms for death of spores Bacterial spores may be in both activated or dormant states or consist of a combination of species with different thermal resistances. Four different models describing the thermal death kinetics of the spores are depicted in Table E.3.26 (Smerage and Teixeira, 1993):

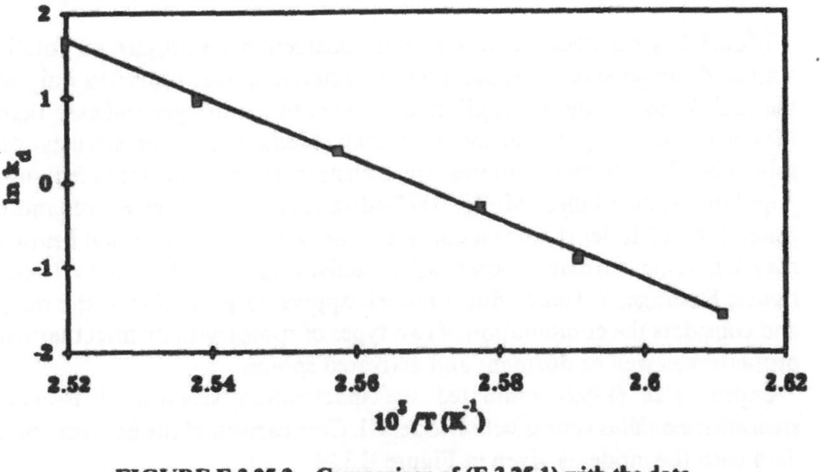

FIGURE E.3.25.2 Comparison of (E.3.25.1) with the data.

Table E.3.26. Mechanism and kinetic models for microbial inactivation of spores.

model	mechanism	kinetics
I	dormant spores (x) $\xrightarrow{k_d}$ dead spores	$\frac{dx}{dt} = -k_d x$
II	dormant spores (x) $\xrightarrow{k_a}$ activated spores (x_a)	$\frac{dx}{dt} = -k_a x$
	activated spores (x_a) $\xrightarrow{k_d}$ dead spores	$\frac{dx_a}{dt} = k_a x - k_d x_a$
III	dormant spores (x) $\xrightarrow{k_{d1}}$ dead spores	$\frac{dx}{dt} = -(k_a + k_{d1}) x$
	dormant spores (x) $\xrightarrow{k_a}$ activated spores (x_a)	$\frac{dx_a}{dt} = k_a x - k_{d2} x_a$
	activated spores (x) $\xrightarrow{k_{d2}}$ dead spores	
IV	dormant spores (x_i) $\xrightarrow{k_{id1}}$ dead spores	$\frac{dx_i}{dt} = -(k_{ia} + k_{id2}) x_i$
	dormant spores (x_i) $\xrightarrow{k_{ia}}$ activated spores (x_{ia})	$\frac{dx_{ia}}{dt} = k_{ia} x_i - k_{id2} x_{ia}$
	activated spores (x_{ia}) $\xrightarrow{k_{id2}}$ dead spores $i = 1$ represents the first population, $i = 2$ represents the second population, such that $x = \sum_{i=1}^{2} x_i$	

Model I is the classical model and considers a homogeneous totally activated, single species population of bacterial spores subjected only to thermal death. Model II (Shull model) refers to a more general case than Model I and is capable of more accurate prediction of the sterilization processes. The dormant spores are transformed into an activated subpopulations, then killed. Model III (Rodriguez-Sapru model) is even more general than Model II since it considers the possibility that lethal heating may kill some dormant spores before activating them. Model IV (Rodriguez, Smerage, Teixeira, Busta model) applies to general case the most and considers the combination of two types of spores with distinct thermal properties as well as dormant and activated spores.

Sapru *et al.* (1992) simulated the inactivation kinetics of *Bacillus stearothermophilus* spores with Model III. Comparison of the experimental data with the model is given in Figure E.3.26.

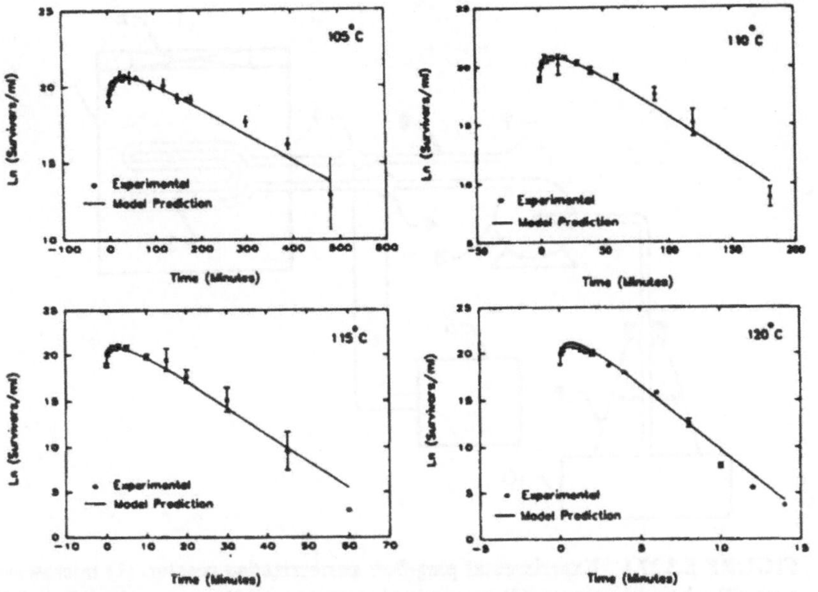

FIGURE E.3.26 Predicted and corresponding experimental survivor curves of *B. stearothermophilus* spores at 105, 110, 115 and 120°C (Sapru *et al.*, 1992; © IFT, reproduced by permission).

Example 3.27. A model for pasteurization with microwaves in a tubular flow reactor (Özilgen and Özilgen, 1991) Polar molecules, i.e. water, try to align with the microwaves in a microwave field. The microwaves of commercial ovens vibrate with frequency of 2450 Hz. This rapid movement of polar molecules heats the medium, which may be used to sterilize liquid fermentation media or for pasteurizing liquid foods. An experimental plug-flow pasteurization reactor shown in Figure E.3.27.1 is used with this purpose.

At a constant temperature, the thermal death rate of the microorganisms is:

$$\frac{dx}{dt} = -k_d x \tag{3.46}$$

Temperature effects on the death rate constant k_d may be expressed with the Arrhenius equation:

$$k_d = k_{d0} \exp\left\{-\frac{E_a}{RT}\right\} \tag{3.5}$$

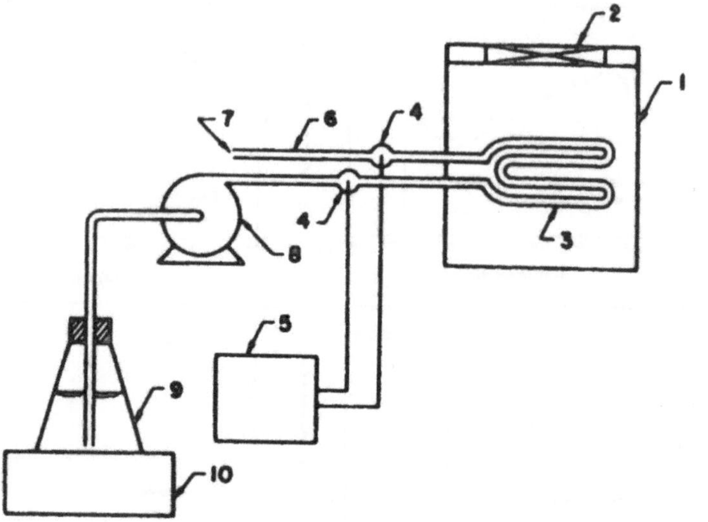

FIGURE E.3.27.1 Experimental plug-flow pasteurization reactor. (1) microwave oven, (2) microwave source, (3) pasteurization reactor, (4) thermocouples, (5) digital temperature read-out, (6) constant temperature section, (7) pasteurized medium outlet, (8) peristaltic pump, (9) inoculated medium reservoir, (10) magnetic stirrer. Reprinted by permission of the publisher from A model for Pasteurization with Microwaves in a Tubular Flow Reactor, Özilgen and Özilgen, *Enzyme and Microbial Technology*, 13(5), 419–423. © 1991 by Elsevier Science Ltd.

Heating of a stagnant fluid via microwave absorption may be expressed as:

$$\not p = \omega \varepsilon_0 E^2 \kappa \qquad \text{(E.3.27.1)}$$

where $\not p$ = absorbed microwave power, ω = angular frequency, ε_0 = dielectric constant of free space, E = electric field intensity coupled by matched load, κ = relative dielectric loss factor. The constant κ is the overall measure of the ability of the material to respond to the microwave field. Temperature effects on κ may be expressed as:

$$\kappa = \phi \exp(-\beta T) \qquad \text{(E.3.27.2)}$$

where β = constant. Heating of a stagnant liquid may be expressed after combining (E.3.27.1) and (E.3.27.2) as:

$$\not p = \alpha \exp(-\beta T) \qquad \text{(E.3.27.3)}$$

where $\alpha = \omega \varepsilon_0 E^2 \phi$ = constant. Thermal energy balance around an infinitely small shell element of the pasteurization reactor as depicted in Figure 3.3

requires:

$$\begin{bmatrix}\text{input rate}\\ \text{into the}\\ \text{shell}\\ \text{element}\end{bmatrix}-\begin{bmatrix}\text{output}\\ \text{rate from the}\\ \text{shell}\\ \text{element}\end{bmatrix}+\begin{bmatrix}\text{generation}\\ \text{rate in the}\\ \text{shell}\\ \text{element}\end{bmatrix}$$

$$=\begin{bmatrix}\text{accumulation}\\ \text{rate in the}\\ \text{shell}\\ \text{element}\end{bmatrix} \quad \text{(E.3.27.4)}$$

After assuming no radial temperature distribution we may convert (E.3.27.4) in mathematical terms as:

$$\frac{\pi D_r^2 \upsilon\rho c}{4}(T-T_{ref})|_z-\frac{\pi D_r^2 \upsilon\rho c}{4}(T-T_{ref})|_{z+\Delta z}-\pi D_r \Delta z\, U(T-T_{env})$$

$$+\frac{\pi D_r^2\,\Delta z\,\alpha e^{-\beta T}}{4}=\frac{\pi D_r^2\,\Delta z\,\rho c}{4}\frac{dT}{dt} \quad \text{(E.3.27.5)}$$

where D_r is the diameter of the reactor, υ, c and ρ are the velocity, specific heat and density of the liquid. The term $(\pi D_r^2\,\upsilon\rho c/4)(T-T_{ref})|_z$ is the enthalpy of the liquid entering into the shell element at distance z from the entrance of the heating section of the reactor. It should be noticed that the enthalpy is calculated with respect to an arbitrarily chosen reference temperature T_{ref}. The term $(\pi D_r^2\,\upsilon\rho c/2)(T-T_{ref})|_{z+\Delta z}$ is the enthalpy of the liquid leaving the shell element. The length of the shell is Δz as depicted in Figure E.3.3. The term $\pi D_r \Delta z_z U(T-T_{env})$ describes the enthalpy loss from the shell element to the environment with convection. Heating rate of the liquid in the shell element with microwaves is described by $\pi D_r^2 \Delta z\, \alpha e^{-\beta T}/4$, unsteady state thermal energy accumulation in the shell element is described by $(\pi D_r^2 \Delta z \rho c/4)(dT/dt)$.

Under steady state conditions the unsteady state thermal energy accumulation term becomes zero and the equation is rearranged as:

$$\frac{dT}{dz}=K_1\exp(-\beta T)-K_2(T-T_{env}) \quad \text{(E.3.27.6)}$$

where $K_1=\alpha/\upsilon\rho c$ and $K_2=U/D_r\upsilon\rho c$. Solutions of (E.3.27.6) simulates the variation of the temperature profiles along the reactor (Fig. E.3.27.2).

Although temperature varies drastically with distance, the flow reactor may be considered as a combination of 1 cm long small segments, and the

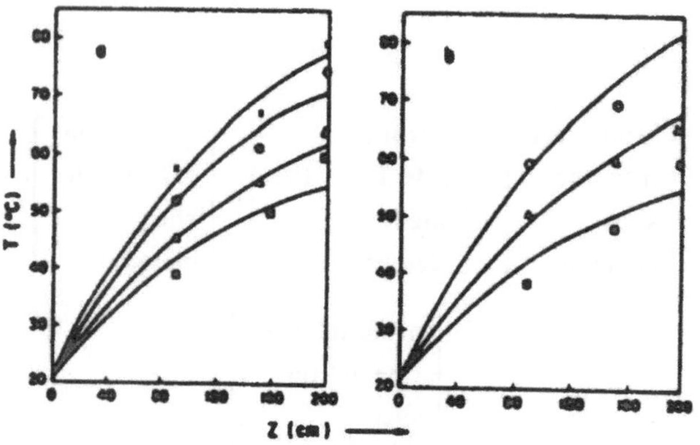

FIGURE E.3.27.2 Typical sample plots for variation of temperature in the reactor with distance from the entrance of the oven. Solid curves (–) are simulations. (a) Diameter of the reactor D_r was constant (7.5 mm) and the experimental data were obtained with the following average velocities: (x) $v = 228$ cm/min, (●) $v = 270$ cm/min, (▲) $v = 360$ cm/min, (■) $v = 402$ cm/min; (b) flow rate was constant (91.8 cm^3/min) and the experimental data were obtained with the following average velocities in the reactors with the given diameters: (○) $v = 186$ cm/min, $D_r = 11.8$ mm; (▲) $v = 228$ cm/min, $D_r = 9.0$ mm; (■) $v = 402$ cm/min, $D_r = 7.5$ mm. Numerical values of the model constants were $K_1 = 2.0 \pm 0.2$, $K_2 = 0.0069 \pm 0.0001$, $\beta = 0.00012$ (1/°C). Reprinted by permission of the publisher from A model for Pasteurization with Microwaves in a Tubular Flow Reactor, Özilgen and Özilgen, *Enzyme and Microbial Technology*, 13(5), 419–423. © 1991 by Elsevier Science Ltd.

temperature of any of these segments may be considered constant. The fraction of biomass surviving each segment may be calculated as:

$$\frac{xi}{x_0} = \exp\left(-\frac{k_{di}}{v}\right) \tag{E.3.27.7}$$

where x_0 and x_i are the viable cell counts at the entrance and exit of the ith segment, respectively. Parameter k_{di} is the death rate constant calculated with (E.3.27.7) at the average temperature of the ith segment T_i.

A simplified flow diagram of the computer program used to simulate viable fraction of the microorganisms at the exit of the system is given in Figure E.3.27.3.

In Figure E.3.27.3 parameters v and v_1 were the average velocity of the medium in the microwave oven and in the constant temperature section, respectively. Parameter T_0 was the temperature of the medium before entering the microwave reactor. The numerical value of z_{i-1} was zero for $i = 1$ indicating the beginning of the tubular reactor in the microwave oven. The program computes the viable biomass concentration surviving after passing each 1-cm section of the microwave oven between the START and the first IF statement. The ratio $1/v$ was the residence time of the medium in

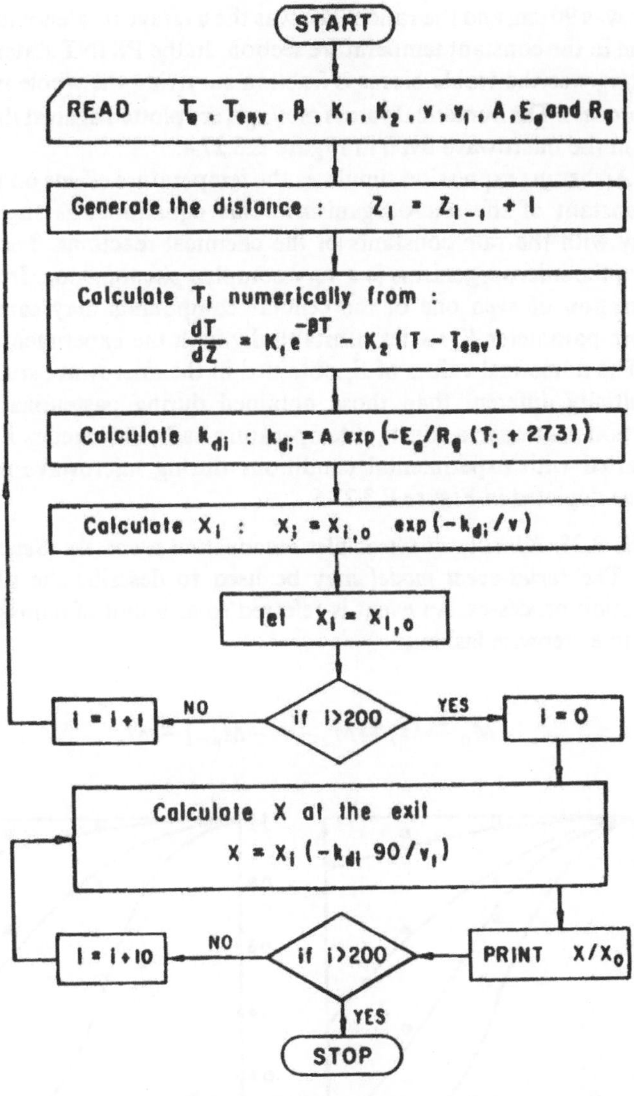

FIGURE E.3.27.3 A simplified flow diagram of the computer program used to simulate viable fraction of the microorganisms at the exit of the system ($90/v_i$ = time spent in the exit section). Reprinted by permission of the publisher from A model for Pasteurization with Microwaves in a Tubular Flow Reactor, Özilgen and Özilgen, *Enzyme and Microbial Technology*, 13(5), 419–423. © 1991 by Elsevier Science Ltd.

each 1-cm section of the reactor. The program computes the viable biomass concentration surviving the constant temperature section between the first and the second IF statements. The length of the constant temperature section was 90 cm, and the ratio $90/v_1$ was the average residence time of the medium in the constant temperature section. In the PRINT statement, the ratio x/x_0 was the viable biomass fraction surviving the whole pasteurization process. The numerical values of x/x_0 were plotted against the reactor length in the microwave oven in Figure E.3.27.4.

The Arrhenius expression simulates the temperature effects on the death rate constant of the microorganisms. This equation was suggested in analogy with the rate constants of the chemical reactions. The thermal death of the microorganisms is a very complex phenomenon. Irreversible denaturation of even one of the cellular compounds may cause death, therefore parameter E_a varies substantially with the experimental conditions. The numerical values of E_a obtained in the microwave studies were substantially different than those obtained during pasteurization with convection heating in a constant temperature bath. Parameters E_a and k_{d0} also varied with experimental conditions during microwave pasteurizations as depicted in Figure E.3.27.5.

Example 3.28. Kinetics of ultraviolet inactivation processes (Severin et al., 1983) *The series-event model* may be used to describe the ultraviolet inactivation processes. An event is referred to as a unit of damage. Event occurs in a stepwise fashion:

$$M_0 \xrightarrow{kI} M_1 \xrightarrow{kI} M_2 \xrightarrow{kI} \cdots M_{n-1} \xrightarrow{kI} M_n \qquad \text{(E.3.28.1)}$$

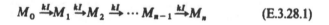

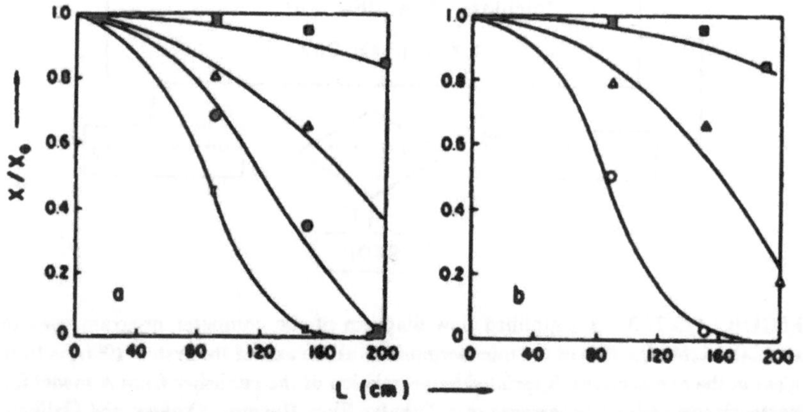

FIGURE E.3.27.4 Comparison of the model (–) with the experimentally determined surviving fractions of the microorganisms (●, ▲, ■, ○, x, △) at the exit of the experimental set-up. Symbols refer to the same experimental conditions as Figure E.3.27.2. Reprinted by permission of the publisher from A model for Pasteurization with Microwaves in a Tubular Flow Reactor, Özilgen and Özilgen, *Enzyme and Microbial Technology*, 13(5), 419–423. © 1991 by Elsevier Science Ltd.

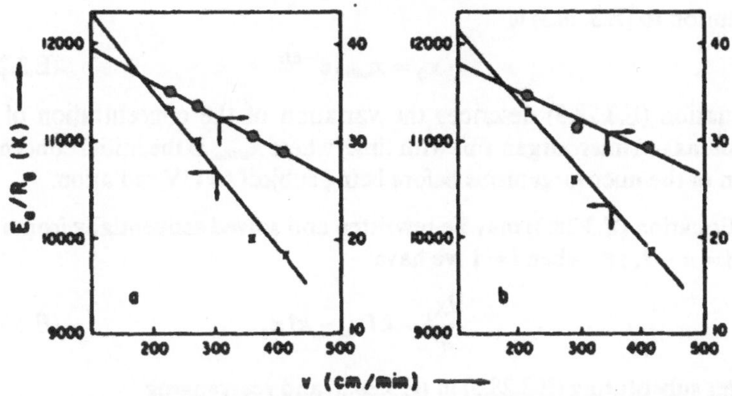

FIGURE E.3.27.5 Variation of the activation energy E_a/R and the pre-exponential constant k_{a0} of the Arrhenius expression with the average velocity of the tubular flow reactor. (a) Diameter of the tubular flow reactor was constant $D_r = 7.5$ mm, (b) flow rate was constant 91.8 cm^3/min. Reprinted by permission of the publisher from A model for Pasteurization with Microwaves in a Tubular Flow Reactor, Özilgen and Özilgen, *Enzyme and Microbial Technology*, 13(5), 419–423. © 1991 by Elsevier Science Ltd.

where k = constant, I = total point ultraviolet intensity and M_i = an organism which has reached event level i. The rate at which an organism passes from one event level to the next is first-order with respect to the ultraviolet intensity and independent of the event level occupied by the organism. As long as an organism is exposed to the ultraviolet it collects damage. An event threshold exists such that an organism which collects damage greater than the tolerable event threshold is deactivated. The threshold may vary depending on the species, strain or culturing conditions, however for a given set of conditions the threshold level is constant.

For a well mixed, flat, thin layered, closed batch reactor the rate at which the organisms pass through event level i is:

$$R_{xi} = kI\,x_{i-1} - kI\,x_i \tag{E.3.28.2}$$

Equation (E.3.28.2) may be incorporated in (3.7) written for species at damage level i in a batch reactor as:

$$\frac{dx_i}{dt} = kI\,x_{i-1} - kI\,x_i \tag{E.3.28.3}$$

With $i = 0$ (x_0 = concentration of the undamaged microorganisms) (E.3.28.3) is rewritten as:

$$\frac{dx_0}{dt} = -kI\,x_0 \tag{E.3.28.4}$$

Solution to (E.3.28.3) is

$$x_0 = x_{\text{initial}}\, e^{-kIt} \qquad \text{(E.3.28.5)}$$

Equation (E.3.28.5) describes the variation of the concentration of the undamaged microorganisms with time, where x_{initial} is the initial concentration of the microorganisms before being subject to UV radiation.

Equation (E.3.28.3) may be rewritten and solved sequentially from $i = 0$ to $i = n - 1$, i.e., when $i = 1$ we have:

$$\frac{dx_1}{dt} = kIx_0 - kIx_1 \qquad \text{(E.3.28.6)}$$

After substituting (E.3.28.5) in (E.3.28.6) and rearranging

$$\frac{dx_1}{dt} + kIx_1 = kI\,x_{\text{initial}}\, e^{-kIt} \qquad \text{(E.3.28.7)}$$

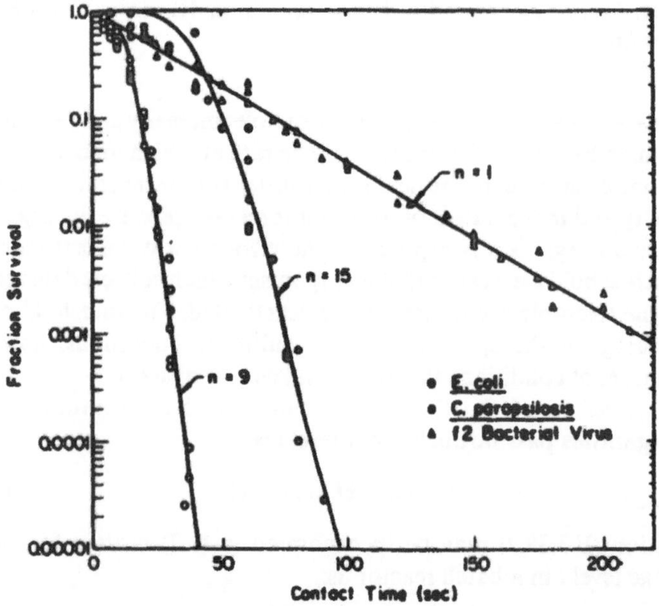

FIGURE E.3.28 Comparison of (E.3.28.10) with the experimental data obtained with a flat-tray reactor. Kinetic parameters for individual species were: *Escherichia coli* $n = 9$, $k = 1.538 \times 10^{-3}\, \text{cm}^2/\text{mWs}$; *Candida parasilosis* $n = 15$, $k = 0.891 \times 10^{-3}\, \text{cm}^2/\text{mWs}$; f2 bacterial virus $n = 1$, $k = 0.0724 \times 10^{-3}\, \text{cm}^2/\text{mWs}$. Reprinted from *Water Research*, volume 17, Severin *et al.*, Kinetic modeling of disinfection of water, pages 1669–1678, © 1983 with permission from Elsevier Science Ltd, Pergamon Imprint, The Boulevard, Langford Lane, Kidlington 0X5 1GB, UK.

The solution to (E.3.28.7) is

$$x_1 = x_{\text{initial}} e^{-kIt} kIt \tag{E.3.28.8}$$

The general expression for time varying density of organisms at any level i is:

$$x_i = \frac{x_{\text{initial}}(kIt)^i}{i!} e^{-kIt} \tag{E.3.28.9}$$

The total fraction of the surviving organisms between any level 0 and $n-1$ is:

$$\frac{x}{x_{\text{initial}}} = \sum_{i=0}^{n-1} \frac{x_i}{x_{\text{initial}}} = \exp(-kIt) \sum_{i=0}^{n-1} \frac{(kIt)^i}{i!} \tag{E.3.28.10}$$

where n = threshold number of the damaged sites required for inactivation. Comparison of (E.3.28.10) with the experimental data is depicted in Figure E.3.28.

REFERENCES

Aguilera, J. M., Oppermann, K. and Sanchez, F. (1987) Kinetics of browning of sultana grapes. *Journal of Food Science*, **52**, 990–993.

Akdoğan, H. and Özilgen, M. (1992) Kinetics of microbial growth, gas production and dough volume increase during leavening. *Enzyme and Microbial Technology*, **14**, 141–143.

Bailey, E. J. and Ollis, D. F. (1986) *Biochemical Engineering Fundamentals*, 2nd ed. McGraw-Hill Book Co. USA.

Barth, M. M., Kerbel, E. L., Perry, A. K. and Schmidt, S. J. (1993) Modified atmosphere packaging affects ascorbic acid, Enzyme activity and market quality of broccoli. *Journal of Food Science*, **58**, 140–143.

Bayindirli, A., Özilgen, M. and Ungan, S. (1991) Kinetic analysis of *Aspergillus oryzae* cultivations on starch. *Biocatalysis*, **5**, 71–78.

Chen, H. C. and Zall, R. R. (1983) Continuous lactose hydrolysis in fixed-bed reactors containing catalytic resins. *Journal of Food Science*, **48**, 1741–1744, 1757.

Doke, S. N. and Ninjoor, V. (1987) Characteristics of an alkaline proteinase and exopeptidase from shrimp (*Penaeus indicus*) muscle. *Journal of Food Science*, **52**, 1203–1208, 1211.

Edelstein-Keshet, L. (1988) *Mathematical Models in Biology*, Random House, New York.

Fennema, O. R., Powrie, W. D. and Marth, E. H. (1973) *Low Temperature Preservation of Foods and Living Matter*. Marcel Dekker, Inc. New York.

Frels, J. and Rupnow, J. H. (1985) Characterization of two α-amylase inhibitors from black bean (*Phaseolus vulgaris*) *Journal of Food Science*, **50**, 72–77, 105.

Goldstein, L., Levin, Y. and Katchalski, E. A. (1964) A water-insoluble polyionic derivative of trypsin. II. Effect of polyelectrolyte carrier on the kinetic behavior of the bound trypsin. *Biochemistry*, **3**, 1913–1919.

Gupta, S. K. and Chand, S. (1994) Bioconversion of sugars to ethanol in a chemostat employing *S. cerevisiae* - Dynamic response to perturbations in process parameters. *Process Biochemistry*, **29**, 343–354.

Hallström, B., Skjöldebrand, C. and Trägradh, C. (1988) *Heat Transfer & Food Products*, Elsevier Appl. Sci. London.

Harada, T., Tirtohusodo, H. and Paulus, K. (1985) Influence of temperature and time on cooking kinetics of potatoes. *Journal of Food Science*, **50**, 459–462, 472.

Hill, C. H. and Grieger-Block, R. A. (1980) Kinetic data: Generation, interpretation, and use. *Food Technology*, **34**(2), 56–66.

Hill, W. M. and Sebring, M. (1990) Desugarization of egg products, in *Egg Science and Technology*, Stadelman, W. J. and Cotterill, O. J. (editors) Food Products Press, New York.

Karel, M. (1975) Radiation Preservation of Foods, in *Physical Principles of Food Preservation*, (Karel, M., Fennema, O. R. and Lund, D.) Marcel Dekker, Inc. New York.

Kuwajima, K. A. (1977) A folding model of α-lactalbumin deduced from the three-state denaturation mechanism. *Journal of Molecular Biology*, **114**, 241–258.

Labuza, T. P. (1980) Enthalpy/entropy compensation in food reactions. *Food Technology*, **34**(2), 67–77.

Leoni, O., Iori, R. and Palmeri, S. (1985) Purification and properties of lipoxygenase in germinating sunflower seeds. *Journal of Food Science*, **50**, 88–92.

Lienhard, G. E. (1973) Enzymatic catalysis and transition state theory. *Science*, **180**, 149–154.

Luedeking, R. and Piret, E. L. (1959) A kinetic study of lactic acid fermentation. Batch process at controlled pH. *Journal of Biochemical and Microbiological Technology and Engineering* **1**, 393–412.

Lund, D. (1977) Maximizing nutrient retention. *Food Technology*, **31**(2), 71–78.

Mulchandani, A. and Luong, J. H. T. (1989) Microbial growth kinetics revisited. *Enzyme and Microbial Technology*, **11**, 66–72.

Özen, S. and Özilgen, M. (1992) Effects of substrate concentration on growth and lactic acid production by mixed cultures of *Lactobacillus bulgaricus and Streptococcus thermophilus*. *Journal of Chemical Technology and Biotechnology*, **54**, 57–61.

Özilgen, M. (1988) Kinetics of amino acid production by over-producer mutant microorganisms. *Enzyme and Microbial Technology*, **10**, 110–114.

Özilgen, M, Ollis, D. F. and Ogrydziak, D. (1988) Kinetics of batch fermentations with *Kluyveromyces fragilis*. *Enzyme and Microbial Technology*, **10**, 165–172 and 640.

Özilgen, S. and Özilgen, M. (1991) A model for pasteurization with microwaves in a tubular flow reactor. *Enzyme and Microbial Technology*, **13**, 419–423.

Özilgen, M., Çelik, M. and Bozoğlu, T. F. (1991) Kinetics of spontaneous wine production. *Enzyme and Microbial Technology*, **13**, 252–256.

Pagliarini, E., Vernille, M. and Peri, C. (1990) Kinetic study on color changes in milk due to heat. *Journal of Food Science*, **55**, 1766–1767.

Regan, D. L., Lilly, M. D. and Dunnill, P. (1974) Influence of intraparticle diffusional limitation on the observed kinetics of immobilized enzymes and on catalyst design. *Biotechnology and Bioengineering*, **16**, 1081–1093.

Rodrigo, M., Martinez, A., Sanchez, T., Peris, M. J. and Safon, J. (1993) Kinetics of *Clostridium sporogenes* PA3679 spore destruction using computer-controlled thermoresistometer. *Journal of Food Science*, **58**, 649–652.

Sapru, V., Teixeira, A. A., Smerage, G. H. and Lindsay, J. A. (1992) Predicting thermophilic spore population dynamics for UHT sterilization process. *Journal of Food Science*, **57**, 1248–1252, 1257.

Sarikaya, A. and Özilgen, M. (1991) Kinetics of peroxidase inactivation during thermal processing of whole potatoes. *Lebensmittel-Wissenschaft und Technologie* **24**, 159–163.

Severin, B. F., Suidan, M. T. and Engelbrecht, R. S. (1983) Kinetic modeling of disinfection of water. *Water Research*, **17**, 1669–1678.

Shuler, M. L. and Kargi, F. (1992) *Bioprocess Engineering*. Prentice Hall, Englewood Cliffs, New Jersey.

Shulz, G. E. and Schirmer, R. H. (1979) *Principles of Protein Structure*. Springer-Verlag. New York, USA.

Smerage, G. H. and Teixeira, A. A. (1993) Dynamics of heat destruction of spores: a new view. *Journal of Industrial Microbiology*, **12**, 211–220.

Tseng, M. M-C. and Phillips, C. R. (1981) Mixed cultures: commensalism and competition with *Proteus vulgaris* and *Saccharomyces cerevisiae*. *Biotechnology and Bioengineering*, **23**, 1639–1651.

Ülgen, N and Özilgen, M. (1991) Kinetic compensation relations for ascorbic acid degradation and pectinesterase inactivation during orange juice pasteurizations. *Journal of the Science of Food and Agriculture*, **57**, 93–100.

Wang, D. I. C., Cooney, C. L., Demain, A. L., Dunnill, P. Humphrey, A. E. and Lilly, M. D. (1979) *Fermentation and Enzyme Technology*, John Wiley and Sons, New York.

Whitaker, J. R. (1972) *Principles of Enzymology for the Food Sciences*. Marcel Dekker Inc. New York.

Whiting, R. C. and Buchanan, R. L. (1994) Microbial modeling. *Food Technology*, **48**(6), 113–120.

Woodward, J. and Arnold, S. L. (1981) The inhibition of β-glucosidase activity in *Trichoderma reesei* C30 cellulase by derivatives and isomers of glucose. *Biotechnology and Bioengineering*, **23**, 1553–1562.

CHAPTER 4

Mathematical Modeling in Food Engineering Operations

4.1. THERMAL PROCESS MODELING

Thermal processing may be applied at different levels (Tab. 4.1). It may aim to kill the microorganisms, destroy the toxins or inactivate the enzymes. Softening of the tissue is desired with some foods, i.e., beans, meat, etc., but undesirable with others, i.e., fruit preserves. Vitamins, flavor or aromatic constituents may be lost with undesirable side reactions. Microorganisms are considered dead when they do not proliferate. With sub-lethal heat treatment they may be injured and became more harmful, because they may escape detection on the selective media, while still being capable of repairing themselves and producing toxins (Hurst, 1977).

Classification of some foods for their acidity is given in Table 4.2. Microorganisms of the *acid foods* are heat labile, where *Clostridium botulinum* cannot grow and spores cannot germinate. Processing at 100°C is sufficient for commercial sterilization and the time required to achieve complete sterilization is usually less than the time required for softening the tissue. Microorganisms of the *low acid foods* may form heat stable spores, therefore processing requires temperatures above 100°C, and the time required to cook the tissue is usually less than the time required to achieve complete sterilization.

Table 4.1 Levels of thermal processing

Total sterilization	microbial growth is not possible on any media
Commercial sterilization	microorganisms do not proliferate in the can, but may grow on a better medium
Pasteurization	mild heat treatment. Only pathogenic microorganisms are destroyed. Inactivation of alkaline phosphatase indicates efficiency of pasteurization, i.e., destruction of tuberculosis causing microorganisms in milk
Blanching	applied mainly to fruits and vegetables prior to freezing to inactivate the enzymes and remove gas and wilt the leafy vegetables. Peroxidase, lipase, phenolase and catalase are among the indicator enzymes of blanching

Table 4.2 Classification of the foods according to their acidity (Jay, 1978; Jackson and Shinn, 1979)

Low acid foods (pH 5.3 and higher)	Medium acid food (pH 5.3 to 4.5)	Acid foods (pH 4.5 to 3.7)	High acid foods (pH 3.7 and lower)
Meats (beef, chicken, duck, ham, veal)	macaroni in tomato sauce	Fruits (peach, pear, plum)	Fruit juices (cranberry, grapefruit, lemon, lime, orange)
Seafood (clams, crabs, fish (most species), shrimp	spaghetti and meat	Fruit and vegetable juices (orange, papaya, prune, tomato, vegetable juice cocktail	Pickles and sauerkraut
Dairy products (butter, cheddar cheese, cream, milk)	ravioli		Fruits (apple, pineapple plum)
Vegetables (asparagus, broccoli, cauliflower, corn, lima beans, soy beans, peas,	Fruit and Vegetables (banana, eggplant fig, green beans, okra, pumpkin)		
Olives (ripe)	mashed potatoes		
Soups (bean, chicken, noodle, mushroom, pea)	Soups (tomato, vegetable)		
mushrooms			

Equations (3.70) and (3.71) may combined and rearranged to obtain:

$$F_{T_{ref}}^{z} \equiv D_{T_{ref}} \log(x_0/x) = \int_0^t 10^{(T - T_{ref})/z} dt \tag{4.1}$$

where x_0 and x are the initial load and the final concentration of the microorganisms, respectively; and t is the duration of the process. Thermal process time required to reduce the initial microbial load x_0 to a final safe microbial concentration x at a constant temperature T_{ref} is:

$$F_{required} = D_{T_{ref}} \log(x_0/x) \tag{4.2}$$

where $\log(x_0/x)$ and $D_{T_{ref}}$ indicates the number of the log cycles of microbial reduction required for a safe product, and the heating time at T_{ref} for one log cycle of reduction, respectively. The can center is considered as the critical point, i.e., the latest sterilized point of conduction heating foods. Right hand side of (4.1) is the actual level of sterilization received by the food:

$$F_{process} = \int_0^t 10^{(T - T_{ref})/z} dt \tag{4.3}$$

$F_{process}$ received at the critical point of the can should be equal or larger than $F_{required}$ to achieve a safe process:

$$F_{process} \geqslant F_{required} \tag{4.4}$$

The C value (cooking value) is equivalent to the F value, but based on the kinetic parameters of the nutrients or other chemicals, instead of the microorganisms.

Example 4.1. Calculation of process time and nutrient loss at a constant temperature A convection heating food has a contaminant microorganism with $D = 75$ s at $T_{ref} = 100$°C and $z = 10$°C.

a) In a constant temperature process how long should this product be treated to get 10^7 cfu/ml of reduction at 100 °C and 121°C?

Solution We may calculate D value at 121°C as $D_T = D_{T_{ref}} 10^{(T_{ref} - T)/z}$, after substituting the numbers $D_{121} = 0.60$ s. It was given in the problem statement that $x_0/x = 10^7$, $D_{T_{ref}} = 75$ s and $T_{ref} = 100$°C. After substituting the numbers in (4.2) we calculate $F_{required} = 525$ s at 100°C and $F_{required} = 4.2$ s at 121°C. Since these are constant temperature processes $t = 525$ s at 100°C, and $t = 4.2$ s at 121°C.

b) The food has a major nutrient with $D = 90$s at $T_{ref} = 121$°C and $z = 25$ °C. What percent of the nutrient will be lost at 100°C and 121°C?

Solution We may calculate D value at 100°C as $D_T = D_{T_{ref}} 10^{(T - T_{ref})/z}$, after substituting the numbers $D_{100} = 623$ s. The C value of the process is $C_{T_{ref}}^{z} \equiv D_{T_{ref}} \log(c_0/c) = \int_0^t 10^{(T - T_{ref})} dt$ after integrating and substituting the numbers we calculate $c/c_0 = 89.8$ % at 121°C and $c/c_0 = 14.4$ %at 100°C

Example 4.2. Implications of the cooking values Consider the following processes for sterilizing a liquid food under following uniform temperatures: i) Rotary pressure cooker: 35 min, 113°C, ii) Intermediate aseptic process: 2 min at 127°C, iii) UHT (ultra high temperature) processing: 4 sec at 149°C

a) Calculate $F_{process}$ at $T_{ref} = 121°C$ for each of these processes ($z = 18°C$).

Solution $F_{process} = \int_0^t 10^{(T-T_{ref})/z}\,dt$, after substituting the numbers we calculate $F_{process} = 12.6$ min with rotary pressure cooker, $F_{process} = 4.3$ min with intermediate aseptic process and $F_{process} = 2.4$ min with UHT.

b) Calculate $C_{process}$ for a nutrient with $D = 15$ min at $T_{ref} = 121°C$ ($z = 15.5°C$) and determine the percentage of the nutrient surviving each processes.

Solution $C^z_{T_{ref}} = \int_0^t 10^{(T-T_{ref})/z}\,dt$ after substituting the numbers $C_{process} =$ 10.7 min with rotary pressure cooker, $C_{process} = 4.9$ min with intermediate aseptic process and $C_{process} = 4.3$ min with UHT. Since $C^z_{T_{ref}} = D_{T_{ref}} \log(c_0/c)$ we may calculate the fraction of the nutrient loss after substituting the numbers: $c/c_0 = 19\%$ with rotary pressure cooker, $c/c_0 = 47.1\%$ with intermediate aseptic process and $c/c_0 = 52\%$ with UHT. With the given criteria UHT appears to be the best process since it gives the maximum nutrient retention.

Example 4.3. Process time calculation with variable time temperature profile A convection heating food was processed in still retort at 120°C. The critical point temperature was measured at 1.5 minutes intervals. In the heating cycle the average critical point temperature at each interval was:

i	1	2	3	4	5	6	7	8	9	10	11
T(°C)	46	56	74	92	102	109	113	115	116	117	118

The critical point temperature remained almost constant at 118°C during the holding period, then the steam was turned off and the cooling period started. In the cooling cycle the critical point temperature was.

j	1	2	3	4	5	6	7
T(°C)	111	66	45	35	30	28	27

The required processing time at 121°C is 6 min. The z value is 10.4°C. Determine the time to turn the steam off for a safe process.

Solution We may assume that the critical point temperature profiles remain the same in the heating and cooling periods under similar processing conditions. We may adjust the holding time to achieve the safe process. Equation (4.13) may be written to cover all the cycles as:

$$F_{process} = \sum_{i=1}^{11} 10^{(T_i - T_{ref})/z}\,\Delta t_i + \sum_{n=1}^{N} 10^{(T_n - T_{ref})/z}\,\Delta t_n + \sum_{j=1}^{7} 10^{(T_j - T_{ref})/z}\,\Delta t_j$$

where i, n and j are the indices of the time increment in the heating, holding and the cooling periods, respectively. After substituting the numbers:

$$6 = 1.5\left\{\sum_{i=1}^{11} 10^{(T_i - 121)/10.4} + \sum_{n=1}^{N} 10^{(118-121)/10.4} + \sum_{j=1}^{7} 10^{(T_j - 121)/10.4}\right\}$$

After substituting the values of T_i, T_n and T_j this equation may be solved to obtain $N = 4$. The time to turn the steam off is $1.5 \times (11 + 4) = 22.5$ min, the total process time is $1.5 \times (11 + 4 + 7) = 33$ min.

In the heating cycle of a retorting process the difference between the retort (T_R) and the critical point (T) temperatures approaches zero logarithmically:

$$\frac{d(\log(T_R - T))}{dt} = -\frac{1}{f_h} \tag{4.5}$$

where f_h is the time required for $(T_R - T)$ to change one log cycle. A similar observation is generally made in the cooling cycle with $(T - T_{CW})$:

$$\frac{d(\log(T - T_{CW}))}{dt} = -\frac{1}{f_c} \tag{4.6}$$

where T_{CW} is the cooling water temperature and f_c is time required for $T - T_{CW}$ to change one log cycle. Equations (4.5) and (4.6) may be solved to relate the critical point temperature to time elapsed since the beginning of each cycle:

$$t = f_h \log\left[j_h \frac{T_R - T_0}{T_R - T}\right] \tag{4.7}$$

$$t = f_c \log\left[j_c \frac{T_h - T_{CO}}{T - T_{CW}}\right] \tag{4.8}$$

where $T = T_0$ at $t = 0$ and $T = T_{CO}$ at $t_c = 0$; $T = T_h$ at the critical point at $t = t_h =$ end of the heating period; j_h and j_c are correction factors to compensate for deviation of integrated forms of (4.5) and (4.6) from linear behavior at the beginning of each cycle. The time to turn the steam off, i.e., end of the heating period is:

$$t_{\text{heating}} = f_h(\log j_h(T_R - T_0) - \log(T_R - T_g)) \tag{4.9}$$

where $T = T_g$ at the end of the heating period. The retort temperature may reach its final value T_R with some time lag called the retort coming-up time (t_{CUT}). A correction term may be added to (4.9) to compensate the effect of the coming-up time (Ball, 1923):

$$t_{\text{heating}} = f_h(\log j_h(T_R - T_0) - \log(T_R - T_g)) + 0.58\, t_{CUT} \tag{4.10}$$

After using (4.5)–(4.8) Ball (1923) presented Tables of $f_h/\mathscr{U}$ as a function of $T_R - T_g$ with parameters $T_R - T_{CW}$ and z to achieve $F_{\text{process}} = F_{\text{required}}$.

National Canners Association (1968) presented Ball's Tables in graphical form (Figs. 4.1–4.4). When parameter f_h is a constant through the entire heating period, parameter $\mathcal{U}$ is defined as:

$$\mathcal{U} = F_{\text{required}}\, 10^{(T_{ref} - T_R)/z} \tag{4.11}$$

In some processes semi-solid foods, i.e., Jell-O, thickened puddings, high fat meats, etc., may be converted into liquid upon heating; or liquid foods, i.e., peaches in syrup, corn in brine, etc., may be converted into semi-solids. Parameter $f_h/\mathcal{U}$ is defined for such processes as:

$$\frac{f_h}{\mathcal{U}} = \frac{f_{h2}}{(F_{\text{required}}) + (10^{(T_{ref} - T_R)/z}) + r((f_{h2} - f_{h1})/(f_{h1}/\mathcal{U}_1))} \tag{4.12}$$

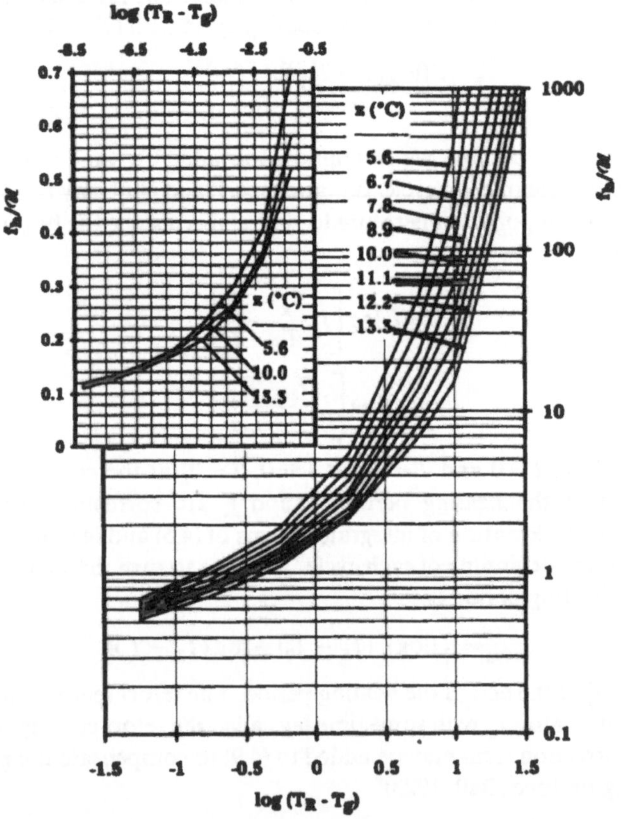

FIGURE 4.1 Graphical relation between $f_h/\mathcal{U}$ and $\log T_g$ when $T_R - T_{CW} = 100°C$ (National Canners Association, 1968; redrawn and published with kind permission of the copyright owner Chapman & Hall Inc.).

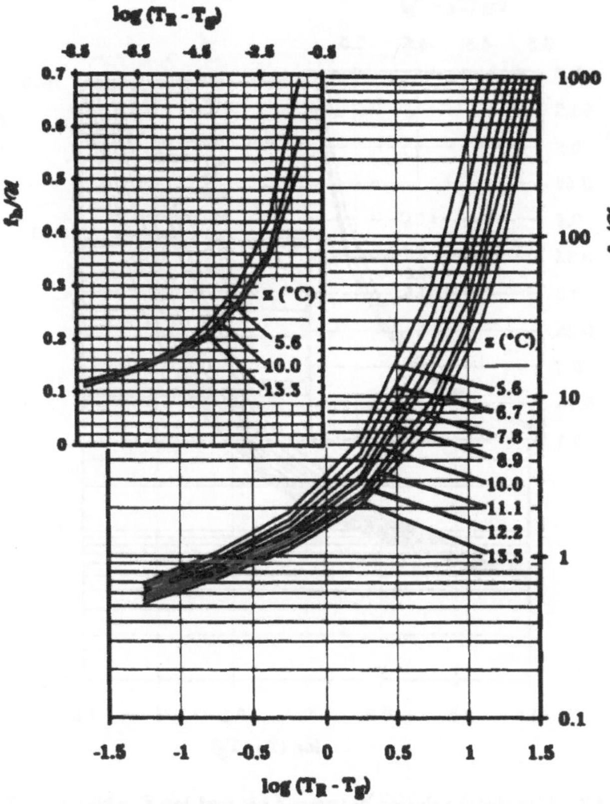

FIGURE 4.2 Graphical relation between $f_h/\mathscr{U}$ and $\log T_g$ when $T_R - T_{CW} = 89°C$ (National Canners Association, 1968; redrawn and published with kind permission of the copyright owner Chapman & Hall Inc.).

where f_{h1} and f_{h2} are values of parameter f_h before and after the phase change, respectively. Parameter r is a correction factor given in Figure 4.4, value of parameter $\mathscr{U}$ at the end of the first phase is $\mathscr{U}_1$. The z values of the chemical reactions are substentially different than those of the microbial death kinetics, therefore Figures 4.1–4.4 may not be used with the processes involving nutrient degradation or enzyme inactivation. The relations between $f_h/\mathscr{U}$ and $T_{RT} - T_g$ at different values of z and j as given in Tables 4.3 and 4.4 are in the closer range to those of the processes involving chemical reactions.

Ball's formula method is based on an empirical model and has some simplifying assumptions, i.e., $f_h = f_c, j_c = 1.41$ and inaccuracies in its derivation; however it served the food industry for more than half a century and is one of the milestones in food process modeling. The Ball's formula method may be used for large variety of foods processed under diverse conditions.

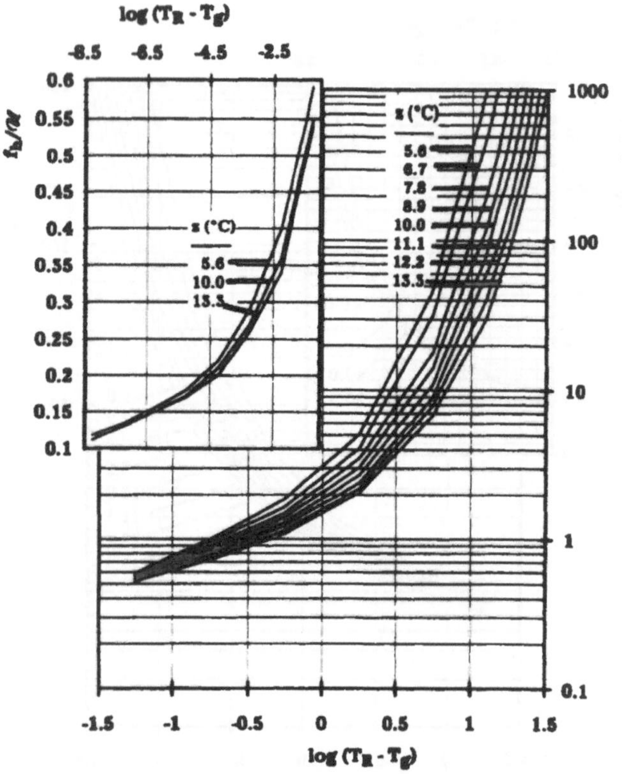

FIGURE 4.3 Graphical relation between $f_h/\mathcal{U}$ and $\log T_g$ when $T_R - T_{CW} = 72°C$ (National Canners Association, 1968; redrawn and published with kind permission of the copyright owner Chapman & Hall Inc.).

Theoretical models are available for the limed conditions, i.e., pure convection and pure conduction, only. Heat transfer by conduction and convection occurs simultaneously in most foods, but one of these mechanisms dominates in most cases. An excellent critical review of the Ball's formula method has been made by Merson *et al.* (1978).

Example 4.4. Perfect convection model of thermal processing When the food in the can is a perfectly mixed homogeneous liquid with uniform temperature and the retort is operating at a constant temperature T_R with no coming up time the thermal energy balance around the can requires

$$UA(T_R - T) = mc\frac{dT}{dt} \tag{E.4.4.1}$$

where A is the total heat transfer area, U is the total heat transfer coefficient, m is the total mass of the food, c is the average specific heat of the food, T is

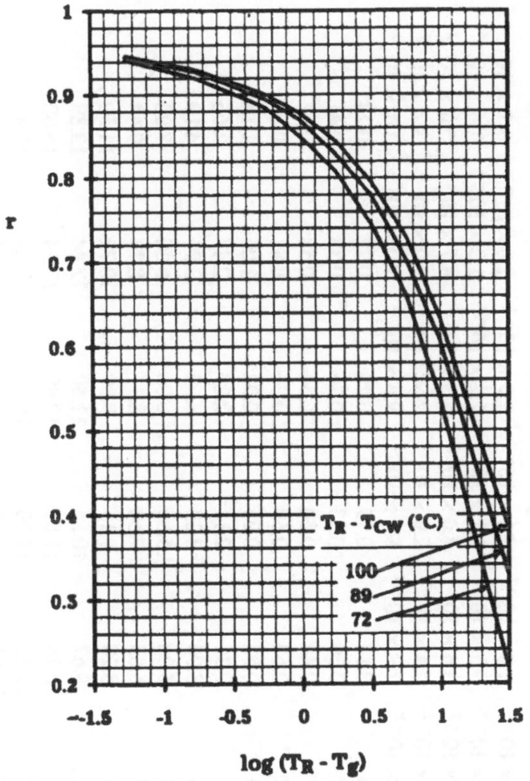

FIGURE 4.4 Graphical relation between r and $\log T_g$ when $z = 10°C$ (National Canners Association, 1968; redrawn and published with kind permission of the copyright owner Chapman & Hall Inc.). Parameter r increases with z within the range of ± 0.01 of the values given in the chart when $5.6°C \leqslant z \leqslant 13.3°C$.

the food temperature and t is time. The total heat transfer coefficient may be expressed in terms of its components:

$$\frac{1}{U} = \frac{1}{h_0} + \frac{\Delta r}{k} + \frac{1}{h_i} \tag{E.4.4.2}$$

where h_0 and h_i are the outside and inside heat transfer coefficients, Δr is the thickness of the can and k is the thermal conductivity of the can. When the food is at a constant temperature T_0 at time $t = 0$ (E.4.4.1) may be solved as (Merson *et al.*, 1978):

$$t = \frac{2.303mc}{UA} \log\left(\frac{T_R - T_0}{T_R - T}\right) \tag{E.4.4.3}$$

Table 4.3 The relations between $f_h/\mathscr{U}$ and $T_R - T_g$ (°C) at $z = 24.5$°C with different values of j (Jen *et al.*, 1971; converted into SI units and published by permission of IFT)

$f_h/\mathscr{U}$ ↓	$T_R - T_g$; j→ 0.40	0.60	0.80	1.00	1.20	1.40	1.60	1.80	2.00
0.1	3.7×10^{-8}	7.1×10^{-8}	1.1×10^{-7}	1.4×10^{-7}	1.8×10^{-7}	2.1×10^{-7}	2.5×10^{-7}	2.8×10^{-7}	3.1×10^{-7}
0.2	4.4×10^{-5}	5.6×10^{-5}	6.8×10^{-5}	8.1×10^{-5}	9.3×10^{-5}	1.1×10^{-4}	1.2×10^{-4}	1.3×10^{-4}	1.4×10^{-4}
0.3	2.2×10^{-3}	2.6×10^{-3}	2.9×10^{-3}	3.3×10^{-3}	3.6×10^{-3}	4.0×10^{-3}	4.3×10^{-3}	4.7×10^{-3}	5.0×10^{-3}
0.4	1.6×10^{-2}	1.8×10^{-2}	1.3×10^{-2}	2.2×10^{-2}	2.4×10^{-2}	2.6×10^{-2}	2.8×10^{-2}	3.0×10^{-2}	3.2×10^{-2}
0.5	5.1×10^{-2}	5.8×10^{-2}	6.5×10^{-2}	7.2×10^{-2}	7.9×10^{-2}	8.7×10^{-2}	9.4×10^{-2}	1.0×10^{-1}	1.1×10^{-1}
0.6	0.11	0.12	0.14	0.16	0.17	0.19	0.20	0.22	0.24
0.7	0.19	0.22	0.25	0.27	0.30	0.33	0.36	0.38	0.41
0.8	0.29	0.33	0.37	0.41	0.46	0.50	0.54	0.58	0.62
0.9	0.40	0.46	0.52	0.58	0.63	0.69	0.76	0.81	0.87
1.0	0.52	0.60	0.68	0.76	0.83	0.91	0.98	1.06	1.14
1.5	1.20	1.38	1.57	1.75	2.31	2.11	2.29	2.47	2.66
2.0	1.89	2.18	2.46	2.75	3.04	3.33	3.62	3.91	4.19
3.0	3.12	3.60	4.09	4.58	5.06	5.55	6.06	6.50	7.00
4.0	4.18	4.84	5.50	6.17	6.83	7.50	8.11	8.77	9.44
5.0	5.08	5.89	6.67	7.50	8.28	9.11	9.89	10.72	11.50
6.0	5.89	6.83	7.78	8.72	9.67	10.56	11.50	12.44	13.39
7.0	6.56	7.61	8.67	9.72	10.78	11.83	12.89	13.94	15.00
8.0	7.22	8.39	9.56	10.67	11.83	13.00	14.17	15.28	16.44
9.0	7.89	9.11	10.33	11.56	12.83	14.06	15.28	16.50	17.72
10.0	8.56	9.89	11.17	12.44	13.78	15.06	16.33	17.67	18.94
12.5	9.94	11.39	12.89	14.39	15.83	17.33	18.78	20.28	21.72
15.0	11.17	12.78	14.39	15.94	17.56	19.11	22.33	22.33	23.89
20.0	13.50	15.22	17.00	18.78	20.56	22.33	24.06	25.83	27.61
25.0	15.27	17.17	19.06	20.94	22.83	24.72	26.61	28.50	30.39

Table 4.4 The relations between $f_h/\mathscr{U}$ and $T_R - T_g$ (°C) at $z = 25.5$°C with different values of j (Jen *et al.*, 1971; converted into SI units and published by permission of IFT)

	$T_R - T_g$								
$f_h/\mathscr{U}$ ↓	j → 0.40	0.60	0.80	1.00	1.20	1.40	1.60	1.80	2.00
0.1	3.7×10^{-8}	7.2×10^{-8}	1.1×10^{-7}	1.4×10^{-7}	1.8×10^{-7}	2.1×10^{-7}	2.5×10^{-7}	2.8×10^{-7}	3.2×10^{-7}
0.2	4.6×10^{-5}	5.8×10^{-5}	7.1×10^{-5}	8.4×10^{-5}	9.7×10^{-5}	1.1×10^{-4}	1.2×10^{-4}	1.4×10^{-4}	1.5×10^{-4}
0.3	2.3×10^{-3}	2.7×10^{-3}	3.1×10^{-3}	3.5×10^{-3}	3.8×10^{-3}	4.2×10^{-3}	4.6×10^{-3}	5.0×10^{-3}	5.4×10^{-3}
0.4	1.7×10^{-2}	1.9×10^{-2}	2.1×10^{-2}	2.3×10^{-2}	2.5×10^{-2}	2.8×10^{-2}	3.0×10^{-2}	3.2×10^{-2}	3.4×10^{-2}
0.5	5.2×10^{-2}	6.0×10^{-2}	6.8×10^{-2}	7.6×10^{-2}	8.4×10^{-2}	9.1×10^{-2}	9.9×10^{-2}	1.1×10^{-1}	1.2×10^{-1}
0.6	0.11	0.13	0.15	0.16	0.18	0.20	0.21	0.23	0.25
0.7	0.20	0.23	0.26	0.29	0.32	0.35	0.38	0.41	0.44
0.8	0.30	0.34	0.39	0.43	0.48	0.52	0.57	0.61	0.66
0.9	0.41	0.48	0.54	0.61	0.67	0.73	0.79	0.86	0.92
1.0	0.54	0.63	0.71	0.79	0.87	0.96	1.04	1.12	1.21
1.5	1.24	1.44	1.64	1.83	2.03	2.22	2.42	2.62	2.81
2.0	1.96	2.27	2.58	2.89	3.20	3.51	3.82	4.13	4.44
3.0	3.22	3.75	4.27	4.80	5.47	5.83	6.39	6.89	7.44
4.0	4.33	5.04	5.78	6.44	7.17	7.89	8.61	9.28	10.00
5.0	5.27	6.11	7.00	7.89	8.72	9.61	10.44	11.33	12.22
6.0	6.11	7.11	8.11	9.11	10.11	11.17	12.17	13.17	14.17
7.0	6.83	7.94	9.11	10.22	11.39	12.50	13.67	14.78	15.94
8.0	7.50	8.72	9.94	11.22	12.44	13.67	14.94	16.17	17.39
9.0	8.22	9.56	10.89	12.16	13.50	14.83	16.17	17.44	18.78
10.0	8.94	10.33	11.72	13.17	14.56	15.94	17.33	18.72	20.11
12.5	10.39	11.94	13.50	15.11	16.67	18.22	19.83	21.39	23.00
15.0	11.67	13.39	15.11	16.83	18.50	20.22	21.94	23.67	25.33
20.0	14.11	15.94	17.83	19.72	21.61	23.50	25.39	27.28	29.11

Comparison of (E.4.4.3) with (4.7) implies that when pure convection is the prevailing mechanism of heat transfer (Merson *et al.*, 1978):

$$f_h = \frac{2.303\,mc}{UA} \tag{E.4.4.4}$$

and

$$j_h = 1 \tag{E.4.4.5}$$

Example 4.5. Perfect conduction model of thermal processing This model is valid when the food in the can is not moving, temperature varies with time in radial and axial directions and the temperature on the can surface is constant. The governing transport equation for heat transfer is obtained after simplifying (2.17) as:

$$\frac{1}{r}\frac{\partial}{\partial r}\left(r\frac{\partial T}{\partial r}\right) + \frac{\partial^2 T}{\partial z^2} = \frac{1}{\alpha}\frac{\partial T}{\partial t} \tag{E.4.5.1}$$

where r and z are distances along the radial and the longitudinal directions, respectively, and parameter α is the thermal diffusivity of the food. Equation (E.4.5.1) may be solved with the following initial and the boundary conditions:

$$T(r, z, 0) = T_0 \tag{E.4.5.1.a}$$

$$T(R, z, t) = T_R \tag{E.4.5.1.b}$$

$$T(r, \pm L, t) = T_R \tag{E.4.5.1.c}$$

$$\frac{dT}{dr} = 0 \quad \text{at } r = 0 \tag{E.4.5.1.d}$$

$$\frac{dT}{dz} = 0 \quad \text{at } z = 0 \tag{E.4.5.1.e}$$

where R and $2L$ are the radius and the height of the can, respectively. Equation (E.4.5.1.a) implies that the food was initially at a constant temperature T_0, (E.4.5.1.b) and (E.4.5.1.c) shows that the food surfaces were at the retort temperature throughout processing. Equations (E.4.5.1.d) and (E.4.5.1.e) indicate the symmetrical temperature profiles around the center line and the horizontal plane between the lower and the upper halves of the can. A new dependent variable may be defined to apply the method of separation of variables:

$$Y_{\text{can}} = \frac{T_R - T}{T_R - T_0} = \mathscr{R}(r, t)\,\mathscr{Z}(z, t) \tag{E.4.5.2}$$

Equation (E.4.5.2) may be arranged as:

$$\mathscr{Z}\left(\frac{\partial^2 \mathscr{R}}{\partial r^2}\right)+\frac{1}{r}\frac{\partial \mathscr{R}}{\partial r}-\frac{1}{\alpha}\frac{\partial \mathscr{R}}{\partial t}\Bigg)+\mathscr{R}\left(\frac{\partial^2 \mathscr{Z}}{\partial z^2}-\frac{1}{\alpha}\frac{\partial \mathscr{Z}}{\partial t}\right)=0 \quad \text{(E.4.5.3)}$$

Equation (E.4.5.3) is satisfied if:

$$\frac{\partial^2 \mathscr{R}}{\partial r^2}+\frac{1}{r}\frac{\partial \mathscr{R}}{\partial r}-\frac{1}{\alpha}\frac{\partial \mathscr{R}}{\partial t}=0 \quad \text{(E.4.5.4)}$$

and

$$\frac{\partial^2 \mathscr{Z}}{\partial z^2}-\frac{1}{\alpha}\frac{\partial \mathscr{Z}}{\partial t}=0 \quad \text{(E.4.5.5)}$$

Equations (E.4.5.4) and (E.4.5.5) are actually thermal energy balances for an infinitely long cylinder and an infinitely long slab, respectively. The functions $\mathscr{R}(r, t)$ and $\mathscr{Z}(z, t)$ are actually the solutions of (E.4.5.4) and (E.4.5.5), respectively; where

$$\mathscr{R}(r, t) = Y_{\text{cylinder}} = \left(\frac{T_R - T}{T_R - T_0}\right)_{\text{cylinder}} \quad \text{(E.4.5.6)}$$

$$\mathscr{Z}(r, t) = Y_{\text{slab}} = \left(\frac{T_R - T}{T_R - T_0}\right)_{\text{slab}} \quad \text{(E.4.5.7)}$$

The initial and the boundary conditions for (E.4.5.4) are

$$\mathscr{R}(r, 0) = 1 \quad \text{(E.4.5.4.a)}$$

$$\mathscr{R}(R, t) = 0 \quad \text{(E.4.5.4.b)}$$

$$\frac{d\mathscr{R}}{dr} = 0 \quad \text{at} \quad r = 0 \quad \text{(E.4.5.4.c)}$$

The initial and the boundary conditions for (E.4.5.5) are

$$\mathscr{Z}(z, 0) = 1 \quad \text{(E.4.5.5.a)}$$

$$\mathscr{Z}(L, t) = 0 \quad \text{(E.4.5.5.b)}$$

$$\frac{d\mathscr{Z}}{dz} = 0 \quad \text{at} \quad z = 0 \quad \text{(E.4.5.5.c)}$$

Equations (E.4.5.4.a) and (E.4.5.5.a) imply that the cylinder and the slab were initially at a constant temperature T_0, (E.4.5.4.b) and (E.4.5.5.b) show that the cylinder and the slab surfaces were at the retort temperature T_R throughout processing. Equations (E.4.5.4.c) and (E.4.5.5.c) indicate the symmetric temperature profiles around the center line of the cylinder and the horizontal plane between the lower and the upper halves of the slab.

Solutions to the cylinder and the slab equations are:

$$Y_{\text{cylinder}} = 2 \sum_{n=1}^{\infty} \frac{J_0\left(B_n \frac{r}{R}\right)}{J_1(B_n) B_n} \exp\left(-B_n (k/\rho c)(t/R^2)\right) \tag{E.4.5.8}$$

where J_0 and J_1 are zero and first order Bessel functions of type one, B_n is the nth eigen value of $J_0(B_n)$, ρ and c are the average density and heat capacity of the food.

$$Y_{\text{slab}} = \sum_{m=1}^{\infty} \left[2 \frac{(-1)^{m+1}}{\beta_m}\right] \cos\left[\beta_m \frac{2z}{L}\right] \exp\left(-\beta_m^2 (k/\rho c)(t/L^2)\right) \tag{E.4.5.9}$$

where $\beta_m = (2m-1)\pi/2$. Values of the Bessel functions $J_0(x)$, $J_1(x)$ and $J_1(B_n)$ for the eigen values B_n of $J_0(B_n) = 0$ are given in Tables 2.10 and 2.14.

Equation (E.4.5.2) requires (Merson *et al.*, 1978):

$$Y_{\text{can}} \sum_{m=1}^{\infty} \sum_{n=1}^{\infty} \left[2 \frac{(-1)^{m+1}}{\beta_m}\right] \cos\left[\beta_m \frac{z}{L}\right] \left[\frac{2J_0\left(B_n \frac{r}{R}\right)}{B_n J_1(B_n)}\right]$$

$$\exp\left[-\left(\frac{\beta_m^2}{L^2} + \frac{B_n^2}{R^2}\right)\left(\frac{k}{\rho c}\right) t\right] \tag{E.4.5.10}$$

Equation (E.4.5.10) describes the unsteady state temperature variations within the can as a function of the radial location r and the longitudinal location z. When the processing times are long, solution of (E.4.5.10) may be approximated with the first term of the series solution, i.e., $m = 1$, $\beta_1 = \pi/2$, $n = 1$, $B_1 = 2.4048$, $J_1(B_1) = 0.5191$. Further simplifications may be done when the long time solution is sought at the can center, i.e., $z = 0$, $\cos(0) = 1$, $r = 0$, $J_0(0) = 1$ and (E.4.5.10) may be rewritten as (Merson *et al.*, 1978):

$$Y_{\text{can center}} = \frac{T_R - T}{T_R - T_0} = 2.04 \exp\left[-\left(\frac{\pi^2}{4L^2} + \frac{B_1^2}{R^2}\right)\left(\frac{k}{\rho c}\right) t\right] \tag{E.4.5.11}$$

Equation (E.4.5.11) may be rearranged as:

$$t = \frac{2.303}{\left(\frac{\pi^2}{4L^2} + \frac{B_1^2}{R^2}\right)\left(\frac{k}{\rho c}\right)} \log\left[2.04 \frac{T_R - T_0}{T_R - T}\right] \tag{E.4.5.12}$$

Comparison of (E.4.5.12) with (4.7) implies that, when pure conduction is the prevailing mechanism of heat transfer (Merson *et al.*, 1978):

$$f_h = \frac{2.303}{\left(\frac{\pi^2}{4L^2} + \frac{B_1^2}{R^2}\right)\left(\frac{k}{\rho c}\right)} \qquad \text{(E.4.5.13)}$$

and

$$j_h = 2.04 \qquad \text{(E.4.5.14)}$$

The method we employed here to solve the partial differential equation is also referred to as *Newman's method* (Newman, 1936).

It should be noticed that (4.7) was based on experimental observations, whereas (E.4.4.3) and (E.4.5.12) have a theoretical basis. Although theoretical analysis is limited to pure convection and to pure conduction only, the experimental approach may be used with almost any food regardless of the heat transfer mechanism, therefore both theoretical and experimental approaches are used together for process calculations. Equations (E.4.4.4) and (E.4.5.13) show that the f_h values are affected by the can size. The mass of the food in the can is $m = \rho\pi R^2 2L$, the total heat transfer area of a cylindrical can (radius $= R$, height $= 2L$) is $A = 2\pi R^2 L(2/R + 1/L)$. It may be assumed that the total heat transfer coefficient U is independent of the can size. After substituting the equations for m and A into (E.4.4.4) the ratio of the f_h values may be expressed as a function of the can size as:

$$f_{h2}/f_{h1} = \frac{R_2 L_2}{R_2 + 2L_2} \Bigg/ \frac{R_1 L_1}{R_1 + 2L_1} \qquad (4.13)$$

A similar conversion factor may be obtained for the conduction heating foods after substituting $B_1 = 2.405$ in (E.4.5.13) as:

$$f_{h2}/f_{h1} = \left(\frac{5.78}{R_1^2} + \frac{\pi^2}{4L_1^2}\right) \Bigg/ \left(\frac{5.78}{R_2^2} + \frac{\pi^2}{4L_2^2}\right) \qquad (4.14)$$

Example 4.6. Calculation of f_h and j values from the physical properties of the food and the use of Ball's formula method for thermal process calculation a) A convection heating product ($c_p = 3.8$ kJ/kg K, $\rho_p = 960$ kg/m^3) is heated in a 307 × 306 size can of 0.03 cm thickness. The can wall has k = 69 W/mK with $h_{inside} = 350$ W/m^{2}K, $h_{outside} = 450$ W/m^{2}K. Calculate the f_h and j_h values for this process.

Solution The first number in the can size indicates the height (first digit = integer number of inches, the second and the third digits = additional height in 1/16 folds of one inch). The second number indicates the can diameter.

$$\text{Can height} = 2L = \left(3 + \frac{7}{16}\right)(2.54) = 8.7 \text{ cm}$$

$$\text{Can radius} = R = \left(3 + \frac{6}{16}\right)(2.54)\left(\frac{1}{2}\right) = 4.3 \text{ cm}$$

where 2.54 is a constant to convert the numbers in cm from inches. $V = \pi R^2 2L = 5.05 \times 10^{-4}\,\text{m}^3$, $m = \rho V = 0.485$ kg and $A = 2\pi R^2 L(2/R + 1/L) = 3.51 \times 10^{-2}\,\text{m}^2$. The total heat transfer coefficient is:

$$\frac{1}{U} = \frac{1}{h_0} + \frac{\Delta r}{k} + \frac{1}{h_i} \tag{E.4.4.2}$$

After substituting the numbers in (E.4.4.2) we will obtain $U = 197$ W/m^{2}K. The f_h and j values for a convection heating food are:

$$f_h = \frac{2.303mc}{UA} \tag{E.4.4.4}$$

and

$$j_h = 1 \tag{E.4.4.5}$$

After substituting the numbers in (E.4.4.4) we will calculate $f_h = 10.2$ min.

b) If F_{required} is 5 min at $T_{\text{ref}} = 100$°C and $z = 10$°C, calculate the time from steam on to steam off when $T_R = 110$ °C, $T_{CW} = 21$°C and $T_0 = 40$°C.

Solution We have $T_R - T_{CW} = 89$°C and $\mathscr{U} = F_{\text{req}} 10^{T(T_{\text{ref}} - T_R)/10} = 0.5$ and $f_h/\mathscr{U} = 20.4$. We will use Figure 4.2 with the given values of, $f_h/\mathscr{U}$ and z to determine $\log(T_R - T_g) = 0.85$, therefore $T_g = 103$°C.

When there is no retort coming-up time, the heating time is:

$$t_{\text{heating}} = f_h(\log j_h(T_R - T_0) - \log(T_R - T_g)) \tag{4.9}$$

After substituting the numbers in (4.9) we will obtain $t_{\text{heating}} = 10.2$ min.

Example 4.7. Use of the Ball's formula method with changing can size A conduction heating product has heat sensitive nutrients with $D = 180$ min at $T_{\text{ref}} = 121$°C, $z = 24.5$°C. It is processed in 211 × 212 size cans with $f_h = f_c = 60$ min, $j_h = j_c = 2.0$, $T_R = 121$°C, $t_{CUT} = 10$ min, $T_0 = 70$°C and $T_{CW} = 21$°C. The processing requirement $F_{\text{required}} = 10$ min at 121°C with $z = 13.3$°C.

a) Use Ball's formula method to calculate the time from steam on to steam off and determine what percent of the nutrients will be retained at the center of this product at the end of processing?

Solution Can height $= 2L = (2 + 12/16)(2.54) = 6.98$ cm

Can radius $= R = (2 + 11/16)(2.54)(1/2) = 3.41$ cm

We have $\mathcal{U} = F_{req} 10^{(T_{ref} - T_R)/10} = 10$ min and $f_h/\mathcal{U} = 6$. Since $T_R - T_{CW} = 100°C$ We will use Figure 4.1 with $z = 13.3°C$ to determine $\log(T_R - T_g) = 0.70$, therefore $T_g = 116°C$.

When there is retort coming-up time, the heating time is calculated as:

$$t_{heating} = f_h(\log j_h(T_R - T_0) - \log(T_R - T_g)) + 0.58\, t_{CUT} \qquad (4.10)$$

After substituting the numbers in (4.10) we will obtain $t_{heating} = 84.5$ min.

We will use Table 4.3 with $j = 2.0$ and $T_R - T_g = 5°C$ and determine $f_h/\mathcal{U} = 2.29$ with interpolation; therefore $\mathcal{U} = f_h/2.29 = 26$ min. Since $\mathcal{U} = C_{process} 10^{(T_{ref} - T_R)/10}$ we calculate $C_{process} = 26$ min. Since we have

$$C_{process} = D_{T_{ref}} \log \frac{c_{final}}{c_{initial}} \text{ we calculate } \frac{c_{final}}{c_{initial}} = 71.4\%.$$

b) What is the time from steam on to steam off, and what percent of the nutrients will be retained at the center of the product if you use a 202 × 308 size can?

Solution Can height $= 2L = (3 + 8/16)(2.54) = 8.89$ cm
Can radius $= R = (2 + 2/16)(2.54)(1/2) = 2.7$ cm
Equation (4.14) gives the variation of f_h with the can size for a conduction heating food as:

$$f_{h2}/f_{h1} = \left(\frac{5.78}{R_1^2} + \frac{\pi^2}{4L_1^2}\right) \Big/ \left(\frac{5.78}{R_2^2} + \frac{\pi^2}{4L_2^2}\right) \qquad (4.14)$$

After substituting the numbers in (4.14) we will calculate $f_{h2} = 45.8$ min. We have $\mathcal{U} = F_{req} 10^{(T_{ref} - T_R)/10} = 10$ min and $f_h/\mathcal{U} = 4.58$. Since $T_R - T_{CW} = 100°C$. We will use Figure 4.1 with $z = 13.3°C$ to determine $\log(T_R - T_g) = 0.67$, therefore $T_g = 116.4°C$. The heating time is:

$$t_{heating} = f_h(\log j_h(T_R - T_0) - \log(T_R - T_g)) + 0.58\, t_{CUT} \qquad (4.10)$$

After substituting the numbers in (4.10) we will obtain $t_{heating} = 67.5$ min.

We will use Table 4.3 with $j = 2.0$ and $T_R - T_g = 4.6°C$ and determine $f_h/\mathcal{U} = 2.15$ with interpolation; therefore $\mathcal{U} = f_h/2.15 = 21.3$ min. Since $\mathcal{U} = C_{process} 10^{(T_{ref} - T_R)/10}$ we calculate $C_{process} = 21.3$ min. Since we have

$$C_{process} = D_{T_{ref}} \log \frac{c_{final}}{c_{initial}} \text{ we calculate } \frac{c_{final}}{c_{initial}} = 76\%.$$

Example 4.8. Wehrle-Merson model to predict critical point temperatures of conduction heating foods for short heating times a) Temperature distribution in an infinite slab may be expressed as (Carslaw and Jaeger, 1959):

$$\frac{T-T_0}{T_R-T_0}=\left[\sum_{n=0}^{\infty}(-1)^n \operatorname{erfc}\left\{\frac{(2n+1)L-z}{2\sqrt{\alpha t}}\right\}+\sum_{n=0}^{\infty}(-1)^n \operatorname{erfc}\left\{\frac{(2n+1)L+z}{2\sqrt{\alpha t}}\right\}\right] \quad \text{(E.4.8.1)}$$

where z is distance from the mid-plane of the slab, α is thermal diffusivity of the food, and $2L$ is the thickness of the slab. A short time solution for the axis of an infinite cylinder may be expressed as (Carslaw and Jaeger, 1959):

$$\frac{T-T_0}{T_R-T_0}=\frac{R}{\sqrt{\pi\alpha t}}\left[\exp\left(-\frac{R^2}{8\alpha t}\right)\right]\left[K_{1/4}\left(\frac{R^2}{8\alpha t}\right)\right] \quad \text{(E.4.8.2)}$$

where $K_{1/4}$ is the modified Bessel function of the second order 1/4. Make a one-term approximation of the first expression, then combine the two expressions to obtain an expression for variation of the temperature at the critical point of the can of a conduction heating food.

Solution With $n=0$ and $z=0$ (E.4.8.1) reduces to the following form:

$$\frac{T-T_0}{T_R-T_0}=2\operatorname{erfc}\left\{\frac{L}{2\sqrt{\alpha t}}\right\} \quad \text{(E.4.8.3)}$$

After following the same procedure as in Example 4.5 we can state that:

$$\left(\frac{T_R-T}{T_R-T_0}\right)_{\text{can}}=\left(\frac{T_R-T}{T_R-T_0}\right)_{\text{slab}}\times\left(\frac{T_R-T}{T_R-T_0}\right)_{\text{cylinder}} \quad \text{(E.4.8.4)}$$

After substituting (E.4.8.2) and (E.4.8.3) in (E.4.8.4) we will obtain

$$\left(\frac{T_R-T}{T_R-T_0}\right)_{\text{can}}=\left(1-2\operatorname{erfc}\left\{\frac{L}{2\sqrt{\alpha t}}\right\}\right)\left(1-\frac{R}{\sqrt{\pi\alpha t}}\left[\exp\left(-\frac{R^2}{8\alpha t}\right)\right]\left[K_{1/4}\left(\frac{R^2}{8\alpha t}\right)\right]\right) \quad \text{(E.4.8.5)}$$

Equation (E.4.8.5) was originally used to simulate can center temperature data (Wehrle, 1980) during heating of tomato paste.

b) Use (E.4.8.5) and (E.4.5.12) to plot the center temperature of a can as a function of time during the heating period, then compare it with the following set of data (Ward *et al.*, 1984):

Time (min)	T (°C)	Time (min)	T (°C)	Time (min)	T (°C)
0	7.2	22	47.8	47	77.8
7	12.8	27	54.4	57	80.0
12	17.8	30	61.1	64	82.2
15	27.2	38	70.0	73	83.9
18	38.9				

$T_R = 87.8°C$, $T_0 = 7.2°C$, $f_h = 42.3$ minutes for processing in 307 × 296 size can.

Solution Can height $= 2L = (2 + 6/16)(2.54) = 6.03$ cm and can radius $= R = (3 + 7/16)(2.54)(1/2) = 4.36$ cm. We may calculate the thermal diffusivity $\alpha = k/\rho c$ from f_h through (E.4.5.13):

$$f_h = 2.303 \Big/ \left\{ \left(\frac{\pi^2}{4L^2} + \frac{B_1^2}{R^2} \right) \alpha \right\} = 2.303 \Big/ \left\{ \left(\frac{(3.14)^2}{(6.03)^2} + \frac{(2.41)^2}{(4.36)^2} \right) \alpha \right\} = 42.3$$

After substituting the numbers we will get $\alpha = 0.094\ \text{cm}^2/\text{min}$. Values of the modified Bessel's function of the second order 1/4 are given in Table 2.15. Variation of critical point temperature is calculated by using (E.4.8.5) in Table E.4.8.1:

Table E.4.8.1 Calculation of the critical point temperature by using (E.4.8.5)

t (min)	$\frac{R}{\sqrt{\pi\alpha t}}$	$\exp\left(-\frac{R^2}{8\alpha t}\right)$	$K_{1/4}\left(\frac{R^2}{8\alpha t}\right)$	$\text{erfc}\left\{\frac{L}{2\sqrt{\alpha t}}\right\}$	$\left(\frac{T_R - T}{T_R - T_0}\right)_{can}$	T(°C)
0	∞	0	0	0	1	7.2
7	3.035	0.027	0.0174	0.009	0.981	8.8
12	2.318	0.012	0.1009	0.045	0.907	14.7
15	2.073	0.185	0.1714	0.073	0.798	23.5
18	1.893	0.246	0.2479	0.098	0.710	30.5
22	1.712	0.317	0.3042	0.139	0.603	39.2
27	1.545	0.392	0.4487	0.179	0.468	50.1
30	1.467	0.431	0.5481	0.203	0.388	56.5
38	1.303	0.514	0.7238	0.258	0.249	67.7
47	1.171	0.584	0.8978	0.312	0.145	76.1
57	1.064	0.642	1.0832	0.378	0.063	82.7
64	1.004	0.674	1.1651	0.386	0.048	83.9
73	0.940	0.707	1.3066	0.417	0.022	86

Equation (E.4.5.12) will be rearranged to obtain

$$\left(\frac{T_R - T}{T_R - T_0}\right)_{can} = 2.04 \exp\left\{ -\left(\frac{\pi^2}{4L^2} + \frac{B_1^2}{R^2} \right) \left(\frac{k}{\rho c} \right) t \right\} \qquad \text{(E.4.8.6)}$$

After substituting the constants (E.4.8.6) will be rewritten as:

$$\left(\frac{T_R - T}{T_R - T_0}\right)_{can} = 2.04 \exp\left\{ -\left(\frac{3.14^2}{6.03^2} + \frac{2.41^2}{4.36^2} \right) 0.094t \right\} = 2.04 \exp(-0.054t) \qquad \text{(E.4.8.7)}$$

Temperature estimates by using (E.4.8.7) are given in Table E.4.8.2. Equations (E.4.8.5) and (E.4.8.7) are compared with the experimental data in Figure E.4.8, which shows that (E.4.8.5) gives better estimates than (E.4.8.7) especially at the beginning of the heating process. This was an expected

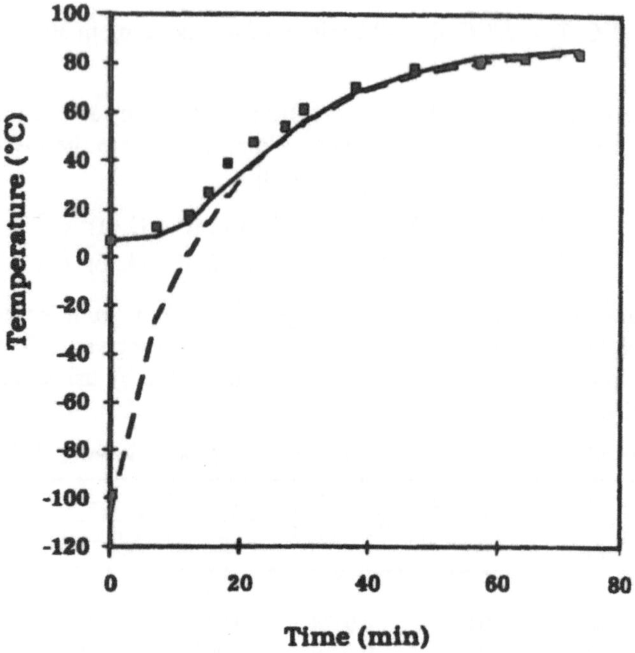

FIGURE E.4.8 Comparison of (E.4.8.5) (---) and (E.4.8.7) (—) with the experimental data (■).

result because it was clearly expressed in Example 4.5 that the solution to (E.4.5.10) may be approximated with the first term of the series solution when the processing times are long; therefore the use of (E.4.5.12), (E.4.8.6) and subsequently (E.4.8.7) are subject to this constraint.

Table E.4.8.2 Calculation of the critical point temperature by using (E.4.8.7)

t (min)	$\left(\frac{T_R - T}{T_R - T_0}\right)_{can}$	T(°C)
0	2.04	− 105.6
7	1.40	− 24.9
12	1.067	1.8
15	0.908	14.7
18	0.772	25.6
22	0.622	37.7
27	0.475	49.5
30	0.404	55.3
38	0.262	66.7
47	0.161	74.8
57	0.094	80.2
64	0.064	82.6
73	0.030	84.6

The *Lethality Fourier Number Method* of thermal process calculation is used with conduction heating foods (Lenz and Lund, 1977). The Arrhenius expression (3.5) may be used to express the death rate constant of the microorganisms in terms of a reference:

$$k = k_{ref} \exp\left[-\left(\frac{E_a}{R_g}\right)\left(\frac{1}{T} - \frac{1}{T_{ref}}\right)\right] \tag{4.14}$$

where k_{ref} is the death rate constant at T_{ref}. The first order death rate expression (3.46) may be integrated after substituting (4.14) as:

$$\ln(x_0/x) = k_{ref} \int_0^t \exp\left[-\left(\frac{E_a}{R_g}\right)\left(\frac{1}{T} - \frac{1}{T_{ref}}\right)\right] dt \tag{4.15}$$

where x_0 and x are the viable microbial counts at times 0 and t, respectively; R_g is the gas constant. Equation (4.15) may be made dimensionless:

$$\frac{\alpha \ln(x_0/x)}{k_{ref} R^2} = \int_0^\tau \exp\left[-\left(\frac{E_a}{R_g}\right)\left(\frac{1}{T} - \frac{1}{T_{ref}}\right)\right] d\tau \tag{4.16}$$

where $\alpha = k/\rho c$ is thermal diffusivity, R is the radius of the can, $\tau = \alpha t/R^2$ is the Fourier number, $\mathcal{L} = \alpha \ln(x_0/x)/k_{ref}R^2$ is the lethality number. The temperature profile in the can depends on the radial and longitudinal locations and time, i.e., $T = T(r, z, \tau)$. r and z describe the radial and longitudinal location in the can with centre ($r = 0$, $z = 0$). Temperature profiles in the can during the heating period may be described by the unsteady state heat conduction equation for a finite cylinder (Carslaw and Jaeger, 1959):

$$\frac{T_R - T}{T_R - T_0} = \left\{\frac{4}{\pi}\sum_{n=0}^{\infty}\frac{(-1)^n}{2n+1}\exp\left[-\left(n+\frac{1}{2}\right)^2 \pi^2 \left(\frac{R}{L}\right)^2 \tau\right]\cos\left[\left(n+\frac{1}{2}\right)\pi\left(\frac{Z}{L}\right)\right]\right\}$$
$$\left\{2\sum_{k=1}^{\infty}\frac{J_0(\beta_k(r/R_{can}))}{\beta_k J_1(\beta_k)}\exp(-\beta_k^2 \tau)\right\} \tag{4.17}$$

where β_k are the roots of $J_0(\beta) = 0$, and L is the half of the height of the can. A similar expression for the cooling period is:

$$\frac{T - T_{CW}}{T_R - T_0} = \left\{4\sum_{j=1}^{\infty}\sum_{i=1}^{\infty}\frac{(-1)^i \exp\left\{-\left[\left(\frac{R}{L}\right)^2\left(i+\frac{1}{2}\right)^2 \pi^2 + \beta_j^2\right]\tau_w\right\}}{\left(i+\frac{1}{2}\right)\pi \beta_j J_1(\beta_j)}\right.$$

$$\left.\cos\left[\left(i+\frac{1}{2}\right)\pi\left(\frac{z}{L}\right)\right]\right\}\left\{\left(\frac{T_R - T_{CW}}{T_R - T_0}\right) - \exp\left[-\left(\beta_j^2\left(i+\frac{1}{2}\right)^2 \pi^2\left(\frac{R_{can}}{L_{can}}\right)^2 \tau_W\right]\right\} \tag{4.18}$$

where τ_w is the cooling period Fourier number. The ratio x_0/x may be calculated after solving (4.15) simultaneously with (4.17) or (4.18). Lenz and Lund (1977) calculated the Lethality number ($\mathscr{L}$) at the critical point ($r = 0$, $z = 0$) for various microorganisms with different activation energies. Figure 4.5 may be used when $T_{ref} = T_R$, $T_{CW} = 15.5°C$, $T_R - T_0 = 55.5°C$, $L/R = 1.11$ and α = constant. It was found that the $\mathscr{L}$ number was not sensitive to the changes in $T_R - T_0$ and T_{CW}, i.e., $\mathscr{L}$ number changes less than 1% for a change of 11 °C in $T_R - T_0$ or 5.5°C in T_{CW}.

Example 4.9. Lethality-Fourier number method for process calculation A conduction heating food with $\alpha = 0.097$ cm^2/min has been packed into 211 × 304 cans, and processed at $T_R = 121°C$, $T_0 = 65.5°C$ and $T_{CW} = 15.5°C$. The process was based on inactivation of *Clostridium botulinum* with $E_a = 2.7 \times 10^5$ J/mole ($z = 10$ °C) and $D = 0.5$ min at 121°C. Commercial sterility is accomplished when the container has been exposed to an equivalent heating time of 12 × D values. Calculate the required heating time.

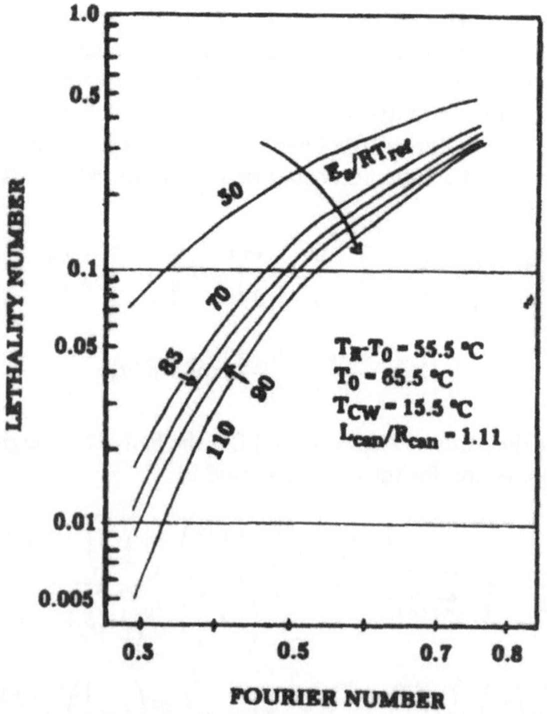

FIGURE 4.5 Lethality-Fourier number chart (Lenz and Lund, 1977; © IFT, reproduced by permission).

Solution With a 211 × 304 can $R=(1/2)\ (2+11/16)\ (2.54)=3.4$ cm, $2L=(3+4/16)\ (2.54)=8.24$ cm, $L/R=1.21$ (satisfies the constraint). $F_{\text{required}}=D_{T_{\text{ref}}}\log(x_0/x)=12\times D_{T_{\text{ref}}}$, $\mathscr{L}=\alpha\ln(x_0/x)/k_{\text{ref}}$ $R^2=0.05$, $E_a/R_g T_{\text{ref}}=83$. With the given $\mathscr{L}$ number and $E_a/R_g T_{\text{ref}}$ we may read the Fourier number from Figure 4.5 as $\tau=0.4=\alpha t_{\text{heating}}/R^2$. We may calculate the heating time from the Fourier number as 48 min.

Example 4.10. A transfer function approach to predict transient internal temperatures during sterilization (*Chiheb et al.*, 1994) The canned food (container + contents) during the sterilization process is simulated as an input-output transfer function with N time constants $\tau_1, \tau_2, \tau_3, \ldots, \tau_N$ and time delay t_d (Fig. E.4.10.1).

The model in Laplace domain is:

$$\frac{y(s)}{x(s)}=\frac{\exp(t_d s)}{\prod_{i=1}^{N}(1+\tau_i s)} \tag{E.4.10.1}$$

the inverse Laplace transformations of $x(s)$ and $y(s)$ are defined as:

$$x(t)=\frac{T_{\text{USR}}-T_R}{T_0-T_R} \tag{E.4.10.2}$$

and

$$y(t)=\frac{T-T_0}{T_R-T_0} \tag{E.4.10.3}$$

where T is the internal temperature, T_{USR} is the unsteady state retort temperature before attaining the final steady state retort temperature T_R and T_0 is the temperature of the food at the beginning of the heating phase. Time delay t_d represents the time needed for an input signal to produce

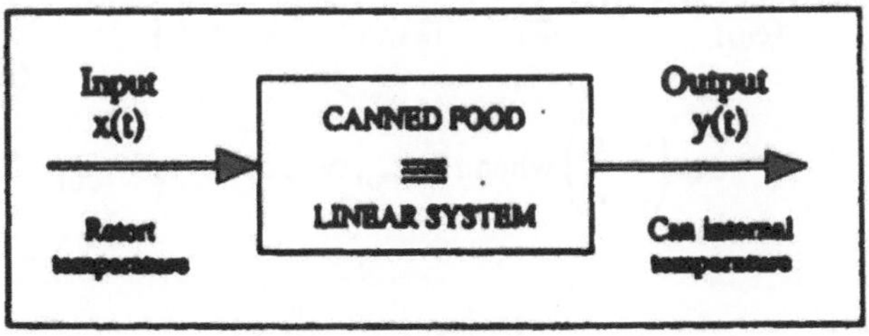

FIGURE E.4.10.1 Representation of a canned food during sterilization as an input-output linear system. The input is the retort temperature, the output is the internal temperature (Chiheb *et al.*, 1994; © IFT, reproduced by permission).

a variation at the output after its propagation within the system. Time constants are in decreasing order such that $\tau_1 > \tau_2 > \tau_3 > \cdots > \tau_N$.

The time constants $\tau_1, \tau_2, \tau_3, \ldots, \tau_N$ may be calculated from the roots of

$$\prod_{i=1}^{N}(1+\tau_i s)=0 \tag{E.4.10.4}$$

With a first order system where $N = 1$ we have

$$\tau_1 = \frac{f}{2.303} \tag{E.4.10.5}$$

where f is the slope index of the semi logarithmic temperature history curve as defined in (4.5) or (4.6). A reduced internal temperature was defined as:

$$T^*(t) = \frac{T - T_R}{T_0 - T_R} = 1 - y(t) \tag{E.4.10.6}$$

When the food is initially in equilibrium with the retort temperature and there is no retort coming-up time we will have

$$T^*(t) = 1 \quad \text{when } t \leqslant t_d \tag{E.4.10.7.a}$$

$$T^*(t) = \sum_{i=1}^{N} J_i \exp\left(-\frac{t}{\tau_i}\right) \quad \text{when } t > t_d \tag{E.4.10.7.b}$$

where

$$j_i = \frac{(\tau_i)^{N-1}\exp(t_d/\tau_i)}{\prod_{k=1\, k\neq i}^{N}(\tau_i - \tau_k)} \tag{E.4.10.7.c}$$

When the retort temperature attains T_R with some delay we will have:

$$T^*(t) = 1 \quad \text{when } t \leqslant t_d \tag{E.4.10.8.a}$$

$$T^*(t) = 1 - \frac{1}{t_{\mathrm{CUT}}}\left[(t - t_d) - \sum_{i=1}^{N}\tau_i + \sum_{i=1}^{N} j_i \tau_i \exp\left(-\frac{t}{\tau_i}\right)\right] \quad \text{when } t_d < t < t_{\mathrm{CUT}} \tag{E.4.10.8.b}$$

$$T^*(t) = \sum_{i=1}^{N} j_i^{t_{\mathrm{CUT}}} \exp\left(-\frac{t}{\tau_i}\right) \quad \text{when } t > t_{\mathrm{CUT}} > t_d \text{ or } t > t_d > t_{\mathrm{CUT}} \tag{E.4.10.8.c}$$

where

$$j_i^{t_{\mathrm{CUT}}} = \frac{j_i \tau_i(\exp(t_{\mathrm{CUT}}/\tau_i) - 1)}{t_{\mathrm{CUT}}} \tag{E.4.10.8.d}$$

The detailed procedure has been described by Chiheb *et al.* (1994) to evaluate the constants of the model. While expressing the internal temperature history for complete sterilization the critical point temperature may not be in equilibrium with the retort temperature, and may keep on increasing during the first moments of cooling. To account for this phenomenon the temperature at the can center is expressed as:

$$T(t) = T_h - (T_h - T_0)\, T_h^*(t) \text{ when } t \leq t_g \qquad \text{(E.4.10.9)}$$

and

$$T(t) = T_c - (T_h - T_0)\, T_h^*(t) - (T_c - T_h)\, T_c^*(t - t_g) \text{ when } t > t_g \qquad \text{(E.4.10.10)}$$

where subscripts c and h indicate the cooling and heating phases, respectively. Parameter t_g is the time length of the heating period. The reduced heating and the cooling temperatures were defined, respectively as:

$$T_h^* = (T - T_h)/(T_0 - T_h) \text{ and } T_c^* = (T - T_c)/(T_h - T_c).$$

Comparison of the model with the experimental data is shown in Figure E.4.10.2. The two phases were modeled by using (E.4.10.8.a) - (E.4.10.8.d).

In *continuous processing* a time-temperature history of the individual liquid pockets is needed in order to calculate the sterilization given to them. When the flow model of the fluid is available, the time spent by each pocket in the sterilization equipment may be determined by the help of the flow model. When such a model is not available, the residence time distribution of the liquid pockets is needed. The time spent by a liquid pocket in the process vessel is called the residence time. Process analysis may be simplified by considering the constant temperature holding tube only. The average

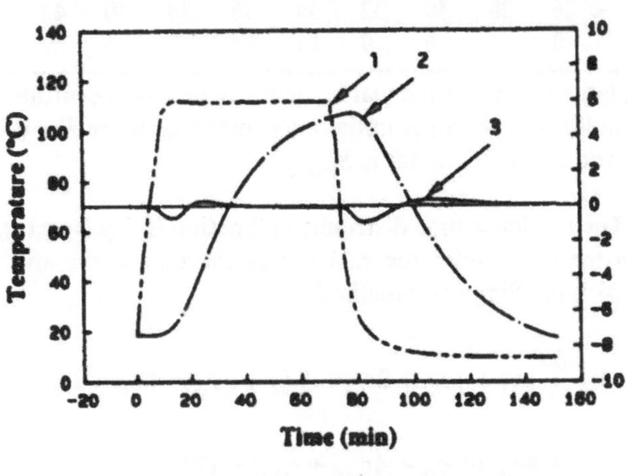

FIGURE E. 4.10.2 The two phases of the process. (1) retort temperature, (2) temperature at the can center, (3) deviation between the predicted - measured temperatures. Model parameters $N = 2$, $j_{j1}^{cut} = 2.22$, $j_1 = 1.87$, $t_d = 4.0$ min, $\tau_1 = 24.0$ min, $\tau_2 = 8.84$ min, $j_2 = -0.917$, standard deviation = 0.13 (Chiheb *et al.*, 1994; © IFT, reproduced by permission).

exit concentration of the microorganisms, $\bar{x}$, from the holding tube may be calculated as:

$$\bar{x} = \int_0^{\infty} x_0 \exp(-kt)E(t)\,dt \tag{4.19}$$

where x_0 is the potential average microbial load of the pockets at the entrance, k is the death rate constant and $E(t)$ is the residence time distribution function. Equation (4.19) considers the pockets with residence times between 0 to ∞. The residence time distribution function $E(t)$ may have values between 0 and 1. When there are no liquid pockets with a specific residence time t_s the residence time distribution function corresponding to t_s will be assigned $E(t_s) = 0$. The residence time distribution function $E(t)$ is different for each process vessel and changes with the flow conditions. It needs to be determined experimentally.

Example 4.11. Thermal process calculation by using the residence time distribution in a holding tube In a HTST process, liquid food enters the heating section at 4°C and leaves at 72°C. Holding temperature is constant at 72°C. The food leaves the cooling section at 12°C. The residence time of the food is very small in the heating and the cooling sections, therefore sterilization calculations may be based on the holding section only. When a tracer was injected into milk at the inlet of the holding tube, the tracer concentrations $c_t(t)$ were recorded at the exit as:

t (s)	< 26	28	30	32	34	36	38	40	42	44	> 46
$c_t(t)$ (g/l)	0	2	6	9	15	14	9	5	1	1	0

a) Calculate the ratio of the viable microorganism concentration at the exit of the holding tube to the initial microbial load when $D = 0.2$ min at $T_{ref} = 72$°C and $z = 6.7$°C. What is $F_{Process}$?

Solution The residence time distribution function is: $E(t_i) = c_t(t_i)/A$ where A is the area under the $c_t(t)$ versus the time curve and may be calculated with the Simpson's method:

$$A = \frac{\Delta t}{3}(c_{t1} + 4c_{t2} + 2c_{t3} + 4c_{t4} + 2c_{t5} + 4c_{t6} + 2c_{t7}$$

$$+ 4c_{t8} + 2c_{t9} + 4c_{t10} + c_{t11}) = 122$$

It should be noticed that $\int_0^{\infty} E(t)\,dt = 1$. Equation (4.19) may be rewritten as:

$$\frac{\bar{x}}{x_0} = \int_0^{\infty} \exp\left(-\frac{2.303\,t}{D}\right) E(t)\,dt.$$

The integral may be calculated using Simpson's method (Table 2.21):

$$\frac{\bar{x}}{x_0}=\frac{\Delta t}{3}\left[\exp\left(-\frac{2.303t_1}{D}\right)E(t_1)+4\exp\left(-\frac{2.303t_2}{D}\right)E(t_2)\right.$$
$$+2\exp\left(-\frac{2.303t_3}{D}\right)E(t_3)+4\exp\left(-\frac{2.303t_4}{D}\right)E(t_4)$$
$$+2\exp\left(-\frac{2.303t_5}{D}\right)E(t_5)+4\exp\left(-\frac{2.303t_6}{D}\right)E(t_6)$$
$$+2\exp\left(-\frac{2.303t_7}{D}\right)E(t_7)+4\exp\left(-\frac{2.303t_8}{D}\right)E(t_8)$$
$$+2\exp\left(-\frac{2.303t_9}{D}\right)E(t_9)+4\exp\left(-\frac{2.303t_{10}}{D}\right)E(t_{10})$$
$$\left.+\exp\left(-\frac{2.303t_{11}}{D}\right)E(t_{11})\right]$$

At time t_i the terms of the equation are:

i	$t_i\,(s)$	$\exp\left(-\frac{2.303\,t_i}{D}\right)$	$E(t_i)$
1	26	0.007	0
2	28	0.005	0.016
3	30	0.003	0.049
4	32	0.002	0.074
5	34	0.001	0.123
6	36	0.001	0.115
7	38	0.001	0.074
8	40	0.000	0.041
9	42	0.000	0.008
10	44	0.000	0.008
11	46	0.000	0

After substituting the numerical values: $\frac{\bar{x}}{x_0}=1.4\times10^{-3}$

$$F_{\text{process}}=D_{T_{\text{ref}}}\log(x_0/x)=0.57\text{ min}$$

b) What percentage of an enzyme with $D=5$ min at $T_{\text{ref}}=72$ °C and $z=33$°C will be destroyed in the process?

Solution Integral may be taken with Simpson's method, similarly as those of the microorganisms. At time t_i terms of the equation are:

i	$t_i(s)$	$\exp\left(-\frac{2.303\,t_i}{D}\right)$	$E(t_i)$
1	26	0.819	0
2	28	0.807	0.016
3	30	0.794	0.049
4	32	0.782	0.074
5	34	0.770	0.123
6	36	0.759	0.115
7	38	0.747	0.074
8	40	0.736	0.041
9	42	0.724	0.008
10	44	0.713	0.008
11	46	0.702	0

Therefore $\bar{c}/c_0 = 0.777$, i.e., 77.7% of the enzyme will survive the process.

Sterilization in the heating and cooling sections is usually neglected and process analysis is simplified by considering the constant temperature holding tube only. ***Axially dispersed plug flow*** may be described as axial dispersion imposed on plug flow. Plug flow in a tube is characterized by the flat velocity profile. Axial dispersion describes the additional random motion of the liquid pockets due to molecular or turbulent diffusion. The intensity of axial dispersion may be described with the axial dispersion coefficient (D_Z) or the Peclet number ($Pe = vL/D_Z$), where v and L are the average velocity of the fluid and the length of the flow vessel, respectively. The limiting case with $Pe \to 0$ (or $D_Z \to \infty$) describes the totally mixed flow, like flow in a CSTR, and $Pe \to \infty$ (or $D_Z \to 0$) describes the plug flow with no mixing. In the holding tube the equation of continuity for spores is:

$$\frac{\partial x}{\partial t} + \frac{\partial N_m}{\partial z} = R_d \tag{4.20}$$

where N_m and R_d are the net flux in the flow direction and the death rate of the microorganisms, respectively. The net flux, including the axial dispersion is:

$$N_m = -D_Z\frac{dx}{dz} + vx \tag{4.21}$$

Assuming steady state and after substituting the first order death rate expression for R_d and (4.21) for N_m (4.20) becomes

$$-D_Z\frac{\partial^2 x}{\partial z^2} + v\frac{\partial x}{\partial z} = -kx \tag{4.22}$$

The boundary conditions for (4.22) are:

$$\text{at } z=0^+ \quad \upsilon x_0 = \left(\upsilon x - D_z \frac{dx}{dz}\right)_{z=0^+} \tag{4.22.a}$$

and

$$\text{at } z=L \quad \left(\frac{dx}{dz}\right)_{z=L} = 0 \tag{4.22.b}$$

Solution of (4.22) with (4.22.a) and (4.22.b) is (Aiba *et al.*, 1973):

$$\frac{X(L)}{X_0} = \frac{4y\exp(Pe/2)}{(1+y)^2\exp(Pe\,y/2) - (1+y)^2\exp(Pe\,y/2)} \tag{4.23}$$

where $y = \left(1 + \frac{4Da}{Pe}\right)^{1/2}$, $Da = k\,L/\upsilon$ = Damköhler number. Solutions of (4.23) are given in graphical form in Figure 4.6.

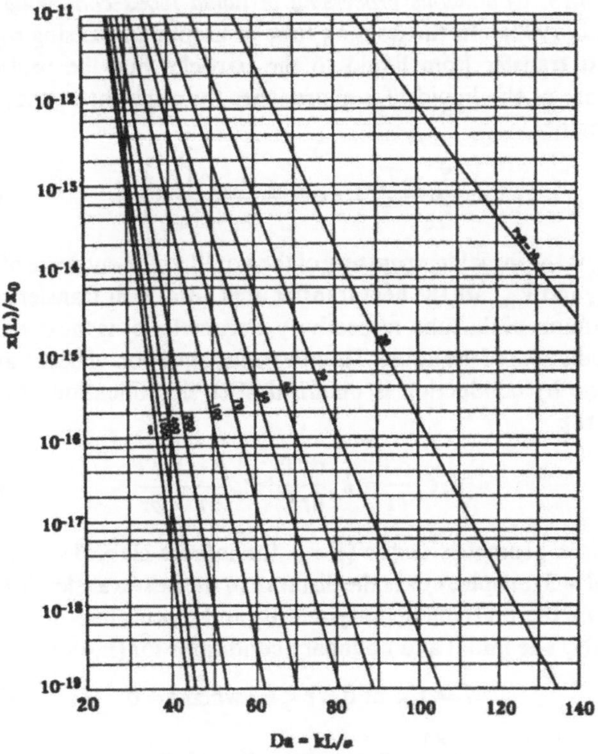

FIGURE 4.6 Effect of axial dispersion on the destruction of the microorganisms at a constant temperature in the holding tube of the continuous pasteurizer (Aiba *et al.*, 1973; © University of Tokyo Press, reproduced by permission).

Example 4.12. Use of axially dispersed plug flow model in continuous thermal process calculations 150 m^3/day of liquid food with potential microbial load of 10^4 cfu/ml will be sterilized at 72°C in a continuous sterilizer. There are 50 parallel tubes of 1 cm diameter in the holding section. D value of the indicator microorganisms is 0.02 min at $T_{ref} = 72$°C. Calculate the required tube length for complete sterilization, i.e., $x_{exit}/x_0 = 10^{-12}$. Consider the following cases:

a) $Pe = 10$
b) Plug flow prevails in the tube.

Solution Flow rate through a single tube is 3 m^3/day, v = flow rate/cross sectional tube area = 6.37 m/min. with $x_{exit}/x_0 = 10^{-12}$ $Da = 100$ when $Pe = 10$ and $Da = 28$ when plug flow prevails (Figure 4.6). Death rate constant is $k_d = (2.303/D_T) = 115\ \text{min}^{-1}$. After substituting the numbers in $Da = kL/v$ we calculate $L = 5.5$ m with $Pe = 10$, and $L = 1.6$ m with plug flow.

Example 4.13. Continuous processing of liquid foods containing particles (*Yang et al.*, 1992) In the holding tube of aseptic processing equipment, when heat transfer from liquid to the particles may be neglected, the temperature of the liquid (T_L) approaches the ambient temperature (T_a) exponentially:

$$\frac{T_L - T_a}{T_{L0} - T_a} = \exp\left\{-\frac{U_{tube} A_{tube}}{\rho_L c_L V_{tube}}\right\} \qquad \text{(E.4.13.1)}$$

where T_{L0} is the initial temperature of the liquid at the entrance of the tube; A_{tube}, U_{tube} and V_{tube} are the heat transfer area, total heat transfer coefficient and the volume of the tube, respectively. Parameter ρ_L is the density and c_L is the specific heat of the liquid. The governing equation of heat transfer into the particle by conduction is obtained after simplification of equations (2.16)–(2.18):

$$\rho_p c_p \frac{\partial T}{\partial t} = k_p \frac{\partial^2 T}{\partial r^2} + k_p \frac{\beta - 1}{r} \frac{\partial T}{\partial r} \qquad \text{(E.4.13.2)}$$

where β is a geometric factor ($\beta = 1$ for infinite slab, $\beta = 2$ for infinite cylinder, $\beta = 3$ for sphere), r is the distance in the heat transfer direction; k_p, ρ_p and c_p are thermal conductivity, density and specific heat of the particles, respectively. The initial and boundary conditions for (E.4.13.2) are:

$$T = T_{p0} \ \text{at}\ 0 < r < r_p \ \text{when}\ t = 0 \qquad \text{(E.4.13.2.a)}$$

$$\frac{dT}{dr} = 0 \ \text{at}\ r = 0 \ \text{when}\ t > 0 \qquad \text{(E.4.13.2.b)}$$

$$-k_p \frac{\partial T}{\partial r} = h_{Lp}(T_L(t) - T) \ \text{at}\ r = r_p \ \text{when}\ t \geq 0 \qquad \text{(E.4.13.2.c)}$$

where r_p is the radius of the particle, h_{Lp} is the convective heat transfer coefficient from liquid to particle. With a spherical particle ($\beta = 3$) after applying the L'Hospital rule at $r = 0$ on the second term (E.4.13.2) gives:

$$\rho_p c_p \frac{\partial T}{\partial t} = 3 k_p \frac{\partial^2 T}{\partial r^2} \text{ at } r = 0 \text{ when } t > 0 \qquad \text{(E.4.13.3)}$$

Equation (E.4.13.3) is the only equation to be solved to evaluate the time temperature history at the critical point of the particle. F values may be calculated at the particle center by using the time-temperature data with:

$$F_i = \int_0^t 10^{(T - T_{ref})/z} dt \qquad \text{(E.4.13.4)}$$

The average $F_{process}$ of the food is:

$$F_{process} = \int_0^\infty F_i(t)\, E(t)\, dt \qquad \text{(E.4.13.5)}$$

The residence time distribution of the particles under each experimental condition was obtained from the experimental data. A typical time-temperature profile and F values of the fluid, particle surface and center are given in Figure E.4.13.1

The model also shows the effect of the convective heat transfer coefficient between the liquid and the particles, and the particle to fluid density ratio, on the thermal processing received by the particles (Figs. E.4.13.2 and E.4.13.3).

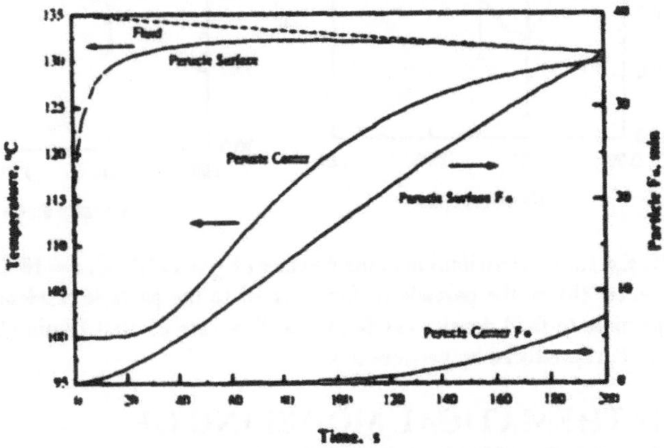

FIGURE E.4.13.1 A typical time-temperature profile and F values ($T_{ref} = 121.1$°C, $z = 10$ °C) of the fluid, particle surface and center; particle with density 1040 kg/m^3 at flow rate 41.5 L/min (Yang *et al.*, 1992 © IFT, reproduced by permission).

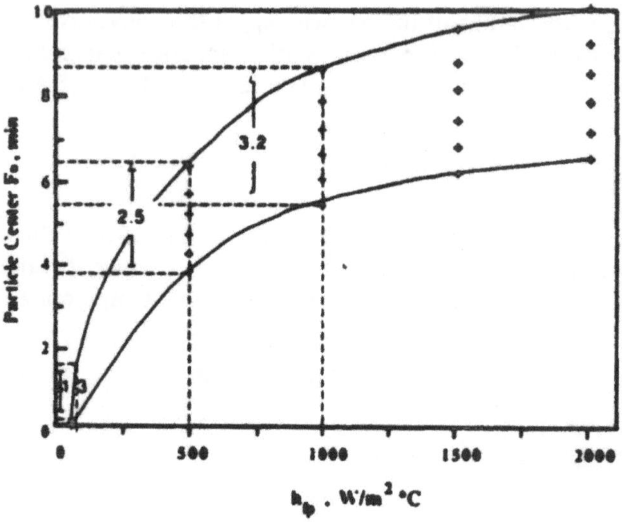

FIGURE E.4.13.2 Effects of the particle residence times and the fluid-to-particle convective heat transfer coefficients (h_{Lp}) on the particle center F values (T_{ref} = 121.1°C, z = 10 °C) in the holding section. Particle density is 1040 kg/m^3 at a flow rate of 41.5 L/min. The points at given h_{Lp} represents minimum to maximum residence times (Yang *et al.*, 1992; ©IFT, reproduced by permission).

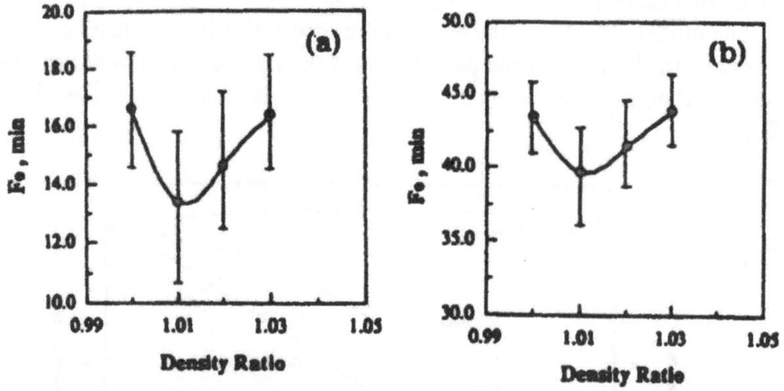

FIGURE E.4.13.3 Distribution of the F values (T_{ref} = 121.1 °C, z = 10°C) (a) at the particle center (b) on the particle surface related to the particle residence times at various particle-to-fluid density ratios for the flow rate of 30.0 L/min (Yang *et al.*, 1992; ©IFT, reproduced by permission).

4.2. MATHEMATICAL MODELING OF EVAPORATION PROCESSES

In an evaporation process water is removed by vaporization to increase the concentration of non-volatile solutes in a liquid food or beverage. It is the most economical and widely used method of concentration (Karel, 1975a). Although the liquid fed into the evaporator may have the physical properties of water, as the concentration increases, the solution behaves differently. The density, viscosity (or consistency), and boiling point may increase with the solid content. Several rate processes are involved in evaporation, but the process may be simply considered as heat transfer from a heater to a boiling liquid. Heat-transfer analysis, combined with mass and energy balance, is usually adequate for modeling the evaporation processes, provided that the boiling point and the physical properties of the liquid are known as a function of the pertinent temperatures and the solute concentrations. Most evaporators are heated by steam condensing in metal tubes. Usually steam is at a low pressure, below 3×10^5 Pa (absolute), and the boiling liquid is under vacuum, up to about 5×10^3 Pa (absolute). The vacuum reduces the boiling temperature of the liquid and increases the temperature difference, and subsequently heat transfer rates, between the steam and the liquid. Many food components are damaged when heated to moderate temperatures for relatively short times. The vacuum also helps to reduce the heat damage to the temperature sensitive food components. In a process consisting of a series of evaporators each evaporator is called an *effect*. When a single evaporator is used, the vapor from the boiling liquid is discarded or used as a steam after recompression. This method is called *single-effect evaporation*. If the vapor from one evaporator is fed into the steam chest of a second evaporator, and the vapor from the second effect to a condenser, the operation becomes double effect. This operational procedure may be generalized by using a higher number of evaporators and referring to it as *multiple-effect evaporation*.

Example 4.14. Basic calculations for a single effect tomato paste concentration process 200 000 kg/day of tomato pulp at 35°C with 5% solids will be converted into tomato paste with 30% solids in a single stage evaporator (Fig. E.4.14) operating under vacuum at 93°C (77 kPa, $H = 2664$ kJ/kg, $h = 387$ kJ/kg, where H and h are the enthalpies associated with the vapor and the liquid water, respectively). Saturated steam was supplied to the heat exchanger under 1250 kPa pressure ($T = 190$°C, $H = 2786$ kJ/kg, $h = 808$ kJ/kg). Specific heat of tomato pulp was 4.01 kJ/kg°C.

a) How much vapor should be removed?

Solution After taking 24 of operation as the basis for our calculations and assuming that negligible amounts of solids are entrained with the vapor stream, the solids balance is $Fx_F = Lx_L$. After substituting $F = 200\,000$ kg, $x_F = 0.05$ and $x_L = 0.30$ we obtain $L = 33\,333$ kg. Total material balance requires $F = V + L$, therefore $V = 166\,667$ kg.

b) What is the steam economy V/S of the process?

Solution We will choose liquid water at 93°C as a reference and make total energy balance around the evaporator:

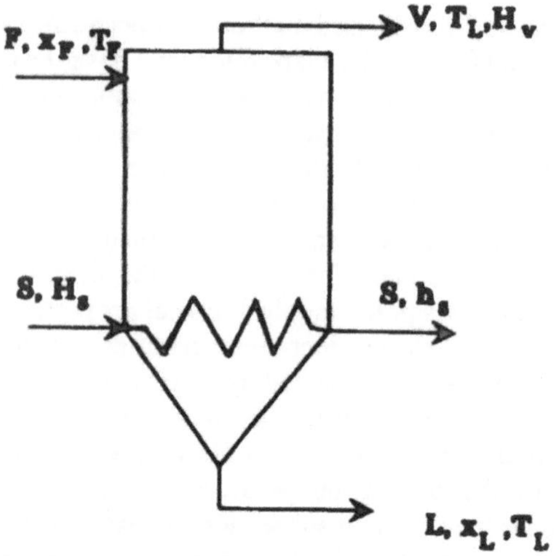

FIGURE E.4.14 Single state evaporator for concentration of the tomato pulp.

$$F\, c_{p,\text{pulp}}(T_F - T_{ref}) + S(H_s - h_{ref}) = Lc_L(T_L - T_{ref}) + V(H_V - h_{ref}) + S(h_s - h_{ref})$$

After substituting $F = 200\ 000$ kg, $c_{p,\text{pulp}} = 4.01$ kJ/kg°C, $T_F = 35$°C, $T_{ref} = T_L = 93$°C, $H_s = 2786$ kJ/kg, $h_{ref} = 387$ kJ/kg, $V = 166\ 667$ kg, $H_V = 2664$ kJ/kg, $h_s = 808$ kJ/kg we calculate $S = 215\ 378$ kg. The steam economy is $V/S = 77.4\%$.

c) If the overall heat transfer coefficient is 3100 W/m^{2}K what is the total heat transfer area?

Solution Total heat transfer from the heat exchanger to the tomato paste is $q = U\,A\,\Delta T = S(H_s - h_s)$. After substituting $S = 2.5$ kg/s, $H_s = 2786$ kJ/kg, $h_s = 808$ kJ/kg, $U = 3100$ W/m^{2}K and $\Delta T = 190 - 93 = 97$ K, the total heat transfer area is calculated as 16.4 m^2.

Example 4.15. Basic calculations for a double effect tomato paste concentration process A double effect forward feed process (Fig. E.4.15) was suggested to achieve the same process described in the previous example. The evaporators will have identical heat transfer areas of 16.4 m^2. The first evaporator will operate at 150°C (476 kPa, $H = 2747$ kJ/kg, $h = 632$ kJ/kg). The second evaporator will operate at 93°C (77 kPa, $H = 2664$ kJ/kg, $h = 387$ kJ/kg). Heat transfer coefficients are expected to be $U_1 = 3750$ W/m^{2}K and $U_2 = 3300$ W/m^{2}K.

a) Make the heat balance around the heat exchanger of the second evaporator to calculate the amount of vapor removed in the first effect.

Solution Total heat transfer from the heat exchanger to the tomato paste in the second evaporator is $q_2 = U_2\,A_2\,\Delta T_2 = V_1\,(H_{v1} - h_{v1})$. After substi-

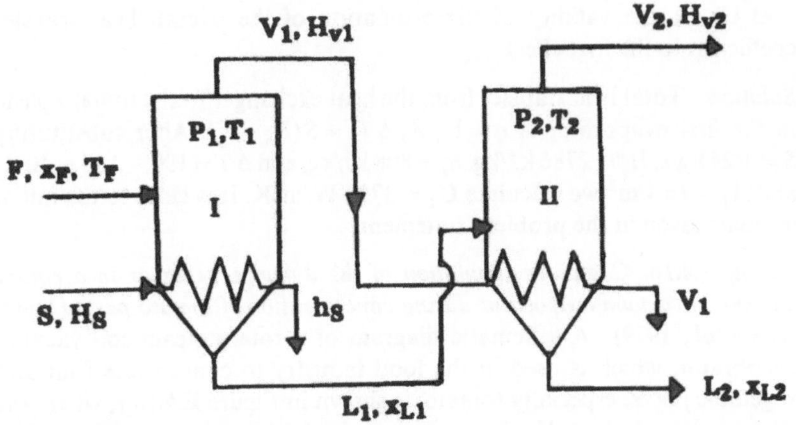

FIGURE E.4.15 Double stage evaporator for concentration of the tomato pulp.

tuting $H_{v1} = 2747$ kJ/kg, $h_{v1} = 632$ kJ/kg, $U_2 = 3300$ W/m^{2}K, $A_2 = 16.4$ m^2 and $\Delta T = 150 - 93 = 57$ K, amount of vapor removed in the first effect is calculated as $V_1 = 1.28$ kg/s $= 110\ 744$ kg/day.

b) What is be the solids content of the concentrated pulp stream leaving the first evaporator?

Solution Total material balance around the first evaporator is $F = V_1 + L_1$. When we base our calculations on a single day of operation with $F = 200\,000$ kg and $V_1 = 110\,744$ kg we calculate $L_1 = 89256$ kg. When we neglect the solids entrained with V_1, solids balance around the first evaporator is $F\,x_F = L_1\,xL_1$. After substituting $F = 200\,000$ kg, $L_1 = 89256$ kg and $x_F = 0.05$ we calculate $x_{L1} = 0.11$.

c) How much steam should be used in the first evaporator?

Solution We will choose liquid water at 150°C as a reference and make the total energy balance around the evaporator:

$$F\,c_{p,\text{pulp}}(T_F - T_{\text{ref}}) + S(H_s - h_{\text{ref}}) = L_1 c_{L1}(T_{L1} - T_{\text{ref}}) + V_1(H_{V1} - h_{\text{ref}}) + S(h_s - h_{\text{ref}})$$

After substituting $F = 200\ 000$ kg, $c_{p,\text{pulp}} = 4.01$ kJ/kg°C, $T_F = 35$°C, $T_{\text{ref}} = T_L = 150$°C, $H_s = 2786$ kJ/kg, $h_{\text{ref}} = 632$ kJ/kg, $V_1 = 110\ 744$ kg, $H_V = 2747$ kJ/kg, $h_s = 808$ kJ/kg we calculate $S = 106\ 786$ kg.

d) What is the steam economy with the new design?

Solution Total material balance around the evaporators requires $F = V_1 + V_2 + L_2$, after substituting $F = 200\ 000$ kg, $V_1 = 110\ 744$ kg and $L_2 = 33\ 333$ kg (from the previous example) we calculate $V_2 = 55\ 923$ kg. Steam economy $= (V_1 + V_2)/S = 1.56 = 156\%$. Using a two stage evaporation process substantially increased the steam economy.

e) Check the validity of the estimation of the overall heat transfer coefficient in the first effect

Solution Total heat transfer from the heat exchanger to the tomato paste in the first evaporator is $q = U_1 A_1 \Delta T_1 = S(H_s - h_s)$. After substituting $S = 1.24$ kg/s, $H_s = 2786$ kJ/kg, $h_s = 808$ kJ/kg, and $\Delta T = 190 - 150 = 40$ K, and $A_1 = 16.4$ m^2 we calculate $U_1 = 3739$ W/m$_2$K. It is close to the initial estimate given in the problem statement.

Example 4.16. Computer simulation of the dynamic behavior in a rotary steam-coil vacuum evaporator during concentration of tomato paste (Lima-Hon et al., 1979) A schematic diagram of a rotary steam-coil vacuum evaporator, which is used in the food industry to concentrate fruit and vegetable juices, especially tomato, is shown in Figure E.4.16.1, where the steam coil is submerged in the product, acting both as a heat transfer surface and an agitation unit.

In modeling the dynamic behavior of vacuum evaporation of tomato paste the liquid phase was assumed to be well mixed and in equilibrium with the vapor phase. No boiling was considered at the coil/liquid interface. Heat losses were assumed to be negligible and no fouling formed.

The mass balance on the liquid inside the evaporator (W) is:

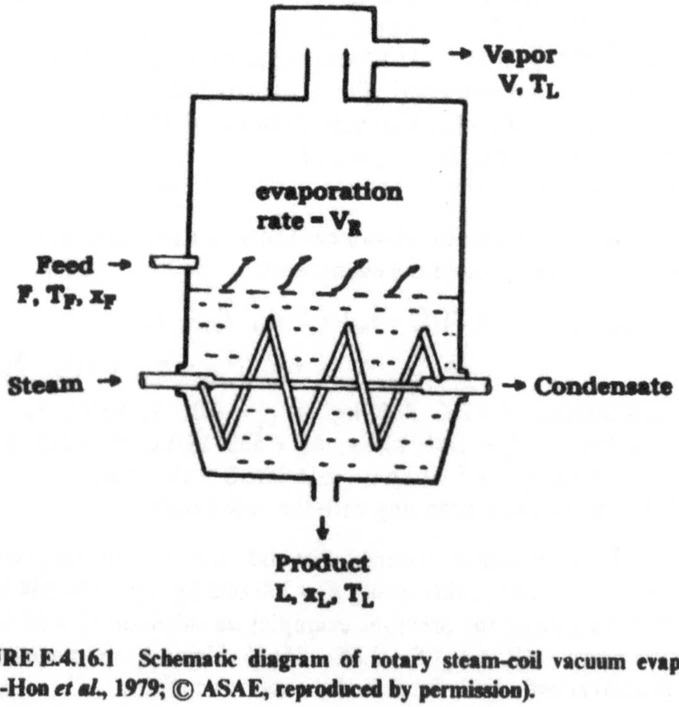

FIGURE E.4.16.1 Schematic diagram of rotary steam-coil vacuum evaporator (Lima-Hon *et al.*, 1979; © ASAE, reproduced by permission).

$$\frac{dW}{dt} = F - V_R - L \qquad \text{(E.4.16.1)}$$

where W is the amount of liquid in the evaporator and V_R is the rate of evaporation. Water and solid balances around the liquid in the evaporator require:

$$F(1 - x_F) = V_R + L(1 - x_L) \qquad \text{(E.4.16.2)}$$

$$F x_F = L x_L \qquad \text{(E.4.16.3)}$$

The mass balance on vapor is:

$$\frac{dM}{dt} = V_R - V \qquad \text{(E.4.16.4)}$$

where M is the amount of vapor in the evaporator. Energy balance on liquid inside the evaporator is:

$$\frac{d(Wc_{WL}T_L)}{dt} = F c_{WF} T_F - V_R(c_{WL}T_L + \Delta H_{vap}) - Lc_{WL}T_L \qquad \text{(E.4.16.5)}$$

where c_{WL} and c_{WF} are the specific heats of water at temperatures T_L and T_F, respectively; ΔH_{vap} is the latent heat of evaporation at temperature T_L. The pressure inside the evaporator was obtained from the ideal gas law:

$$PV_{vapor} = \frac{M}{MW_W} RT_L \qquad \text{(E.4.16.6)}$$

where MW_W is the molecular weight of water. Volume of the vapor phase is:

$$V_{vapor} = V_{evaporator} - \frac{W}{\rho} \qquad \text{(E.4.16.7)}$$

where $V_{evaporator}$ is the volume of the evaporator and ρ is density of the liquid phase. The temperature is obtained from the pressure/boiling point relationship:

$$T = -\frac{5210}{\log(P) - 13.96} \qquad \text{(E.4.16.8)}$$

The heat flux Q is obtained from

$$Q = U_0 A_0 (T_{steam} - T_L) \qquad \text{(E.4.16.9)}$$

where U_0 is the over-all heat transfer coefficient, based on the outside coil surface, and A_0 is the outside surface area of the coil. The over-all heat transfer coefficient was derived from the local heat transfer coefficient equation as follows:

$$\frac{1}{U_0}=\frac{1}{h_0}+\frac{A_0\Delta x_w}{A_m k_w}+\frac{A_0}{A_i h_i} \quad \text{(E.4.16.10)}$$

where h_0 and h_i are the inside and outside film heat transfer coefficients, A_i and A_m are the inside and the mean surface area of the coil, Δx_w and k_w are the thickness and the thermal conductivity of the coil wall.

The effect of physical properties on the local heat transfer coefficient was correlated by the following equation:

$$\left(\frac{h_0 D_t}{k}\right)=0.60\,x^{-2.51}\left(\frac{\rho N D_e^2}{\mu}\right)^{0.62}\left(\frac{c\mu}{k}\right)^{1/3}\left(\frac{\mu}{\mu_w}\right)^{0.14} \quad \text{(E.4.16.11)}$$

where c = specific heat, D_e = outside diameter of the coil, D_t = inside diameter of the evaporator, N = rotational speed of the coil, x = percentage of the total solids in the liquid, μ = viscosity and μ_w = viscosity at outside coil wall temperature. Equation (E.4.16.11) is valid when $1 \leqslant \text{Re} \leqslant 200$ and $20\% \leqslant x \leqslant 50\%$.

The following correlations were used for the physical properties

$$c=4186\,(1-0.67\,x) \quad \text{(E.4.16.12)}$$

$$\rho=99.6\,(0.44\,x+0.997) \quad \text{(E.4.16.13)}$$

$$k=\frac{(5.75-4.8x)(993+500x-0.57(T-273))\,10^{-4}}{1.11-0.0036(T-273)} \quad \text{(E.4.16.14)}$$

$$\mu=0.362\,x^{2.4}\left(\frac{\dot{\gamma}}{500}\right)^{-0.58}\exp\left(\frac{950}{T}\right) \quad \text{(E.4.16.15)}$$

where temperature T and shear rate $\dot{\gamma}$ appear in K and 1/s, respectively. Equations (E.4.16.12) – (E.4.16.15) predicts the specific heat c in J/Kg, density ρ in kg/m^3, thermal conductivity k in W/mK and viscosity μ in N/sm^2.

The computer flow diagram of the mathematical model has been depicted in Figure E.4.16.2. Variation of the overall heat transfer coefficient, concentration of the solutes in the evaporator, evaporation rates and the wall temperature during the operation are simulated in Figures E.4.16.3 and E.4.16.4.

Example 4.17. Modeling the operation of the multiple effect evaporators used in the sugar industry (*Radovic et al.*, 1979) Operation of a five-effect evaporator system used in the sugar industry (Fig. 4.17.1) has been simulated by using the following equations:
Enthalpy balance:

$$F(h(T_F,x_F)-h(T_{L1},x_1))+(S-E_0)\,\Delta H_{vap}=$$
$$(F-L_1)[H(T_{L1})-h(T_{L1},x_1)] \quad \text{(E.4.17.1)}$$

$$L_{i-1}(h(T_{Li-1},x_{i-1})-h(T_{Li},x_i))+(L_{i-2}-L_{i-1}-E_{i-1}+R_{i-1})$$

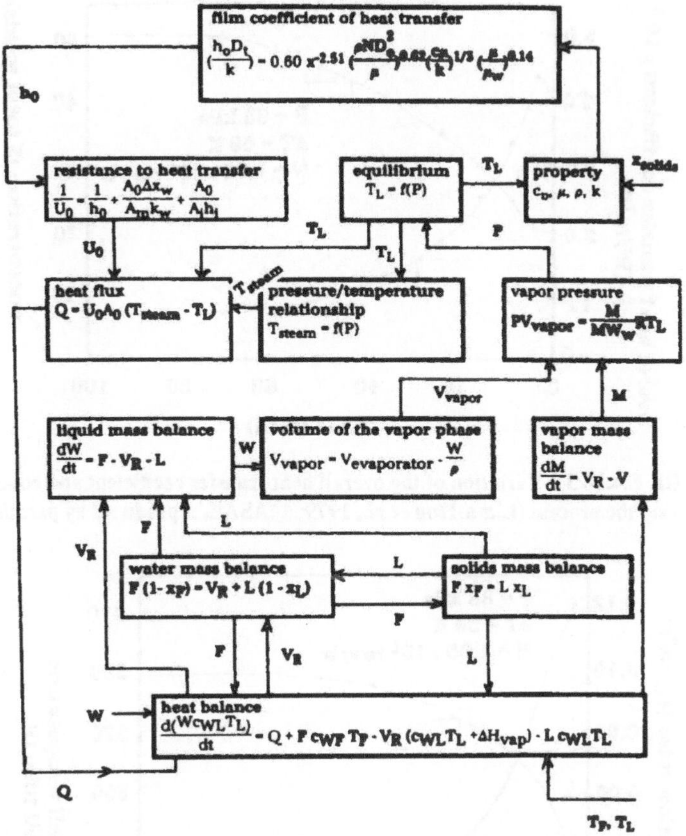

FIGURE E.4.16.2 Mathematical model of the dynamic evaporation process (Lima-Hon *et al.*, 1979; ©ASAE, reproduced by permission).

$$[H(T_{Li-1}) - h(T_{i-1})] = (L_{i-1} - L_i)[H(T_{Li}) - h(T_{Li}, x_i)]$$

$$i = 2, 3, \ldots \qquad \text{(E.4.17.2)}$$

Heat transfer rate:

$$U_1 A_1 (T_S - T_{L1}) = (S - E_0)\Delta H_{vap} \qquad \text{(E.4.17.3)}$$

$$U_i A_i (T_{i-1} - T_{Li}) = (L_{i-2} - L_{i-1} - E_{i-1} + R_{i-1})$$

$$[H(T_{Li-1}) - h(T_{i-1})] \quad i = 2, 3, \ldots \qquad \text{(E.4.17.4)}$$

Phase equilibrium:

$$B_1 + B_2 x_i + B_3 T_i + B_4 x_i^2 + B_5 x_i T_i + B_6 T_i^2 + B_7 x_i^3$$

$$+ B_8 x_i^2 T_i + B_9 x_i T_i^2 + B_{10} T_i^3 = T_{Li} \qquad \text{(E.4.17.4)}$$

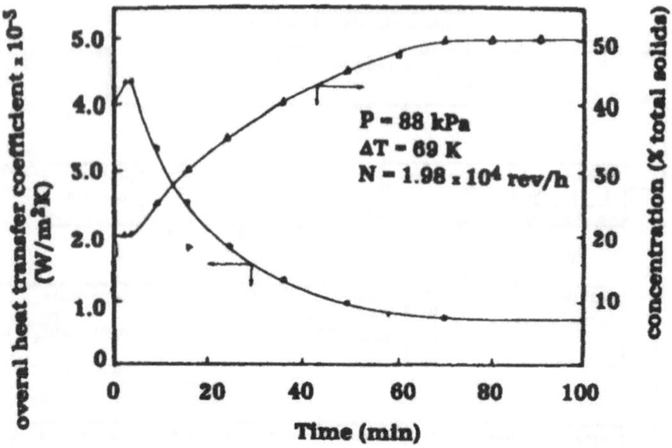

FIGURE E.4.16.3 Variation of the overall heat transfer coefficient and concentration during the process (Lima-Hon *et al.*, 1979; ©ASAE, reproduced by permission).

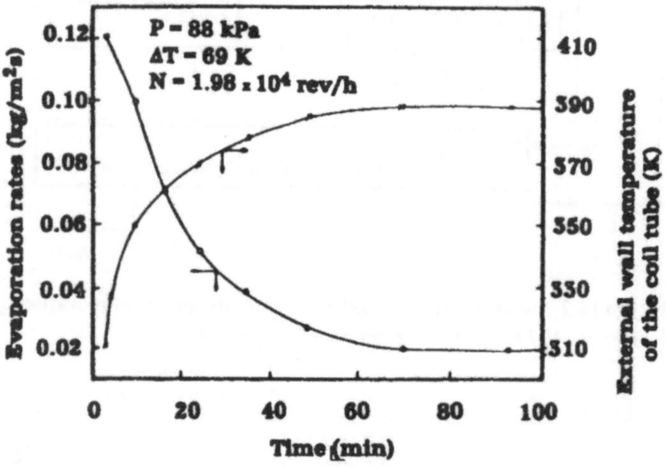

FIGURE E.4.16.4 Variation of the evaporation rates and external wall temperature during the process (Lima-Hon *et al.*, 1979; ©ASAE, reproduced by permission).

Solute balance:

$$F x_F = L_i x_i \quad i = 1, 2, 3, \ldots \tag{E.4.17.5}$$

The overall heat transfer coefficient for each effect may be calculated as a function of inlet and outlet solution concentrations. Although various

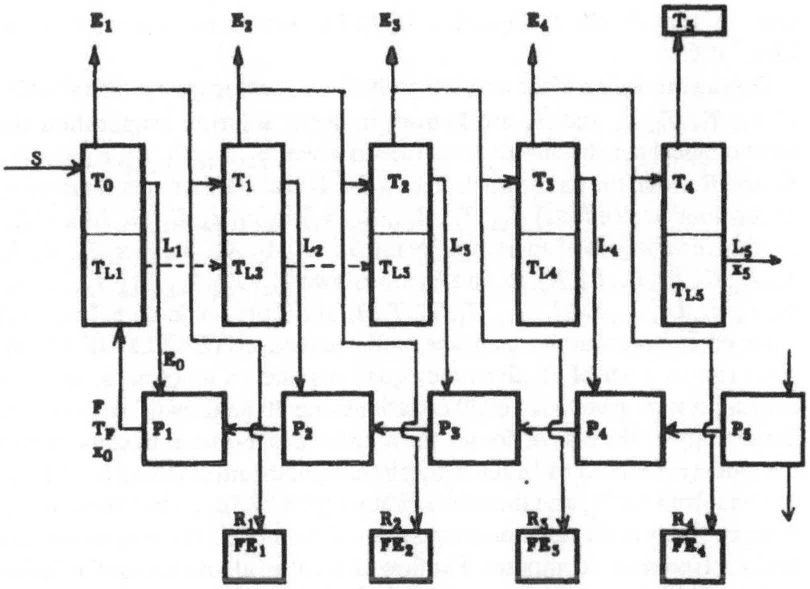

FIGURE E.4.17.1 A typical five-effect forward feed sugar solution evaporator plant including vapor bleeds, solution preheaters, flash evaporators, and condenser. The solution preheaters (P) are used to heat up the feed by using the vapor leaving the evaporators, then the vapor stream is passed through a flash evaporator (FE) to reduce its pressure and temperature and also to remove the liquid in equilibrium with vapor (Reprinted with permission from Radovic *et al.*, © 1979 ACS).

correlations are available only the Baloh equation is given here:

$$U_i = \frac{18.84 \times 10^6}{(x_{i-1}^2 + x_i^2 + 800)} \tag{E.4.17.6}$$

where $h(T, x)$ and $H(T, x)$ are the temperature and solute concentration dependent enthalpies of the liquid and the vapor phases; T_{Li} is the temperature of the boiling liquid leaving the ith effect, T_F is the temperature of the feed, T_i is the saturation temperature at pressure P_i of the vapor leaving the ith effect, T_S is the temperature of the steam entering the first evaporator, x_F is the mass fraction of solute in the solution entering the first effect, x_i is the mass fraction of the solute in the solution leaving i^{th} effect, F is the flow rate of the solution at the inlet of the plant, S is the flow rate of the steam to the first effect, E_0 is flow rate of non-condensable steam in the first effect, E_i is flow rate of vapor bleeds from the ith evaporator, ΔH_{vap} is the latent heat of vaporization of the steam entering the first evaporator, R_i is the flow rate of vapor returned from a flash evaporator to $i + 1^{st}$ effect, $B_1 - B_{10}$ coefficients of the polynomial for the phase equilibrium relationship, U_i is the overall heat transfer coefficient for the ith effect and A_i is the heat transfer area in the ith effect. Equation (E.4.17.6) estimates values of U_i in kJ/m^2 h°C.

During the design of the above five effect evaporator system the variables F, x_F, T_F, T_S, x_5 and T_5 are known; in sugar solution evaporation the technological constraints also sets the variables, E_0, E_1, E_2, E_3, E_4, R_1, R_2, R_3 and R_4, then the unknown variables are A (heat transfer area assumed to be the same in all effects), T_{L1}, T_{L2}, T_{L3}, T_{L4}, x_1, x_2, x_3, x_4, L_1, L_2, L_3 and L_4. During the analysis of an existing plant A_1, A_2, A_3, A_4, A_5, F, x_F, T_F, T_S, T_5, E_0, E_1, E_2, E_3, E_4, R_1, R_2, R_3 and R_4 are known T_{L1}, T_{L2}, T_{L3}, T_{L4}, T_{L5}, x_1, x_2, x_3, x_4, x_5, L_1, L_2, L_3, L_4, L_5, T_1, T_2, T_3, T_4 and S are not known. In case of a five effect evaporator system the model (equations (E.4.17.1)–(E.4.17.6)) gives rise to a set of 18 algebraic equations and 18 unknowns, and this existing system gives a set of 20 equations and 20 unknowns. It should be noted that, for the case of design, the number of simultaneous equations of the model is reduced to 18 since the phase equilibrium equation ((E.4.17.4)) can be solved for T_{L_5} and the solute balance ((E.4.17.5)) can be solved for x_5, independently of the remaining equations of the model. These equations are easily solved with a computer. The flow chart of an algorithm for the design and analysis of operation of a multiple-effect evaporator system is depicted in Figure E.4.17.2.

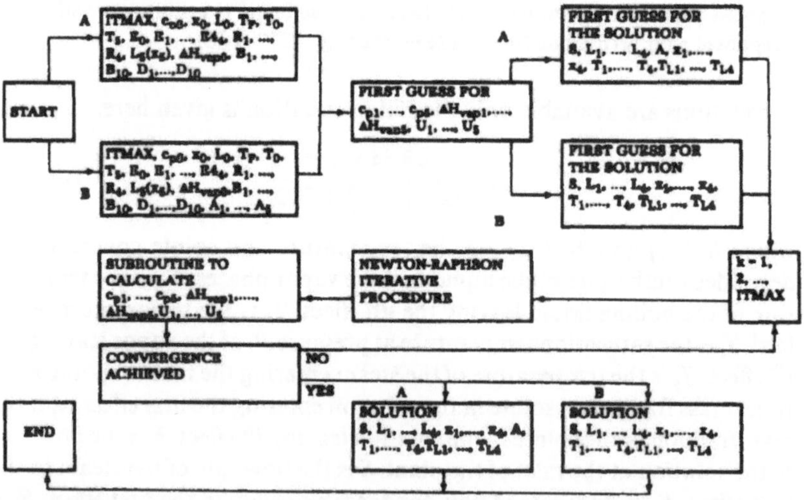

FIGURE E.4.17.2 Algorithm for the design (A) and analysis (B) of a multiple-effect evaporator system. c_{pS} = specific heat of steam, $D_1, \ldots, D_{10}$ = coefficients of the polynomial for the specific heat of the solution, ITMAX = maximum number of iterations, ΔH_{vap0} = latent heat of evaporation of primary water vapor in the first effect (Reprinted with permission from Radovic *et al.*, © 1979 ACS).

4.3. MATHEMATICAL MODELING OF CRYSTALLIZATION PROCESSES

There are two basic types of crystallization processes involved in food processing: *Crystallization from a solution* and *crystallization from a melt.* Typical examples crystallization from a solution are crystal sugar production, crystallization of sugar in jams, potassium acid tartarate precipitation in wines, crystallization of high melting point glycerides in vegetable oils, etc. The most common examples of crystallization from a melt are freezing and freeze concentration processes.

Crystallization occurs in two distinct steps: *Nucleation* and *crystal growth.* Various kinds of nucleation have been identified. Spontaneous or primary nucleation occurs due to high supersaturation in the absence of crystals. The approximate supersolubility curve for sucrose is shown in Figure 4.7. Nucleus formation starts above the supersolubility curve. In the metastable zone, crystals may grow but nucleation cannot occur. Crystals dissolve when the concentration is below the solubility curve.

Secondary nucleation is caused by presence of particles of a material that is being crystallized. Adding seed crystals to a solution in the metastable zone is a typical application of the secondary nucleation process. *Contact nucleation,* a special case of the secondary nucleation, is the formation of new nuclei from a parent crystal of the material being crystallized. Crystals may collide with each other or with the surface of the crystallization equipment and give birth to new crystals. A solution may be brought into

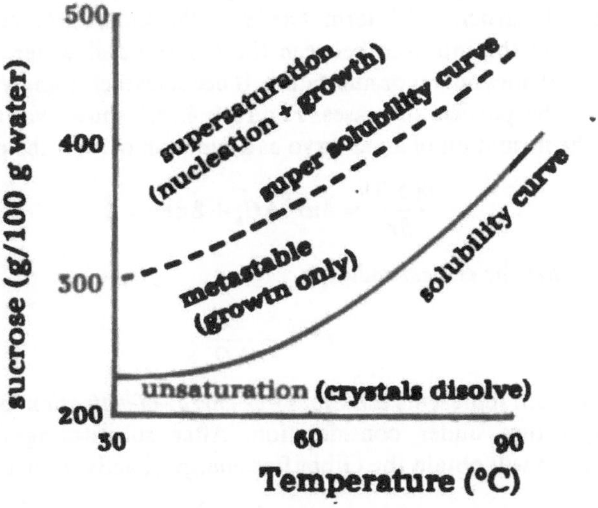

FIGURE 4.7 The approximate supersolubility curve for sucrose.

the labile or metastable zone after cooling at a constant concentration or concentrating at a constant temperature. Evaporators are used to achieve the later process.

During *homogeneous nucleation* molecules combine spontaneously, without the presence of any foreign molecules, to form an embryo, which may grow further and form a nucleus:

$$A + A \rightleftarrows A_2$$
$$A_2 + A \rightleftarrows A_3$$
$$\vdots$$
$$A_{n-1} + A \rightleftarrows A_n$$

Embryos smaller than a critical size do not grow and dissociate. During *heterogeneous nucleation*, the embryos are formed on the catalytic surfaces of foreign particles.

Example 4.18. Critical embryo size for stable ice nucleation (*Franks*, 1982) The total Gibbs free energy of formation of an embryo (ΔG) is:

$$\Delta G = \frac{4}{3}\pi r^3 \Delta G_v + 4\pi r^2 \sigma \qquad \text{(E.4.18.1)}$$

where r = radius of the embryo, ΔG_v = Gibbs energy of forming the bulk solid from the liquid phase and σ is the interfacial tension between water and embryo. The term $4/3\pi r^3 \Delta G_v$ is always negative below 0°C, therefore favors nucleation. It becomes increasingly more negative as the radius of the particle increases. This term relates mostly to the decrease in Gibbs energy when hydrogen bonds form during association of water molecules in an ordered particle. The term $4\pi r^2 \sigma$ is the Gibbs free energy of the formation of the interface between the embryo and water. It is always positive and does not favor nucleation. It becomes increasingly larger as the radius of the particle increases. Figure E.4.18.1 shows variation of ΔG during the formation of an embryo as a function of r. At the maximum:

$$\frac{\partial(\Delta G)}{\partial r} = 4\pi r^2 \Delta G_v + 8\pi r \sigma = 0 \qquad \text{(E.4.18.2)}$$

implying that the *critical radius* is:

$$r^* = -\frac{2\sigma}{\Delta G_v} \qquad \text{(E.4.18.3)}$$

The maximum represents the *Gibbs free energy of activation* for nucleation at temperature under consideration. After substituting (E.4.18.3) in (E.4.18.1) we will obtain the Gibbs free energy of activation as:

$$\Delta G^* = \frac{16\pi\sigma^3}{3\Delta G_v^2} \qquad \text{(E.4.18.4)}$$

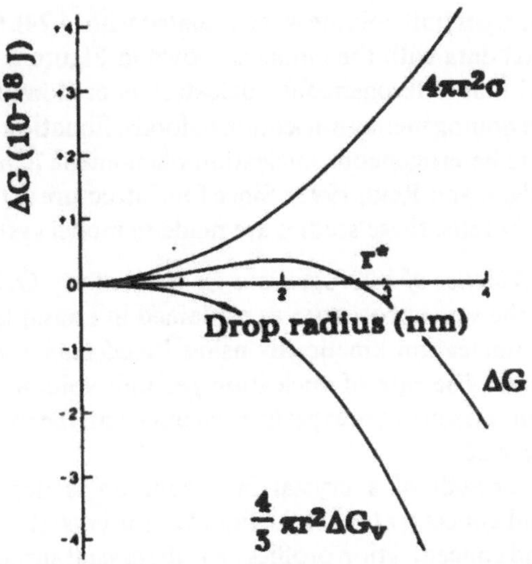

FIGURE E.4.18 Thermodynamics of homogeneous nucleation of ice in supercooled water at −40°C. The critical radius $r^* = 1.85$ nm and contains approximately 200 molecules (Franks, 1982; ©Plenum Press, reproduced by permission).

At a critical cluster size, by gaining a further molecule the free energy of transfer from bulk to a cluster becomes negative and the nucleation process becomes spontaneous. The homogeneous nucleation rate is determined by the rate by which the clusters of critical size gain a further molecule. Hoffman (1958) and Buckle (1961) have shown that the homogeneous nucleation rate may be expressed as:

$$J = A \exp(B\tau_\theta) \tag{4.24}$$

where τ_θ is a function of reduced temperature θ.

$$\tau_\theta = \frac{1}{\theta^3(1-\theta)^2} \tag{4.25}$$

in which $\theta = T/T_m$, where T = process temperature, T_m = equilibrium melting point.

Example 4.19. Kinetics of homogeneous ice nucleation Charoenrein and Reid (1989) studied homogeneous ice nucleation in water and sucrose solutions. Water or sucrose solution were dispersed in continuous phase (silicone oil) to obtain 10^6–10^7 drops per unit volume. Growth of the ice crystals is rapid compared to nucleation, therefore the rate of the nucleation process was assumed to be the same as droplet freezing, and determined after measuring the heat released in a differential scanning calorimeter. The rate of nucleation per unit volume was simulated with (4.24). Comparison of the experimental data with the model is shown in Figure E.4.19.

It is believed that heterogeneous nucleation is predominant and more important than homogeneous nucleation in foods. Equation (4.24) was also used to simulate heterogeneous nucleation phenomena (Charoenrein and Reid,1989; Özilgen and Reid, 1993). Since food structure is too complex to obtain decisive results, these studies are made in model systems.

Example 4.20. Kinetics of heterogeneous ice nucleation Özilgen and Reid (1993) applied the same procedure as explained in Example 4.19 to study heterogeneous nucleation kinetics by using *Pseudomonas syringae* as the nucleating agents. The rate of nucleation per unit volume was simulated with (4.24). Comparison of the experimental data with the model is exemplified in Figure E.4.20.

The rate of growth of a crystal in a solution is dependent on the temperature and concentration of the liquid at the crystal face. Qualitative temperature and concentration profiles near the crystal surface are depicted in Figure 4.8.

Example 4.21. Modeling crystal growth with combined solute diffusion and surface reaction During crystal growth in a solution solute molecules diffuse through the laminar film surrounding the crystal and join the lattice with a surface reaction. Crystal growth rate may be expressed as a function

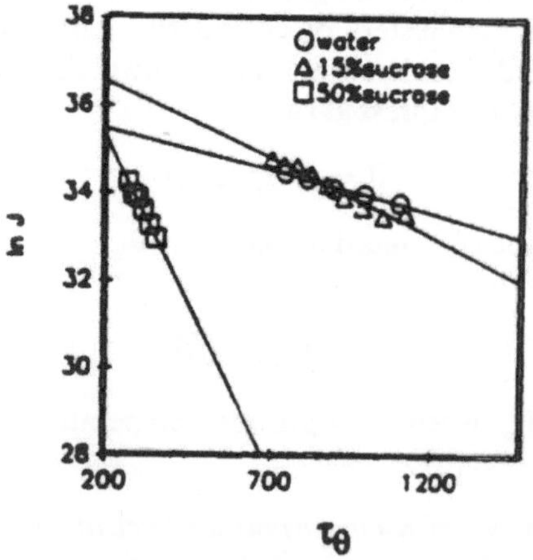

FIGURE E.4.19 **Homogeneous ice nucleation rate in water and sucrose solution as a function of τ_θ. The constants were: $A = 3.32 \times 10^{27}$ ($s^{-1}m^{-3}$), $B = -0.396$ with water; $A = 3.10 \times 10^{26}$ ($s^{-1}m^{-3}$), $B = -0.397$ with 15% sucrose solution; $A = 2.24 \times 10^{22}$ ($s^{-1}m^{-3}$), $B = -0.382$ with 50% sucrose solution (Charoenrein and Reid, 1989; © Elsevier Science Publishers B.V., reproduced by permission).**

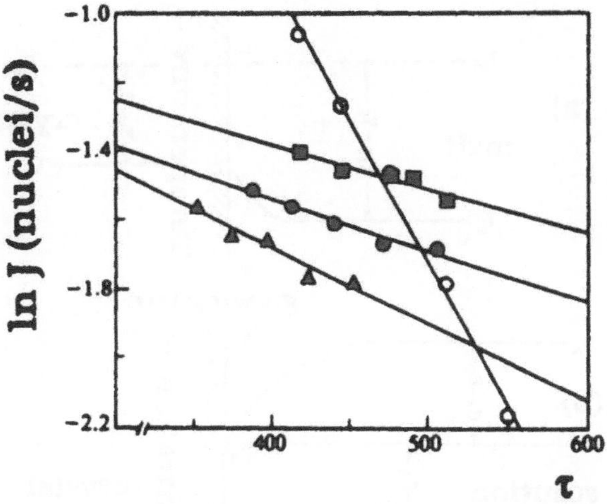

FIGURE E.4.20 **Heterogeneous ice nucleation rate in *P. syringae*/D-alanine solutions as a function of τ_g. Experimental data were shown in symbols and obtained with the following D-alanine concentrations; (○) no D-alanine, $A = 5.0 \times 10^{15}$ ($s^{-1}m^{-3}$), $B = -8.3 \times 10^{-3}$; (■): 1.0 mole/L D-alanine, $A = 6.4 \times 10^{14}$ ($s^{-1}m^{-3}$), $B = -1.3 \times 10^{-3}$; (●): 1.5 mole/L D-alanine, $A = 1.84 \times 10^{14}$ ($s^{-1}m^{-3}$), $B = -1.5 \times 10^{-3}$; (▲): 2.0 mole/L D-alanine $A = 4.8 \times 10^{14}$ ($s^{-1}m^{-3}$), $B = -2.2 \times 10^{-3}$, (Özilgen and Reid, 1993; ©Academic Press, reproduced by permission).**

of the mass transfer as (Coulson and Richardson, 1968):

$$\frac{dm}{dt} = kA(c_A - c_B) \tag{E.4.21.1}$$

where A = interfacial area. Mass transfer coefficient k may be expressed by using the film model (Example 2.16) as:

$$k = \frac{D}{\delta} \tag{E.4.21.2}$$

where D = diffusivity of the solute, δ = hypothetical film thickness. The surface reaction rate on the interface is directly proportional to the supersaturation at the interface (Coulson and Richardson, 1968):

$$\frac{dm}{dt} = K_s A(c_B - c_S)^n \tag{E.4.21.3}$$

where K_s = surface reaction rate constant and n = constant. When $n = 1$ after combining (E.4.21.1), (E.4.21.2) and (E.4.21.3) we will obtain an expression for the crystal growth rate in a solution when both the surface

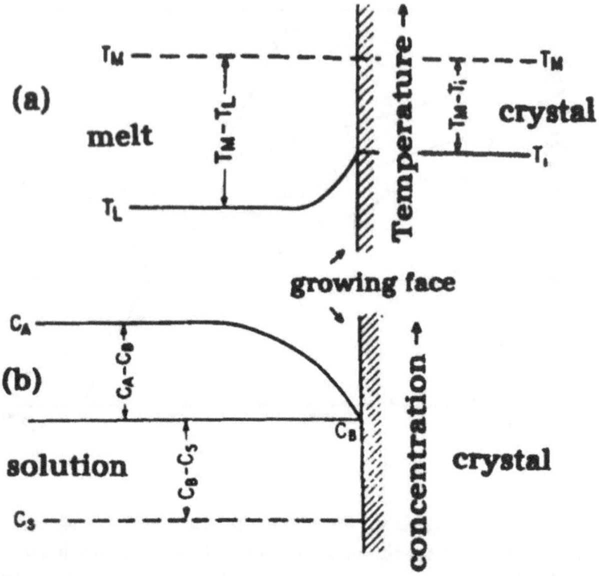

FIGURE 4.8 (a) Qualitative temperature profile for crystallization from a melt and (b) qualitative concentration profile for crystallization from a solution. c_A = concentration of the solute in the bulk liquid, c_B = concentration of the solute on the crystal surface, c_S = concentration of the saturated liquid; T_i = temperature of the interface, T_L = temperature of the liquid, T_M = melting point.

reaction and the mass transfer rates are not negligible:

$$\frac{dm}{dt} = \frac{A(c_A - c_S)}{\dfrac{\delta}{D} + \dfrac{1}{K_s}} \qquad \text{(E.4.21.4)}$$

Example 4.22. Kinetics of potassium bitartarate crystallization from wines (Dunsford and Boulton, 1981) When the mass transfer on the crystal surface is the rate limiting step, the rate of potassium bitartarate removal from wines with crystallization was described as:

$$-\frac{dm}{dt} = k_D \frac{DA}{\delta}(c - c^*) \qquad \text{(E.4.22.1)}$$

where $-dm/dt$ is the rate of solute disappearance, k_D is a constant, D is the diffusivity of the solute, A is the surface area of the crystal δ is the film thickness and $(c - c^*)$ is the degree of supersaturation. When the rate determining step is the reaction of the solute molecules joining the lattice, the rate of solute removal from the solution would be:

$$-\frac{dm}{dt} = K_s(c - c^*)^2 \tag{E.4.22.2}$$

It was concluded that crystallization of potassium bitartarate from table wines generally displays rate limiting steps controlled by nucleation, surface reaction and mass transport at various points in time. The length of time for which each process is controlling is determined by the particle size, crystal loading and to a lesser extent, the level of agitation. With Pinot noir wine the test for the mass transport limiting stage (Fig. E.22.a) shows that with 400-μm and 200-μm particles the surface reaction rate controlling regime starts after approximately four and eight minutes, respectively. With 20-μm particles the linear plot was not observed even up to 50 minutes (Fig. E.22.a). The crystallization rates were controlled by nucleation rates before the beginning of the mass transfer rate controlling step; therefore the longest nucleation rate controlling step was observed with 20-μm particles. The surface reaction rate controlling regions with Pinot noir wine are depicted in Figure E.22.b.

Example 4.23. Nucleation and growth kinetics of lactose (*Shi et al.*, 1990)

Nucleation and growth rates of sodium monohydrate in a continuous crystallizer were expressed as:

$$J_N = 1.99 \times 10^{18} \exp(-17.0/RT)\, M_t^i (S - 1)^{1.9} \tag{E.4.23.1}$$

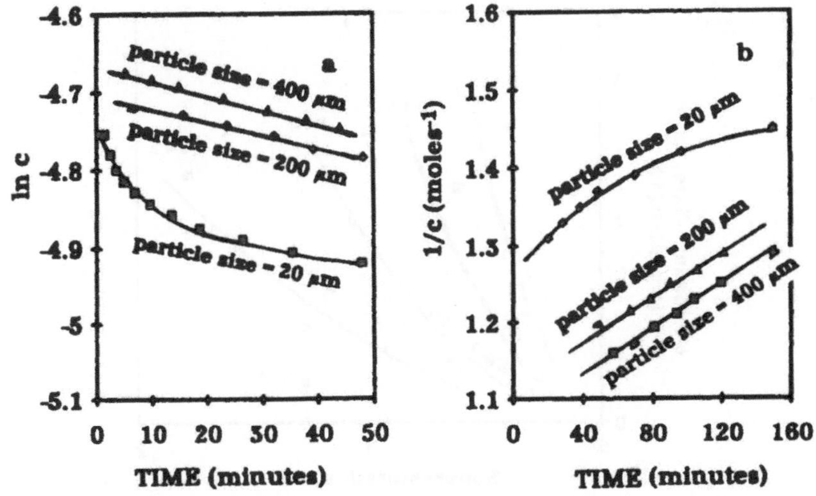

FIGURE E.4.22 The first (a) and the second (b) order regions of potassium bitartarate crystallization process in Pinot noir with particles of different size (Dunsford and Boulton, 1981; © American Society of Enology and Viticulture, reproduced by permission).

$$J_G = 6.1 \times 10^{15} \exp(-22.1/RT)(S-1)^{2.5} \qquad \text{(E.4.23.2)}$$

where J_N and J_G are the nucleation (number of nuclei/mL min) and linear growth (μm/min) rates, respectively and M_t is the suspension density (g crystal/mL suspension). Parameter S is the supersaturation ratio, defined as:

$$S = \frac{c}{c_s - FK_m(c - c_s)} \qquad \text{(E.4.23.3)}$$

where c = total crystal concentration; c_s = equilibrium solubility of lactose; F = temperature dependent factor for derepression of solubility of α-lactose by β-lactose and K_m = ratio of β/α lactose in mutarotation equilibrium. Comparison of (E.4.23.1) and (E.4.23.2) with experimental data is depicted in Figure E.4.23.

The equilibrium between the crystals and the surrounding liquid is dynamic. Some ions leave the crystal, while new ones join it. Since the net number of the ions leaving the crystal is the same as the ones joining, there is no net change between the phases. The solubility, S, of a solid sphere of an ionic compound with the general formula $M_m X_x$ and diameter D is related to that of a flat surface, S_0, by the *Kelvin equation* (Hiemenz, 1977) as:

$$S = S_0 \exp\left\{\frac{\gamma MW}{(m+x)RT\rho D}\right\} \qquad (4.26)$$

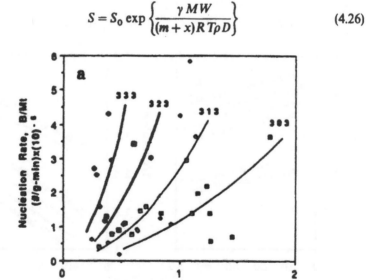

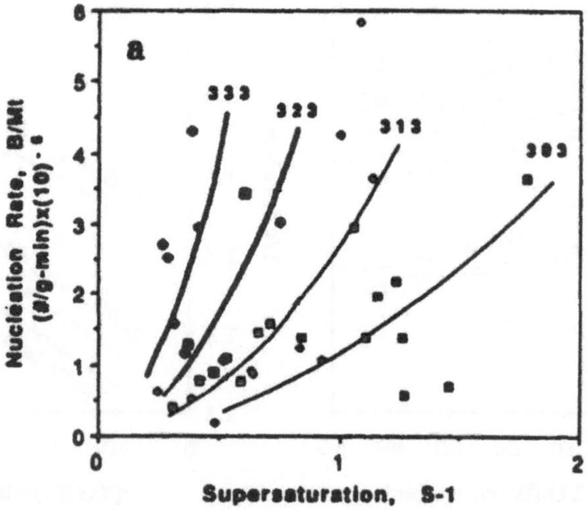

FIGURE E.4.23 (a) Nucleation (normalized by suspension density) and (b) crystal growth rates as a function of supersaturation at various temperatures (K), lines represent best fit as determined by (E.4.23.1) or (E.4.23.2) (Shi *et al.*, 1990; © IFT, reproduced by permission).

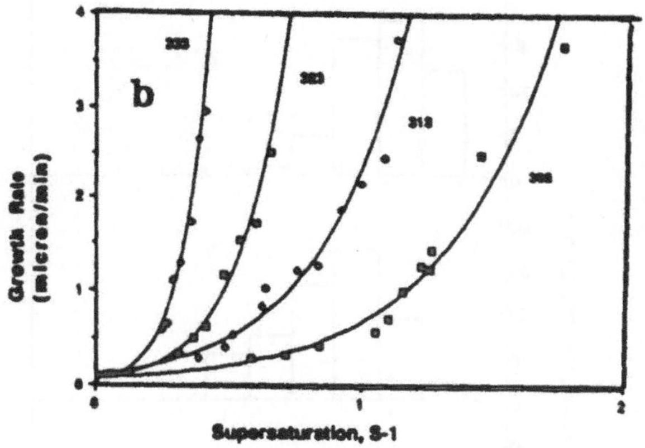

FIGURE E.4.23 (Continued).

where γ = interfacial tension, MW = molecular weight of the solid, ρ = density of the solid, R = gas constant and T = absolute temperature. Equation (4.26) implies that solubility increases exponentially with decreasing particle diameter, therefore in dynamic equilibrium the smaller particles eventually disappear while the large ones get larger. This phenomena is called *recrystallization* and associated with loss in quality of frozen meats during storage. Recrystallization is enhanced with temperature fluctuations. Martino and Zaritzky (1988) reported that both the mean and the variance of the equivalent crystal diameters increased during storage of meat tissue (Fig. 4.9).

Example 4.24. Kinetics of recrystallization of ice in frozen beef tissue (*Martino and Zaritzky*, 1988) Analysis of the experimental data indicates that crystal size may increase up to a limiting value D_∞ attainable in the system. The crystal size growth rates were expressed as a function of the difference of the mean curvature $(1/D_{eq})$ and the limiting curvature $(1/D_\infty)$:

$$\frac{dD_{eq}}{dt} = k\left(\frac{1}{D_{eq}} - \frac{1}{D_\infty}\right) \quad \text{(E.4.24.1)}$$

where k = rate constant. After integrating (E.4.24.1) we will obtain:

$$\ln\left[\frac{D_\infty - D_0}{D_\infty - D_{eq}}\right] + \frac{D_0 - D_{eq}}{D_\infty} = \frac{k}{D_\infty^2}t \quad \text{(E.4.24.2)}$$

where D_0 = mean equivalent initial diameter. Comparison of (E.4.24.2) with the experimental data is shown in Figure E.4.24.1.

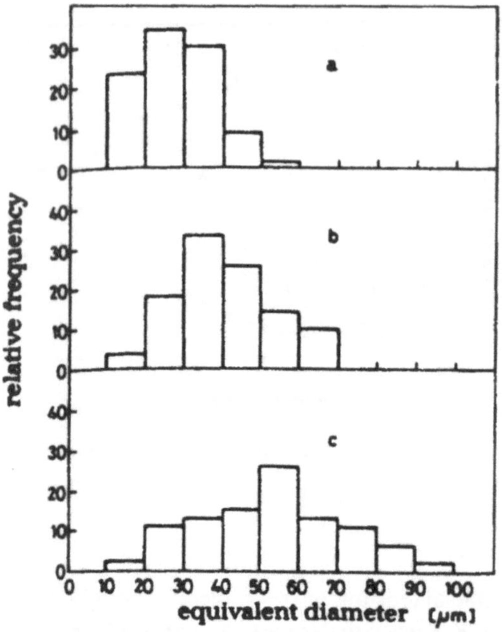

FIGURE 4.9 Histograms of the crystal size distribution in the meat tissue during storage at −5 °C. (a) Immediately after freezing, (b) After 5 days of storage, (c) After 40 days of storage (Martino and Zaritzky, 1988; © IFT, reproduced by permission).

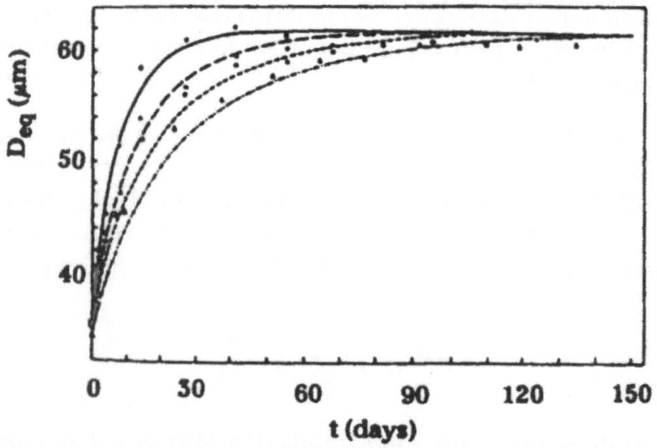

FIGURE E.4.24.1 Effect of recrystallization on mean equivalent diameter at different storage temperatures (maximum standard deviation of D_{eq} is 3.22 μm). Experimental data are shown in symbols, the model are shown with the lines: –o– $T = -5$°C; - -+- - $T = -10$°C; - — - $T = -15$°C; - · - · Δ- . - $T = -20$°C (Martino and Zaritzky, 1988; © IFT, reproduced by permission).

Effect of temperature on the crystallization rate constant k may be expressed in terms of the Arrhenius expression:

$$k = k_0 \exp\left\{-\frac{E_a}{RT}\right\} \tag{3.5}$$

Comparison of the linearized form of (3.5) with experimental data is depicted in Figure E.4.24.2.

Example 4.25. Kinetics of recrystallization of ice during ripening in freeze concentration In a freeze concentration process a liquid food or beverage is partially frozen, then the ice crystals (almost pure water) are removed to obtain a liquor with concentrated solutes. A simplified flow diagram of a two-stage freeze concentration process is shown in Figure E.4.25. The crystallizer of Figure E.4.25 actually consist of two sections. The first section is usually a heat exchanger, where nucleation and growth of ice crystals is achieved. The second section is a ripening tank, where large crystals become larger and take on a round shape, while the small crystals disappear to make ice removal easier and more efficient.

Freezing temperatures for small ice crystals are slightly decreased because the surface free energy per unit volume is larger for small crystals than

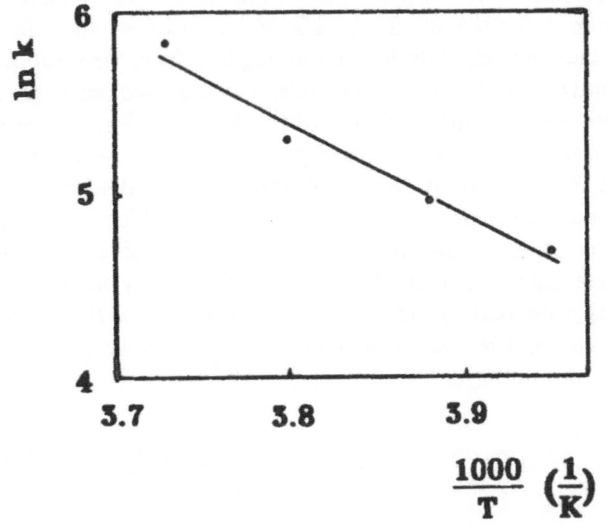

FIGURE E.4.24.2 Effect of temperature on recrystallization kinetic constant (k) in meat tissue (Martino and Zaritzky, 1988; © IFT, reproduced by permission). $E_a = (42.37 \pm 4.75) \times 10^3$ J/mol, $\ln k_0 = 24.78 \pm 2.20$ (parameter k was expressed in μm^2/days).

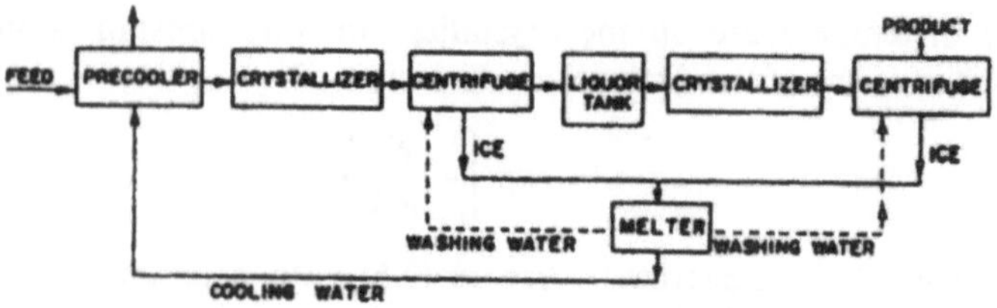

FIGURE E.4.25 A two-stage freeze concentration process (Karel, 1975; © Marcel Dekker, Inc., Reprinted from Principles of Food Science, Part II, Physical Principles of Food Preservation, page 291 by courtesy of Marcel Dekker Inc.).

for large crystals. Crystals of a radius r are in equilibrium when the total free energy is minimum. This occurs at solution temperatures ΔT lower than the equilibrium T for very large crystals (Schwartzberg, 1990):

$$\Delta T = \frac{2\sigma T}{\rho_s \Delta H r} \tag{E.4.25.1}$$

where σ is the interfacial tension between the ice crystals and the surrounding medium, ΔH is the latent heat of crystallization and ρ_s is the density of ice. During the ripening process the solution contains slurried ice crystals of mixed sizes, and the temperature T adjusts to a value higher than the equilibrium T for the small crystals, and to a lower value for the large crystals. The small crystals melt, removing heat from the solution, which in turn, removes heat from large crystals, which consequently grow. Similar effects also occur in different parts of the dendrites. Radii of curvature are very small and positive at the tips of the dendrites and dendrite branches, so they melt. Radii of curvature are very small, but negative, at the base of the clefts between the dendrite branches, so they fill in (Schwartzberg, 1990). If ripening is adiabatic and frictional heat production and secondary nucleation are negligibly small, then the heat taken up by melting crystals with $D_p < D_n$ is generated by growth of crystals with $D_p > D_n$, where D_n is the equilibrium particle size. After equating the rates of heat released and absorbed, we will have

$$2\pi \int_0^{D_n} h D_p^2 (T_p - T_n) f(D_p)\, dD_p = 2\pi \int_{D_n}^{D_{max}} h D_p^2 (T_n - T_p) f(D_p)\, dD_p \tag{E.4.25.2}$$

where h is the heat transfer coefficient, $f(D_p)$ is the number of crystals in the diameter range between D_p and $D_p + dD_p$, T_p and T_n are the respective equilibrium temperatures for particle sizes D_p and D_n. We also have the mathematical identity $(T_p - T_n) = (T + \Delta T_p - T + \Delta T_n) = (\Delta T_p - \Delta T_n)$.

Equation (E.4.25.1) requires

$$\Delta T_p - \Delta T_n = \frac{2\sigma T}{\rho_s \Delta H}\left(\frac{1}{D_p} - \frac{1}{D_n}\right) \qquad \text{(E.4.25.3)}$$

The total energy involved in the phase change of the particles in the diameter range between D_p and $D_p + dD_p$ is:

$$\frac{\pi D_p^2}{4}\frac{D_p}{dt}\rho_s \Delta H = h D_p^2 \frac{2\sigma T}{\rho_s \Delta H}\left(\frac{1}{D_n} - \frac{1}{D_p}\right) \qquad \text{(E.4.25.4)}$$

After rearranging (E.4.25.4) we will obtain an expression for the rate of particle size increase in the ripening process as:

$$\frac{D_p}{dt} = \frac{8\sigma h T}{(\rho_s \Delta H)^2}\left(\frac{1}{D_n} - \frac{1}{D_p}\right) \qquad \text{(E.4.25.5)}$$

4.4. FREEZING TIME CALCULATIONS

The calculation of freezing times for foods, involves a complex problem of heat transfer with simultaneous phase change, variable thermal properties and, in many cases, anisotropy problems. In Figure 4.10 we have a slab of food subject to heat removal by convection symmetrically from both sides. Initially the food is assumed to be at its freezing temperature, $T_{freezing}$, but not frozen.

The rate of heat removal from the surface by convection is:

$$q = hA(T_{surface} - T_{coolant}) \qquad (4.27)$$

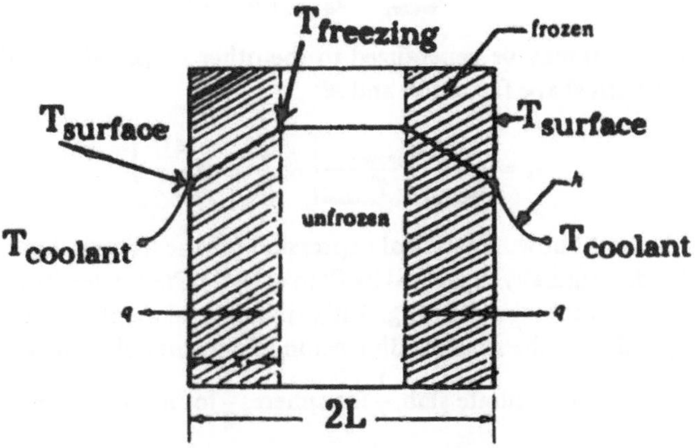

FIGURE 4.10 Temperature profile during freezing.

where A is the surface area. Under steady state conditions the heat being conducted through the frozen layer of thickness x is:

$$q = \frac{kA}{x}(T_{\text{freezing}} - T_{\text{surface}}) \tag{4.28}$$

Equations (4.27) and (4.28) may be combined to obtain:

$$q = \frac{(T_{\text{freezing}} - T_{\text{coolant}})A}{x/k + 1/h} \tag{4.29}$$

The rate of heat removal may also be calculated as:

$$q = \frac{dm}{dt}\Delta H_{\text{product}} = A\rho\,\Delta H_{\text{product}}\frac{dx}{dt} \tag{4.30}$$

The latent heat of fusion of the food $\Delta H_{\text{product}}$ may be estimated as (Heldman, 1983):

$$\Delta H_{\text{product}} = \omega_{\text{Food}}\,\Delta H_{\text{F, water}} \tag{4.31}$$

where ω_{Food} = weight fraction of water in the food, $\Delta H_{\text{F, water}}$ = specific freezing enthalpy of water. Equations (4.29) and (4.30) may be combined as:

$$A\rho\,\Delta H_{\text{product}}\frac{dx}{dt} = \frac{(T_{\text{freezing}} - T_{\text{coolant}})A}{x/k + 1/h} \tag{4.32}$$

After rearranging and integrating between $t = 0$ and $x = 0$ to $t = t_F$ and $x = L$ we will obtain:

$$t_{\text{F}} = \frac{\rho\,\Delta H_{\text{product}}}{T_{\text{freezing}} - T_{\text{coolant}}}\left[\frac{L}{h} + \frac{L^2}{2k}\right] \tag{4.33}$$

Equation (4.33) may be generalized to the other shapes after multiplying each term with shape factors $\mathscr{P}$ and $\mathscr{R}$:

$$t_{\text{F}} = \frac{\rho\,\Delta H_{\text{product}}}{T_{\text{freezing}} - T_{\text{coolant}}}\left[\mathscr{P}\frac{2L}{h} + \mathscr{R}\frac{4L^2}{2k}\right] \tag{4.34}$$

This is the best known analytical expression for the freezing time calculation of foods, originally suggested by Plank (1941). Parameter R denotes the thickness of an infinite slab (Fig. 4.10), diameter of a sphere, diameter of a long cylinder, or the smallest dimension of a rectangular block or brick.

Also, $\mathscr{P} = \frac{1}{2}$ for infinite slab, $\frac{1}{6}$ for sphere, $\frac{1}{2}$ for infinite cylinder

$\mathscr{R} = \frac{1}{8}$ for infinite slab, $\frac{1}{24}$ for sphere, $\frac{1}{16}$ for infinite cylinder

For a rectangular brick by dimensions $2Lx\beta_1 2Lx\beta_2 2L$ Ede (1949) prepared a chart to determine the shape factors $\mathscr{P}$ and $\mathscr{R}$ (Fig. 4.11). Equation (4.34) may also be used for calculation of thawing times by using the k of the thawed material.

Example 4.26. Freezing time calculations with Planck's equation a) Beef carcass (average dimensions 150 cm × 60 cm × 7 cm, $k_{average} = 1.04$ W/mK, $\rho = 1060$ kg/m^3, $\omega = 75\%$, $c_p = 3.5$ kJ/kg K and initially at the freezing temperature of beef $= -2.8°C$) is being frozen in an air-blast freezer ($h = 22$ W/m^{2}K) with air at $-30°C$, where the latent heat of fusion of water to ice is about 334 kJ/kg. Determine the freezing time of the carcass.

Solution With an infinite slab the shape factors are $\mathscr{P} = 1/2$ and $\mathscr{R} = 1/8$. The latent heat of freezing of the product is $\Delta H_{product} = \omega\,\Delta H_{water} = 250.5$ kJ/kg. The characteristic length $L = 0.035$ m. After substituting the data in (4.34) we will obtain

$$t_F = \frac{(1060\ \mathrm{kg/m^3})(250.5\ \mathrm{kJ/kg})}{(270.2-243)\mathrm{K}}\left[\frac{1}{2}\frac{2(0.035\,\mathrm{m})}{(22\ \mathrm{W/m^2K})}+\frac{1}{8}\frac{4(0.035\ \mathrm{m})^2}{(1.04\ \mathrm{W/mK})}\right]$$

$$\times\left(\frac{\mathrm{Ws}}{\mathrm{J}}\right)\left(\frac{1000\,\mathrm{J}}{\mathrm{kJ}}\right)\left(\frac{1\mathrm{h}}{3600\,\mathrm{s}}\right) = 5.9\mathrm{h}$$

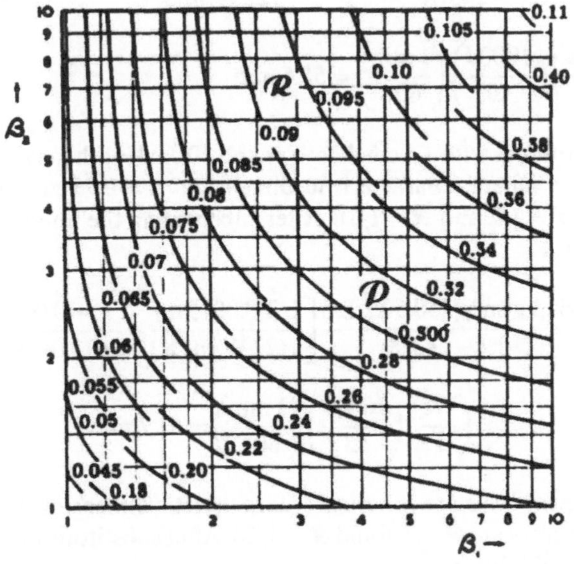

FIGURE 4.11 Shape factors for a rectangular brick by dimensions $2L \times \beta_1 2L \times \beta_2 2L$ (Ede, 1949; © EMAP Maclaren, Ltd., reproduced by permission of RAC).

b) Determine the freezing time of the individual units after the carcass is converted into the following products (you may assume that the properties of the products are almost the same as that of the carcass):

i. Steaks (2 cm × 7 cm × 10 cm)
ii. Sausages (diameter = 1.5 cm, length = 10 cm)
iii. Meat balls (diameter = 3 cm)
iv. Hamburger patties (thickness = 1 cm, diameter = 7 cm)
v. Fifty hamburger patties are packaged with paper (thickness = Δx = 3 mm, $k_{apparent}$ = 0.14 W/m K). Individual patties are separated with paper. The effective convective heat transfer coefficient may be calculated as:

$$\frac{1}{h_{eff}} = \frac{1}{h} + \frac{\Delta x}{k_{apparent}} \qquad \text{(E.4.26.1)}$$

Solution i) The steaks with the dimensions of 2 cm × 7 cm × 10 cm have the shape of a rectangular brick with $\beta_1 = 3.5$ and $\beta_2 = 5$. The shape factors are determined from Figure 4.11 as $\mathcal{P} = 0.336$ and $\mathcal{R} = 0.094$. The characteristic dimension $L = 0.01$ m. After substituting the numbers in (4.34) we will obtain

$$t_F = \frac{(1060\ \text{kg/m}^3)(250.5\ \text{kJ/kg})}{(270.2-243)\text{K}}\left[0.336\frac{2(0.01\ \text{m})}{(22\ \text{W/m}^2\text{K})} + 0.094\frac{4(0.01\ \text{m})^2}{(1.04\ \text{W/mK})}\right]$$
$$\times\left(\frac{\text{Ws}}{\text{J}}\right)\left(\frac{1000\ \text{J}}{\text{kJ}}\right)\left(\frac{1\ \text{min}}{60\ \text{s}}\right) = 55.6\ \text{min}$$

ii) The sausages with 1.5 cm diameter and 10 cm length may be regarded as infinite cylinders (characteristic dimension $L = 0.75$ cm). The shape factors are $\mathcal{P} = 1/2$ and $\mathcal{R} = 1/16$. After substituting the numbers in (4.34) we will obtain

$$t_F = \frac{(1060\ \text{kg/m}^3)(250.5\ \text{kJ/kg})}{(270.2-243)\text{K}}\left[\frac{1}{2}\frac{2(0.0075\ \text{m})}{(22\ \text{W/m}^2\text{K})} + \frac{1}{16}\frac{4(0.0075\ \text{m})^2}{(1.04\ \text{W/mK})}\right]$$
$$\times\left(\frac{\text{Ws}}{\text{J}}\right)\left(\frac{1000\ \text{J}}{\text{kJ}}\right)\left(\frac{1\ \text{min}}{60\ S}\right) = 57.7\ \text{min}$$

iii) The meat balls are spherical (characteristic dimension $L = 1.5$ cm). The shape factors are $\mathcal{P} = 1/6$ and $\mathcal{R} = 1/24$. After substituting the numbers in (4.34) we will obtain

$$t_F = \frac{(1060\ \text{kg/m}^3)(250.5\ \text{kJ/kg})}{(270.2-243)\text{K}}\left[\frac{1}{6}\frac{2(0.015\ \text{m})}{(22\ \text{W/m}^2\text{K})} + \frac{1}{24}\frac{4(0.015\ \text{m})^2}{(1.04\ \text{W/mK})}\right]$$

$$\times\left(\frac{\text{Ws}}{\text{J}}\right)\left(\frac{1000\,\text{J}}{\text{kJ}}\right)\left(\frac{1\,\text{min}}{60\,\text{s}}\right)=42.8\text{ min}$$

iv) The hamburger patties may be considered as infinite slabs with characteristic length $L = 0.5$cm. The shape factors are $\mathscr{P} = 1/2$ and $\mathscr{R} = 1/8$. After substituting the data in (4.34) we will obtain

$$t_F=\frac{(1060\text{ kg/m}^3)(250.5\text{ kJ/kg})}{(270.2-243)\text{K}}\left[\frac{1}{2}\frac{2(0.005\,\text{m})}{(22\,\text{W/m}^2\text{K})}+\frac{1}{8}\frac{4(0.005\,\text{m})^2}{(1.04\,\text{W/mK})}\right]$$

$$\times\left(\frac{\text{Ws}}{\text{J}}\right)\left(\frac{1000\,\text{J}}{\text{kJ}}\right)\left(\frac{1\,\text{min}}{60\,\text{s}}\right)=38.9\text{ min}$$

v) The package may be regarded as an infinite cylinder (characteristic dimension $L = 3.5$ cm). The shape factors are $\mathscr{P} = 1/2$ and $\mathscr{R} = 1/16$. The effective convective heat transfer coefficient is calculated from (E.4.26.1)

$$\frac{1}{h_{eff}}=\frac{1}{h}+\frac{\Delta x}{k_{apparent}}=\frac{1}{22}+\frac{0.003}{0.14}$$

therefore $h_{eff} = 14.95$ W/m²K.
After substituting the numbers in (4.34) we will obtain

$$t_F=\frac{(1060\text{ kg/m}^3)(250.5\text{ kJ/kg})}{(270.2-243)\text{K}}\left[\frac{1}{2}\frac{2(0.035\,\text{m})}{(14.95\,\text{W/m}^2\text{K})}+\frac{1}{16}\frac{4(0.035\,\text{m})^2}{(1.04\,\text{W/mK})}\right]$$

$$\times\left(\frac{\text{Ws}}{\text{J}}\right)\left(\frac{1000\,\text{J}}{\text{kJ}}\right)\left(\frac{1\,\text{h}}{3660\,\text{s}}\right)=6.5\text{ h}$$

Planck's equation does not account for the time required to remove the heat to bring the food from its initial temperature to the initial freezing temperature $T_{freezing}$ and the time required to bring the frozen food from $T_{freezing}$ to its final temperature. It also considers a constant thermal conductivity k. Numerous simplified models were suggested after Planck to overcome this equation's shortcomings and some of them are given here:

i) Cleland and Earle equation (1982a)

$$t_F=\frac{\Delta H_{beginning-final}}{(T_{beginning}-T_{coolant})(\text{EHTD})}\left[P\frac{2L}{h}+R\frac{4L^2}{k_{solid}}\right] \tag{4.35}$$

where P and R the empirical modification constants.
$c_{unfrozen}$ = volumetric specific heat of the unfrozen material (J/m³°C)
c_{solid} = volumetric specific heat of the fully frozen material (J/m³°C)
EHTD = number of equivalent heat transfer dimensions (EHTD = 1 for an infinite slab, EHTD = 2 for an infinite cylinder, EHTD = 3 for a sphere)
ΔH = change of enthalpy between $T_{beginning}$ (temperature where freezing begins) to the temperature where freezing is complete k_{solid} = thermal conductivity of the fully frozen product

$2L$ = characteristic dimension measured through the slowest cooling point
With brick shaped foods

$$P = 0.5(1.026 + 0.5808\,Pk + Ste(0.226\,Pk + 0.1050)) \qquad (4.35.a)$$

$$R = 0.125\,(1.202 + Ste(3.410\,Pk + 0.7336)) \qquad (4.35.b)$$

$$Pk = \frac{c_{unfrozen}(T_{initial} - T_{beginning})}{\Delta H} = \text{Plank number} \qquad (4.35.c)$$

$T_{initial}$ = initial temperature of the food

$$Ste = \frac{c_{solid}(T_{beginning} - T_{coolant})}{\Delta H} = \text{Stefan number} \qquad (4.35.d)$$

$$EHTD = 1 + W_1 + W_2 \qquad (4.35.e)$$

where

$$W_1 = \left(\frac{Bi}{Bi+2}\right)\left(\frac{5}{8\beta_1^3}\right) + \left(\frac{2}{Bi+2}\right)\left(\frac{2}{\beta_1(\beta_1+1)}\right) \qquad (4.35.f)$$

$$W_2 = \left(\frac{Bi}{Bi+2}\right)\left(\frac{5}{8\beta_2^3}\right) + \left(\frac{2}{Bi+2}\right)\left(\frac{2}{\beta_2(\beta_2+1)}\right) \qquad (4.35.g)$$

$$Bi = \frac{2hL}{k_{solid}} \qquad (4.35.h)$$

Cleland and co-workers suggested improved versions of their studies (Cleland, 1990) and extended the use of these equations with other shapes (Cleland *et al.*, 1987).

ii) Mascheroni and Calvelo equation (1982)

$$t_F = t_{pre\text{-}cooling} + t_{phase\ change} + t_{tempering} \qquad (4.36)$$

The precooling and tempering periods were evaluated by solving the heat transfer equation without phase change for a symmetric homogeneous slab of $2L$ thickness without end effects, with constant initial and external temperatures. The pre-cooling and tempering times may be evaluated from Figures 4.12 a and b, respectively. The phase change time may be calculated as:

$$t_{phase\ change} = \frac{\rho_{unfrozen}\,\Delta H_{water}\,\omega L^2}{(T_{beginning} - T_{coolant})\,k_{frozen}}\left[\frac{1}{Bi_{frozen}} + \frac{1}{2}\right] \qquad (4.37)$$

$Bi_{frozen} = 2hL/k_{frozen}$
$Bi_{average} = 2hL/k_{average}$
ΔH_{water} = heat of solidification of water at 0°C
$k_{average} = (k_{unfrozen} + k_{frozen})/2$
k_{frozen} = thermal conductivity evaluated at T_{frozen}

$k_{unfrozen}$ = thermal conductivity of the unfrozen food
T_{final} = final temperature of the frozen product
$T_{frozen} = (T_{beginning} + T_{coolant})/2$
$\alpha_{average} = (\alpha_{unfrozen} + \alpha_{frozen})/2$
α_{frozen} = thermal diffusivity of the frozen food evaluated at T_{frozen}
$\alpha_{unfrozen}$ = thermal diffusivity of the unfrozen food
x = water content of the food on wet basis
$\rho_{unfrozen}$ = density of the unfrozen product
ω = average of the ice content (weight of ice/initial weight of water) of the product at temperatures T_{final} and $T_{coolant}$.

With meat containing 74% water the ice content (weight of ice/initial weight of water) of the meat may be calculated from $\omega = 0.9309 - ((3.466 \times 10^{-3}T)/(273.16 - T))$

iii) De Michelis and Calvelo equation (1982)

In industrial freezing processes symmetric conditions can rarely be achieved. It is a common occurrence that there is considerable difference between the plates of a plate freezer, or there may be poor thermal contact between the food and the refrigerated plates. In some cases the food may be frozen in boxes, where the air space at the top of the box may also be a source of the nonsymmetric behavior. The position of the thermal center may be evaluated from Figures E.4.13.a and E.4.13.b when the cause of the asymmetry is the difference in Biot numbers, and in temperature, respectively. When different temperatures ($T_{coolant} \neq T_{coolant}$) and Biot numbers on both sides of a slab ($Bi_1 \neq Bi_2$) are combined to cause asymmetry, the final position of the thermal center may be obtained as (De Michelis and Calvelo, 1982):

$$\left(\frac{L_{tc}}{2L}\right) = \left(\frac{L_{tc}}{2L}\right)_t \pm \left[\left(\frac{L_{tc}}{2L}\right)_h - 0.5\right] \tag{4.38}$$

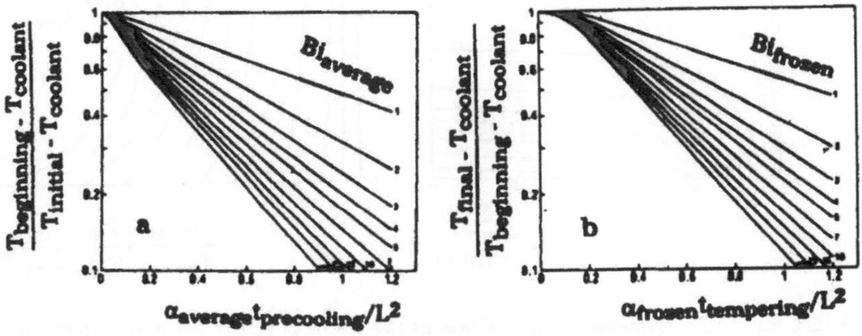

FIGURE 4.12 Curves for (a) pre-cooling and (b) tempering time calculations (Mascheroni and Calvelo, 1982; © IFT, reproduced by permission).

where L_{tc} is the location of the thermal center and $2L$ is the thickness of the slab; $(L_{tc}/2L)_t$ is the dimensionless position of the thermal center estimated in Figure 4.13.b and $(L_{tc}/2L)_h$ is the dimensionless position of the thermal center estimated in Figure 4.13.a The $\pm$ sign must be replaced with + when both causes of asymmetry cause displacement of the thermal center in the same direction; it is replaced with - when causes are on different sides and tend to compensate each other.

Freezing times were calculated with (4.36). The precooling and tempering periods were evaluated by solving the heat transfer equation without phase change for a symmetric homogeneous slab of thickness $2L$ without end effects, with constant initial and external temperatures. The solutions are presented in Figures 4.14.a and b. The phase change time may be calculated as given by (4.39) (De Michelis and Calvelo, 1982):

$$t_{\text{phase change}} = \frac{\rho_{\text{unfrozen}}\, x \Delta H_{\text{water}}\, \omega L_m^2}{(T_{\text{beginning}} - T_{\text{coolant}})\, k_{\text{frozen}}} \left[\frac{1}{Bi_{\text{frozen}}} + \frac{1}{2} \right] \tag{4.39}$$

where $L_m = (L + L_{\text{tempering}})/2$.

Example 4.27. Cleland and Earle's model to predict heating and cooling rates of solids including those with irregular shapes The cooling rate at the slowest cooling point of the food may be described as (Cleland and Earle, 1982b):

$$\frac{T_{\text{center}} - T_{\text{coolant}}}{T_{\text{initial}} - T_{\text{coolant}}} = K_1 \exp(-K_2\, Fo) \tag{E.4.27.1}$$

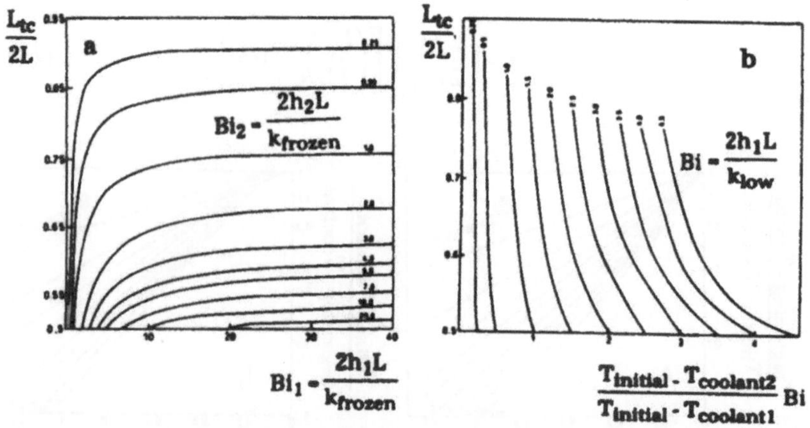

FIGURE 4.13 **Dimensionless position of the thermal center for asymmetry originated in (a) different heat transfer coefficients and (b) different temperatures of the cooling media. Thermal conductivity k_{low} was evaluated at the lowest cooling medium temperature (De Michelis and Calvelo, 1982; © IFT, reproduced by permission).**

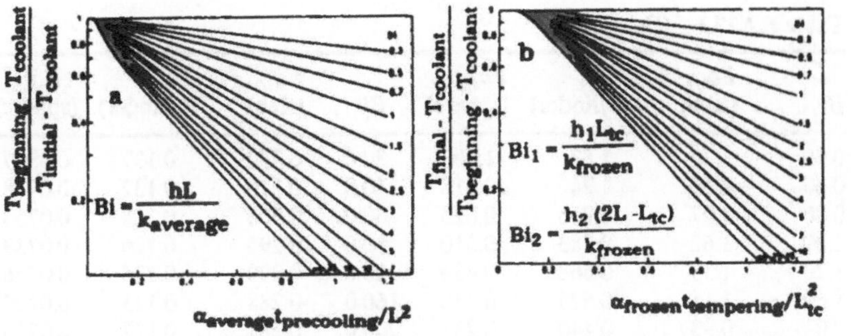

FIGURE 4.14 Curves for (a) precooling and (b) tempering time calculations (De Michelis and Calvelo, 1982; © IFT, reproduced by permission). Where $L_{tempering}$ is the positions of the thermal center in the tempering period. In Figure E.4.13.b either Bi_1 or Bi_2 may be used to calculate Bi.

where K_1 and K_2 are constants, $Fo = kt/c\rho L^2$ is the Fourier number. The Fourier number required to reduce $(T_{center} - T_{coolant})/(T_{initial} - T_{coolant})$ by half is:

$$Fo_{1/2} = \frac{k}{c\rho L^2} t_{1/2} \qquad \text{(E.4.27.2)}$$

where $Fo_{1/2}$ is determined by the Biot number and the geometry of the object to be frozen. After combining (E.4.27.1) and (E.4.27.2) we will obtain:

$$\frac{T_{center} - T_{coolant}}{T_{initial} - T_{coolant}} = K_1 \exp(-K_2 N Fo_{1/2}) \qquad \text{(E.4.27.3)}$$

where N = number of the $Fo_{1/2}$ values elapsed. Values of $Fo_{1/2}$ as a function of the Biot number for infinite slabs, infinite cylinders and spheres are given in Table E.4.27.1. The $Fo_{1/2}$ values for finite objects may be described in terms of those of infinite objects or spheres as described in Table E.4.27.2. An alignment chart has been given in Figure E.4.27 to relate EHTD, Biot number, N and $(T_{center} - T_{coolant})/(T_{initial} - T_{coolant})$.

Table E.4.27.1 Values of $Fo_{1/2}$ as a function of the Biot number ($Bi = 2hL/k$) for infinite slabs, infinite cylinders and spheres (Cleland and Earle, 1982b; Table 1)

Bi	$Fo_{1/2}$ (slab)	$Fo_{1/2}$ (cylinder)	$Fo_{1/2}$ (sphere)	Bi	$Fo_{1/2}$ (slab)	$Fo_{1/2}$ (cylinder)	$Fo_{1/2}$ (sphere)
0.02	69.8	34.9	23.3	6.0	0.487	0.217	0.132
0.04	34.9	17.4	11.6	8.0	0.433	0.190	0.115
0.08	17.6	8.75	5.82	10.0	0.402	0.175	0.105
0.12	11.8	5.86	3.90	12.0	0.381	0.165	0.0984
0.16	8.90	4.42	2.93	16.0	0.355	0.153	0.0906
0.20	7.16	3.55	2.36	20.0	0.339	0.146	0.0862

Table E.4.27.1 (Continued)

Bi	$Fo_{1/2}$ (slab)	$Fo_{1/2}$ (cylinder)	$Fo_{1/2}$ (sphere)	Bi	$Fo_{1/2}$ (slab)	$Fo_{1/2}$ (cylinder)	$Fo_{1/2}$ (sphere)
0.40	3.70	1.82	1.20	30.0	0.320	0.137	0.0807
0.60	2.55	1.24	0.818	40.0	0.310	0.132	0.0778
0.80	1.97	0.956	0.625	60.0	0.300	0.128	0.0753
1.00	1.62	0.783	0.510	80.0	0.295	0.126	0.0739
1.20	1.39	0.668	0.434	120.0	0.290	0.124	0.0726
1.60	1.11	0.525	0.338	160.0	0.288	0.123	0.0720
2.0	0.937	0.440	0.281	200.0	0.287	0.122	0.0716
3.0	0.710	0.327	0.205	∞	0.281	0.120	0.0702
4.0	0.598	0.271	0.168				

Table E.4.27.2 The $Fo_{1/2}$ values for finite objects in terms of the $Fo_{1/2}$ values for the infinite objects or sphere

brick shaped foods	$Fo_{1/2} = \frac{Fo_{1/2(\text{infinite slab})}}{EHTD}$	(E.4.27.4)
finite cylinder foods with height > diameter (L = radius)	$Fo_{1/2} = \frac{2Fo_{1/2(\text{infinite cylinder})}}{EHTD}$	(E.4.27.5)
finite cylinder foods with height < diameter (L = half of the height)	$Fo_{1/2} = \frac{Fo_{1/2(\text{infinite slab})}}{EHTD}$	(E.4.27.6)
oval shapes	$Fo_{1/2} = \frac{3Fo_{1/2(\text{sphere})}}{EHTD}$	(E.4.27.7)

The use of Tables E.4.27.1 and 2 and Figure E.4.27 will be explained by summarizing an example solved by Cleland and Earle (1982b) with slight modifications. An irregular cut of beef (weight 136 kg, $T_{\text{initial}} = 38°\text{C}$, thickness $\cong 20$ cm, $k = 0.50$ W/mK and $c\rho = 3.78 \times 10^6$ J/m^3 K) is cooled with air (velocity = 2 m/s, $T_{\text{coolant}} = 0°\text{C}$, $h = 14.6$ W/m^2 K). The slowest cooling point temperature becomes 9°C after 16 h of cooling. How long it would take to achieve $T_{\text{center}} = 5.6°\text{C}$ with a geometrically similar 68 kg cut of beef? We may calculate the followings with the data available:

$$Bi = \frac{2hL}{k} = 5.8, \quad Fo = \frac{kt}{c\rho L^2} = 0.77, \quad (T_{\text{center}} - T_{\text{coolant}})/(T_{\text{initial}} - T_{\text{coolant}}) = 0.24$$

$EHTD$ for an irregular object is:

$$Fo_{1/2} = \frac{3Fo_{1/2(\text{sphere})}}{EHTD} \quad \text{(E.4.27.7)}$$

$Fo_{1/2(\text{sphere})}$ corresponding to $Bi = 5.8$ will be obtained with interpolation from Table E.4.27.1 as 0.136. It is also known that

$$Fo = Fo_{1/2}N \quad \text{(E.4.27.8)}$$

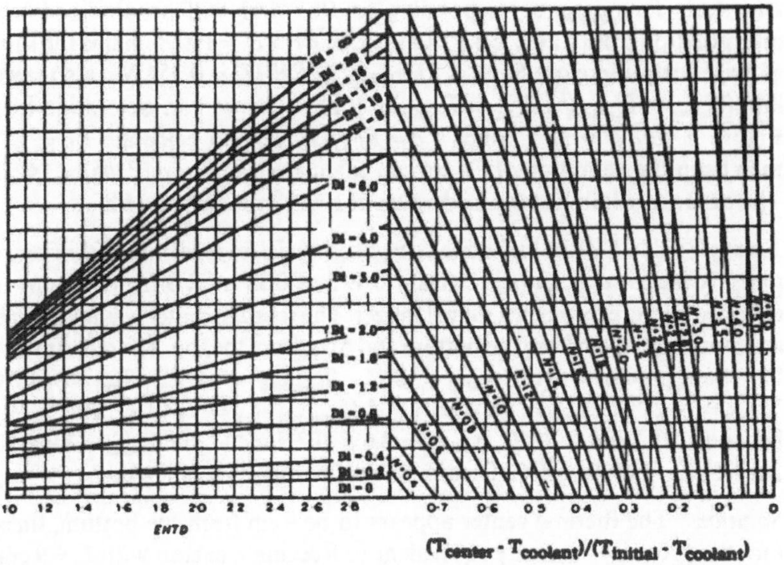

FIGURE E.4.27 Alignment chart to relate EHTD, Biot number, N and $(T_{center} - T_{coolant})/(T_{initial} - T_{coolant})$. Reprinted by permission of the publisher from a simple method for prediction of heating and cooling rates in solids of various shapes, Cleland and Earle, International Journal of Refrigeration, volume 5, pages 98–106, © 1982 Elsevier Science Ltd.

By using (E.4.27.7), (E.4.27.8) and Figure E.4.27 and applying a trial and error procedure we will find $N = 2.51$ and $EHTD = 1.33$. We will apply the following procedure during the trial and error solution:

i. Make an initial guess of $EHTD$
ii. Calculate $Fo_{1/2}$ from (E.4.27.7)
iii. Calculate N from (E.4.27.8)
iv. Locate the $EHTD$ in the horizontal axis of Figure E.4.27, then proceed vertically until reaching $Bi = 5.8$. Proceed in horizontal direction to find the calculated value of N, then proceed vertically to read $(T_{center} - T_{coolant})/(T_{initial} - T_{coolant})$. If the value you read is 0.24 the guess of $EHTD$ is correct.
v. Repeat steps i–iv until finding the correct answer.

If we assume that the cuts of beef have a cube-like structure the thickness of the second cut will be

$$2L_{second} = 2L_{first}\left(\frac{\text{weight of the second cut}}{\text{weight of the first cut}}\right)^{1/3} \qquad \text{(E.4.27.9)}$$

after substituting the numbers in (E.4.27.9) we will calculate $2L_{second} = 15.8$ cm. We may calculate the Biot number with the data available as

$Bi = 4.61$. $Fo_{1/2(sphere)}$ corresponding to $Bi = 4.61$ will be obtained with interpolation from Table E.4.27.1 as 0.157. $EHTD = 1.33$ remains the same as previously, therefore $Fo_{1/2} = 3Fo_{1/2(sphere)}/EHTD = 0.354$. We also calculate $(T_{center} - T_{coolant})/(T_{initial} - T_{coolant}) = 0.15$, then we can determine from Figure E.4.27 $N = 3.15$. Since $Fo = Fo_{1/2}N$ we can calculate $Fo = 1.12$, then from the definition of Fo we can calculate $t = (Fo c \rho L^2)/(k) = 14.5$ h. The experimentally determined value of t was reported as 16 h.

Example 4.28. Cleland and Earle's simplified model for prediction of the freezing times (Cleland and Earle, 1982a) 0.59 m × 0.38 m × 0.15 m cartons of deboned beef are frozen in a tunnel freezer. The thermal center was found to be approximately 9 cm from the bottom and 6 cm from the top. The relevant data for the process were $T_{initial} = 10°C$, $T_{beginning} = -2°C$, $T_{final} = -15°C$, $T_{coolant} = -21°C$, $k_{solid} = 1.6$ W/m°C, $\Delta H_{beginning-final} = 240 \times 10^{-6}$ J/m^3 between 10°C and −15°C, $c_{unfrozen} = 3.6 \times 10^6$ J/m^{3}°C and $c_{frozen} = 1.9 \times 10^6$ J/m^{3}°C, $h = 12.9$ W/m^2°C. Determine the freezing time of the beef.

Solution The thermal center appears to be 9 cm from the bottom, therefore the process is actually equivalent to freezing a carton with $L = 9$ cm.

$$Pk = \frac{c_{unfrozen}(T_{initial} - T_{beginning})}{\Delta H} = 0.180,\ Ste = \frac{c_{solid}(T_{beginning} - T_{coolant})}{\Delta H} = 0.15$$

$$Bi = \frac{2hL}{k_{solid}} = 1.45,\ \beta_1 = \frac{0.38}{0.18} = 2.11,\ \beta_2 = \frac{0.59}{0.18} = 3.28$$

$$P = 0.5(1.026 + 0.5808\,Pk + Ste(0.226\,Pk + 0.1050)) = 0.576$$

$$R = 0.125(1.202\,Ste(3.410\,Pk + 0.7336)) = 0.176$$

$$W_1 = \left(\frac{Bi}{Bi+2}\right)\left(\frac{5}{8\beta_1^3}\right) + \left(\frac{2}{Bi+2}\right)\left(\frac{2}{\beta_1(\beta_1+1)}\right) = 0.20$$

$$W_2 = \left(\frac{Bi}{Bi+2}\right)\left(\frac{5}{8\beta_2^3}\right) + \left(\frac{2}{Bi+2}\right)\left(\frac{2}{\beta_2(\beta_2+1)}\right) = 0.09,$$

$$EHTD = 1 + W_1 + W_2 = 1.29$$

$$t_F = \frac{\Delta H_{beginning-final}}{(T_{beginning} - T_{coolant})(EHTD)}\left[P\frac{L}{h} + R\frac{L^2}{k_{solid}}\right] = 31.6\,\text{h}$$

Example 4.29. Mathematical model for nonsymmetric freezing of beef A slab of meat (water content = 74%, fat content = 5%, $\rho = 1060$ kg/m^3, thickness = 5 cm, initially at 8°C) is placed in a blast freezer ($h_1 = 36$ W/m^{2}K, $T_1 = -30°C$, $h_2 = 32$ W/m^{2}K, $T_2 = -25°C$). Determine the time required to reduce the thermal center to −10°C).

Solution Since there is an asymmetric freezing process we will use the De Michelis and Calvelo equation. We will start the solution by calculating the required parameters:

$$T_{\text{coolant}} = \left(\frac{(-30)+(-25)}{2}\right) = -27.5°\text{C}$$

$$T_{\text{frozen}} = \frac{T_{\text{beginning}} + T_{\text{coolant}}}{2} = \left(\frac{(-2.8)+(-27.5)}{2}\right) = -15.15°\text{C}$$

Thermal conductivity will be calculated by Pham and Willix (1989) correlations:
Thermal conductivity is commonly related to the moisture content of nonfat meat when heat flow is non perpendicular to the fibers as:

$$k_{\text{unfrozen}} = 0.080 + 0.52x \tag{2.47.a}$$

$$k_{\text{frozen}} = -0.28 + 1.9x - 0.0092T \tag{2.47.b}$$

where x = moisture content (kg water/kg dry matter), thermal conductivity k has the units of W/mK.

$$x = \frac{74\,\text{kg}\,\text{H}_2\text{O}}{26\,\text{kg dry matter}} = 2.486$$

After substituting x in (2.47.a) and (2.47.b) we will calculate $k_{\text{unfrozen}} = 1.56$ W/mK and $k_{\text{frozen}} = 2.76$ W/mK. At the lowest medium temperature ($T_1 = -30$°C) thermal conductivity was $k_{\text{low}} = 2.89$ W/mK. The average thermal conductivity is defined as:

$$k_{\text{average}} = \frac{k_{\text{frozen}} + k_{\text{unfrozen}}}{2} = 2.16\,\text{W/mK}$$

The Biot numbers are:

$$Bi_1 = \frac{2h_1L}{k_{\text{frozen}}} = 0.65, \quad Bi_2 = \frac{2h_1L}{k_{\text{frozen}}} = 0.58 \quad \text{and} \quad Bi = \frac{2h_1L}{k_{\text{low}}} = 0.62$$

Since $Bi_1 = 0.65$ and $Bi_2 = 0.58$ we will read $(L_{tc}/2L)_h = 0.5$ from Figure 4.12.a. We have $Bi = 0.62$ and calculate $(T_{\text{initial}} - T_{\text{coolant2}})/(T_{\text{initial}} - T_{\text{coolant1}})$ $Bi = (8-(-25))/(8-(-30)) \times 0.62 = 0.54$, therefore we will read $(L_{tc}/2L)_t = 0.65$ from Figure 4.12.b. The actual location of the thermal center is:

$$\left(\frac{L_{tc}}{2L}\right) = \left(\frac{L_{tc}}{2L}\right)_t \pm \left[\left(\frac{L_{tc}}{2L}\right)_h - 0.5\right] \tag{4.38}$$

After substituting the numbers we will obtain $(L_{tc}/2L) = 0.65$ or $L_{tc} = 2(0.025)(0.65) = 0.033$ m.

Siebel's equation will be used to assume the specific heat as:

$$c_{\text{unfrozen}} = 1674.72\,\omega_f + 837.36\,\omega_s + 4186.8\,\omega_w \tag{2.45.a}$$

$$c_{\text{frozen}} = 1674.72\,\omega_f + 837.36\,\omega_s + 2093.4\,\omega_w \tag{2.45.b}$$

where $\omega_f = 0.05$ (mass fraction of fat), $\omega_s = 0.21$ (mass fraction of non-fat solids), $\omega_w = 0.74$ (mass fraction of water) in the food. After substituting the numbers we will calculate $c_{unfrozen} = 3357.8$ J/kg K and $c_{frozen} = 1808.7$ J/kg K. We also calculate that $c_{average} = (c_{frozen} + c_{unfrozen})/2 = 2583.3$ J/kg K.

We will use Figure 4.13.a to determine the precooling time. The required parameters are:

$$\frac{T_{beginning} - T_{coolant}}{T_{initial} - T_{coolant}} = 0.7, \; Bi = \frac{hL}{k_{average}} = 0.39,$$

$$\alpha_{average} = \frac{k_{average}}{\rho\, c_{average}} = 7.89 \times 10^{-7}\,\mathrm{m^2/s}$$

After using Figure 4.13.b we will determin $\alpha_{average}\, t_{precooling}/L^2 = 1.18$, then substitute values of L and average in this equation and calculate $t_{precooling} = 948$ s.

We will use Figure 4.13.b to determine the tempering time. The required parameters are:

$$\frac{T_{final} - T_{coolant}}{T_{beginning} - T_{coolant}} = 0.71, \; Bi = \frac{h_1 L_{tc}}{k_{frozen}} = 0.43,$$

$$\alpha_{frozen} = \frac{k_{frozen}}{\rho\, c_{frpzen}} = 1.44 \times 10^{-6}\,\mathrm{m^2/s}$$

After using Figure 4.13.b we will determine $\alpha_{frozen}\, t_{tempering}/L_{tc}^2 = 1.14$, then substitute values of L and average in this equation and calculate $t_{tempering} = 862$ s.

We will use (4.39) to calculate $t_{phase\ change}$.

$$t_{phase\ change} = \frac{\rho_{unfrozen}\, x\, \Delta H_{water}\, \omega L_m^2}{(T_{beginning} - T_{coolant})\, k_{frozen}} \left[\frac{1}{Bi_{frozen}} + \frac{1}{2}\right] \qquad (4.39)$$

where $L_m = (L + L_{tempering})/2$ and ω = average of the ice content (weight of ice/initial weight of water) of the product at temperatures T_{final} and $T_{coolant}$ and defined as $\omega = 0.9309 - (3.466 \times 10^{-3} T)/(273.16 - T)$. After substituting the numbers in (4.39) we will calculate $t_{phase\ change} = 9328$ s.

The total freezing time is:

$$t_F = t_{pre\text{-}cooling} + t_{phase\ change} + t_{tempering} \qquad (4.36)$$

After substituting the numbers in (4.36) we will calculate $t_F = 11\,138$ s $= 3.1$ h

Equation of energy, i.e., (2.16)–(2.18), is the starting point of the numerical freezing time calculation methods. In rectangular coordinates the equation of energy is:

$$\frac{\partial T}{dt}+v_x\frac{\partial T}{dx}+v_y\frac{\partial T}{dy}+v_z\frac{\partial T}{dz}+\frac{\dot{\psi}_G}{\rho c_p}+\frac{\partial}{dx}\left(\alpha\frac{\partial T}{\partial x}\right)+\frac{\partial}{dy}\left(\alpha\frac{\partial T}{\partial y}\right)+\frac{\partial}{\partial z}\left(\alpha\frac{\partial T}{\partial z}\right) \tag{2.16}$$

during freezing of an infinite slab with heat transfer in x direction only $v_x = v_y = v_z = \dot{\psi}_G = dT/dy = dT/dz = 0$, therefore (2.16) becomes

$$\frac{\partial T}{\partial t}=\frac{\partial}{\partial x}\left(\alpha\frac{\partial T}{\partial x}\right) \tag{4.40}$$

Equation (4.40) may be expressed in various finite difference schemes, some of the commonly used ones are given in Table 4.5.

Table 4.5 Finite difference schemes for (4.40) (Cleland and Earle, 1984)

i. Lees Scheme

$$(\rho c)^j_n\frac{T^{j+1}_n-T^{j-1}_n}{2\Delta t}=\frac{1}{3(\Delta x)^2}\{k^j_{n+1/2}[(T^{j+1}_{n+1}-T^{j+1}_n)+(T^j_{n+1}-T^j_n)+(T^{j-1}_{n+1}-T^{j-1}_n)]$$
$$-k^j_{n-1/2}[(T^{j+1}_n-T^{j+1}_{n-1})+(T^j_n-T^j_{n-1})+(T^{j-1}_n-T^{j-1}_{n-1})]\} \tag{4.40a}$$

ii. Modified Crank-Nicholson Scheme

$$(\rho c)^j_n\frac{T^{j+1}_n-T^j_n}{\Delta t}=\frac{1}{2(\Delta x)^2}\{k^j_{n+1/2}[(T^{j+1}_{n+1}-T^{j+1}_n)+(T^j_{n+1}-T^j_n)]$$
$$-k^j_{n-1/2}[(T^{j+1}_n-T^{j+1}_{n-1})+(T^j_n-T^j_{n-1})]\} \tag{4.40.b}$$

iii. Fully Implicit Scheme

$$(\rho c)^j_n\frac{T^{j+1}_n-T^j_n}{\Delta t}=\frac{1}{(\Delta x)^2}[k^j_{n+1/2}(T^{j+1}_{n+1}-T^{j+1}_n)-k^j_{n-1/2}(T^{j+1}_n-T^{j+1}_{n-1})] \tag{4.40c}$$

iv. Fully Explicit Scheme

$$(\rho c)^j_n\frac{T^{j+1}_n-T^j_n}{\Delta t}=\frac{1}{(\Delta x)^2}[k^j_{n+1/2}(T^j_{n+1}-T^j_n)-k^j_{n-1/2}(T^j_n-T^j_{n-1})] \tag{4.40d}$$

v. Enthalpy Transformation (explicit) Scheme

$$\frac{H^{j+1}_n-H^j_n}{\Delta t}=\frac{1}{(\Delta x)^2}[k^j_{n+1/2}(T^j_{n+1}-T^j_n)-k^j_{n-1/2}(T^j_n-T^j_{n-1})] \tag{4.40e}$$

where H^{j+1}_n and T^{j+1}_n are related after each time step

Table 4.5 (Continued)

vi. Modified Crank-Nicholson Scheme Using Thermal Diffusivity

$$\frac{T_n^{j+1}-T_n^j}{\Delta t}=\frac{1}{2(\Delta x)^2}\left\{\left(\frac{k}{\rho c}\right)_{n+1/2}^j [(T_{n+1}^{j+1}-T_n^{j+1})+(T_{n+1}^j-T_n^j)] - \left(\frac{k}{\rho c}\right)_{n-1/2}^j [(T_n^{j+1}-T_{n-1}^{j+1})+(T_n^j-T_{n-1}^j)]\right\} \quad (4.40.f)$$

The numerical solution using explicit finite difference scheme is straightforward, but the solution may become unstable because of the oscillations caused by the sharp peak of apparent specific heat function at the initial phase change temperature, and when large time increments are used. Fully implicit and Crank-Nicholson type finnite difference schemes result in sets of non-linear equations, which require time consuming iterative methods of solution. The Lees scheme uses the average of the temperature variations at three time levels on the right hand side of (4.40.a), therefore smoothes the oscillations (Mannapperuma and Singh, 1988). Equation (4.40.a) may be rewritten as:

$$c_n^j\frac{T_n^{j+1}-T_n^{j-1}}{2\Delta t}=\frac{1}{\Delta x}\left[k_{n+1/2}^j\frac{\bar{T}_{n+1}-\bar{T}_n}{\Delta x}-k_{n-1/2}^j\frac{\bar{T}_n-\bar{T}_{n-1}}{\Delta x}\right] \quad (4.41)$$

where $\bar{T}_n - 1/3(T_n^{j+1}+T_n^j+T_n^{j-1})$, c_n^j = volumetric specific heat evaluated at T_n^j; $k_{n+1/2}^j$ and $k_{n-1/2}^j$ are the thermal conductivity evaluated at $T_{n+1/2}^j$ and $T_{n-1/2}^j$, respectively.

Example 4.30. Prediction of freezing times with an enthalpy-based numerical method (Mannapperuma and Singh, 1988) The heat balance for a slab, cylinder or sphere element of unit cross section area at the interior node n between time levels j and $j+1$ may be written as:

$$V_n\left[\frac{H_n^{j+1}-H_n^j}{\Delta t}\right]=A_{n+1/2}k_{n+1/2}\left[\frac{T_{n+1}^j-T_n^j}{\Delta r}\right]-A_{n-1/2}k_{n-1/2}\left[\frac{T_n^j-T_{n-1}^j}{\Delta r}\right] \quad (E.4.30.1)$$

where A is the heat transfer area. The nodal representation of the food is depicted in Figure E.30.1.

Equation (E.4.30.1) was rearranged as:

$$H_n^{j+1}=H_n^j+\frac{\Delta t}{v_n(\Delta r)^2}\left\{a_{n+1/2}k_{n+1/2}((T_{n+1}^j-T_n^j)) - a_{n-1/2}k_{n-1/2}\left[\frac{T_n^j-T_{n-1}^j}{\Delta r}\right]\right\} \quad (E.4.30.2)$$

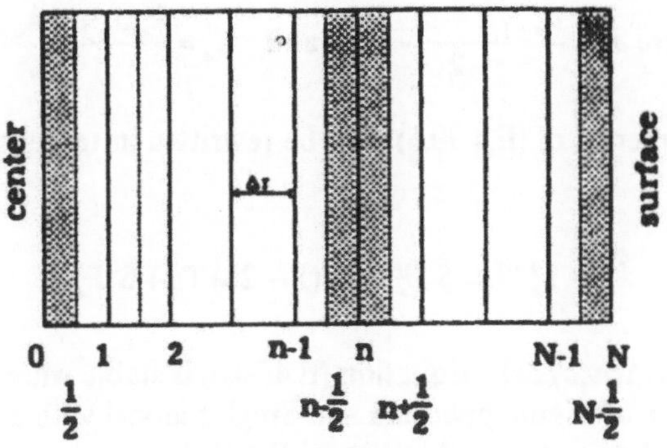

FIGURE E.4.30.1 The nodal representation of the food (Mannapperuma and Singh, 1988; © IFT, reproduced by permission).

where $a_n = A_n \Delta r / V_n$ = dimensionless area factor. Values of the volume and the area factors are given in Table E.4.30.

Table E.4.30 Values of the area and volume factors in different coordinate systems

	rectangular	*cylindrical*	*spherical*
v_0	1/2	1/8	1/24
v_n	1	n	$n^2 + 1/12$
v_N	1/2	$(1/2)(N - 1/4)$	$(1/2)(N^2 - N/2 + 1/12)$
$a_{1/2}$	1	1/2	1/4
$a_{n \pm 1/2}$	1	$n \pm 1/2$	$(n \pm 1/2)^2$
a_N	1	N	n^2

A finite difference equations at $n = 0$ (center) may be obtained as:

$$H_0^{j+1} = H_0^j + \frac{\Delta t}{v_0(\Delta r)^2}\{a_{1/2}k_{1/2}(T_1^j - T_0^j)\} \quad \text{(E.4.30.3)}$$

A similar equation may be obtained at $n = N$ (surface) as:

$$H_N^{j+1} = H_N^j + \frac{\Delta t}{v_N(\Delta r)^2}\left\{a_N h(T_{\text{coolant}}^j - T_N^j) - a_{N-1/2}k_{N-1/2}\left[\frac{T_N^j - T_{n-1}^j}{\Delta r}\right]\right\} \quad \text{(E.4.30.4)}$$

For studying the stability criteria the values of k and a on the right hand of (E.4.30.2) are averaged to obtain

$$H_n^{j+1} = H_n^j + \frac{a_n k_n \Delta t}{v_n(\Delta r)^2}\left\{(T_{n+1}^j - T_n^j) - \left[\frac{T_n^j - T_{n-1}^j}{\Delta r}\right]\right\} \quad \text{(E.4.30.5)}$$

$$\text{where } a_n = \frac{a_{n+1/2} + a_{n-1/2}}{2} \quad \text{and} \quad k_n = \frac{k_{n+1/2} + k_{n-1/2}}{2}.$$

The enthalpy terms of (E.4.30.5) may be rewritten in terms of'the specific heats as:

$$\frac{c_n^{j+1}}{c_n^j} T_n^{j+1} = S\, T_{n+1}^j + (1 - 2S)\, T_n^j + S\, T_{n-1}^j \qquad \text{(E.4.30.6)}$$

where $S = a_i k_i \Delta t / v_i c_i^j (\Delta r)^2$. Equation (E.4.30.6) is stable when $S \leqslant 1/2$.

Comparison of Mannapperuma and Singh's model with a sample set of experimental data is depicted in Figure E.4.30.2.

4.5. MATHEMATICAL MODELING OF DRYING AND FRYING PROCESSES

Drying phenomena have also been discussed in Chapter 2 within the context of modeling mass transfer in food systems; in Examples 2.3, 2.4, 2.12, 2.15 and 2.27 different aspects of drying phenomena were worked out. In a drying process almost all water present in the food is removed through vaporization. It is usually the final step before packaging to improve the microbial and chemical stability of the food. In a drying process the phase change of water may be achieved by heat supplied by the heated air (convective drying), direct contact with a hot surface (conduction heating)

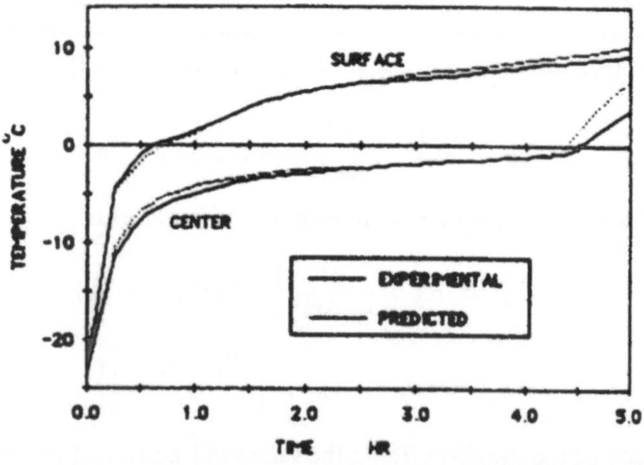

FIGURE E.4.30.2 Comparison of the model with the experimental data during thawing of a slab of Tylose (Mannapperuma and Singh, 1988; © IFT, reproduced by permission).

radiation or microwaves. Water molecules in the food might be bound to the ionic groups or make hydrogen bonds with the food solids. Unbound free water molecules make hydrogen bonds with each other. In a drying process the energy requirement for evaporation of unbound free water is less than that required for evaporation of hydrogen bound water. Water molecules which are bound to the ionic groups require the largest energy and it may be so large that destructive reactions occur in the food before they are removed. Drying data is usually expressed as the amount of water carried per unit dry mass of a process stream:

$$x = \frac{\text{mass of water}}{\text{mass of dry solids}} \quad \text{or} \quad \mathscr{H} = \text{humidity} = \frac{\text{mass of water}}{\text{mass of dry air}}$$

Drying rate may be defined as:

$$R_d = -\frac{1}{A}\frac{dx_f}{dt} \tag{4.42}$$

where R_d = drying rate per unit weight of the dry solids, A = surface area of the food per unit dry weight of the solids, t = time. Free moisture content per unit weight of the dry solids may be calculated as:

$$x_f = x - x^* \tag{4.43}$$

where $x = (W - W_s)/W_s$ = total moisture content of the food per unit dry weight of the solids, W = total weight of the food, W_s = weight of the dry solids in the food, x^* = equilibrium moisture content of the food per unit weight of the dry solids. It should be noticed that (4.42) was originally defined in non-food applications and may not be used in food processing. Most food materials contain more than 70% water and experience drastic shape changes upon drying which make it almost impossible to evaluate A as a function of time easily; then bulk parameter drying rate expressions may be used after combining R and A in the same bulk parameter.

i) The basic model of drying This is among the oldest models used since the 1920's and is discussed in detail in the literature (McCabe *et al.*, 1985; Porter *et al.*, 1973). In a drying process first the solids are charged into the dryer and heated-up if the temperature of the dryer is higher than that of the wet solids. If the surface is covered with a continuous film of water it attains the wet-bulb temperature at the end of the heating period. The evaporation rate from the surface at the wet-bulb temperature is constant. When the surface remains fully covered with the water film and evaporation is the rate determining step, the process is referred to be in the constant rate drying period. Water movement during the constant rate period is assumed to be controlled by several factors, including capillary suction and diffusion. The capillaries extend from small reservoirs of water in the solid to the drying surface. In a drying process, first moisture moves by capillary action to the surface rapidly enough to maintain a uniform wetted surface. Surface tension of the liquid is a controlling factor on the flow rate of water flow due to capillary suction (Labuza and Simon, 1970). When the food attains the critical moisture content, dry spots start appearing on the surface indicating the beginning of the first falling rate period. In this stage of the drying process the water supply rate to the surface is not sufficient, therefore the dry spots enlarge and the water film decreases continuously. When the water film on the surface disappears the second falling rate period starts, where evaporation occurs below the food surface and diffusion of vapor occurs from the place of vaporization to the surface (Treybal, 1980). The drying process stops when equilibrium is established between the food and the ambient air. Variation of the drying rate during the drying process of nixtamal (lime solution treated, cooked and dried raw maize) is depicted in Figure 4.15.

With the depletion of the sub-surface water reservoirs the mechanical structure collapses and the food undergoes shrinkage. In foods where shrinkage occurs, the shortening of the path length for water migration may offset the reduction in matrix porosity (Labuza and Simon, 1970). If the food is subject to too high temperatures at the beginning of the second falling rate period, the entrances of the partially empty capillaries shrink rapidly and prevent removal of the remaining water at deeper locations. This is called case hardening. This entrapped moisture may diffuse to the surface during storage and cause microbial spoilage. Case hardening may be prevented by using moderately moist or warm air for drying.

The mechanism of drying in slow drying non-porous materials, such as soap, gelatin, glue, and in the later stages of drying of clay, wood, textiles, leather, paper, foods and starches is liquid diffusion (Geankoplis, 1983). Diffusion of liquid water may result because of concentration gradients between the depths of the solid, where the water concentration is high, and the surface, where it is low. These gradients are set up during drying from the surface (Treybal, 1980).

ii) Hallström and co-worker's regular regime model The *regular regime model* of drying has been proposed by Schoeber (1976) to enable calculations drying processes where diffusivity is highly dependent on water content. The drying process is divided into three periods described in Figure 4.16. The *constant rate period* has already been discussed. The *penetration period* begins when the moisture profile starts to develop in the product and prevails until a stable moisture profile is developed. The following equation describes the drying rates in a slab during the penetration period (Hallström *et al.*, 1988; Hallström, 1992):

$$R_d = \frac{1}{\kappa \rho_s L} \frac{x_0 - \bar{x}}{x_0 - x^*} \tag{4.44}$$

where R_d is kg moisture transferred/m^2 s, κ is a constant, ρ_s is the density of solids, L is the half-thickness of the slab and $\bar{x}$ is the average moisture content of the slab. The *regular regime period* starts when a stable moisture

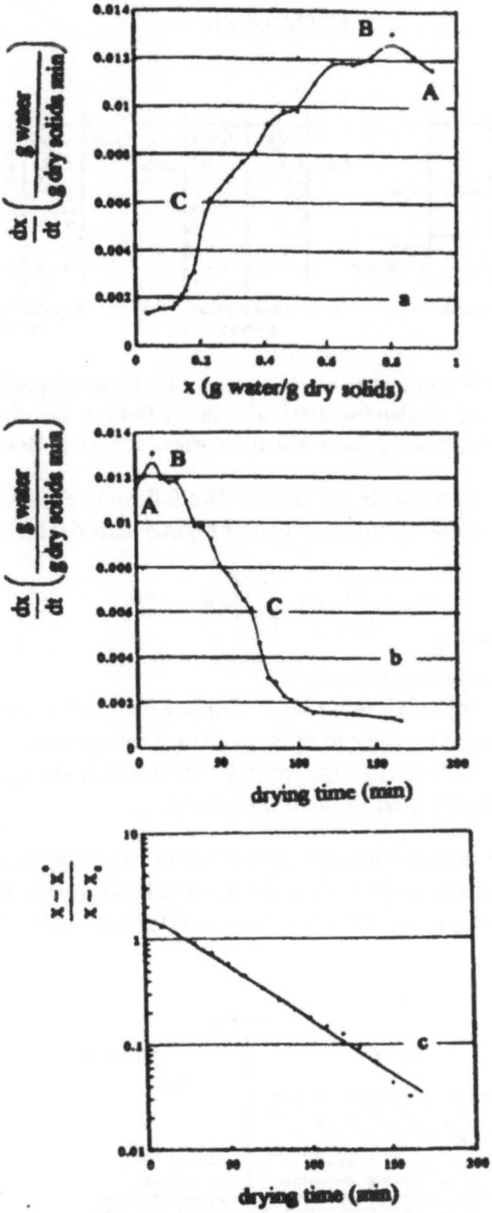

FIGURE 4.15 Variation of the drying rate of nixtamal with (a) moisture content and (b) drying time. Drying was reported to occur esentially in the falling rate period. The section AB may actually represent the initial heating period. A change in trend during the falling rate period is shown at point C. (c) A falling rate drying model $(x-x^*)/(x_c-x^*) = -0.73\,t$ simulates the data (x_c = critical moisture content, x^* = equilibrium moisture content, t = time in minutes). temperature was 85°C during the first one hour of drying, then decreased to 60°C (Gasson-Lara, 1993). Reproduced with permission of the American Institute of Chemical Engineers.

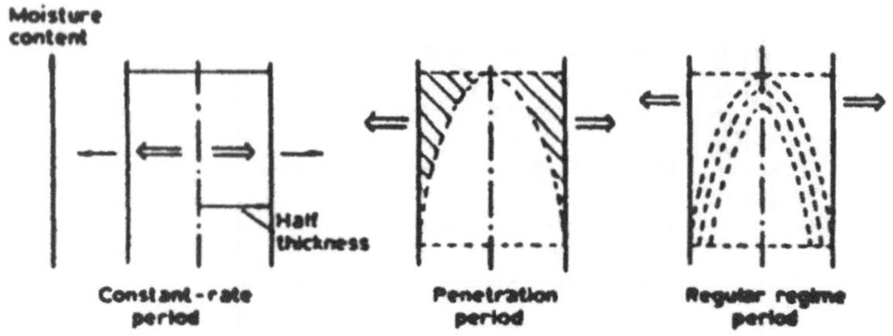

FIGURE 4.16 Hallström and co-workers' model for moisture profiles inside the material during drying (Hallström, 1992; © Marcel Dekker, Inc. Reprinted from *Handbook of Food Engineering*, page 330, by courtesy of Marcel Dekker Inc.).

profile starts moving towards the center. The following equation describes the drying rates in a slab during the regular regime period (Hallström, 1992):

$$R_d = \frac{\kappa_1}{\rho_s L} \exp\left\{\frac{\kappa_2}{T}\right\}(x - x^*)^{\kappa_3} \tag{4.45}$$

where κ_1, κ_2 and κ_3 are constants.

iii) Shrinking-core model (King, 1968) Surface water film disappears and a dry shell is formed after the end of the constant rate period. The heat and mass transfer mechanisms through the dry shell are depicted in Figure 4.17. The basic assumptions used in the model were:

i. Mass transfer occurs with diffusion of water vapor in the outer shell.
ii. The vapor pressure in the shell is assumed to be in equilibrium with the local moisture content at the local temperature of the food

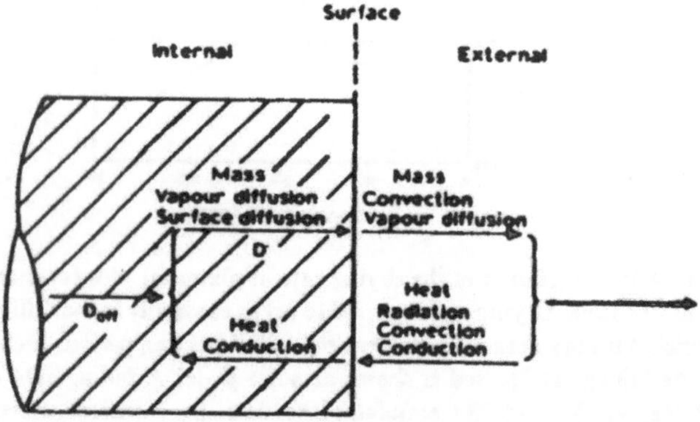

FIGURE 4.17 Mechanisms of heat and mass transfer in the shrinking-core model (King, 1968; © IFT, reproduced by permission).

iii. Heat of sorption is constant with respect to moisture content and temperature
iv. Shrinkage of the product is negligible
v. The effect of heat of sorption is large in comparison to the sensible heat of the product

King (1968) suggested the vapor transport rate coupled with heat transfer as:

$$\frac{\rho_s}{MW_w}\frac{\partial c}{\partial t}=\frac{\partial}{\partial z}\left[\frac{D\varepsilon}{\tau^2}\frac{P_w^o}{RT}\frac{P}{P-p_w}\left(\frac{\partial a_w}{\partial c_w}\right)_T\left(\frac{\gamma}{\gamma+1}\right)\frac{\partial c}{\partial z}\right] \tag{4.46}$$

where

$$\gamma=\frac{kR^2T^3}{(\Delta H_s)^2a_wP_w^o}\frac{\tau^2}{D\varepsilon}$$

a_w = water activity, D = diffusivity of water vapor in the air, ΔH_s = heat of sorption, k = thermal conductivity, MW_w = molecular weight of water, P_w^o = vapor pressure of water, R = gas constant, τ = tortuosity and ε = void fraction. The left hand side of (4.46) gives the transport of the condensed phase, while the right hand side gives the transport of the vapor phase in the dry shell. After comparing (4.46) with the equation of continuity we will obtain the effective diffusivity as:

$$D_{eff}=\frac{MW_wP_w^oD\varepsilon}{\rho_sRT}\frac{P}{\tau^2}\frac{P}{P-p_w}\left(\frac{\partial a_w}{\partial c_w}\right)_T\left(\frac{\gamma}{\gamma+1}\right) \tag{4.47}$$

where the term $MW_w\,P_w^o(\partial a_w/\partial c_w)_T$ represents the ratio of moisture density in the vapor to the moisture density in the condensed phase. The term $\gamma/\gamma+1$ gives the limiting mechanism for heat and mass transport. When thermal conductivity is small and the mass diffusion is large γ becomes smaller than 1 indicating that the process is heat transfer controlled. When $\gamma \ll 1$ we will have $\gamma/(\gamma+1)\cong\gamma$. High values of heat of sorption, high relative humidity and high temperature creates a tendency towards heat transfer control (King, 1968).

Example 4.31. Expressions for the diffusivity of moisture in pasta and bakery products. Expressions for the apparent diffusivity of water in food systems have been given in Chapter 2.10 and Example 2.12. The temperature effects on diffusivity is generally expressed with an Arrhenius expression:

$$D_{eff}=D_0\exp\left(-\frac{E}{RT}\right) \tag{E.4.31}$$

Constants of (E.4.31) may be expressed as a function of the moisture content. The major operation during production of noodles is drying and the majority of the process occurs during the falling rate period. The drying

process is achieved under careful control because the quality of the product is affected by the drying behavior, therefore the drying behavior of pasta products has been studied by many research groups and the following constants for (4.31) were reported:
$D_0 = 7.8 \times 10^{-2}x^2 + 1.67 \times 10^{-2}x + 1.5 \times 10^{-3}\,\mathrm{m^2/s}$, $E = 48\ 580$ J/mole (Andrieu *et al.*, 1988); $D_0 = 3.1 \times 10^{-7}x - 9.1 \times 10^{-9}\,\mathrm{m^2/s}$, $E = 21600$ J/mole (Andrieu and Stamatopoulos, 1986).

The following expressions were reported for the apparent diffusivities of bakery products in the temperature range of $293 \leqslant T \leqslant 373$ K (Tong and Lund, 1990):
Bread ($0.10 \leqslant x \leqslant 0.75$):
$D_{eff} = 28945 \times 10^{-4} \exp(1.26x - 2.76x^2 + 4.96x^3 - 6117.4/T)(\mathrm{m^2/s})$
Biscuit ($0.10 \leqslant x \leqslant 0.60$):
$D_{eff} = 9211.4 \times 10^{-4} \exp(0.45x - 6104.5/T)(\mathrm{m^2/s})$
Muffin ($0.10 \leqslant x \leqslant 0.95$):
$D_{eff} = 61672.9 \times 10^{-4} \exp(0.39x - 6664.0/T)(\mathrm{m^2/s})$

Example 4.32. Modeling simultaneous heat and mass transfer in foods with dimensional changes and variable transport parameters (*Balaban and Pigott*, 1988). Air drying of ocean perch (*Sebates marinus*) fillets was studied. The model was based on the following observations: **i.** Shrinkage is not negligible and ranges up to 50% during drying. **ii.** Transport parameters and physical properties were assumed to be functions of local moisture content and temperature, **iii.** The mass and flux of water vapor was negligible in comparison to those of liquid water. **iv.** The fish fillets were assumed to be continuous infinite slabs. The mass and the heat balance equations were:

$$\rho L \frac{\partial x}{\partial t} = \frac{\partial}{\partial z}\left\{D_L \rho_L \frac{\partial x}{\partial z}\right\} \tag{E.4.32.1}$$

$$\rho_b c_b \frac{\partial T}{\partial t} = \frac{\partial}{\partial z}\left\{k \frac{\partial T}{\partial z}\right\} \tag{E.4.32.2}$$

where c_b = specific heat of the bulk medium, D_L = effective diffusivity of liquid, k = thermal conductivity, x = volume fraction of water in fish, z = space dimension in the direction of thickness, ρ_b = bulk density and ρ_L = density of liquid. The initial and the boundary conditions were:

$$IC1\colon x = x_0 \quad \text{when } t = 0 \tag{E.4.32.1.a}$$

$$IC2\colon T = T_0 \quad \text{when } t = 0 \tag{E.4.32.2.a}$$

$$BC1\colon \frac{\partial x}{\partial z} = 0 \quad \text{at} \quad z = L \text{ (at the center, thickness of the slab} = 2L) \tag{E.4.32.1.b}$$

$$BC2: \frac{\partial T}{\partial z} = 0 \quad \text{at} \quad z = L \text{ (at the center)} \qquad \text{(E.4.32.2.b)}$$

$$BC3: K_c(c_{surface} - c_{air}) = D_L \rho_L \frac{\partial x}{\partial z} \quad \text{at} \quad z = 0 \qquad \text{(E.4.32.1.c)}$$

$$BC4: \Delta H_{vap} D_L \rho_L \frac{\partial x}{\partial z} - k \frac{\partial T}{\partial z} = h(T_{air} - T_{surface}) \quad \text{at} \quad z = 0 \qquad \text{(E.4.32.2.c)}$$

where $c_{surface}$ and c_{air} are concentration of water at the surface and in the air, respectively. The following expressions were used for the physical and transport constants:

$$K_c = 8.92 \times 10^{-3} v_0 (Re)^{-0.2} \frac{c_{surface}}{c_{surface} - c_{air}} \qquad \text{(E.4.32.3)}$$

$$h = 1.145 \times 10^{-3} (v_0 \rho_{air})^{0.8} \qquad \text{(E.4.32.4)}$$

$$k = -8.16 \times 10^{-2} + 6.49 \times 10^{-4} T + 1.012 \times 10^{-1} x \qquad \text{(E.4.32.5)}$$

$$c_b = 0.9615 + 3.6872 \times 10^{-1} x - 2.237 \times 10^{-3} T \qquad \text{(E.4.32.6)}$$

where v_0 is the velocity of air, $c_{surface}$ was taken from the isotherm relation, assuming that liquid and vapor are in equilibrium at the surface. It was assumed that the vapor pressure on the surface may be calculated from the expression

$$a_w = \frac{P_v}{P_{sat}} \qquad \text{(E.4.32.7)}$$

where P_{sat} is the vapor pressure of pure water and

$$a_w = \exp\left\{\frac{\dfrac{1}{(94x)^2} - 9.5018 \times 10^{-5}}{3.114 \times 10^{-3} - 1.3103 \times 10^{-4} T}\right\} \qquad \text{(E.4.32.8)}$$

$$P_{sat} = \exp\left\{54.51 - \frac{6887}{T_{abs}} - 5.3331 \log T_{abs}\right\} \qquad \text{(E.4.32.9)}$$

The diffusion coefficient of moisture in the tissue did not change with the water content between the initial conditions and the critical moisture content. Below this level (after 9 h) a step change decrease was observed and the diffusivity became one seventh of its initial value. Temperature effects on the diffusion coefficient were expressed as:

$$D_L = 2.72 \times 10^{-5} \exp\left(-\frac{3620}{T_{abs}}\right) \qquad \text{(E.4.32.10)}$$

Temperature T has the units of °C in (E.4.32.5)–(E.4.32.8) and K in (E.4.32.9)–(E.4.32.10). It should be noticed that x is dimensionless and its value changes between 0 and 1. Other variables are made dimensionless after substituting $\theta = (T - T_0)/(T_{air} - T_0)$ and $\zeta = z/2L$.

In classical finite difference applications the differences between the nodes are equal and constant. The picture of the drying product is shown in Figure E.4.32.1. During the drying process $\Delta\zeta_i$ (distance between the nodes denoted by i and $i+1$) changes at each time step due to shrinkage. The first and second order derivatives at node i were calculated as:

$$\frac{\partial x}{\partial \zeta} \text{ at node } i = \frac{x_{i+1} - x_{i-1}}{\Delta\zeta_i + \Delta\zeta_{i-1}} \tag{E.4.32.11}$$

$$\frac{\partial^2 x}{\partial \zeta^2} \text{ at node } i = \frac{\dfrac{x_{i+1} - x_i}{\Delta\zeta_i} - \dfrac{x_i - x_{i-1}}{\Delta\zeta_{i-1}}}{\dfrac{\Delta\zeta_i + \Delta\zeta_{i-1}}{2}} \tag{E.4.32.12}$$

The same procedure was also applied to calculate $\partial\theta/\partial\zeta$ and $\partial^2\theta/\partial\zeta^2$. The dimensionless numerical forms of (E.4.32.1) and (E.4.32.2) were solved by computer program. The inputs of the program were the air properties (relative humidity, temperature, velocity and viscosity); initial product dimensions; shrinkage (after assuming the volume change of material is known as a function of moisture); density (as a function of moisture); heat capacity (as a function of moisture and temperature); thermal conductivity (as a function of moisture and temperature); diffusion coefficient of water (as a function of moisture and temperature); heat and mass transfer coefficients on the surface (as a function of air properties); product water activity (as a function of moisture and temperature); initial product moisture and temperature; number of the nodes in the half thickness; total drying time

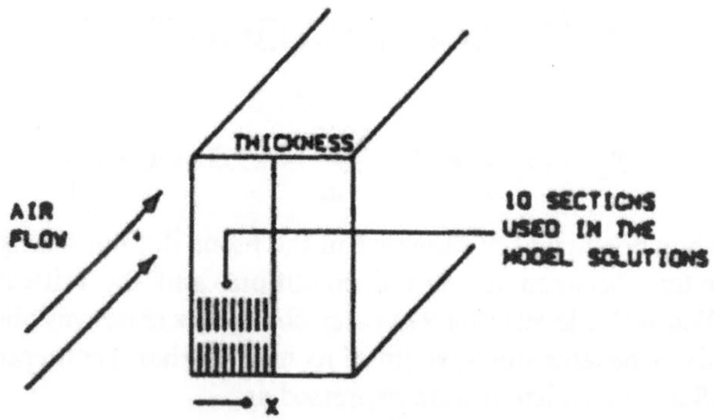

FIGURE E.4.32.1 Cross section of the infinite rectangular slab (Balaban and Pigott, 1988; © IFT, reproduced by permission).

and the time steps to be taken during solution. Based on these inputs the model determines values of temperature and x at each node at each time step; shrinkage at each section (calculated from the average moisture content of each section at each time steps). The calculated and the experimental values of the percentage change in the values of x during the drying process are depicted in Figure E.4.32.2. Prediction of the moisture gradient along the slab is depicted in Figure E.4.32.3.

Example 4.33. Drying behavior of thin biscuits during baking process (Turhan and Özilgen, 1991). Biscuits with 7 cm × 5 cm × 0.25 cm dimensions are baked in an oven at a constant temperature. Drying phenomena during baking in the falling rate period may be simulated with the simplified form of the equation of continuity in rectangular coordinate system as:

$$\frac{\partial x}{\partial t} = D\frac{\partial^2 x}{\partial z^2} \qquad \text{(E.4.33.1)}$$

where D = diffusivity of water in the biscuit and z = distance along the shortest dimension of the biscuit. Solution to (E.4.33.1) is:

$$\frac{x - x^*}{x_0 - x^*} = \frac{8}{\pi^2}\sum_{n=0}^{\infty}\frac{1}{(2n+1)^2}\exp\left[-\frac{(2n+1)^2\pi^2}{4}\frac{Dt}{L^2}\right] \qquad \text{(E.4.33.2)}$$

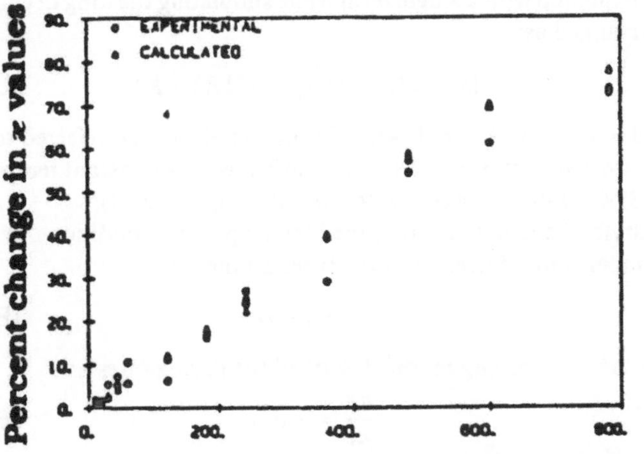

FIGURE E.4.32.2 Comparison of the experimental and the calculated values of the percentage x with 30 samples of ocean perch slabs during the drying process. (▲) calculated (○) experimental (Balaban and Pigott, 1988; © IFT, reproduced by permission).

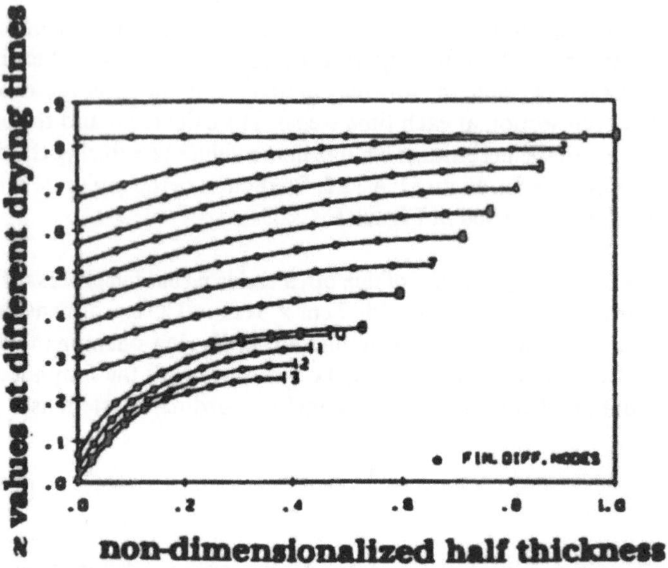

FIGURE E.4.32.3 Prediction of the moisture gradient in a slab of ocean perch during the drying process. Each curve represents one hour, (○) location of the finite difference nodes (Balaban and Pigott, 1988; © IFT, reproduced by permission).

where L = thickness of the biscuit. When D/L^2 is a constant, (E.4.33.2) may be approximated with a single term while simulating the long drying times, and rearranged as:

$$\ln (x - x^*) = \ln ((x_0 - x^*) B) - Kt \qquad \text{(E.4.33.3)}$$

where $B = 8/\pi^2$ and $k = \pi^2 D/4L^2$. Parameter K may be referred to as the drying rate constant. When a biscuit was baked at a constant temperature, (E.4.33.3) was found to agree with the data (Fig. E.4.33.1).

The drying behavior is determined by the process conditions. When the oven temperatures decreased linearly with time as:

$$T = T_0 - \alpha t \qquad \text{(E.4.33.4)}$$

the constant rate drying model described the drying rates as:

$$\frac{dx}{dt} = -\phi \qquad \text{(E.4.33.5)}$$

Equation (E.4.33.5) was integrated as:

$$x = x_0 - \phi t \qquad \text{(E.4.33.6)}$$

Comparison of (E.4.33.6) with the experimental data is shown in Figure E.4.33.2.

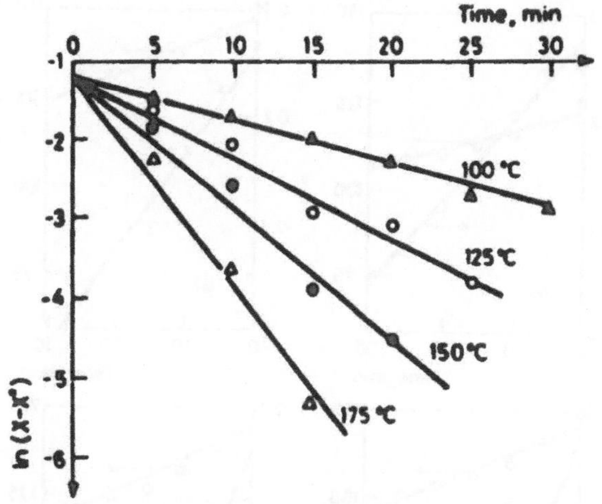

FIGURE E.4.33.1 Comparison of the falling rate drying model (—) with the data (▲, ○, ●, Δ) obtained in the constant oven temperature experiments. Oven temperatures and the numerical values of constants were as follows: (▲) 100°C, $K = 0.9 \times 10^{-3}s^{-1}$, (○) 125°C, $K = 1.67 \times 10^{-3}s^{-1}$, (●) 150°C, $K = 2.82 \times 10^{-3}s^{-1}$, (Δ) 175°C, $K = 4.48 \times 10^{-3}s^{-1}$ (Turhan and Özilgen, 1991; © Akadémiai Kiadó, Budapest, reproduced by permission).

Example 4.34. Modeling heat and mass transfer to cookies undergoing commercial baking (Hayakawa and Hwang, 1982). Baking of two different types (cracker type and chocolate base) of circular disc cookies was studied in a 90 m long oven subdivided into seven different zones. Heat balance around the cookies was:

$$c_{\text{cookie}} V\bar{\rho}\frac{dT}{dt} = (A_{\text{upper}} + A_{\text{sides}})h_{\text{upper and sides}}(T_{\text{air}} + T_{\text{side surfaces}})$$
$$+ A_{\text{bottom}} h_{\text{bottom}}(T_{\text{conveyor}} - T_{\text{bottom surface}})$$
$$+ (A_{\text{upper}} + A_{\text{sides}})\gamma_{\text{air}}(T^4_{\text{air}} - T^4_{\text{side surfaces}})$$
$$+ \gamma_{\text{wall}}(T^4_{\text{wall}} - T^4_{\text{side surfaces}}) + V\Delta H_{\text{vap}}\bar{\rho}\frac{d\bar{x}}{dt} \qquad \text{(E.4.34.1)}$$

where A and T represent the surface area of the cookies and the temperature of the cookies (or the oven), respectively (areas and temperatures pertinent to different locations of the cookies or the oven are depicted by subscripts); c_{cookie}, V and $\bar{\rho}$ are the specific heat, volume and average density of the cookies, respectively; parameter h is the convective heat transfer coefficients on the surface of the cookies at the subscripted locations, γ_{air} and γ_{wall} are constants of the radiative heat exchange equal to the product of the absorbance of oven walls or air geometric correction factor and Stefan-

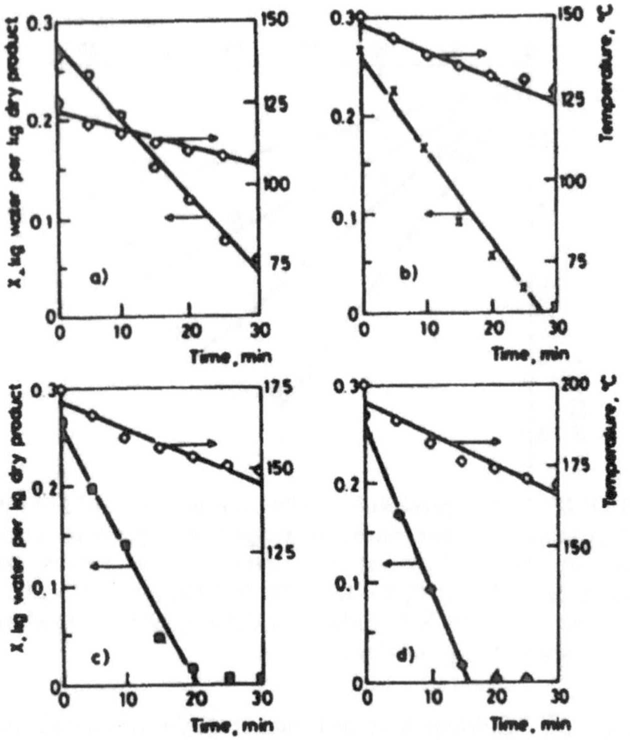

FIGURE E.4.33.2 Oven temperature profiles (◇) and comparison of the constant rate during model (—) with the data (○, x, ■, ◆) obtained in the constant oven temperature experiments. Oven temperatures and the numerical values of constants were as follows: (a) $T = 123 - 0.009\,t\,°C$, $\phi = 1.25 \times 10^{-4} s^{-1}$, (b) $T = 147 - 0.012\,t\,°C$, $\phi = 1.5 \times 10^{-4} s^{-1}$, (c) $T = 171 - 0.013\,t\,°C$, $\phi = 2.0 \times 10^{-4} s^{-1}$, (d) $T = 195 - 0.016\,t\,°C$, $\phi = 2.7 \times 10^{-4} s^{-1}$ (Turhan and Özilgen, 1991; © Akadémiai Kiadó, Budapest, reproduced by permission).

Boltzmann constant. Parameter ΔH_{vap} is the apparent latent heat of vaporization and $\bar{x}$ is the average moisture content (g water/g dry matter) of the cookies. Each term of (E.4.34.1) represents the followings:

$c_{cookie} V \bar{\rho} \dfrac{d\bar{T}}{dt}$	rate of the enthalpy change of a cookie
$(A_{upper} - A_{sides})\, h_{upper\ and\ sides}\, (T_{air} - T_{side\ surfaces})$	convective heat transfer on the upper and side surfaces of a cookie
$A_{bottom}\, h_{bottom}\, (T_{conveyor} - T_{bottom\ surface})$	heat transfer at the bottom surface of a cookie in contact with the conveyor band

	(radiation heat transfer from the bottom of the cookie are also implicitly included into this term, therefore h_{bottom} is actually an apparent constant)
$(A_{\text{upper}} + A_{\text{sides}})\,\gamma_{\text{air}}\,(T_{\text{air}}^4 - T_{\text{side surfaces}}^4)$	Radiative heat transfer from the hot air
$\gamma_{\text{wall}}\,(T_{\text{wall}}^4 - T_{\text{side surfaces}}^4)$	Radiative heat transfer from the side wall of the oven located in the upper space of the conveyor
$V \Delta H_{\text{vap}} \bar{\rho} \dfrac{d\bar{x}}{dt}$	Energy consumed to evaporate water in the cookie

The moisture balance around a cookie is:

$$-V\bar{\rho}_{\text{dry matter}}\frac{d\bar{x}}{dt} = (A_{\text{upper}} + A_{\text{sides}})\,k_{\text{upper and sides}}(\bar{x} - x^*)$$
$$+ A_{\text{bottom}}\,k_{\text{bottom}}(\bar{x} - x^*) \qquad \text{(E.4.34.2)}$$

where $\bar{\rho}_{\text{dry matter}}$ is the average density of the dry matter of the biscuits, and k and x are the mass transfer coefficient and the moisture content (g water/g dry matter) at the locations indicated by the subscripts. In these equations we also have $(A_{\text{upper}} + A_{\text{sides}})/V \cong 1/\ell$ and $(A_{\text{bottom}})/V \cong 2/r$. The latent heat of evaporation of water from the cookies was:

$$\Delta H_{\text{vap}} = \Delta H_{\text{water}}(1 + \alpha \exp(-\beta \bar{x})) \qquad \text{(E.4.34.3)}$$

where α and β are constants; ΔH_{water} is the latent heat of evaporation of pure water, expressed as:

$$\Delta H_{\text{water}} = 2257.9\left[\frac{647.16 - \bar{T}}{274.0}\right]^{0.38} \qquad \text{(E.4.34.4)}$$

where $\bar{T}$ is the average temperature of cookie. Equation (E.4.34.4) is called Watson's equation, where $\bar{T}$ and ΔH_{water} are expressed in K and kJ/kg, respectively. Estimates of the thermophysical constants $h_{\text{upper and sides}}$, h_{bottom}, k_{bottom}, $k_{\text{upper and sides}}$, α, β, γ_{air}, γ_{wall} and experimentally measured values of c_{cookie}, ℓ, $\bar{T}$, T_{air}, $T_{\text{bottom surface}}$, T_{conveyor}, T_{wall}, r, $\bar{x}$, $\bar{\rho}_{\text{dry matter}}$ and $\bar{\rho}$ were obtained. Parameter x^* was assigned to be zero. The techniques used to estimate the thermophysical properties are explained by Hayakawa and Hwang (1982) and beyond the scope of this text. Equations (E.4.34.1) and (E.4.34.2) were solved numerically by using the initial conditions:

$$\bar{x} = x_0 \quad \text{at} \quad t = 0 \qquad \text{(E.4.34.5)}$$

and

$$\bar{T} = T_0 \quad \text{at} \quad t = 0 \qquad \text{(E.4.34.6)}$$

Agreement of the model with the experimental data is depicted with a typical plot in Figure E.4.34.

Example 4.35. Modeling of wheat drying in fluidized beds (Giner and Calvelo, 1987) Early harvest of wheat with about 25% of moisture requires artificial drying of the grain to reduce the water activity to inhibit microbial growth and spoilage reactions. Drying of the grains was modeled after assuming an internal control of water movement and short drying times as:

$$\frac{\bar{x}-x^*}{x_0-x^*}=1-2a_v\sqrt{\frac{Dt}{\pi}}+0.33\,a_v^2\,Dt \qquad \text{(E.4.35.1)}$$

where a_v is the area per unit volume of the wheat. The temperature effect on water diffusivity, D, was expressed with the Arrhenius expression:

$$D=D_0\exp\left(-\frac{E}{RT}\right) \qquad \text{(E.4.31)}$$

After assuming that a perfect mixing of the grains takes place in the fluidized bed and that air leaves the bed in thermal equilibrium with the solids, the

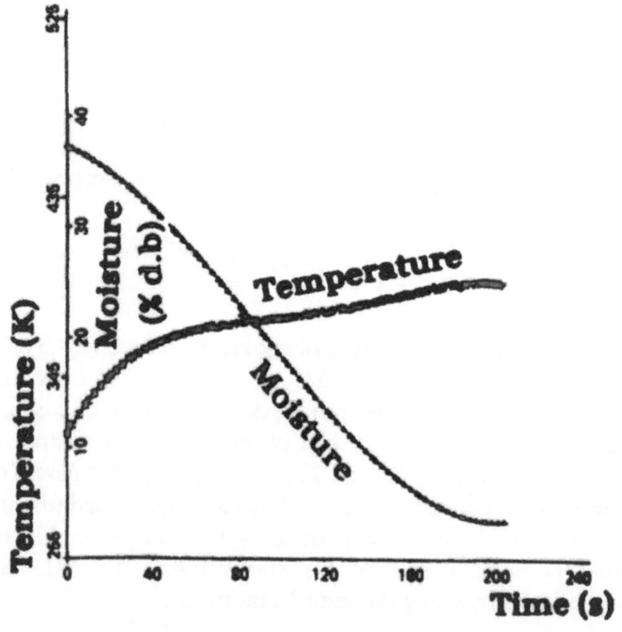

FIGURE E.4.34 Agreement of the solutions of (E.4.34.1) and (E.4.34.2) with the experimental data with cracker type cookies. Experimental data and model fall on the same lines. Abscissa for moisture curves = 10 x% dry basis (Hayakawa and Hwang, 1981; © Academic Press, reproduced by permission).

energy balance is:

$$m_s c_{grain} \frac{dT}{dt} = m_s \Delta H_d \frac{d\bar{x}}{dt} + \rho_{air} v_0 c_{air} (T_{air1} - T) \qquad \text{(E.4.35.2)}$$

where $m_s = \rho_s(1 - \varepsilon_0) S H_0$ = mass of the dry solid, c_{air} = specific heat of air, c_{grain} = specific heat of the partially dried grains, H_0 = bed height under fixed bed conditions, ΔH_d = heat of desorption, S = cross sectional area of the bed, T_{air1} = inlet air temperature, v_0 = superficial air velocity, ρ_{air} = density air, ρ_s = density of the dry solids. Specific heat of the partially dried grains were:

$$c_{grain} = c_{solids} + \bar{x} c_{water} \qquad \text{(E.4.35.3)}$$

where c_{solids} and c_{water} are the specific heats of the dry solids and water, respectively. The balance of water flowing through the bed is:

$$\rho_{air} v_0 S (\mathscr{H}_2 - \mathscr{H}_1) = -m_s \frac{d\bar{x}}{dt} \qquad \text{(E.4.35.4)}$$

where $\mathscr{H}_1$ and $\mathscr{H}_2$ are the inlet and exit air humidities, respectively. We define the following dimensionless groups $\eta = a_v^2 D_0 t$, $\psi = (x_0 - \bar{x})/(x_0 - x_s)$, $\phi = (T_{air1} - T)/(T_{air1} - T_0)$, where x_s is the average equilibrium solids moisture, T_{air1} is the inlet air temperature and T_0 is the initial temperature of the grains. There are also constants defined as: $K_1 = c_0(T_{air1} - T_0)/(\Delta H_d (x_0 - x_s))$, $K_2 = \rho_{air} v_0 S c_s (T_{air1} - T_0)/(m_s \Delta H_d a_v^2 D_0 (x_0 - x_s))$ and $K_3 = c_{water} (x_0 - x_s)/x_0$. When we substitute (E.4.35.3) in (E.4.35.2) and convert in dimensionless form, we will obtain

$$K_1(1 - K_3\psi)\frac{d\phi}{d\eta} + K_2\phi = \frac{d\psi}{d\eta} \qquad \text{(E.4.35.5)}$$

Equations (E.4.35.1) and (E.4.31) are converted into the dimensionless forms as:

$$\psi = \frac{2a_v}{\sqrt{\pi}} \sqrt{\int_0^t Dt} - 0.33\, a_v^2 \int_0^t D\, dt \qquad \text{(E.4.35.6)}$$

and

$$D = D_0 \exp\left[K_5 \frac{1 - \phi}{K_4 - \phi}\right] \qquad \text{(E.4.35.7)}$$

where $K_4 = T_{air1}/(T_{air1} - T_0)$ and $K_5 = E/RT_0$. After replacing (E.4.35.7) in (E.4.35.6) we will obtain:

$$\frac{d\psi}{d\eta} = \frac{1}{U\psi} \exp\left[K_5 \frac{1 - \phi}{K_4 - \phi}\right] \qquad \text{(E.4.35.8)}$$

where $U(\psi) = 1 - Z/0.33Z$ and $Z = (1 - 0.33\pi\psi)^{1/2}$. Equation (E.4.35.8) may be rewritten as:

$$\frac{d\eta}{d\psi} = U(\psi)\exp\left[-K_5\frac{1-\phi}{K_4-\phi}\right] \qquad \text{(E.4.35.9)}$$

Equation (E.4.35.5) is combined with (E.4.35.8) and rearranged as:

$$\frac{d\phi}{d\psi} = \frac{1 - K_2\phi U(\psi)\,\exp(K_5(1-\phi)/(K_4-\phi))}{K_1(1-K_3\psi)} \qquad \text{(E.4.35.10)}$$

In (E.4.35.9) and (E.4.35.10) the dimensionless water content ψ is the independent variable with the initial conditions:

$$\eta = 0 \quad \text{at} \quad \psi = 0 \qquad \text{(E.4.35.11)}$$

and

$$\phi = 1 \quad \text{at} \quad \psi = 0 \qquad \text{(E.4.35.12)}$$

Equations (E.4.35.9) and (E.4.35.10) were solved numerically after using the appropriate initial conditions to calculate the times and the corresponding temperatures required to attain a given water content in the grain. The values of ψ and ϕ obtained at each time can be used to evaluate the humidity of the outgoing air by means of the dimensionless form of (E.4.35.4):

$$\mathscr{H}_2 = \mathscr{H}_1 + \frac{K_6}{U(\psi)}\exp(K_5(1-\phi)/(K_4-\phi)) \qquad \text{(E.4.35.13)}$$

where $K_6 = m_s(x_0 - x_s)D_0 a_v^2/\rho_{air} v_0 S$. The model has been tested for various experimental conditions. Agreement of the model with the data is depicted in Figures E.4.35.1 and 2 with various inlet air temperatures and bed heights.

In a freeze drying process a food is frozen, than water is removed by transfer from the solid state to the vapor state by sublimation. A vapor pressure versus temperature phase diagram for free water is shown in Figure 4.18. The process $1 \rightarrow 2$ is conventional drying under constant pressure, where liquid water is evaporated. The process $3 \rightarrow 4$ represents freeze drying, where ice is converted directly into vapor. It should be noticed that the freeze drying requires a vacuum and is conducted at lower temperatures, therefore expensive equipment and long processing times are needed, which increases the cost of the process.

Example 4.36. Heat and mass transfer in freeze drying Schematic representation of two common cases of freeze drying is given in Figures E.4.36.a and b:

i) Calculation of the freeze drying time of an infinite slab when heat and mass transfer occurs through the dry layer

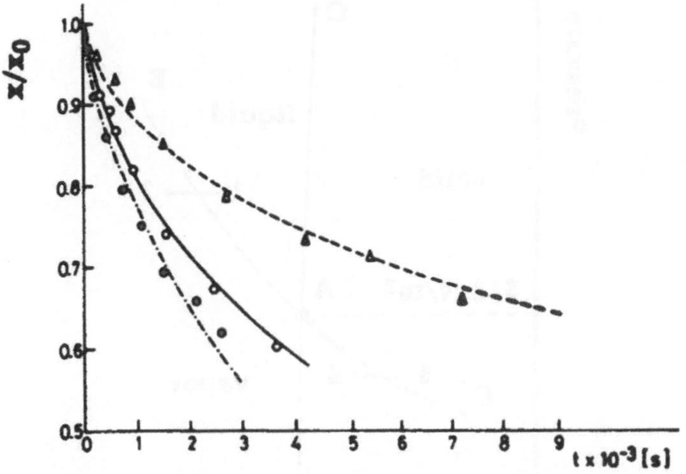

FIGURE E.4.35.1 Comparison of the model with the experimental data at different inlet air temperatures. Lines describes the model, symbols shows the experimental data: (– Δ –) $T_{air1} = 40°C$; (– O –) $T_{air1} = 60°C$; (.–●.–.) $T_{air1} = 75°C$. Other experimental parameters are: $v_e = 120$ cm/s; $H = 10$ cm, $T_e = 20°C$ (Giner and Calvelo, 1987; © IFT, reproduced by permission).

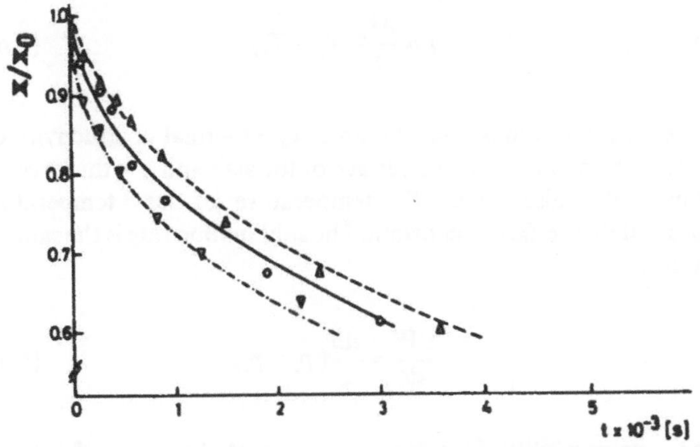

FIGURE E4.35.2 Comparison of the model with the experimental data at different initial bed heights. Lines describes the model, symbols shows the experimental data: (.–Δ.–.) thin layer; (– o –) $H_e = 4$ cm; (–▲–) $H_e = 10$ cm. Other experimental parameters are: $v_e = 120$ cm/s; $T_{air1} = 60°C$ (Giner and Calvelo, 1987; © IFT, reproduced by permission).

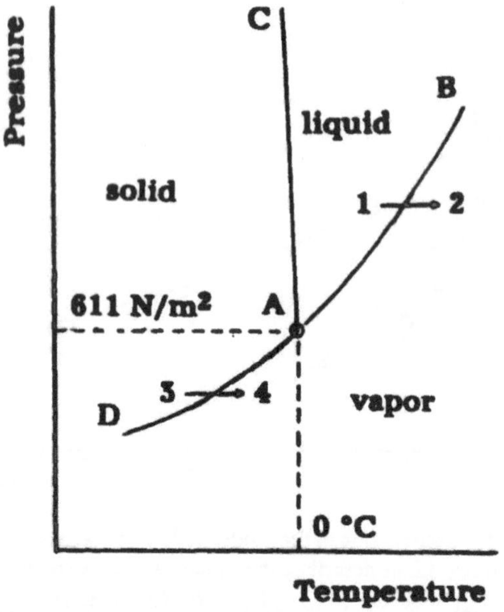

FIGURE 4.18 A qualitative vapor pressure versus temperature phase diagram for free water. AC: melting line, AD: sublimation line, AB: vaporization line.

Heat transfer rate at any instant is given by:

$$q = \frac{Ak_d}{z}(T_s - T_i) \tag{E.4.36.1}$$

where A = heat and mass transfer area, k_d = thermal conductivity of dry layer, T_s = temperature on the surface of the slab and z = thickness of the dry layer. We also have T_i = temperature of ice = temperature of frozen/dry slab interface = constant. The sublimation rate is the same as the weight loss:

$$-\frac{dW}{dt} = \frac{Ab}{z}(p_i - p_s) \tag{E.4.36.2}$$

where b = permeability of the dry layer, p_i = partial pressure of water vapor on ice, p_s = partial pressure of water vapor on the surface of the slab and W is the weight of the slab. It was assumed that the slab surface attains temperature T_s and pressure p_s as soon as the freeze drying process starts. At the ice/dry layer interface when the moisture content (expressed in kg water/kg dry solids) drops from the initial value of x_0 to final value x_f,

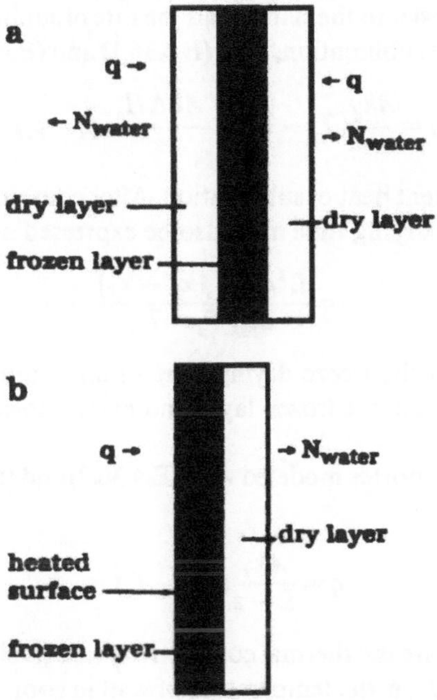

FIGURE E.4.36 Schematic representation of freeze drying of a slab with (a) Heat and mass transfer through the dry layer, (b) Heat transfer through the frozen layer and mass transfer through the dry layer. Figure E.4.36.a is observed when the latent heat of sublimation is supplied by the air. Figure E.4.36.b is observed when the latent heat of sublimation is supplied by the tray from the bottom of the food. N_{water} = water transport rate.

weight loss of the sample may be expressed as:

$$-\frac{dW}{dt} = A\rho(x_0 - x_f)\frac{dz}{dt} \quad \text{(E.4.36.3)}$$

where ρ is the density of the drying slab. When p_i is constant (E.4.36.2) and (E.4.36.3) may be combined and rearranged to obtain:

$$\int_0^L z\,dz = \frac{b(p_i - p_s)}{\rho(x_0 - x_f)}\int_0^{t_d} dt \quad \text{(E.4.36.4)}$$

where $2L$ is the slab thickness and t_d is the drying time. Partial pressure of water vapor p_i on dry/frozen slab interface is assumed to be a constant as a consequence of the constant interfacial temperature T_i. Equation (E.4.36.4) requires

$$t_d = \frac{L^2\rho(x_0 - x_f)}{2b(p_i - p_s)} \quad \text{(E.4.36.5)}$$

The rate of heat transfer to the slab equals the rate of sublimation multiplied by the latent heat of sublimation, then (E.4.36.1) and (E.4.36.2) will require

$$q = \frac{Ak_d}{z}(T_s - T_i) = \frac{Ab\Delta H_s}{z}(p_i - p_s) \qquad \text{(E.4.36.6)}$$

where ΔH_s is the latent heat of sublimation. After combining (E.4.36.5) and (E.4.36.6) the freeze drying time may also be expressed as (Karel, 1975b):

$$t_d = \frac{L^2 \rho \Delta H_s (x_0 - x_f)}{2k_d (T_s - T_i)} \qquad \text{(E.4.36.7)}$$

ii) Calculation of the freeze drying time of an infinite slab when heat transfer occurs through the frozen layer and mass transfer occurs through the dry layer.

Water vapor transport is modeled with (E.4.36.2) and (E.4.36.3). The heat transfer rate is:

$$q = \frac{Ak_f}{L - z}(T_w - T_i) \qquad \text{(E.4.36.8)}$$

where k_f and $L - z$ are the thermal conductivity and thickness of the frozen layer, respectively. T_w is the temperature of wall in contact with the frozen layer. After assuming that the rate of heat transfer to the slab equals the rate of sublimation multiplied by the latent heat of sublimation and combining (E.4.36.2) and (E.4.36.8) we will obtain

$$p_i = p_s + \left(\frac{k_f}{b\Delta H_s}\right)\left(\frac{z}{L - z}\right)(T_w - T_i) \qquad \text{(E.4.36.9)}$$

Since thickness of the dry layer z varies with time, neither p_i nor T_i are constants. After combining (E.4.36.2) and (E.4.36.3) we may obtain (Karel, 1975b):

$$t_d = \frac{\rho(x_0 - x_f)}{b}\int_0^L \frac{z}{p_i(z) - p_s}\,dz \qquad \text{(E.4.36.10)}$$

Freeze drying time t_d may be calculated after substituting an appropriate expression for $p_i(z)$ in (E.4.36.10).

Example 4.37. Heat and mass transport in convection oven frying (Skjldebrand, 1980; Skjöldebrand and Hallström, 1980) Frying in a convection oven is a drying operation at high air temperatures. Although the whole product is dried in the conventional drying processes, only the surface is dried in convective oven frying. During convection oven frying, three different stages are recognized. The first stage is the pre-heating of the product surface from the initial temperature to the wet-bulb temperature. The second stage is the constant rate period, where the surface is covered with a continuous water film and retains the wet-bulb temperature. When the surface water film starts shrinking and exposes the dry product, the temperature of the dry zones starts increasing over the wet-bulb temperature. This third step of the convection oven frying process is the falling rate period. Frying of meat products is signified by the formation of a crust. The crust has remarkably lower moisture content than the initial moisture content of the product and its temperature exceeds 100°C (Fig. E.4.32). During the convection oven frying process the main water evaporation takes place in the vicinity of the 100°C-isotherm.

The following propositions and approximations were made while calculating the profiles given in Figure E.4.37.1:

i. Water transport in the crust occurs in the vapor phase. The mass transfer rate of the vapor is proportional to the vapor pressure and inversely proportional to the thickness of the crust.

ii. Vapor transport is assumed to occur from the location of the 100°C isotherm. Loss of vapor from the other locations in the crust is neglected.

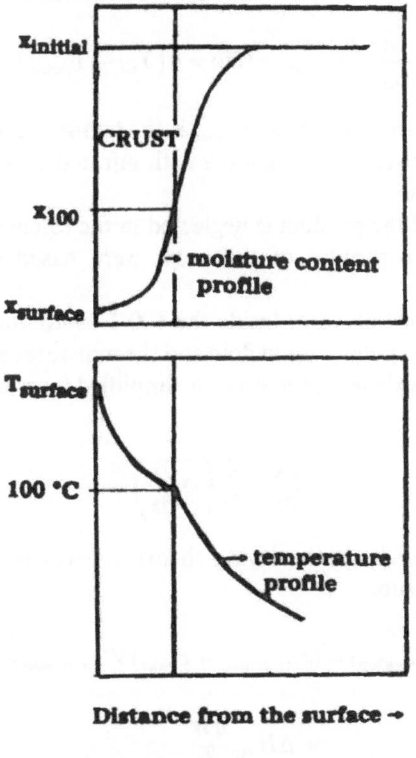

FIGURE E.4.37.1 Fully developed moisture and temperature profiles in the product during convective oven frying (Skjöldebrand and Hallström, 1980; © IFT, reproduced by permission).

iii. The partial vapor pressure in the crust at any moment is supposed to be at equilibrium with the water content of the crust at actual local temperature. This equilibrium is defined by the sorption isotherms:

$$a_w = \exp(-e^{0.033T-2.92}x^{0.0043(T-1.99)}) \tag{E.4.37.1}$$

The vapor pressure at 100°C isotherm is supposed to equal the saturated vapor pressure at this temperature.

iv. Product properties in the crust (including those of sorption) changed because of heat denaturation. For each distance from the surface the change in the sorption properties were supposed to be proportional with the following integral:

$$\ln\frac{x}{x_{100}} = k_{\text{denaturation}}\int_0^t (T(t)-100)\,dt \tag{E.4.37.2}$$

where x_{100} is the moisture content at 100°C, $k_{\text{denaturation}}$ is the denaturation rate constant.

v. Thermal conductivity of the crust is calculated from:

$$\frac{k}{L_{\text{crust}}}(T_{\text{surface}}-100) = h(T_{\text{air}}-T_{\text{surface}}) \tag{E.4.37.3}$$

where L_{crust} is the crust thickness, T_{surface} is the temperature at the surface of the product (measured experimentally with infrared sensing device), T_{air} is the air temperature.

vi. Shrinkage of the product is neglected in the calculations.

vii. Temperature profiles of the crust were based on experimental measurements.

viii. The transport of water inside the 100°C isotherm was assumed to take place only as capillary liquid flow and the moisture content profile was calculated with *Fick's second law*, i.e., a simplified form of the equation of continuity:

$$\frac{\partial x}{\partial t} = \frac{\partial}{\partial z}\left(D\frac{\partial x}{\partial z}\right) \tag{E.4.32.4}$$

ix. According to the basic drying theories h was estimated from the following relationship:

$$Ah(T_{\text{air}}-T_{\text{surface}}) = (c_{\text{dry product}}+c_{\text{water}})W_{\text{dry product}}\frac{dT_{\text{mean}}}{dt} + \Delta H_{\text{vap}}\frac{dW}{dt} \tag{E.4.37.5}$$

where A is the heat and mass transfer area, W and T_{mean} are the weight and the mean temperature of the product, respectively.

x. Interfacial mass transfer coefficient k_G was estimated from the following relationship:

$$\frac{dW}{dt} = \frac{k_G A}{RT}(p_{ws} - p_{wair}) \tag{E.4.37.6}$$

where p_{ws} and p_{wair} are the partial pressures of water vapor on the surface and in the air, respectively. Variation of the transport parameters h and k_G during the experiments is depicted with sample plots in Figure E.4.37.2.

I. Air velocity $= 9$ m/s, $T_{\text{wet bulb}} = 45°\text{C}$, $h_{\text{mean}} = 53\ \text{W/m}^2{}°\text{C}$,

$k_{G\text{mean}} = 0.051$ m/s

II. Air velocity $= 5$ m/s, $T_{\text{wet bulb}} = 45°\text{C}$, $h_{\text{mean}} = 47\ \text{W/m}^2{}°\text{C}$,

$k_{G\text{mean}} = 0.054$ m/s

III. Air velocity $= 5$ m/s, $T_{\text{wet bulb}} = 97°\text{C}$, $h_{\text{mean}} = 30\ \text{W/m}^2{}°\text{C}$,

$k_{G\text{mean}} = 0.062$ m/s

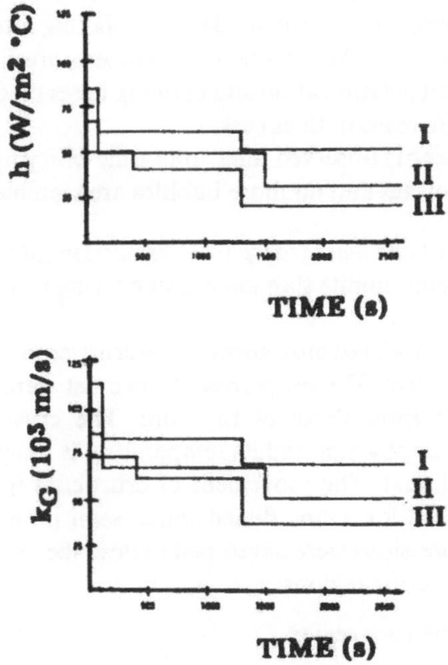

FIGURE E.4.37.2 Variation of the transport parameters (a) h and (b) k_G during the experiments with $T_{air} = 225°\text{C}$ (Skjöldebrand, 1980; © IFT, reproduced by permission).

Heat and mass transfer in immersion-fat frying process has been reviewed by Singh (1995), where conductive heat transfer occurs in the food and convective heat transfer occurs on the interface between the solid food and the surrounding oil. The water vapor bubbles escaping from the surface of the food cause considerable turbulence in the oil, on the other side their accumulation on the under side of the food may prevent effective heat transfer. Moisture transfer from the food to the surroundings is the principle physical phenomena in both drying and frying processes; where the surrounding media are air and oil in drying and frying processes, respectively, therefore these processes have a lot of common characteristics and are discussed in a similar manner

A frying process is composed of four distinct stages (Singh, 1995):

i. **Initial heating period** lasts until the surface of the food is elevated to the boiling temperature of the oil. This stage lasts a few seconds and the mode of heat transfer between the oil and the food is natural convection. No vaporization of water occurs from the surface of the food.
ii. **Surface boiling period** is signalled by evaporation on the surface of the food. The crust begins to form at the surface and the dominant mode of heat transfer on the surface is forced convection due to the considerable turbulence on the surface.
iii. **Falling rate period** in frying is similar to the falling rate period observed in drying processes. The internal core temperature rises to the boiling point, then start gelatinization and cooking takes place. The crust layer continues to increase in thickness.
iv. **Bubble end-point** is observed after long time of frying, when moisture removal diminishes and no more bubbles are seen leaving the surface.

Example 4.38. Heat and mass transport in immersion-fat frying Schematic description of a semi-infinite slab undergoing frying is described in Figure E.4.38.1.

Farkas (1994) considered movement of the crust/core interface as a moving-boundary problem. The properties of the crust were considered to be uniquely different from those of the core. The crust region contains a negligible amount of water and its temperature is much higher than the boiling point of liquid. The movement of crust/core interface of Figure 4.38.1 was modelled for a one dimensional, semi infinite slab. Separate mathematical expressions were developed to describe heat and mass transfer in the core and crust regions:

Heat transfer in the core region:

$$k^{II}\frac{\partial^2 T}{\partial x^2}+N_{\beta x}c_{p\beta}\frac{\partial T}{\partial x}=(\varepsilon_\beta\rho_\beta c_{p\beta}+\varepsilon_\sigma\rho_\sigma c_{p\sigma})\frac{\partial T}{\partial t} \qquad \text{(E.4.38.1)}$$

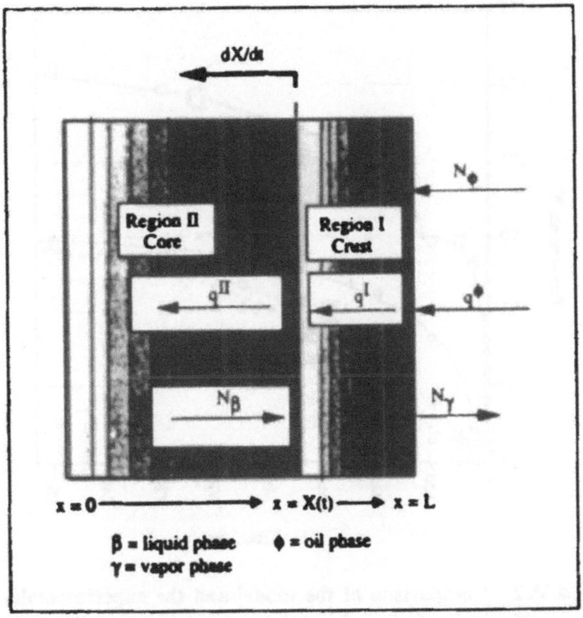

FIGURE E.4.38.1 Semi-infinite slab undergoing frying. The center of the slab is at $x = 0$, total thickness $= L$ (Singh, 1995; © IFT, reproduced by permission).

Mass transfer in the core region:

$$\frac{\partial c_\beta}{\partial g} = D_{\beta\sigma} \frac{\partial^2 c_\beta}{\partial x^2} \tag{E.4.38.2}$$

Heat transfer in the crust region:

$$k^I \frac{\partial^2 T}{\partial x^2} + N_{\gamma x} c_{p\gamma} \frac{\partial T}{\partial x} = (\varepsilon_\gamma \rho_\gamma c_{p\gamma} + \varepsilon_\sigma \rho_\sigma c_{p\sigma}) \frac{\partial T}{\partial t} \tag{E.4.38.3}$$

Mass transfer in the crust region:

$$\frac{\partial}{\partial x}\left[\rho_\gamma \frac{\partial P_\gamma}{\partial x}\right] = 0 \tag{E.4.38.4}$$

where c_{pi} = specific heat of species i, c_β = concentration of β, k^i = effective thermal conductivity of region i, N_{ix} = flux of species i in x direction, P = pressure, ε_i = volume fraction of species i and ρ_i = density of species i. A typical plot for the comparison of the model and the experimentally determined temperature profiles during frying of a semi-infinite potato mix is shown in Figure E.4.38.2.

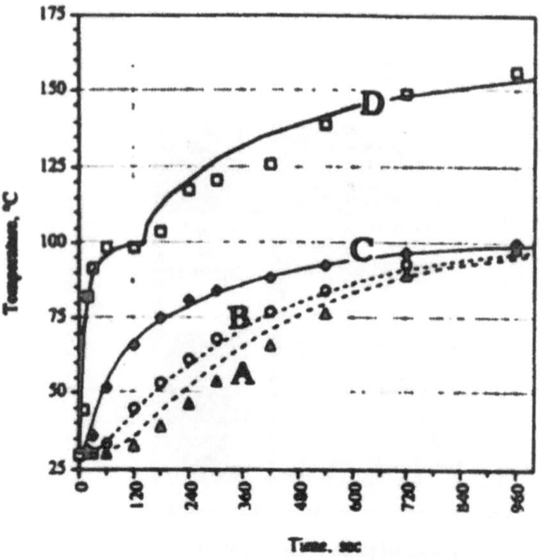

FIGURE E.4.38.2 Comparison of the model and the experimentally determined temperature profiles during frying of a semi-infinite potato mix (thickness = 2.54 cm, diameter = 8.26 cm). Frying oil temperature is 180°C. A is center point; B is 1/3 out; C is 2/3 out; D is 0.05 cm below the surface (Singh, 1995; © IFT, reproduced by permission). The model and the data are shown with the lines and the symbols, respectively.

Osmotic dehydration is a method of concentrating fruits and vegetables prior to air drying, freeze drying or freezing, where part of the natural water contents of fruits and vegetables is removed by immersing the materials in aqueous solutions of osmotic agents. The most commonly used osmotic agents are sucrose for fruits and sodium chloride for vegetables. Other osmotic agents such as lactose, maltodextrin, ethanol, glucose, glycerine and corn syrups have also been used. Osmotic drying is a multicomponent mass transfer process, where water is removed from the fruits and vegetables and osmotic agents are taken up (Biswal and Bozorgmehr, 1992). The detailed discussion of osmosis is given in Chapter 4.6.

Example 4.39. Mass transfer in mixed solute osmotic dehydration of apple rings Mass flux N was previously expressed empirically as:

$$N = k_c(c_i - c_\infty) \tag{2.43.e}$$

It was shown in Example 2.16 that when an unsteady state concentration profile occurs near the interface the mass transfer coefficient may be expressed as $k_c = \sqrt{D/\pi t}$. The loss of any substrate from an apple slice may be described as:

$$-V\frac{dc}{dt} = A\sqrt{\frac{D}{\pi t}}(c_i - c_\infty) \tag{E.4.39.1}$$

where A is the mass transfer area and V is the volume of the fruit slices. When A, c_i, c_∞, D and V do not change much, (E.4.39.1) may be rearranged as:

$$\frac{dc}{dt} = -\left\{\frac{A(c_i - c_\infty)}{V}\sqrt{\frac{D}{\pi}}\right\}\frac{1}{\sqrt{t}} \tag{E.4.39.2}$$

when the term $\{(A(c_i - c_\infty)/V)\sqrt{D/\pi}\}$ is constant. Equation (E.4.39.2) may be rearranged and integrated as

$$c = c_0 - \kappa\sqrt{t} \tag{E.4.39.3}$$

where $\kappa = 2\{(A(c_i - c_\infty)/V)\sqrt{D/\pi}$.

When there is an uptake of subsrate the negative sign disappears from (E.4.39.1)–(E.4.39.3). Comparison of the model with the data is depicted in Figures E.4.39.1 and E.4.39.2.

4.6. MATHEMATICAL MODELING OF FILTRATION AND MEMBRANE SEPARATION PROCESSES

The particle size range of the filtration and membrane separation processes is shown in Figure 4.19 Particles within the ionic range, i.e., aqueous salts, sugars, etc. are separated with *reverse osmosis*. *Ultrafiltration* is used within the molecular range, i.e., virus, albumin protein, etc. *Microfiltration* is used to separate particles within the macromolecule range; and particle *filtration* techniques are employed for the micro and macro particles.

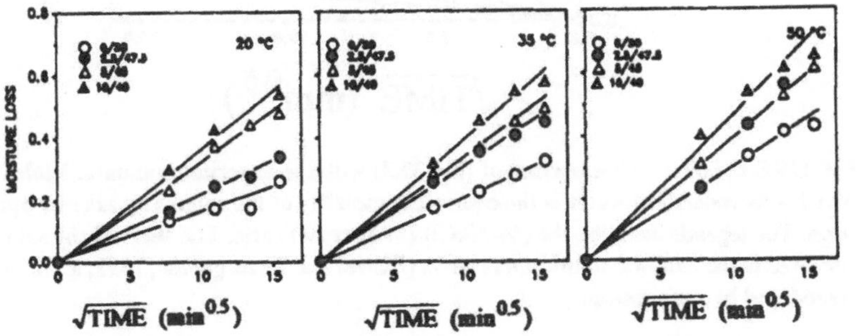

FIGURE E.4.39.1 Comparison of (E.4.39.3) with the experimental data. Moisture loss was defined as $(c_0 - c(t))/c_0$, therefore the slope of the lines were κ/c_0. The legends describe the (% NaCl)/(% sucrose) ratio. The total of the solutes dissolved in the osmotic solution was 50% (Biswal and Bozorgmehr, 1992; © ASAE, reproduced by permission).

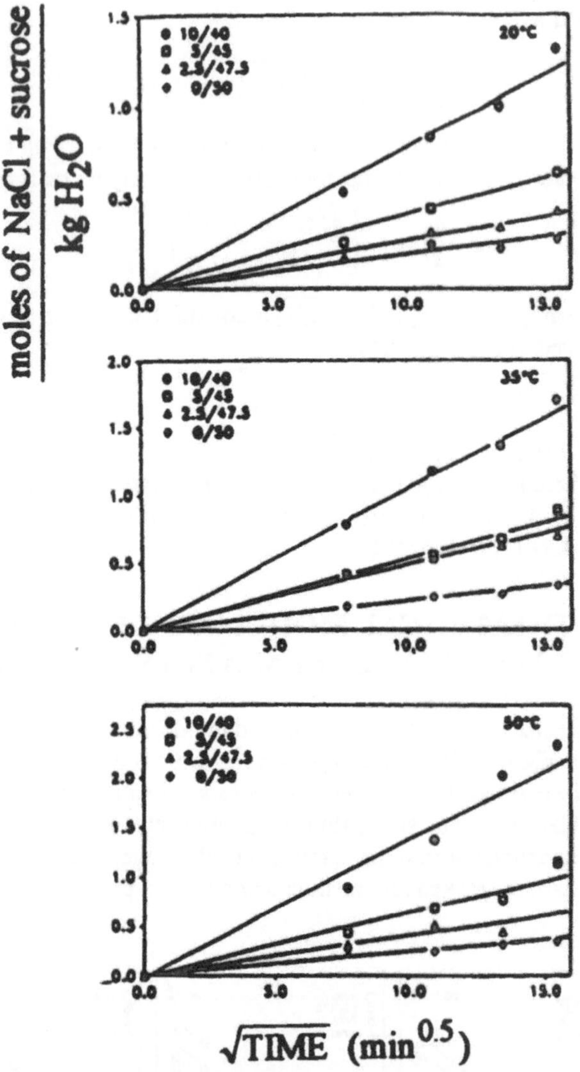

FIGURE E.4.39.2 Comparison of (E.4.39.3) with the experimental data. Moles of NaCl + sucrose/kg of water is the equivalent molality of the solutes up take by apple rings. The legends describe the (% NaCl)/(% sucrose) ratio. The total of the solutes dissolved in the osmotic solution was 50 % (Biswal and Bozorgmehr, 1992; © ASAE, reproduced by permission).

In a particle filtration process the solids are separated from a liquid by means of a medium which retains the solids, but allows the liquid to pass through. Accumulation of the particles on the medium clogs its holes and prevents the flow. *Filter aids*, i.e., inert particles of diatomaceous earth,

particle size (log scale, μm)	0.001	0.01	0.1	1.0	10	100 1000
particle range	ionic	molecular	macro-molecular		micro particle	macro particle

effective particle size range of some well known particles	← salts → ← yeast → ←virus→ ← bacteria → ←sugars→ a typical ←globular→ protein ← milled flour →
separation process and the pressure (kPa) range of the operation	←REVERSE OSMOSIS→ $(20\times10^2 \le \Delta P \le 50\times10^2)$ ←MICRO-FILTRATION→ $(1\times10^2 \le \Delta P \le 2\times10^2)$ ←ULTRAFILTRATION→ $(1\times10^2 \le \Delta P \le 7\times10^2)$ ← PARTICLE FILTRATION → $(0.65\times10^2 \le \Delta P \le 15\times10^2)$

FIGURE 4.19 **Particle size range of the membrane separation and filtration processes.**

cellulose, etc., are used in filtration processes to keep these holes open. In a cake filtration process filtration actually occurs through the cake (Fig. 4.20). The filter medium is precoated with a layer of filter aid before the beginning of filtration. The liquid food is also mixed with the filter aid and fed in to the filter. Filtration occurs through the cake, the food particles and the filter aid are retained in the filter, while the liquid passes through. The cake builds up continuously through the process, the food particles are dispersed among the filter aid particles, therefore clogging is prevented.

Sperry's equation is the most common model used to simulate filtration (Example 1.4):

$$\frac{dV}{dt}=\frac{A\,\Delta P}{\mu}\frac{1}{R_m+\alpha cV/A} \tag{4.48.1}$$

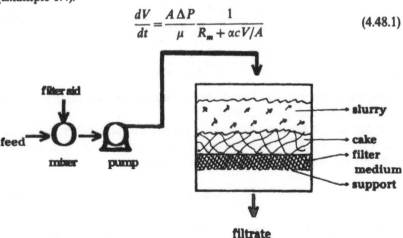

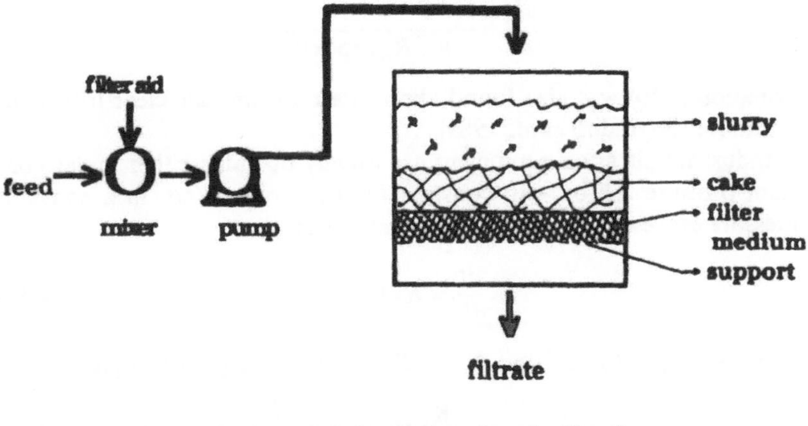

FIGURE 4.20 **Schematic description of a cake filtration process.**

The medium resistance R_m often varies with time. This behavior results when some of the solids penetrate the medium as it compress under applied pressure. The specific cake resistance α is the most troublesome parameter and it changes with ΔP in most practical cakes due to the compressibility of the cake, this is often expressed as (Svarovsky, 1985):

$$\alpha = \alpha_0 (\Delta P_c)^n \tag{4.48.2}$$

where α_0 and n are constants, ΔP_c is the pressure drop across the cake. The average value of the cake resistance α_{av} may be calculated as (Svarovsky, 1985):

$$\alpha_{av} = \frac{\Delta P_c}{\int_0^{\Delta P_c} \frac{d(\Delta P_c)}{\alpha}} = (1-n)\alpha_0 (\Delta P_c)^n \tag{4.48.3}$$

For high feed concentrations, the volumes of the feed slurry and the filtrate differ significantly, therefore correction should be made in (4.48.1) by substituting the effective concentration $c_{corrected}$ for c (Svarovsky, 1985):

$$c_{corrected} = \frac{1}{\dfrac{1}{c} + \dfrac{1}{\rho_s} + \dfrac{\gamma - 1}{\rho}} \tag{4.48.4}$$

where ρ_s and ρ are the densities of the solids and the liquid, respectively; γ is the mass ratio of the wet to dry filter cake. Foods are usually highly complex systems. DeLagarza and Boulton (1984) modified (4.48.1) to model wine filtrations (Example 1.4):

$$\frac{dV}{dt} = \frac{A \Delta P}{\mu} \frac{1}{R_m + \alpha c (V/A)^n} \tag{4.49}$$

$$\frac{dV}{dt} = \frac{A \Delta P}{\mu} \frac{1}{R_m \exp(\kappa V/A)} \tag{4.50}$$

Equation (4.50) was also found appropriate to simulate clear fruit juice filtrations (Bayindirli *et al.*, 1989).

Industrial filtration equipment are usually operated either under constant pressure ΔP or with constant filtration rate dV/dt. In a constant pressure process (4.48.1) may be rearranged as:

$$\frac{dt}{dV} = K_1 V + B_1 \tag{4.51}$$

where $K_1 = \mu \alpha c / \Delta P A^2$ and $B_1 = \mu R_m / \Delta P A$, or integrated to give:

$$t = \frac{K_1}{2} V^2 + B_1 V \tag{4.52}$$

In a constant rate process (4.48.1) may be rearranged as:

$$\Delta P = K_2 V + B_2 \tag{4.53}$$

$$\text{where } K_2 = \frac{\mu\alpha c}{A^2}\frac{dV}{dt} \quad \text{and} \quad B_2 = \frac{\mu R_m}{A}\frac{dV}{dt}.$$

In a constant pressure process, if (4.48.1) is valid, a plot of dt/dV versus V may be used to evaluate constants K_1 and B_1 from the slope and the intercept of (4.51), respectively. A plot of ΔP versus V may be employed in a constant rate process to evaluate K_2 and B_2 as implied by (4.53).

Example 4.40. Constant pressure filtration process of poultry chiller water (Chang et al., 1989) In poultry processing plants the broilers are dipped in cold water to achieve reasonable pre-cooling prior to further processing. The chiller water is contaminated by some post slaughtering residues and soluble constituents of the broilers. Recycling of the poultry chiller water after filtration may reduce ice and water usage and liquid waste effluent. During constant pressure filtration process of poultry chiller water (4.52) failed to simulate the filtration process, and modified to express time dependent cake resistance as:

$$t = (\gamma + \beta t)V^2 + \frac{1}{q}V \tag{E.4.40.1}$$

and

$$t = (\varepsilon + \phi(1 - \exp(-\zeta t))V^2 + \frac{1}{q}V \tag{E.4.40.2}$$

where q = initial volumetric flow rate of the filtrate when only the medium and the precoat of the filter aid in place, γ, β, ε, ϕ and ζ are constants. Equations (E.4.40.1) and (E.4.40.2) were solved to simulate accumulation of the filtrate as:

$$V = \frac{-\frac{1}{q} + \sqrt{\left(\frac{1}{q}\right)^2 + 4(\beta t + \gamma)t}}{2(\beta t + \gamma)} \tag{E.4.40.3}$$

and

$$V = \frac{-\frac{1}{q} + \sqrt{\left(\frac{1}{q}\right)^2 + 4t(\varepsilon + \phi(1 - \exp(-\zeta t)))}}{2(\varepsilon + \phi(1 - \exp(-\zeta)))} \tag{E.4.40.4}$$

Comparison of (E.4.40.3) and (E.4.40.4) with the experimental data are shown in Figure E.4.40.

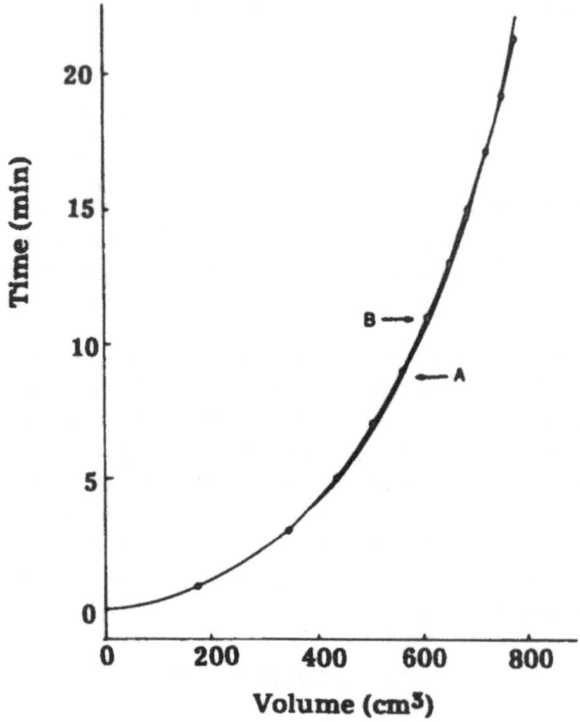

FIGURE E.4.40 Comparison of the mathematical models (—) with the experimental data (•). Curve A represents (E.4.40.3), curve B represents (E.4.40.4) (Chang *et al.*, 1989; © Food & Nutrition Press, Inc., reproduced by permission).

The rate of filtration of an incompressible liquid through a filter cake may be related with the permeability of the medium by using Darcy's equation:

$$\frac{dV}{dt} = \frac{A\kappa\Delta P}{\mu H} \tag{4.54}$$

where V is the volume of the fluid flowing through the medium, μ is the fluid viscosity, ΔP is the pressure drop across the medium, A and H are the cross sectional area and the depth of the medium, respectively. Parameter κ is empirically defined as the permeability of the medium related to the properties of the cake (Ingmanson, 1953; Harvey *et al.*, 1988). When the cake has uniform, straight, cylindrical voids the permeability may be expressed as:

$$\kappa = \frac{d_p^2 \varepsilon}{32} \tag{4.55}$$

where d_p is the diameter of the voids, and ε is the void fraction. For non-cylindrical voids κ may be defined as:

$$\kappa = \frac{d_e^2 \varepsilon}{K''} \tag{4.56}$$

where d_e = (void volume)/(solid surface area), and K'' is an empirical shape factor (also called the Kozeny constant). For the compressible mycelial microbial cakes Oolman and Liu (1991) suggested that

$$\kappa = \frac{H}{H_0} \frac{(\varepsilon_0 - \beta)^3 d_h^2}{(1-\beta)^2 16K''(1-\varepsilon_0)^2} \tag{4.57}$$

where $\beta = \alpha(\Delta P/g\rho_c H)^n$, α = constant, ρ_c = grams of dry hyphae/cm^3 cake, g = gravitational acceleration, n = positive constant, ε_0 = void fraction of a developed cake when ΔP equals zero, H_0 = height of a developed cake when ΔP equals zero and d_h = diameter of hyphae. Equations (4.54)–(4.57) relates the filtration rates to the cake characteristics, therefore require more detailed characterization of the cake than with those of (4.48)–(4.50).

When a pure solvent and a solution (solvent + solute) are separated with a solute impermeable membrane and no external pressure is applied ($\Delta P = 0$), solvent tends to diffuse through the membrane towards the solution side to equalize the solvent chemical potential on both sides (Fig. 4.21). When the solution is dilute and consists of a single solute the

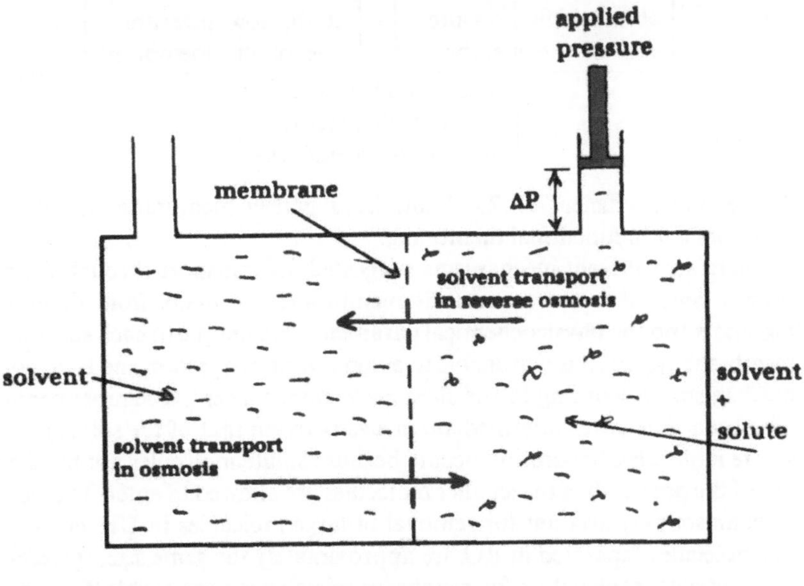

FIGURE 4.21 Solvent flux in osmosis and reverse osmosis processes.

osmotic pressure (π) may be calculated as:

$$\pi = cRT \tag{4.58}$$

where c = concentration of the solute, R = gas constant and T = absolute temperature. In the concentrated solution of a single solute, the osmotic pressure may be calculated as:

$$\pi = A_1 c + A_2 c^2 + A_3 c^3 + \ldots \tag{4.59}$$

where A_1, A_2 and A_3 are constants. When there is more than one solute in a dilute solution, the osmotic pressure may be calculated as:

$$\pi = RT \sum_{i=1}^{n} c_i \tag{4.60}$$

where c_i = concentration of each solute. If we should apply pressure ΔP to the solute side, osmosis prevails as long as $\pi > \Delta P$. Solvent flux stops when $\pi = \Delta P$. If we should increase ΔP such that $\Delta P > \pi$ the solvent flux is reversed towards the pure solvent side. This is called reverse osmosis (RO). Ultrafiltration (UF) and microfiltration (MF) are essentially similar to reverse osmosis, but the particle size range of the processes is different (Tab. 4.19).

The rejection coefficient is used to describe the efficiency of a membrane system for the separation of a specified solute from a well defined solution and is defined as:

$$R = \frac{\begin{bmatrix}\text{solute concentration} \\ \text{at the high pressure} \\ \text{side of the membrane}\end{bmatrix} - \begin{bmatrix}\text{solute concentration} \\ \text{at the low pressure} \\ \text{side of the membrane}\end{bmatrix}}{\begin{bmatrix}\text{solute concentration} \\ \text{at the high pressure} \\ \text{side of the membrane}\end{bmatrix}} \tag{4.61}$$

A rejection coefficient of $R = 1$ implies a perfect membrane and $R = 0$ describes a non functional membrane.

There are different mechanisms suggested for transport through membranes. Since the reverse osmosis membranes are made from different materials and the physicochemical parameters are unique to each solution-membrane system, it is reasonable to assume various rejection and transport mechanisms. According to the *sieve mechanism* concept, a semipermeable membrane has pores intermediate in size between that of the solvent and solute molecules. Separation occurs because solute molecules are blocked out of the pores, while the smaller molecules are allowed to enter. The sieve mechanism may account for removal of large molecules in UF, however the molecules separated in RO are approximately the same size, therefore a purely steric explanation for membrane rejection is not possible (Cheryan, 1992). *Hydrogen-Bonding mechanism* was developed specifically for cellulose acetate membranes. Permeation is suggested to occur as the molecules migrate through the membrane from one hydrogen-bond site to the other. Water molecules are also absorbed on the membrane surface by making hydrogen-bonds, and prevent solute ions from entering. Solute molecules may move through the membrane by displacing the solvent molecules from the absorption sites. Since displacement requires large amounts of energy, relatively few solute molecules are absorbed by the membrane (Cheryan, 1992). The *solution-diffusion mechanism* attributes the solute rejection to large differences in the solubility and diffusivity of the solute and the solvent materials in the membrane material. According to the *preferential sorption-capillary flow mechanism*, partial sorption on the membrane surface is achieved with repulsive or attractive force gradients at the membrane surface. The solvent and the solute molecules move through the membrane pores with the kinetic effects governed by both potential force gradients, and steric effects associated with the structure and size of the solute and solvent molecules relative to the membrane pores. A membrane transport model is needed to relate the solvent or solute fluxes to the operating conditions or the other measurable properties of the system. An extensive review of the membrane transport models has been presented by Cheryan (1992). Some of these models which are suitable for practical calculations are summarized in Table 4.6.

Table 4.6 Models for membrane transport (Cheryan, 1992)

Kadem and Katchalsky's model	$J_{solvent} = L_{solvent}(\Delta P - \sigma \Delta \pi)$	(4.62)
	$J_{solute} = L_{solute}\Delta \pi + (1-\sigma) J_{solvent} c_{ln}$	(4.63)
Pusch's model	$J_{solvent} = L_{solvent}(\Delta P - \sigma \Delta \pi)$	(4.64)
	$J_{solute} = \left[\frac{L_{osmotic}}{L_{solvent}} - \sigma^2\right] L_{solvent}\bar{c}\Delta \pi + J_{solvent}(1-\sigma)\bar{c}$	(4.65)
	$R_{max} = \sigma$	(4.66)
	$R = \frac{R_{max} J_{solvent}}{J_{solvent} + \left(\frac{L_{osmotic}}{L_{solvent}} - R_{max}^2\right) L_{solvent}\pi_{high}}$	(4.67)
solution-diffusion model	$J_{solvent} = A(\Delta P - \Delta \pi)$	(4.68)
	$J_{solute} = B \Delta c$	(4.69)
	where	
	$A = -\frac{D_{solvent.M}\, c_{solvent.M}\, \bar{\upsilon}_{solvent}}{RT\ell}$	(4.70)
	$B = -\frac{D_{solute.M}\, K_{solute}}{\ell}$	(4.71)

Table 4.6 (Continued)

	$R_{max} = 1$	(4.72)
	$R = \dfrac{A(\Delta P - \Delta\pi)}{A(\Delta P - \Delta\pi) + Bc_{solvent/high}}$	(4.73)
pore flow model	$J_{solvent} = \varepsilon v_D$	(4.74)
	$J_{solute} = \dfrac{v}{\varepsilon}\dfrac{K_{low}c_{low} - K_{high}c_{high}\exp(v\tau\ell/\varepsilon D_{solute/solvent})}{1 - \exp(v\tau\ell/\varepsilon D_{solute/solvent})}$	(4.75)
	where	
	$v_p = \dfrac{r^2 n}{8\mu}\dfrac{\Delta P}{\tau\ell}$	(4.76)
	$R_{max} = 1 - \dfrac{K_{high}}{\varepsilon}$	(4.77)
	$R = 1 - \dfrac{K_{high}\exp(v\tau\ell/\varepsilon D_{solute/solvent})}{K_{low} - \varepsilon + \varepsilon\exp(v\tau\ell/\varepsilon D_{solute/solvent})}$	(4.78)
finely porous model	$J_{solvent} = \varepsilon v$	(4.79)
	$J_{solute} = \dfrac{v}{\varepsilon b}\dfrac{K_{low}c_{low} - K_{high}c_{high}\exp[(v\tau\ell/\varepsilon)(f_{solute/solvent}/R_g T)]}{1 - \exp[(v\tau\ell/\varepsilon)(f_{solute/solvent}/R_g T)]}$	(4.80)
	where	
	$v = \dfrac{H}{\varepsilon}\left[\dfrac{1}{1 + (Hf_{solute/membrane}/MW\varepsilon)c_{low}}\right]\dfrac{\Delta P}{\tau\ell}$	[4.81]
	$b = 1 + (f_{solute/membrane}/f_{solute/solvent})$	(4.82)
	$R_{max} = 1 - \dfrac{K_{high}}{\varepsilon}$	(4.83)
	$R = 1 - \dfrac{K_{high}\exp[(v\tau\ell/\varepsilon)(f_{solute/solvent}/R_g T)]}{K_{low} - b\varepsilon + b\varepsilon\exp[(v\tau\ell/\varepsilon)(f_{solute/solvent}/R_g T)]}$	(4.84)

where

b = friction factor

Δc = solute concentration difference across the membrane

c_{ln} = log mean solute concentration difference across the membrane

$\bar{c}$ = $\Delta\pi/RT\Delta\ln a$ (where $\Delta\ln a$ is the difference of the natural logarithms of the solute activity on both sides of the membrane)

c_{low} = solute concentration at the low pressure side solution/membrane interface

c_{high} = solute concentration at the high pressure side solution/membrane interface

$c_{solvent/high}$ = solvent concentration at the high pressure side solution/membrane interface

$c_{solvent,M}$ = concentration of the solvent in the membrane

$D_{solute/solvent}$ = diffusivity of solute in solvent

$D_{solvent,M}$ = diffusivity of solvent in the membrane

$D_{solute.M}$ = diffusivity of solute in the membrane
$f_{solute/membrane}$ = friction coefficient between the solute and membrane
$f_{solute/solvent}$ = friction coefficient between the solute and solvent
$H = \varepsilon r^2/8\mu$ = hydraulic permeability of the membrane
$J_{solvent}$ = solvent flux through the membrane
J_{solute} = solute flux through the membrane
K_{high} = solute distribution coefficient at the high pressure side of the membrane
K_{low} = solute distribution coefficient at the low pressure side of the membrane
K_{solute} = generalized solute distribution coefficient
ℓ = thickness of the membrane
$L_{solvent}$ = hydrodynamic (solvent) permeability coefficient
L_{solute} = solute permeability coefficient
$L_{osmotic}$ = osmotic permeability coefficient
MW = molecular weight of the solute
n = number of pores per unit membrane area
r = pore radius
R_{max} = maximum attainable value of the rejection coefficient
$\bar{v}_{solvent}$ = average transverse velocity of the solvent
ΔP = pressure difference across the membrane
R_g = gas constant
T = absolute temperature
v_p = center-of-mass pore fluid velocity
ε = fractional pore area
μ = viscosity
π_{high} = osmotic pressure on the high pressure side of the membrane
$\Delta\pi$ = osmotic pressure difference across the membrane
σ = reflection coefficient
τ = tortuosity factor

Kadem-Katchalsky and Pusch models are based on the principles of irreversible thermodynamics (Kadem and Katchalsky, 1958; Pusch, 1977). The solution-diffusion model assumes that the membrane is a non-porous diffusive barrier in which all components dissolve in accordance with phase equilibrium considerations and diffuse through by the same mechanism that governs diffusion through liquids and solids (Longsdale, 1972). Pore flow model considers the flux through the membrane as viscous flow through a highly porous membrane (Merten, 1966). Separation of the solute and solvent are assumed to occur because the solid concentration in the pore fluid is not the same as that in the feed solution. After assuming chemical equilibria at the solution-membrane interfaces, solute concentration in the pore fluid at each interface is related to the solute concentration in the permeate and feed solutions through equilibrium or partition coefficients (Cheryan, 1992). Pore flow model includes a tortuosity factor (τ), which accounts for twisting of the pores and an increase in the effective pore length. Solute flux for this model is the sum of the solute flux due to convective or bulk movement within the pores, and diffusion of solute through the membrane pores (Cheryan, 1992). Finely porous model is developed for reverse osmosis membranes whose transport properties are intermediate between those of the solution-diffusion and the pore flow models (Menten, 1966). Finely porous model is a frictional transport model. The frictional transport model assumes that the thermodynamic driving forces for transport of the particles across the membrane are balanced by the frictional forces between the particle and others in the surrounding medium. There are also other models published in the literature which use a combination of the models listed in Table 4.6. A detailed discussion of these models is beyond the scope of the text and an interested reader is recommended to refer to Cheryan (1992).

Constants of the models given in Table 4.6 varies with the kind of membrane, composition and pH of the solution and the operating conditions of the membranes. Fouling of the membranes is the major problem in membrane separation processes, which may also change the efficiency of the separation processes and the value of the model constants. In some processes solute molecules tend to accumulate on the surface of the membrane after preferential transport of the solvent molecules, then a solute concentration gradient is established between the bulk solution and the solution/membrane interface. This is called *concentration polarization*, and increases the effective solute concentration exposed to the membrane. Higher effective solute concentration and subsequently higher osmotic pressure difference is experienced between the sides of the membrane as a result of concentration polarization, and the efficiency of the separation process may decrease. Concentration polarization may be minimized by supplying liquid flow parallel to the membranes, which removes the accumulating solute molecules.

Example 4.41. Effect of the tube diameter on the ultrafiltration rates of apple juice in a process utilizing single pass metallic ultrafiltration tubes (Thomas et al., 1986 *and* 1987) A process utilizing a single pass metallic ultrafiltration is described in Figure E.4.41 to produce apple juice. Pureed apples were treated with cellulase and pectinase, then pumped through a single-pass tubular membrane system consisting of a metallic oxide membrane formed-in-place on the porous structure of sintered stainless steel tubes, juice yields of 75–86% were obtained with a single pass.

In this system the volumetric flow rate of the puree decreases in the tubes due to permeation through the membrane as:

$$dF = -J\pi D\,dx \qquad \text{(E.4.41.1)}$$

where F is volumetric flow rate of the puree in the tube, D is the diameter of the tube, dx is the incremental length of the tube through with permeation $-dF$ occurs through the membrane surface area $\pi D\,dx$; J is the permeation

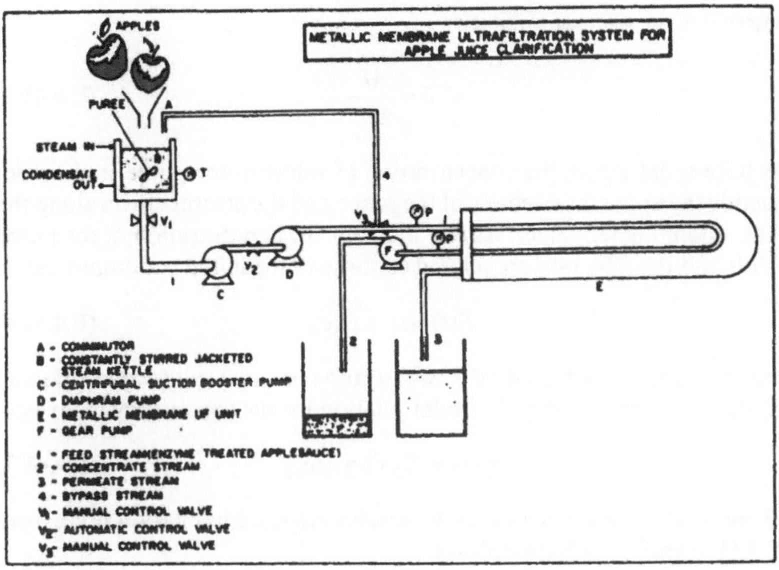

FIGURE E.4.41 Flow diagram for single pass, single stage, metallic membrane ultrafiltration process of apple puree (Thomas *et al.*, 1986 ©; IFT, reproduced by permission).

rate of the juice:

$$J = AP \tag{E.4.41.2}$$

where A is a constant. It was assumed in (E.4.41.2) that the osmotic pressure of the juice and the pressure of the low pressure side of the membrane are negligible in comparison with the pressure P of the high pressure side of the membrane

The viscous properties of the apple puree were described with the power law model:

$$\tau = K\dot{\gamma}^n \tag{E.4.41.3}$$

where τ is shear stress, K is consistency, $\dot{\gamma}$ is shear stress and n is the flow behavior index. In most situations laminar flow of the apple puree is expected in the tube, associated with the following pressure drop rate in flow direction:

$$\frac{dP}{dx} = -\frac{4K}{D^{1+n}}\left(\frac{6n+2}{n}\right)^n \bar{v} \tag{E.4.41.4}$$

where $\bar{v}$ is the average velocity:

$$\bar{v} = \frac{4F}{\pi D^2} \quad \text{(E.4.41.5)}$$

As juice is extracted, the concentration of solute in the puree, and subsequently the apparent viscosity of the puree and the pressure drop along the tube, attain higher values. At any moment the concentration of the puree solids and the flow rate are related to those of the initial conditions as:

$$F(t)c(t) = F_0 c_0 \quad \text{(E.4.41.6)}$$

where F_0 and c_0 are the initial flow rate and the initial solids concentration of the puree, respectively. A similar relation for the consistency index is:

$$K(t) = K_0 f(c(t)/c_0) \quad \text{(E.4.41.7)}$$

where $f(c(t)/c_0)$ is a function with variable $c(t)/c_0$. Elimination of dx from (E.4.41.1) and (E.4.41.4) results in:

$$(K_0 f(c(t)/c_0)\bar{v}^n)dF = \left(\frac{A\pi D^{2+n}}{4}\right)\left(\frac{n}{6n+2}\right)^n PdP \quad \text{(E.4.41.8)}$$

$$(K_0 f(c(t)/c_0)\bar{v}^n)dF = \left(\frac{A\pi D^{2+n}}{4}\right)\left(\frac{D^{2+3n}A}{F_0^{n+1}}\right)\left(\frac{n}{6n+2}\right)^n PdP \quad \text{(E.4.41.9)}$$

After combining (E.4.41.5), (E.4.41.6) and (E.4.41.9) to eliminate F in favor of $c(t)/c_0$ we will obtain:

$$\left(\frac{K_0 f(c(t)/c_0)}{(c(t)/c_0)^{n+2}}\right)d(c(t)/c_0) = -\frac{\pi^{1+n}}{4^{1+n}}\left(\frac{D^{2+3n}A}{F_0^{n+1}}\right)\left(\frac{n}{6n+2}\right)^n PdP \quad \text{(E.4.41.10)}$$

The left hand side of (E.4.41.10) depends on the way the concentration affects the rheological behavior of the retentate. The designer may choose the tube diameter D and the initial flow rate F_0 and, within limits, the operating pressure. The left hand side of (E.4.41.10) may be evaluated at limits for c_{exit}/c_0, e.g., 5 or 10, c_{exit} = suspended solids concentration in the puree at the exit of the tube), then (E.4.41.10) integrates to:

$$-P_e^2 + P_0^2 = \kappa_1 \frac{F_0^{1+n}}{A} D^{-(2+3n)} \quad \text{(E.4.41.11)}$$

where P_0 and P_e are the inlet and the exit pressures to the tube and κ_1 is a constant. Equation (E.4.41.11) may be rearranged as:

$$P_0 - P_e = 2\kappa_1 \frac{F_0^{1+n}}{J_{average}} D^{-(2+3n)} \quad \text{(E.4.41.12)}$$

where $J_{average} = A\ (P_0 + P_e)/2$ is the average apple juice flux through the membrane. Equation (E.4.41.12) implies that the pressure drop along the tube is a function of the tube diameter when all the other experimental parameters remains constant. Thomas *et al.* (1986 and 1987) performed their experiments by using two different tube arrangements. In the first design the tube diameter was constant and 1.56 cm; in the second design the equipment shown in Figure E.4.41 was modified and converted into a two-stage single pass system with 3.12 cm diameter tubes. With 1.56 cm diameter tubes and 86% juice yield the pressure drop $P_0 - P_e$ was calculated as 4830 kPa (experimentally determine pressure drop was also 4830 kPa). With 3.12 cm diameter tubes and 86% juice yield the total pressure drop $P_0 - P_e$ along the two stages was calculated in 1987 to 828 kPa (experimentally determined pressure drop was 1725–2415 kPa). Due to the high solute concentration there was up to 690 kPa pressure fluctuations in the second stage of the equipment with 3.12 cm diameter tube. The large range reported for this tube is actually the limits of the pressure drop caused by the fluctuations.

Example 4.42. Unsteady-state permeate flux of crossflow microfiltration (Chang and Hwang, 1994) In a cross flow microfiltration process a cake layer forms on the membrane and decreases the permeate flux. A schematic description of a cross flow microfiltration process is shown in Figure E.4.42.1. The permeation velocity in a constant pressure microfiltration process is expressed according to Darcy's equation as:

$$v_p = \frac{\Delta P}{\mu(R_m + R_c)} \qquad \text{(E.4.42.1)}$$

where ΔP is the pressure difference across the membrane, R_m and R_c are the membrane and the cake resistances, respectively. There is a back transport of the particles away from the vicinity of the cake with back transport velocity (v_t) which is the sum of the velocities due to lateral migration (v_l), shear-induced diffusion (v_s), Brownian diffusion (v_b), gravity and electric double layer repulsion (v_e):

$$v_t = v_l + v_s + v_b + v_e \qquad \text{(E.4.42.2)}$$

where

$$v_1 = \frac{0.43\,\rho\, v_0^2 d_p^3}{\mu h^2} \qquad \text{(E.4.42.2.a)}$$

$$v_s = \frac{v_0 d_p^2}{20\, h^2} \qquad \text{(E.4.42.2.b)}$$

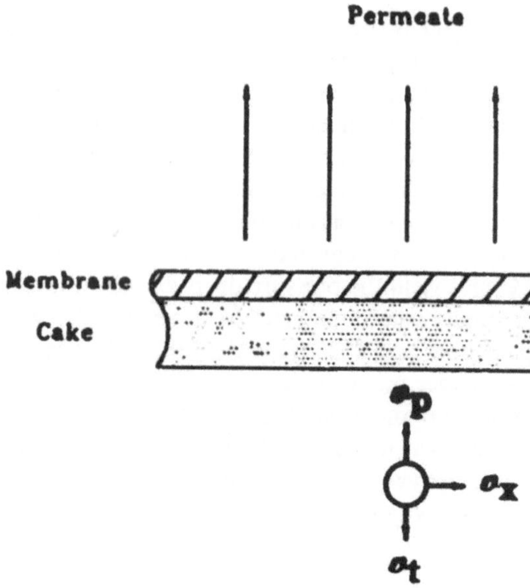

FIGURE E.4.42.1 Schematic representation of the cross flow microfiltration process. (v_x = axial velocity of particle) (Chang and Hwang, 1994; © Marcel Dekker, Inc. Reprinted from *Separation Science and Technology*, Volume 29, pages 1539—1608 by courtesy of Marcel Dekker Inc.).

$$v_b = \frac{2kT}{3\pi\mu d_p h} \tag{E.4.42.2.c}$$

$$v_e = \frac{1}{\mu}\left(\frac{2\varepsilon_0\zeta^2}{45\pi\ell_D} - \frac{A}{36\pi\ell_D^2}\right) \tag{E.4.42.2.d}$$

where A = Hamaker constant, d_p = particle diameter, h = clearance of the cross flow channel, k = Boltzmann constant, ℓ_D = Debye length, T = absolute temperature, v_0 = cross flow velocity, ε_0 = dielectric constant of the suspension, μ = viscosity of the filtrate and ζ = zeta potential. The contribution of gravitational settling to (E.4.42.2) was neglected since the velocity caused by this effect was much smaller than the velocities caused by other effects. The cake resistance was defined as:

$$R_c = \rho_p \alpha c \int \{v_p - (v_l + v_s + v_b + v_e)\}\, dt \tag{E.4.42.3}$$

where c is the particle concentration in the suspension, α is the specific cake resistance and ρ_p is the particle density. Equation (E.4.42.3) may be substituted into (E.4.42.1) to obtain

$$v_p = \frac{\Delta P}{\mu(R_m + \rho_p \alpha c \int \{v_p - (v_1 + v_s + v_b + v_e)\}\, dt)} \tag{E.4.42.4}$$

Integration of (E.4.42.4) gives

$$\int v_p\, dt = v_t t + \frac{\Delta P}{\rho_p \mu \alpha c v_p} - \frac{R_m}{\rho_p \alpha c} \tag{E.4.42.5}$$

During each experiment the volumetric flow rate of suspension was kept constant, but the effective channel height of the filtration module decreased due to the growth of the cake under the surface of the membrane; therefore Chang and Hwang (1994) corrected (E.4.42.2.a)–(E.4.42.2.c) by replacing h with $h - h_{ct}$, where h_{ct} is defined as:

$$h_{ct} = \left(\frac{\Delta P - \mu v_{pi} R_m}{\mu v_{pi}}\right)\left(\frac{d_p^2 \varepsilon_2^3}{180(1-\varepsilon_c)^2}\right) \tag{E.4.42.6}$$

where ε_c is the void fraction of the cake. The crossflow velocity v_0 and the concentration of the suspended particles were also corrected for the changing effective channel height:

$$v_{0(\text{corrected})} = \left[v_0 + \frac{A_m(v_{p0} - v_{pl})}{A_b}\right]\left[\frac{h}{h - h_{ct}}\right] \tag{E.4.42.7}$$

$$c_{\text{corrected}} = c - \frac{(1-\varepsilon_c) h_{ct} A_m}{V_{\text{suspension}}} \tag{E.4.42.8}$$

where A_b and A_m are the cross section area of the channel and the filtration area of the membrane, respectively; $V_{\text{suspension}}$ is the volume of the suspension and $v_{0(\text{corrected})}$ is the corrected crossflow velocity. A stepwise iterative procedure was employed by using the corrected equations and (E.4.42.5) to calculate the permeation velocity v_p. Agreement of the model with the experimental data was confirmed with varying particle size, membranes pore size, cross flow velocity (v_0), pressure difference across the membrane, particle concentration, electrolyte concentration and pH values. A typical plot for the comparison of the model with the experimental data is depicted in Figure E.4.42.2.

Example 4.43. A model for decreasing flux via gel and cake formation and membrane fouling in cheese whey ultrafiltration (Kaiser and Glatz, 1994) During ultrafiltration of cheese whey, the permeation flux undergoes an immediate initial decline, followed by a continuous moderate decline. The flux decline has been attributed to added resistances from fouling (as a result of protein and salt deposition) of the membrane surface and its pores; concentration polarization and cake buildup. The flux $J(t)$ at

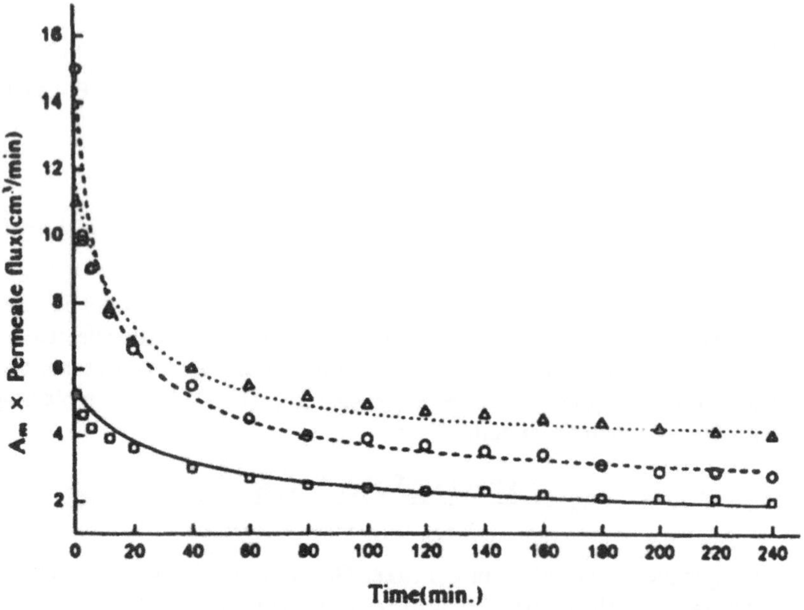

FIGURE E.4.42.2 Comparison of the model with the experimental data when ΔP is (–□–) 1.7×10^4 N/m^2, (...Δ...) 3.8×10^4 N/m^2, (--o--) 5.2×10^4 N/m^2. Symbols shows the experimental data the lines describe the model. Other experimental conditions were: membrane pore size = 0.1 μm, $d_p = 0.6$ μm, $T = 30$ °C, pH = 6.6, $v_0 = 1.8$ m/s and $c = 50$ ppmv (Chang and Hwang, 1994; © Marcel Dekker, Inc. Reprinted from *Separation Science and Technology*, volume 29, pages 1593—1608 by courtesy of Marcel Dekker Inc.).

any instant t may be related to the volume of the permeate $V(t)$ as:

$$J(t) = J_0 V(t) \tag{E.4.43.1}$$

where J_0 is the initial flux. Equation (E.4.43.1) leads to an integrated expression for $V(t)$:

$$V(t) = M\, t^{1/(1+b)} \tag{E.4.43.2}$$

where M and b are the short term and long term parameters, respectively. Gel polarization is attributed to the soluble macromolecules, cake formation is caused by accumulation of the precipitates. Fouling is caused by the molecules adsorbed on the membrane surface. When gel and cake formation and accumulation occurs together the permeate flux may be expressed as:

$$J(t) = \frac{\Delta P}{\mu(R_m + R_g + R_c + R_a(t))} \quad \text{(E.4.43.3)}$$

where subscripts m, g, c and a refer to the resistances attributable to the membrane, gel, cake and the adsorbed foulants, respectively. When the gel layer is limiting (E.4.43.2) is stated as:

$$V(t) = M_g^{0.5} t^{0.5} \quad \text{(E.4.43.4)}$$

where

$$M_g = A_m^2 \left(\frac{c_g - c_0}{c_g - c_p}\right)^2 \left(\frac{c_g - c_p}{c_0 - c_p}\right)\left(\frac{2nD}{n+1}\right) \quad \text{(E.4.43.5)}$$

where A_m is the membrane area, c_g is the concentration where gel formation begins, c_p is the solute concentration in the permeate, c_0 is the bulk solute concentration, D is the solute diffusion coefficient and n is a parameter for the assumed boundary layer profile. When the cake resistance is limiting, (E.4.43.2) is stated as:

$$V(t) = M_c^{0.5} t^{0.5} \quad \text{(E.4.43.6)}$$

where

$$M_c = D_p^2 A_m^2 \left(\frac{2\Delta P}{\mu}\right) \frac{\varepsilon^3\left[\rho_s - \left(\frac{1}{1-\varepsilon}\right)c_s\right]}{150(1-\varepsilon)c_s} \quad \text{(E.4.43.6)}$$

where c_s is the concentration of the solid particles, D_p is particle diameter, ε is the void fraction of the cake and ρ_s is solids density.

Equation (E.4.43.4) is compared with the experimental data after taking the square of the both sides of equation in Figure E.4.43. The slope M has changed every time when ΔP is increased.

4.7. MATHEMATICAL MODELING OF EXTRACTION PROCESSES

In a liquid/solid extraction process the food solids are brought in to contact with a liquid solvent. The solvent dissolves the solute in the solid phase, then the two phases are separated. The liquid/solid extraction process is also referred to as leaching. Solid/liquid extraction may be employed to remove or separate some components from food materials, i.e., crystal sugar production from sugar cane, or it may occur unintentionally in other processes, i.e., blanching. The biological materials have cellular structure and their soluble constituents are usually inside the cell walls, which may make the extraction process difficult. Grinding or drying the food raw materials may help to improve the extraction process by breaking the cell walls.

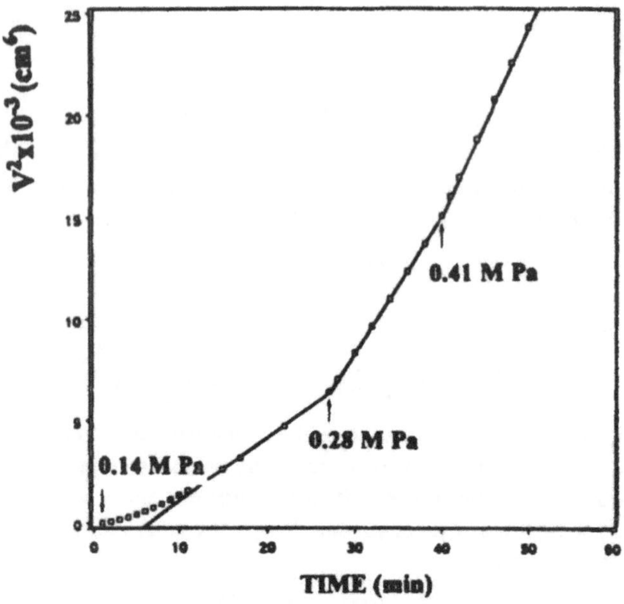

FIGURE E.4.43 Cumulative volume versus time for ultrafiltration of washed acid whey precipitates in an unstirred batch cell. Arrows indicate the points at which ΔP is increased to the levels indicated. Experimental data are shown in symbols, solid lines describes (E.4.43.4) (Kaiser and Glatz, 1988). Reproduced with permission of the American Institute of Chemical Engineers. ©1988 AIChE. All rights reserved.

Extraction is a stage contacting process. *Stage* is a unit of equipment, in which two or more phases are brought into contact for a predetermined period, then separated from each other. When these phases are contacted for a long period to attain thermodynamic equilibrium the stage is referred to as an *equilibrium stage*. Theoretical stage or ideal stage are other names for an equilibrium stage.

Example 4.44. Extraction of sugar from beets (*Yang and Brier*, 1958) The extraction of beet sugar may be resolved in two steps: (1) Diffusion of sugar beet solution through the permeable membrane of beet cells toward the interface between the beets and the extracting solution, (2) Mass transfer of sugar through a liquid film at the interface in to the extracting solution. Diffusion of sugar inside the beet may be modeled with the equation of continuity after simplifying (2.12) as:

$$\frac{\partial c}{\partial t} = D\frac{\partial^2 c}{\partial z^2} \qquad \text{(E.4.44.1)}$$

where c is the sugar concentration in the beet, D is the diffusion coefficient of sugar, t is time and z distance of diffusion within the solid slab extracted.

When we consider a slab of finite thickness $2L$ the following initial and boundary values exist:

$$\text{IC. } c = c_0 \text{ for } -L \leqslant z \leqslant L \text{ at } t = 0 \qquad \text{(E.4.44.1.a)}$$

$$\text{BC. } c = c^* \text{ at } z = -L \text{ and } z = L \qquad \text{(E.4.44.1.b)}$$

where c^* is the sugar concentration of the beets in equilibrium with sugar concentration in bulk liquid. The solution of (E.4.44.1) with the given IC and BC is:

$$\frac{\bar{c} - c^*}{c_0 - c^*} = \frac{8}{\pi^2} \sum_{m=0}^{\infty} \frac{1}{(2m+1)^2} \exp\left[\frac{-D(2m+1)^2 \pi^2 t}{L^2}\right] \qquad \text{(E.4.44.2)}$$

where $\bar{c}$ is the average sugar concentration in the beet slice. All the concentrations were defined as weight of sugars/weight of wet beets. Mass transfer through the liquid film at the solid/liquid interface may be expressed as:

$$-\frac{d\bar{c}}{dt} = \ell A(c_e - c_b) \qquad \text{(E.4.44.3)}$$

where c_e is defined as the sugar concentration in the beet juice in equilibrium with the bulk concentration c_b (units = weight of sugars/weight of extraction liquid). Parameter ℓ is the mass transfer coefficient per unit mass transfer area, corrected for the difference of the units of $\bar{c}$, c^*, c_e and c_b. And A is the total mass transfer area. Yang and Brier (1958) calculated ℓA in the experiments performed in the continuous diffuser (Fig. E.4.44.1) with different liquid phase flow rates and temperatures. Variation of parameter ℓA with time was found to be unaffected by the experimental conditions (Fig. E.4.44.2), implying that the resistance of the liquid film around the beet slices were negligible in comparison with that of the internal mass transfer. Variation of the dimensionless average beet sugar contents with time is depicted in Figure E.4.44.3.

Example 4.45. Countercurrent desalting of pickles (Bomben *et al.*, 1974) Salt stock pickles have about 15–18% salt. They need to be extracted with water to reduce the salt content to about 4% before sent to market. When these pickles are flushed with successive washing with fresh water "weak brine" with 2–3% salt is produced. The recovery of salt from such a low concentration solution is prohibitively expensive, on the other hand the weak brine is difficult to dispose because of the environmental considerations. With countercurrent extraction, however, the salt stock and water are countercurrently contacted in successive tanks. This results in a proportionally lower volume of high concentration spent brine with 13–14% salt, from which it may be economically feasible to recover salt, or it may be easier to dispose the smaller volumes of the spent brine. A schematic drawing of a countercurrent desalting process is shown in Figure E.4.45.1.

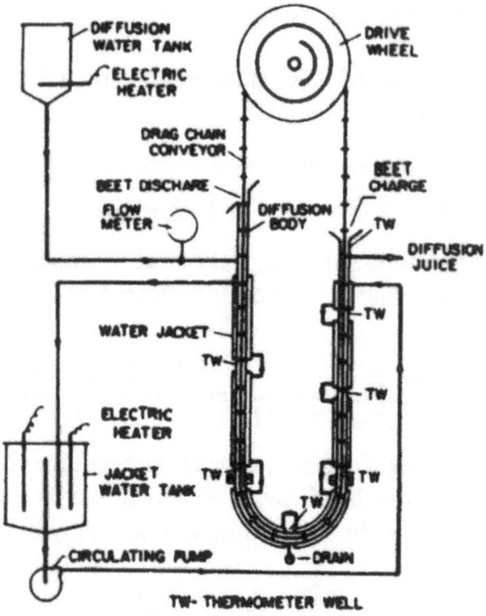

FIGURE E.4.44.1 Flow diagram of the continuous diffusion equipment (Yang and Brier, 1958). Reproduced with permission of the American Institute of Chemical Engineers. © 1958 AIChE. All rights reserved.

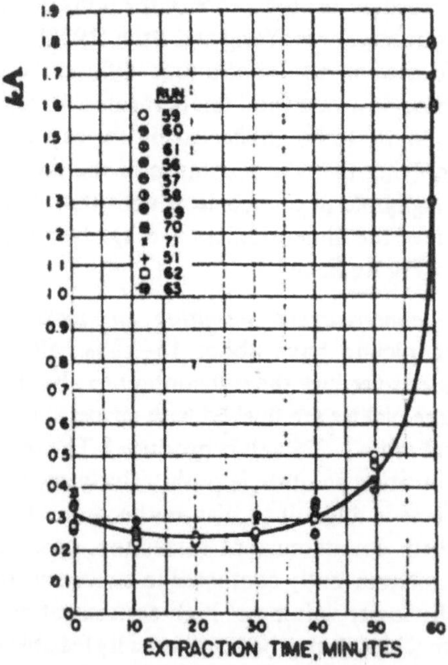

FIGURE E.4.44.2 Variation of kA with extraction time (Yang and Brier, 1958). Reproduced with permission of the American Institute of Chemical Engineers. © 1958 AIChE. All rights reserved.

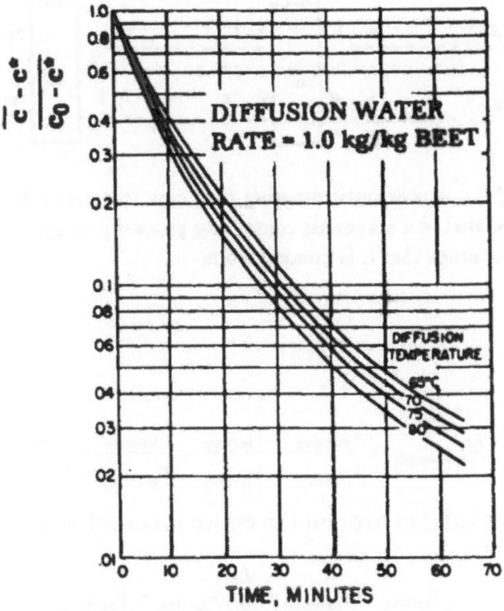

FIGURE E.4.44.3 Variation of the dimensionless average beet sugar contents with extraction time. Solid lines describe (E.4.44.2). Experimental data were reported to be within ±5% deviation from the model. Flow rate of the liquid phase was 1 kg water/1 kg beet (Yang and Brier, 1958). Reproduced with permission of the American Institute of Chemical Engineers. © 1958 AIChE. All rights reserved.

When the viscosity and density changes of salt solutions are not large enough to affect the calculations, and flow rates of the overflow and underflow streams are different, the number of theoretical stages required to achieve a required separation may be calculated as:

$$N_{\text{theoretical}} = \frac{\log\left(\dfrac{y_{\text{entering}} - x_{\text{leaving}}}{y_{\text{leaving}} - x_{\text{entering}}}\right)}{\log\left(\dfrac{y_{\text{entering}} - y_{\text{leaving}}}{y_{\text{entering}} - x_{\text{entering}}}\right)} \tag{E.4.45.1}$$

where x and y indicate the salt concentration in the underflow and overflow, respectively; the subscripts indicate whether the concentrations are pertinent to leaving or entering streams. In a solid/liquid extraction process the solid phase is referred to as underflow and the liquid phase is referred to as overflow by convention. In a special case when the flow rates of overflow and underflow are the same, a simpler equation is substituted

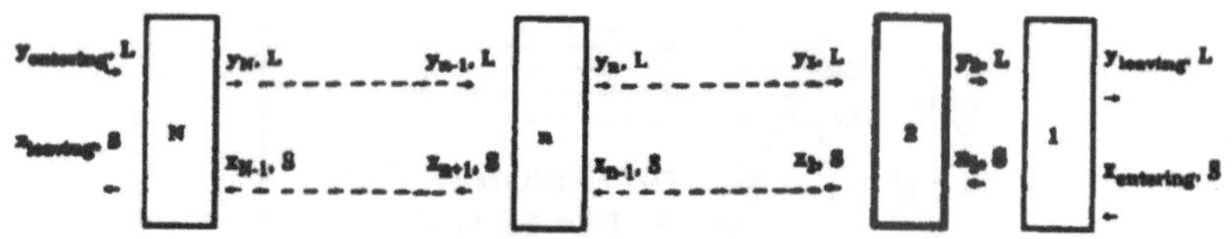

FIGURE E.4.45.1 A schematic drawing of a countercurrent desalting process. It should be noticed that in a stagewise contacting process a stream is subcripted with the number of the stage that it is coming from.

for (E.4.45.1):

$$N_{\text{theoretical}} = \frac{y_{\text{entering}} - y_{\text{leaving}}}{y_{\text{entering}} - x_{\text{leaving}}} = \frac{x_{\text{leaving}} - x_{\text{entering}}}{y_{\text{entering}} - x_{\text{leaving}}} \quad \text{(E.4.45.2)}$$

The overall salt balance around the entire process leads to:

$$x_{\text{leaving}} = x_{\text{entering}} - \frac{L}{S}(y_{\text{leaving}} - y_{\text{entering}}) \quad \text{(E.4.45.3)}$$

where L and S are the flow rates of the overflow and underflow streams, respectively.

The salt stock pickles are long circular cylinders with equivalent radius:

$$R = \sqrt{\frac{m}{\rho \pi \ell}} \quad \text{(E.4.45.4)}$$

where ℓ, m and ρ are the length, mass and density of the salt stock pickle, respectively. When the contacting vessel is well mixed, the rate limiting step is diffusion of salt in the pickles and diffusivity is a constant, the extraction process may be simulated with the equation of continuity in cylindrical coordinate system as:

$$\frac{\partial c}{\partial t} = D\left[\frac{1}{r}\frac{\partial}{\partial r}\left(r\frac{\partial c}{\partial r}\right)\right] \quad \text{(E.4.45.5)}$$

where c is salt concentration in the pickles. Solution to (E.4.45.5) may be expressed as:

$$\frac{M(t)}{M_\infty} = 1 - \sum_{n=1}^{\infty} \frac{4\alpha(1+\alpha)}{4 + 4\alpha + \alpha^2 \lambda_n^2} \exp\left(-\frac{D\lambda_n^2 t}{R}\right) \quad \text{(E.4.45.6)}$$

where $M(t)$ and M_∞ are the amounts of salt having left the salt stock after time t and at equilibrium, respectively; α is the ratio of the rate of overflow to underflow (L/S) and λ_n are the non-zero roots of

$$\alpha \lambda_n J_0(\lambda_n) + 2J_1(\lambda_n) = 0 \quad \text{(E.4.45.7)}$$

The right hand side of (E.4.45.6) is almost equal to the Murphree stage efficiency η for a solid/liquid extraction process:

$$\eta = \frac{M(t)}{M_{\infty}} \tag{E.4.45.8}$$

where Murphree stage efficiency η is defined as the ratio of the number of the theoretical stages ($N_{\text{theoretical}}$) to those of the actual stages (N_{actual}):

$$\eta = \frac{N_{\text{theoretical}}}{N_{\text{actual}}} \tag{E.4.45.9}$$

Variation of $M(t)/M_{\infty}$ with $\sqrt{Dt/R^2}$ is depicted in Figure E.4.45.2. Bomben *et al.* (1974) determined the diffusion coefficient D after comparing the experimentally determined values of $M(t)/M_{\infty}$ with (E.4.45.6). They chose the experimental conditions such that $S = L$, then after combining

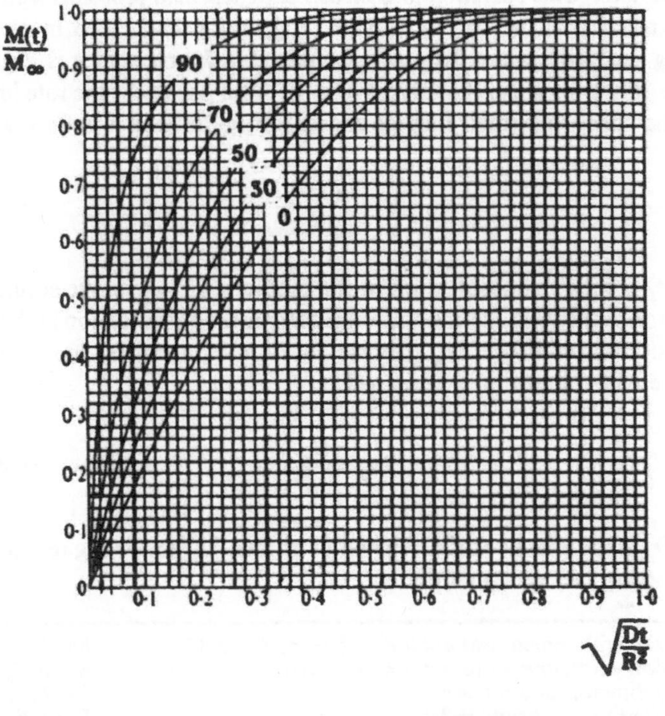

FIGURE E.4.45.2 Variation of $M(t)/M_{\infty}$ with $\sqrt{\frac{Dt}{R^2}}$. Numbers on curves show percentages of total solute finally taken up by the cylinder. Reprinted by permission of the Oxford University Press from Crank, *Mathematics of Diffusion*, ©1956 Oxford University Press.

(E.4.45.2), (E.4.45.3) and (E.4.45.9) they obtained

$$y_{\text{leaving}} = \frac{y_{\text{entering}} + \eta N_{\text{actual}} x_{\text{entering}}}{1 + \eta N_{\text{actual}}} \tag{E.4.45.10}$$

In a 6 stage experimental procedure ($N_{\text{actual}} = 6$), when contacting time of each stage was 8 hours at 49°C, with $L/S = 1$, $x_{\text{entering}} = 16.7\%$ NaCl and $y_{\text{entering}} = 0\%$ NaCl they calculated $y_{\text{leaving}} = 14.3\%$ NaCl, corresponding to $x_{\text{leaving}} = 2.5\%$ NaCl. The experimental results under these conditions were $y_{\text{leaving}} = 13.2–13.6\%$ NaCl and $x_{\text{leaving}} = 2.0–2.8\%$ NaCl, implying good agreement between the model and the data.

There are also well established graphical solution procedures for the stagewise contacting problems, an interested reader is recommended to refer to Treybal (1980), Geankoplis (1983) and McCabe *et al.*, 1985) for the details.

Example 4.46. Chart solution to unsteady state leaching problems with negligible external mass transfer resistance Figure E.4.46. is used in unsteady leaching problems when the external mass transfer resistance is negligible and the diffusion rate of the solvent into the solid particle is the rate limiting step. The unaccomplished average concentration fraction is defined as:

$$E = \frac{c_1 - c_{av}}{c_1 - c_0} \tag{E.4.46.1}$$

where c_1 = solute concentration in the medium, c_{av} = average solute concentration in the solid at time t, c_0 = initial solute concentration in the solid. Dimensionless number Fo'_d (similar to Fourier number) based on the characteristic dimension d is defined as:

$$Fo'_d = \frac{D_{ef} t}{d^2} \tag{E.4.46.2}$$

where D_{ef} is the effective diffusivity and t is time. The following relations for E are used in different geometries:

infinite slab with dimensions $a \times b \times c$ ($b \to \infty$, $c \to \infty$)	$E = E_a$
infinite slab with dimensions $a \times b \times c$ ($c \to \infty$)	$E = E_a E_b$
slab with dimensions $a \times b \times c$	$E = E_a E_b E_c$
cylinder (radius = a, height = $2c$)	$E = E_r E_c$
sphere (radius = a)	$E = E_s$

Olives (which may be assumed cylinders with radius $r = 0.7$ cm and height $2c = 2.5$ cm) lose 80% of their initial salt content when leached in water for 24 h. Calculate the time needed to remove 65% of the initial salt.

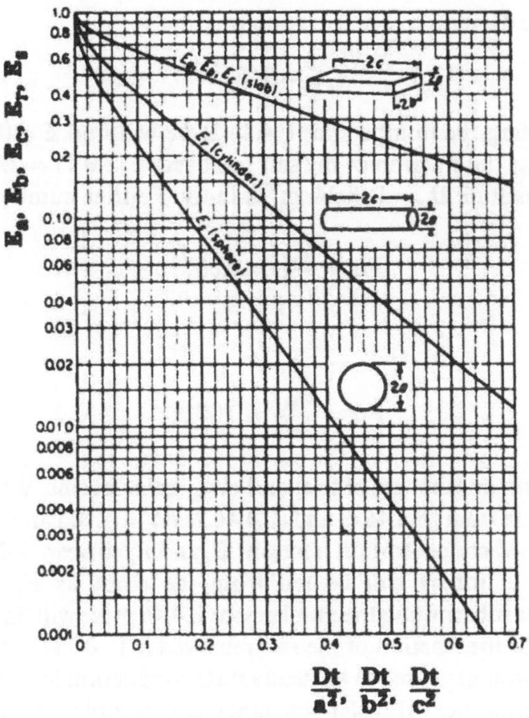

FIGURE E.4.46 Variation of the unaccomplished average temperature fractions with dimensionless numbers during unsteady-state diffusion with negligible surface resistance (Reproduced from Treybal, R. E., Mass Transfer Operations, 3rd ed. ©1980, by permission of McGraw-Hill, companies).

Solution The dimensionless numbers based on the characteristic dimensions are:

$$Fo_r^{\cdot} = \frac{D_{eff}t}{r^2} = 49\,D_{eff} \tag{E.4.46.3}$$

and

$$Fo_c^{\cdot} = \frac{D_{eff}t}{c^2} = 15.4\,D_{eff} \tag{E.4.46.4}$$

We will make trial and error solution to find out D_{eff} required to remove 80% of the salt in 24 hours. When $D_{eff} = 0.0034\,\text{cm}^2/\text{h}$, we will calculate from (E.4.46.3) and (E.4.46.4) $Fo_r^{\cdot} = 0.166$ and $Fo_c^{\cdot} = 0.052$. Unaccomplished concentration profiles E_r and E_c corresponding to these Fourier numbers will be read from the chart as: $E_r = 0.28$ and $E_c = 0.75$. The overall unaccomplished average concentration fraction is:

$$E = E_r E_c \tag{E.4.46.5}$$

After substituting the numbers in (E.4.46.5) we will find $E = 0.21 \cong 0.220$, confirming that the fraction of the salt removed is $1-0.21 = 79\% \cong 80\%$.

We will substitute $D_{eff} = 0.0034\,\mathrm{cm^2/h}$ in the Fourier numbers:

$$Fo_r^* = \frac{D_{eff}t}{r^2} = 0.007\,t \tag{E.4.46.6}$$

and

$$Fo_c^* = \frac{D_{eff}t}{c^2} = 0.0022\,t \tag{E.4.46.7}$$

We will perform another set of trial and error calculations. When $t = 15\,\mathrm{h}$, we will calculate from (E.4.46.6) and (E.4.46.7) $Fo_r^* = 0.105$ and $Fo_c^* = 0.033$. Unaccomplished concentration fractions E_r and E_c corresponding to these dimensionless numbers will be red from the chart as $E_r = 0.395$ and $E_c = 0.87$. After substituting the numbers in (E.4.46.5) we will find $E = 0.344$, confirming that the fraction of the salt removed is $1-0.344 = 65.6 \cong 65\%$.

Figure E.4.46. is also used in unsteady state conduction heating problems when the external heat transfer resistance is negligible after substituting α for D_{eff} in (E.4.46.2) (α = thermal diffusivity):

$$Fo_d = \frac{\alpha t}{d^2} \tag{E.4.46.8}$$

where Fo_d is the Fourier number and defining

$$E = \frac{T_1 - T_{av}}{T_1 - T_0} \tag{E.4.46.9}$$

where T_1 = temperature of the medium, T_{av} = average temperature of the solid at time t and T_0 = initial temperature of the solid.

Example 4.47. Mathematical modeling of blanching potatoes (*Kozempel et al.*, 1981) During production French fries the potatoes are cut in the required shape, then blanched in water. Blanching is also a major step in production of other potato products, including instant potato flakes. Kozempel *et al.* (1981) used the experimental set-up shown in Figure E.4.47.1 to study extraction of the potato constituents, i.e., glucose, potassium, magnesium and phosphorus, during blanching. French fry cuts, 0.95 cm, were processed in the precooker in perforated metal baskets. The precooker had five baskets. The input of the potatoes was achieved by sequentially replacing the baskets. At every 4 minutes a basket of blanched

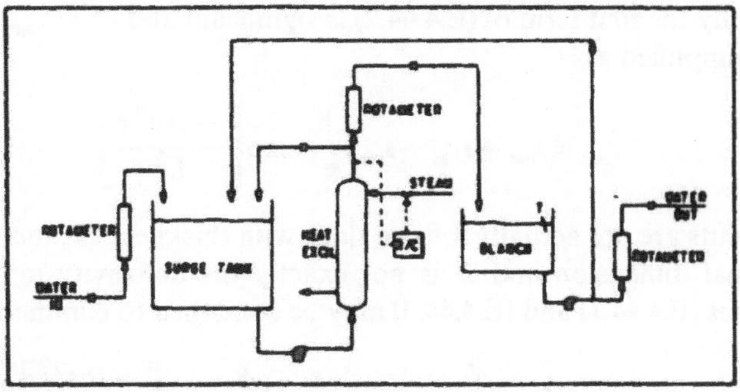

FIGURE E.4.47.1 Equipment for simulating potato precooking process (Kozempel *et al.*, 1981; ©Academic Press, reproduced by permission).

potatoes came out and a new basket came in. Residence time of each basket was 20 minutes in the precooker.

Solute balance around the continuous precooker requires:

$$\begin{bmatrix}\text{input rate}\\ \text{into the}\\ \text{precooker}\end{bmatrix} - \begin{bmatrix}\text{output}\\ \text{rate from the}\\ \text{precooker}\end{bmatrix} = \begin{bmatrix}\text{accumulation}\\ \text{rate in the}\\ \text{precooker}\end{bmatrix} \quad \text{(E.4.47.1)}$$

Equation (E.4.47.1) may be written in mathematical terms as:

$$((W_{ps}x)c_{in} + W_w s_{in}) - ((W_{ps}x)c_{out} + W_w s_{out}) = V\rho\frac{ds}{dt} + \frac{W_{ps}x\tau}{2}\frac{d(c_{in}+c_{out})}{dt} \quad \text{(E.4.47.2)}$$

where W_{ps} is the input rate of the dry potato solids into the precooker, x is the moisture content of the potatoes, expressed in kg water/dry potato solids, and c_{in} is the solute concentration, i.e. potassium, in the water contents of the potato; W_w is flow rate of water in to the precooker and s_{in} is the solute concentration in the input water. The output concentrations of the same streams were denoted as c_{out} and s_{out}. Volume and density of the blanching water in the precooker were V and ρ, respectively. Residence time of the potatoes in the precooker was τ. The term $(c_{in} + c_{out})/2$ was the average solute concentration in water in the potatoes.

When the French cut potatoes were considered as a slab of thickness $2L$, then the solute concentration profile along the slab was simulated similarly as in Example 4.44:

$$\frac{c_{out} - c^*}{c_{in} - c^*} = \frac{8}{\pi^2}\sum_{m=0}^{\infty}\frac{1}{(2m+1)^2}\exp\left[\frac{-D(2m+1)^2\pi^2\tau}{L^2}\right] \quad \text{(E.4.44.2)}$$

When only the first term of (E.4.44.2) is significant and $c^* = s_{out}$, (E.4.44.2) will be simplified as:

$$c_{out} = s_{out} + (c_{in} - s_{out})\frac{8}{\pi^2}\exp\left[\frac{-D\pi^2\tau}{L^2}\right] \quad \text{(E.4.44.3)}$$

French cuts are not actually infinite slabs with thickness $2L$, therefore L is a nominal dimension and D is not exactly the diffusivity in (E.4.44.3). Equations (E.4.44.1) and (E.4.44.3) may be combined to eliminate c_{out}:

$$(W_{ps}x)c_{in} + W_w s_{in} - (W_{ps}x)\left(s_{out} + (c_{in} - s_{out})\frac{8}{\pi^2}\exp\left[\frac{-D\pi^2\tau}{L^2}\right]\right) - W_w s_{out}$$

$$= V\rho\frac{ds_{out}}{dt} + \frac{W_{ps}x\tau}{2}\frac{d}{dt}\left(s_{out} + (c_{in} - s_{out})\frac{8}{\pi^2}\exp\left[\frac{-D\pi^2\tau}{L^2}\right]\right) \quad \text{(E.4.47.4)}$$

Equation (E.4.47.4) may be rearranged as:

$$W_{ps}c_{in}\left\{1 - \frac{8}{\pi^2}\exp\left[\frac{-D\pi^2\tau}{L^2}\right]\right\} - W_{ps}s_{out}$$

$$\left\{1 - \frac{8}{\pi^2}\exp\left[\frac{-D\pi^2\tau}{L^2}\right]\right\} + W_w(s_{in} - s_{out})$$

$$= \left[V\rho + \frac{W_{ps}x\tau}{2}\left\{1 - \frac{8}{\pi^2}\exp\left[\frac{-D\pi^2\tau}{L^2}\right]\right\}\right]\frac{ds_{out}}{dt} \quad \text{(E.4.47.5)}$$

Kozempel *et al.* (1981) solved (E.4.47.5) with Runge-Kutta method. Under steady state conditions, when $ds_{out}/dt = 0$, the solution is:

$$s_{out} = \frac{W_{ps}c_{in}\left\{1 - \frac{8}{\pi^2}\exp\left[\frac{-D\pi^2\tau}{L^2}\right]\right\} + W_w s_{in}}{W_{ps}\left\{1 - \frac{8}{\pi^2}\exp\left[\frac{-D\pi^2\tau}{L^2}\right]\right\} + W_w} \quad \text{(E.4.47.6)}$$

Kozempel *et al.* (1981) compared the model to their experimental data under various experimental conditions and found good agreement. A typical plot is presented in Figure E.4.47.2.

Example 4.48. Fractionation of citrus oils by using a membrane based extraction process (*Brose et al.*, 1995) Citrus oils are obtained from the pressed peels of citrus fruits, such as oranges, lemons and limes. Typically, the pressed oils contain 96–98 wt% terpene hydrocarbons and 1–3 wt% oxygenates (aldehydes, alcohols and esters). The terpenes contribute only slightly to the flavor or fragrance of the oil, and they rapidly oxidize to form undesirable components; the oxygenates are the main flavor components of

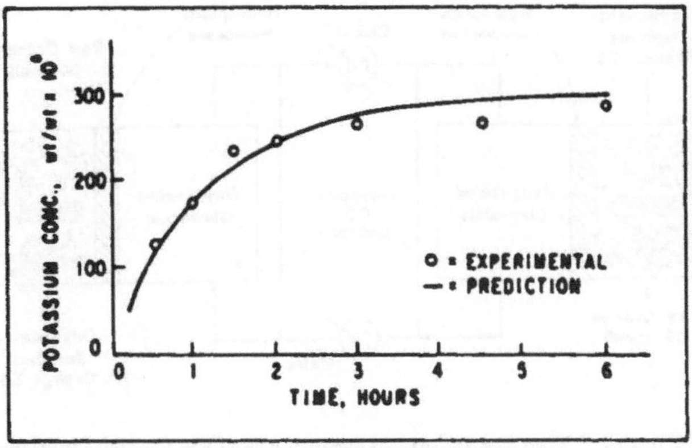

FIGURE E.4.47.2 Comparison of the predicted concentration of potassium in precook water with the experimental data during blanching process for Maine Russet Burbank Potatoes ($D = 0.423$ cm²/h). (Kozempel *et al.*, 1981; ©Academic Press, reproduced by permission).

the oil. It is common industrial practice to remove some or all of the terpenes to concentrate the oxygenates in the oil. A membrane-based extraction process is developed by Brose *et al.* (1995) for the fractionation of the citrus oils. The process relies on highly selective cyclodextrins (CDs) to preferentially bind the desirable oxygenates. In this process, low temperature citrus oil flows on one side of a non-porous membrane and an aqueous CD solution flows on the other side of the membrane (Figure E.4.48.1). Oxygenates diffuse through the water-swollen hydrophobic membrane and preferentially partition into the aqueous CD solution. The aqueous CD solution is then heated and the liberated oxygenates diffuse into a citrus oil solution. Dissociated CD is recycled to the first membrane.

The solute flux J through a membrane is expressed as:

$$J = L\Delta c \tag{E.4.48.1}$$

where L is the permeability and $\Delta c = c^*_{oil} - c_{aq}$; c^*_{oil} is the oxygenate concentration in the orange oil feed c_{aq} and is the oxygenate concentration in the aqueous phase. Concentration c^*_{oil} is defined as an aqueous concentration that would be in equilibrium with the oil phase. The equilibrium relationship used to define c^*_{oil} in terms of the actual solute concentration in the oil phase (c_{oil}) is:

$$c^*_{oil} = \frac{c_{oil}}{K} \tag{E.4.48.2}$$

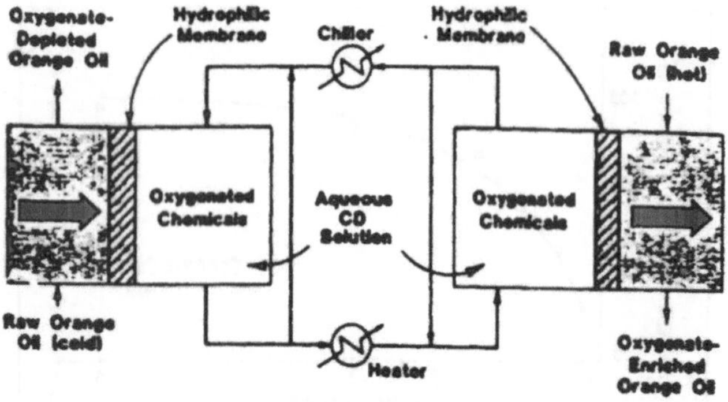

FIGURE E.4.48.1 Membrane-based extraction process for the production of oxygenate-enriched orange oil (Reprinted with permission from Brose et al., © 1995 ACS).

where K is the distribution coefficient. Equations (E.4.48.1) and (E.4.48.2) may be combined to obtain a flux expression as:

$$J = L\left(\frac{c_{oil}}{K} - c_{aq}\right) \tag{E.4.48.3}$$

The overall oxygenate balance equation is:

$$c_{oil,t=0} V_{oil} = c_{oil} V_{oil} + c_{aq} V_{aq} \tag{E.4.48.4}$$

where $c_{oil,t=0}$ is the oxygenate concentration in the oil at time $t = 0$ and V_{oil} is volume of the raw oil reservoir. The oxygenate balance in the oil reservoir is:

$$V_{oil}\frac{dc_{oil}}{dt} = -JA \tag{E.4.48.5}$$

where A is the mass transfer area of the membrane. Equations (E.4.48.3)–(E.4.48.5) were combined to result in

$$c_{oil} = c_{oil,t=0}\left[1 - \frac{1}{\varepsilon} + \frac{\exp(-\varphi\varepsilon t)}{\varepsilon}\right] \tag{E.4.48.6}$$

and

$$c_{aq} = c_{oil,t=0}\left(\frac{V_{oil}}{V_{aq}\varepsilon}\right)(1 - \exp(-\varphi\varepsilon t)) \tag{E.4.48.7}$$

where $\varepsilon = 1 + V_{oil}K/V_{aq}$ and $\varphi = LA/V_{oil}K$.

When the transport of solute is from an aqueous feed solution to an oil product solution, as in the case during regeneration of the aqueous CD solution by stripping with an oil solution, a similar procedure is used to derive the following equations that relate solute concentrations in the oil and aqueous reservoirs as a function of operating time:

$$c_{\text{oil}} = c_{\text{aq},t=0}\frac{V_{\text{aq}}}{V_{\text{oil}}\xi}(1-\exp(-\phi\xi t)) \qquad \text{(E.4.48.8)}$$

and

$$c_{\text{aq}} = c_{\text{aq},t=0}\left(1-\frac{1}{\xi}(1-\exp(-\phi\xi t))\right) \qquad \text{(E.4.48.9)}$$

where $\xi = 1 + V_{\text{aq}}/V_{\text{oil}}K$ and $\phi = LA/V_{\text{aq}}$. The variation of the oxygenate concentrations with time in the aqueous CD feed and product reservoirs are depicted in Figures E.4.48.2.a and b.

Solvents used in the food industry must be non-toxic and not leave any undesirable residues. The solvent is also desired to be non-explosive, inexpensive, readily available, plentiful in nature, easily separable from the substances extracted and easy to regenerate or dispose. Water is a preferred

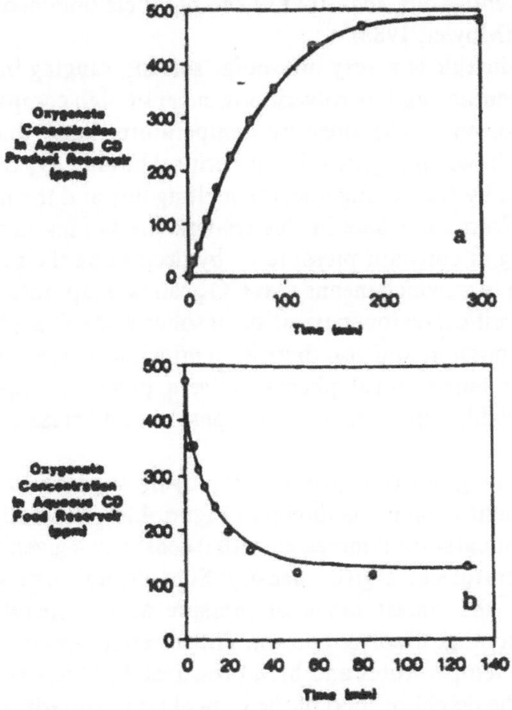

FIGURE E.4.48.2 Time course of the variation of the oxygenate concentrations in the aqueous CD feed (a) and product (b) reservoirs. Experimental data are shown in symbols. Lines describe [E.4.48.7] and [E.4.48.9]. The model and the data are shown with the lines and the symbols, respectively (Reprinted with permission from Brose *et al.*, © 1995 ACS).

solvent for the extraction of polar or ionic substances when its use is feasible in a process. Some organic solvents, i.e., hexane, heptane, thrichloroethylene, isopropyl alcohol, and ethyl alcohol, were also used in the past in the food industry for recovery of mostly non-polar constituents, but extraction with supercritical carbon dioxide is preferred these days. Supercritical carbon dioxide exhibits a polarity classifiable inbetween that of dichloromethane and ethyl ether, consequently non-polar compounds, like fats, oils and aroma compounds, such as terpenes, dissolve quite readily (Hierro and Santa-Maria, 1992). Other compounds displaying a certain degree of solubility in supercritical carbon dioxide are low molecular weight compounds with average polarity, such as caffeine, nicotine, cholesterol and alcohols (Hierro and Santa-Maria, 1992). Supercritical carbon dioxide extraction was reported to be used for extraction of fats, oils, hops, flavor and perfume ingredients (Moyler, 1988). Liquid carbon dioxide rejects protein, waxes, sugars, chlorophyll and pigments to yield an extract which more closely resembles the aroma and taste of the botanical starting material of the essential oils, than that of the products obtained with the steam distillation (Moyler, 1988).

Liquid carbon dioxide is a very non-polar solvent, ranging in polarity near hexane and pentane and its solvent power is not high compared with ordinary liquid solvents. The pressure–temperature phase diagram of carbon dioxide is shown in Figure 4.22. Subcritical liquid CO_2 is found in the triangle formed by the boiling line, the melting line and the line of the critical pressure. From anywhere in this triangle the boiling line may be reached by heating at constant pressure or by decreasing the pressure at a constant temperature, which means that CO_2 can be evaporated (Brogle, 1982). In the supercritical region, separation of solute and solvent by simple evaporation is not possible and also there is no principle difference between the liquid and the supercritical phases. Solvent power of supercritical carbon dioxide is highly dependent on its temperature and pressure (Brogle, 1982).

Schematic flow diagram of a process employing supercritical carbon dioxide as an extracting solvent is shown in Figure 4.23. Generally solvent power of a supercritical solvent increases with i) density at a given temperature and ii) temperature at a given density. Supercritical area shown in Figure 4.22 spans the widest range of pressure and temperature, and therefore the widest range of solvent power. The solvent power reaches very high levels at high temperatures and high pressures, but there is very low solvent power in the neighborhood of the critical point (Brogle, 1982).

Example 4.49. Supercritical extraction of fish oils (*Krukonis*, 1989) Objectionable components, such as odor compounds, free fatty acids and polychlorinated biphenyls (PCB's) are present in fish oils. The PCB's may be removed from the fish oils by supercritical carbon dioxide extraction.

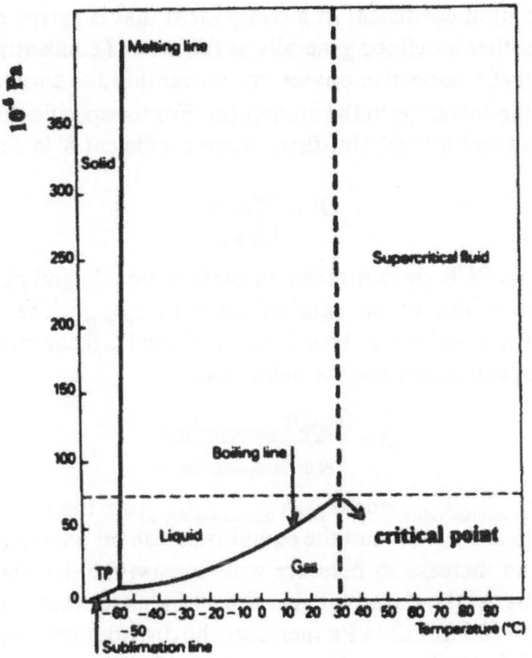

FIGURE 4.22 The pressure – temperature phase diagram of carbon dioxide (Brogle, 1982; © SCI, reproduced by permission).

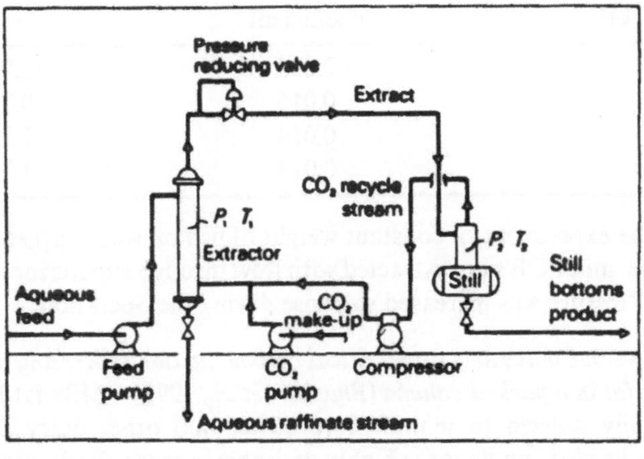

FIGURE 4.23 Flowsheet for a process employing supercritical carbon dioxide as an extracting solvent, and including distillation of the solvent De Filippi, 1982; © SCI, reproduced by permission).

The distribution coefficient of a component that is extracted from one phase into another is defined generally as the ratio of concentrations of the component in the respective phases; by convention the component being extracted by the solvent is in the numerator. For the specific case of carbon dioxide, PCB's and fish oil, the distribution coefficient K is defined as:

$$K = \frac{c_{PCB/gas}}{c^*_{PCB/oil}} \tag{E.4.49.1}$$

where $c_{PCB/gas}$ is PCB concentration in carbon dioxide and $c^*_{PCB/oil}$ is PCB concentration in fish oil in equilibrium with $c_{PCB/gas}$. The distribution coefficient is supposed not to be a function of the PCB concentration. The selectivity β of extraction may be defined as:

$$\beta = \frac{(c_{PCB}/c_{triglycerides})_{extract}}{(c_{PCB}/c_{triglycerides})_{fish\ oil}} \tag{E.4.49.2}$$

where $(c_{PCB}/c_{triglycerides})_{extract}$ and $(c_{PCB}/c_{triglycerides})_{fish\ oil}$ are the ratio of PCB's to triglycerides in the extract and the equilibrium fish oil phases, respectively.

Generally an increase in pressure will increase the distribution coefficient. Referring to the data in Table E.4.49 indicates that an increase in pressure from 1685 to 2224 kPa increases the distribution coefficient from 0.008 to 0.015, but subsequent increases in pressure does not increase the distribution coefficient any further. Table E.4.49 indicates that the selectivity of the process decreases with pressure.

Table E.4.49. Variation of the distribution coefficient and the selectivity with pressure.

extraction pressure (kPa)	distribution coefficient	selectivity
1685	0.008	12.9
2224	0.015	9.7
3370	0.014	2.1
3707	0.014	1.3

During the experiments a constant weight of fish oil was charged into the extraction unit. PCB's are extracted with flow through supercritical carbon dioxide. Pressure was increased stepwise during the operation.

Example 4.50. Continuous supercritical carbon dioxide processing of anhydrous milk fat in a packed column (Bhaskar, et al., 1993) Milk fat has been traditionally utilized to manufacture butter and other dairy products. Although its pleasing flavor is highly desirable in many foods, recognition of its limited dietary value and functional properties in the native form has dramatically reduced its consumption in the recent years. A continuous packed bed column (Fig. E.4.50) was used by Bhaskar, *et al.* (1993) to

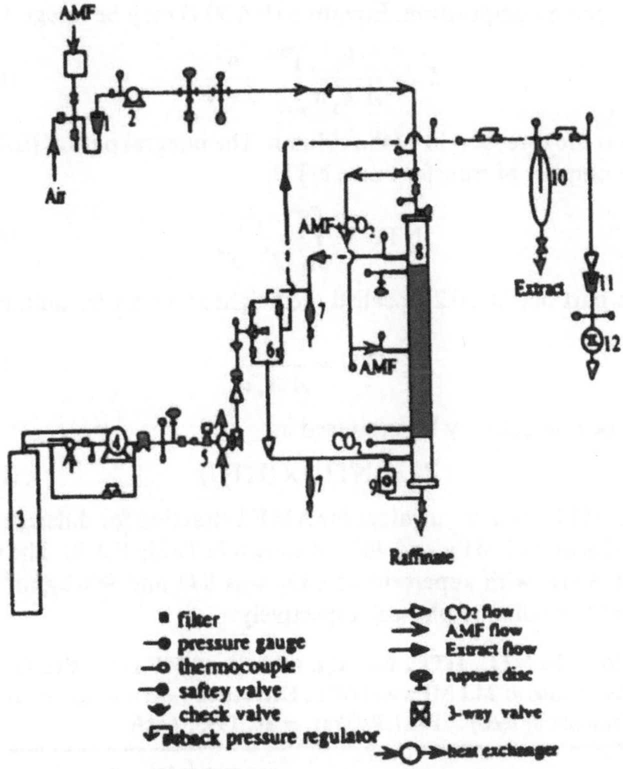

FIGURE E.4.50 A simplified schematic diagram of the continuous pilot-scale system for supercritical carbon dioxide processing of anhydrous milk fat. (1) feed flow meter, (2) feed pump, (3) liquid CO_2 tank, (4) solvent pump, (5) micrometering valve, (6) solvent flow loop, (7) entrainment vessel, (8) packed column, (9) view cell, (10) sampling vessel, (11) solvent flowmeter, (12) dry test meter (Reprinted with permission from Bhaskar *et al.*, ©1993 ACS).

process the continuous phase consisting of anhydrous milk fat (AMF) with supercritical carbon dioxide.

When we consider a differential height dz of the packed column of Figure E.4.50, (cross sectional area of the column = A) mass balance around the differential volume element of Adz requires:

$$-V\,dy = \mathcal{K}_y a(y - y^*)\,A\,dz \tag{E.4.50.1}$$

where V is the mass flow rate of gas, $\mathcal{K}_y a$ is the overall mass transfer coefficient, a is the interfacial area per unit volume, y is the weight fraction in the extract, y^* is the weight fraction corresponding to equilibrium with the liquid phase composition. Equation (E.4.50.1) may be integrated as:

$$Z_T = \frac{V}{A\mathscr{K}_y a}\int_{in}^{out}\frac{dy}{y-y^*} \quad \text{(E.4.50.2)}$$

where Z_T is the total height of the column. The integral part of (E.4.50.2) is called the number of transfer units, NTU:

$$\text{NTU} = \int_{in}^{out}\frac{dy}{y-y^*} \quad \text{(E.4.50.3)}$$

The other part of (E.4.50.2) is called the height of a transfer unit, HTU:

$$\text{HTU} = \frac{V}{A\mathscr{K}_y a} \quad \text{(E.4.50.4)}$$

The total bed height may be expressed as:

$$Z_T = (\text{NTU}) \times (\text{HTU}) \quad \text{(E.4.50.5)}$$

The NTU, HTU, and $\mathscr{K}_y a$ values for AMF extraction for different solvent to feed ratios at 24.1 MPa and 40°C is shown in Table E.4.50. The mixture density of AMF with supercritical CO_2 was 873 and 908 kg/m^3 for the extract and the raffinate phases, respectively.

Table E.4.50 The NTU, HTU, and $\mathscr{K}_y a$ values for AMF extraction for different solvent to feed ratios at 24.1 MPa and 40°C. Extraction yield was defined as (amount of extract/amount of feed) × 100 (Bhaskar, *et al.*, 1993; Table 1)

	Experimental run		
parameters	I	II	III
solvent flow rate (g/h)	9224	9022	9098
AMF processed (g/h)	297	185	124
solvent-to-feed ratio	31	49	73
extract (g/h)	9370	9155	9202
raffinate (g/h)	126	49	19
NTU	1.34	1.17	0.75
HTU (cm)	45	52	82
$\mathscr{K}_y a$ (g/cm^{2}s)	0.0030	0.0026	0.0017
extraction yield (wt%)	49	72	84

4.8. MATHEMATICAL ANALYSIS OF THE PROCESSES FOR DISTILLED BEVERAGE PRODUCTION

Distillation is the major unit operation in distilled alcoholic beverage process. The first step in the distilled alcoholic beverage production is i.e., brandy, whiskey, vodka, rum, tequila, raki, etc. production fermentation of an appropriate carbon source to produce alcohol. Brandy is produced by distillation of white wines, where the carbon source is the grape sugars. Bourbon whiskey is produced by fermenting a mash containing 60–70% corn and small amounts of other grains including rye, barley and wheat for development of unique flavors imparted to distillates (Bluhm, 1983). The mash constituent changes when producing other whiskeys, i.e., Irish whiskey mash bill is composed of 100% malted barley (Brandt, 1982). Vodka can be produced from white potatoes, cassava, sugar cane, sugar beets, etc. Rum is produced from molasses, a by-product of cane or beet sugar refineries. Tequila is made from a cactus plant, *Agave tequilana*. The distilled aniseed-flavored spirits of the Balkan and the Eastern Mediterranean countries, i.e. raki, arrack or ouzo is made from raisins. Although ethyl alcohol and water are the two major components of any distilled spirit, aroma and flavor character depend on a multitude of minor compounds usually referred to as congeners or congenerics. The most abundant congeners of brandy are minor products of alcoholic fermentation derived primarily from sugars, but in part from components of the fruit. Distillation conditions affect the relative quantities of minor compounds recovered in distillate, and some arise from chemical reactions during distillation such as heat induced degradative changes (Guymon, 1974).

The earliest version of industrial distillation equipment are pot stills (Fig. 4.24.a), where direct heat is supplied to the fermented mixture in the pot, vapors rise into the head, pass through the coiled tube, then condensed and drained into a product tank. Distillation is based on separation of components of a liquid mixture by the help of the relative volatility differences. Relative volatility is defined as:

$$\alpha_{AB} = \frac{y_A/x_A}{y_B/x_B} \tag{4.85}$$

where y and x are the mole fractions in the vapor and liquid phases, respectively. The ratio y_i/x_i is called the volatility of component i. Easy separation is achieved when $\alpha_{AB} \ll 1$ or $\alpha_{AB} \gg 1$. Components with higher vapor pressure, i.e., alcohol, will have higher mole fraction in the vapor phase. It should be noticed that components with similar relative volatility to ethanol will also appear in the distillate and contribute to the flavor. Appropriate cuts of distillates from the same or different fermentation products are blended to obtain the optimum product. The coffey still (Fig. 4.24.b) was invented in 1830s in Ireland and is in use today both in Europe and the United States. It is composed of two columns, a beer still and a rectifier. The fermented mash is referred to as beer in the alcoholic beverage literature. Further improvements were made by 'beer still and doubler' design (Fig. 4.24.c). Sketches of the various column arrangements for distillation of brandy are depicted in Figure 4.25. Alcohol is stripped of aldehydes and other low boiling point components, i.e.,

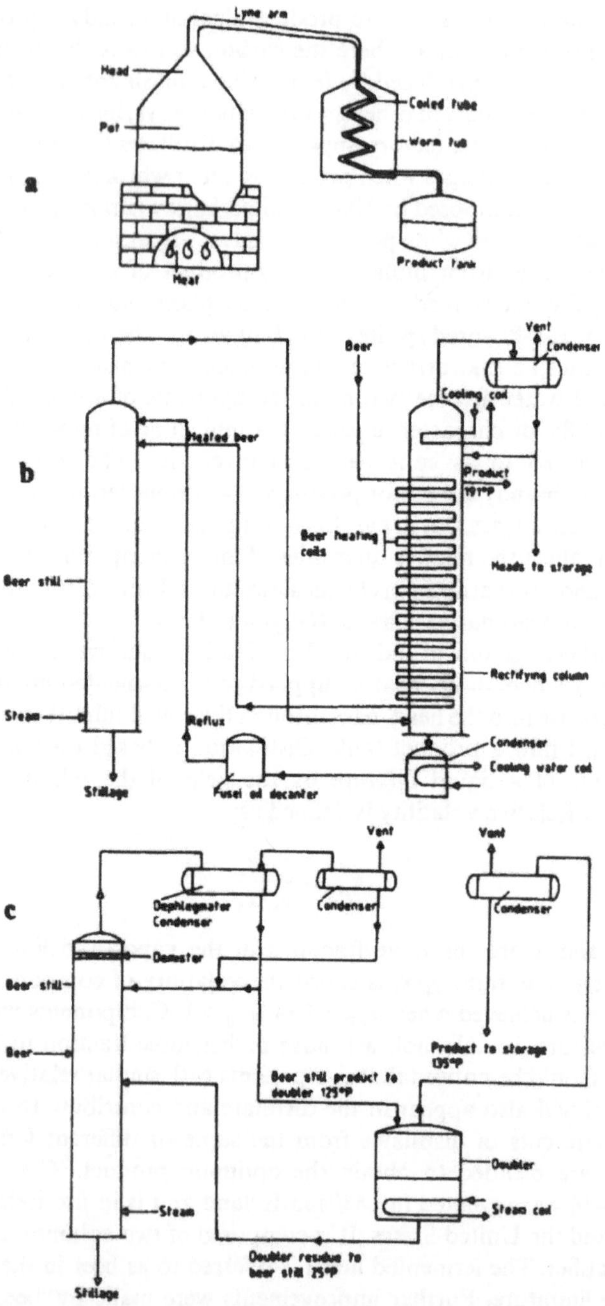

FIGURE 4.24 Drawings of the equipment used for distillation of the alcoholic beverages. (a) An early pot still. (b) Coffey still. (c) Whiskey beer still and doubler (Bluhm, 1983; ©Verlag Chemie GmbH, reproduced by permission).

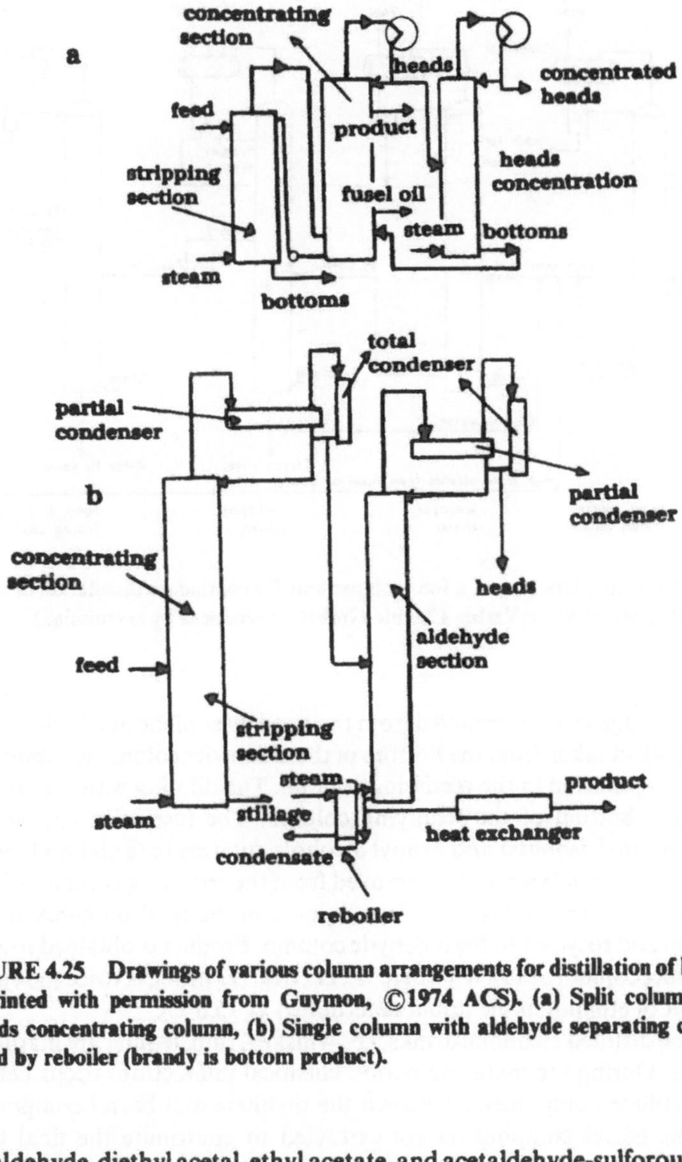

FIGURE 4.25 Drawings of various column arrangements for distillation of brandy (Reprinted with permission from Guymon, ©1974 ACS). (a) Split column with a heads concentrating column, (b) Single column with aldehyde separating column heated by reboiler (brandy is bottom product).

acetaldehyde, diethyl acetal, ethyl acetate, and acetaldehyde-sulforous acid, in the aldehyde column of the designs depicted in Figures 4.25.a and b.

The neutral spirits are essentially alcohol, free of any congeners. The distillation columns used for the production of neutral spirits are different than those of the whiskey production process, where flavor development is of prime concern. Design of a multi-column distillation unit used in neutral spirits production is depicted in Figure 4.26. A more detailed sketch of such a process is also available (Anonymous, 1942). Heads, containing low

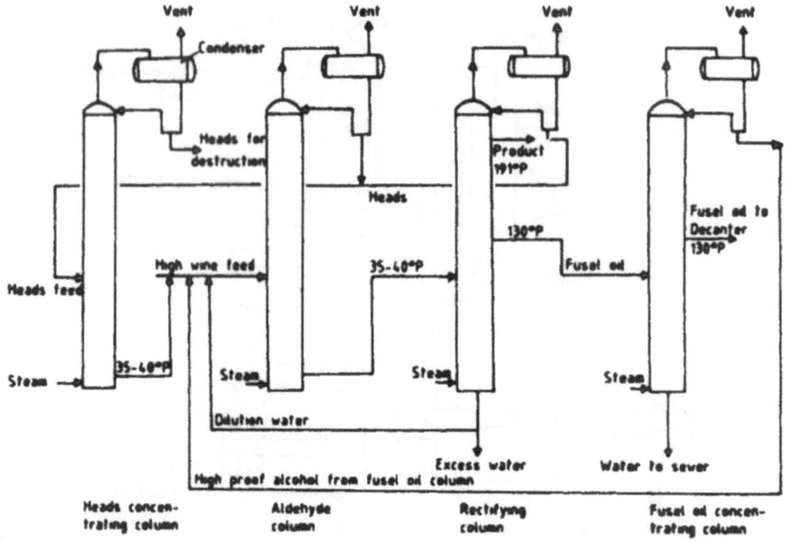

FIGURE 4.26 Drawing of a four column unit for continuous distillation of neutral sprits (Bluhm, 1983; ©Verlag Chemie GmbH, reproduced by premission).

boiling congeners are removed from the condenser of the aldehyde column. The product taken from the bottom of the aldehydes column is subjected to further separation in the rectifying column. The diluting water is removed from the bottom of the rectifying column. The fusel oils, i.e., isoamyl, d-active amyl, isobutyl and propyl alcohols, intermediate and high boiling esters, and phenolics are also removed from the rectifying column. Alcohol carried out with the fusel oils is reclaimed in the fusel oil concentration column and recycled to the aldehyde column. Product is obtained from the rectifying column at 191 °P (where °P, i.e., degrees proof, is twice the volume percent of ethanol in the liquid determined at 15.6°C).

Some distilled alcoholic drinks, i.e., whiskey, rum, tequila are matured in barrels. During the maturing period chemical interactions occur between the distillate components, between the distillate and barrel components, and the barrel components are extracted to contribute the final flavor (Bluhm, 1983).

The mathematical models for the design of the distillation equipment are essentially based on material balances. Mass balance in differential distillation process described in Figure 4.24.a. for an early pot still requires:

$$xL = (L - dL)(x - dx) + y\,dL \tag{4.86}$$

where x and y are the mole fractions of alcohol in the liquid and the vapor phases, respectively and L is the number of moles of the fermentation product. The term xL represents the original amount of the fermentation product, $(x-dx)(L-dL)$ is ethanol remaining in the liquid phase after differential amounts are distilled; and ydL is the amount of ethanol evaporated. Equation (4.86) may be rearranged and integrated as:

$$\ln\left(\frac{L_F}{L_0}\right)=\int_{x_0}^{x_F}\frac{dx}{y-x} \quad (4.87)$$

where subscripts $_0$ and $_F$ describe the initial and the final conditions, respectively. Equation (4.87) may be modified as:

$$L_0x_0=L_Fx_F+(L_0-L_F)y_{av} \quad (4.87)$$

then the average concentration of ethanol in a given cut may be calculated as:

$$y_{av}=\frac{L_0x_0-L_Fx_F}{L_0-L_F} \quad (4.88)$$

Example 4.51. Analysis of whiskey production in a pot still During whiskey production in a pot still the distillation product is distilled two or three times to attain the required ethanol content.

a) In a study made in a Turkish distillery with a steam heated pot still it was reported (Yazicioğlu, 1974) that the initial product for the second distillation operation contained 30% alcohol. The first cut of distillate was 80 L with 83% average alcohol content, the middle cut was 1750 L with 74% average alcohol content and the final cut of distillate was 1170 L with 20% average alcohol content. The remaining liquid in the pot still contains 2.2% ethanol. Calculate the initial volume of the fermentation product charged to the pot still.

Solution Total liquid balance requires:

$$V_0=80+1750+1170+V_R \quad (E.4.51.1)$$

where V_0 and V_R are the initial and the final remaining volumes of the liquid in the pot still. Ethanol balance requires:

$$0.30\ V_0=(80)(0.83)+(1750)(0.74)+(1170)(0.20)+(0.022)\ V_R \quad (E.4.51.2)$$

After solving (E.4.51.1) and (E.4.51.2) together we will obtain $V_0=5500\ L$ and $V_R=2500\ L$.

b) Use the following data (Yazicioğlu, 1974) to calculate the average ethanol content of the middle cut.

Time (h)	T (°C)	Fi (L/h)	y_i (volume fraction)	Time (h)	T (°C)	F (L/h)	y_i (volume fraction)
1	84.5	50	0.83	17	85.5	50	0.789
2	82.5	60	0.845	18	85.5	60	0.792
3	82.5	60	0.845	19	86.5	50	0.771
4	82.6	60	0.835	20	87.5	60	0.73
5	83	60	0.835	21	88	60	0.73
6	83	60	0.829	22	88.5	60	0.723
7	83.5	60	0.845	23	88.5	60	0.718
8	83.6	60	0.835	24	89	60	0.68
9	83.8	50	0832	25	89	60	0.68
10	83.8	60	0.83	26	89.5	60	0.65
11	84.5	50	0.834	27	90	60	0.63
12	84.5	50	0.834	28	90.5	60	0.62
13	84.5	50	0.84	29	90.5	50	0.61
14	85	50	0.83	30	90.5	50	0.60
15	85	50	0.83				
16	85	50	0.83				

F_i = distilled product flow rate from the pot stills during the time period $\Delta t = 1$ h.

Solution The average ethanol content y_{av} of the middle cut is:

$y_{av} = \Sigma_{i=2}^{30} F_i y_i \Delta t / \Sigma_{i=2}^{30} F_i \Delta t = 0.77$, this is in close agreement with the previous statement that the average ethanol content of the middle cut was 74%.

c) Would it be reasonable to simulate such a system by using equilibrium data obtained with the pure alcohol water mixtures?

Solution We will convert the data into mass fractions:

$$y = \frac{y \rho_{ethanol}}{\rho_{ethanol} + (1 - y)\rho_{water}}$$

where y and y are the experimentally determined volume and the corresponding mass fractions of ethanol in the liquid respectively; $\rho_{rethanol} = 0.79\ \mathrm{g/cm^3}$ and $\rho_{water} = 1\ \mathrm{g/cm^3}$ are the densities of ethanol and water, respectively. Equilibrium ethanol mass fractions under 101.3 kPa pressure at the indicated temperatures were y^* and x^* in the vapor and the corresponding liquid phases, respectively.

Time (h)	*T* (°C)	*y*	y^*	x^*	Time (h)	*T* (°C)	*y*	y^*	x^*
1	84.5	0.79	0.72	0.30	17	85.5	0.75	0.70	0.27
2	82.5	0.81	0.76	0.44	18	85.5	0.75	0.70	0.27
3	82.5	84.5	0.76	0.44	19	86.5	0.73	0.67	0.27
4	82.6	83.5	0.76	0.45	20	87.5	0.68	0.65	0.20
5	83	83.5	0.77	0.48	21	88	0.68	0.64	0.18
6	83	82.9	0.77	0.48	22	88.5	0.67	0.62	0.17
7	83.5	84.5	0.79	0.52	23	88.5	0.67	0.62	0.17
8	83.6	83.5	0.78	0.53	24	89	0.63	0.61	0.16
9	83.8	83.2	0.78	0.55	25	89	0.63	0.61	0.16
10	83.8	83	0.78	0.55	26	89.5	0.59	0.59	0.15
11	84.5	83.4	0.80	0.61	27	90	0.57	0.58	0.14
12	84.5	83.4	0.80	0.61	28	90.5	0.56	0.56	0.13
13	84.5	84	0.80	0.61	29	90.5	0.55	0.56	0.13
14	85	0.79	0.71	0.29	30	90.5	0.54	0.56	0.13
15	85	0.79	0.71	0.29					
16	85	0.79	0.71	0.29					

Initial volume of ethanol in the liquid was reported to be about 30%, which corresponds to mass fraction $y = 0.32$ and is in good agreement with the value reported at time $t = 1$ h. Values of x^* during 2–13 hours of operation are between 0.44 and 0.61. They exceed the initial alcohol content of the liquid and are not reasonable. Some discrepancy is also observed between y and y^* during the rest of the process. They became reasonably close only after 24 h of operation. The data indicates clearly that the pot still did not operate under equilibrium distillation conditions and using (4.87) with the equilibrium data pertinent to the pure alcohol water mixtures may cause unacceptable error. There are numerous other compounds in the liquid, therefore the equilibrium data obtained with pure alcohol-water mixtures may not be actually pertinent to the system under consideration. Heat losses through the equipment and variations of the pressure in the pot still may also actually cause the system to deviate from equilibrium conditions.

Mass balances may be performed to obtain the following equations for a continuous fractionating column with rectifying and stripping sections (Fig. 4.27):

$$F = D + B \tag{4.89}$$

where F, D and B are the feed, distillate and bottom product rates, respectively. Relation between the composition of the phases in the enriching (above the feed plate) section of the column is:

$$y_{n+1} = \frac{L_n}{V_{n+1}} x_n + \frac{D x_D}{V_{n+1}} \tag{4.90}$$

and the relation between the composition of the phases in the stripping section (below the feed plate) of the column is:

$$y_{m+1} = \frac{L_m}{L_m - B} x_m - \frac{B x_B}{L_m - B} \tag{4.91}$$

where L and V are the flow rates of downward and upward flowing streams, respectively. Variables x and y denotes the ethanol concentrations in L and V, respectively. The subscripts describe the plate number that the stream is leaving. Generally heat is applied at the base of a tower by means of a reboiler (heat exchanger) as shown in Figure 4.27, but when a water solution is fractionated to give non-aqueous solute as the distillate, and water is removed as the residue product, the heat may be provided by the use of open steam (saturated at the pressure of the column) at the bottom of

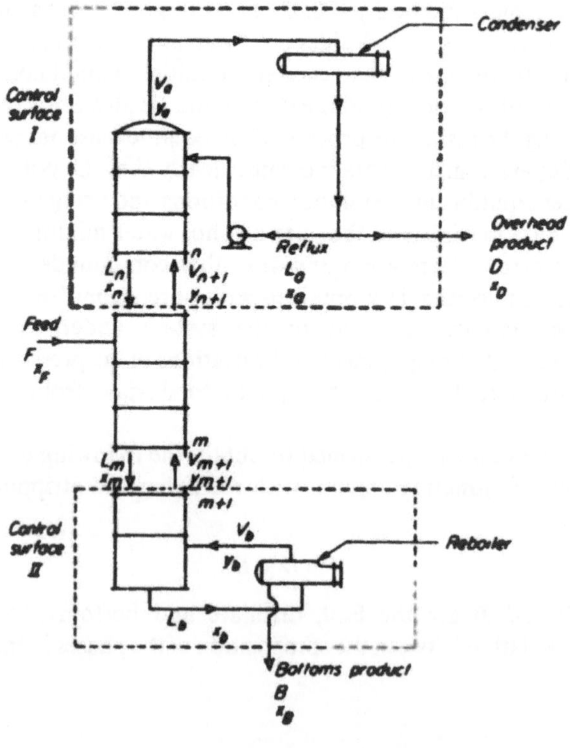

FIGURE 4.27 Material balance diagram for a continuous fractionating column.

the tower; then the reboiler is not used, but more trays are needed in the stripping section of the tower. In such a column the total material balance and the operating line for the stripping section are:

$$F + S = D + B \tag{4.92}$$

$$y_{m+1} = \frac{L_m}{L_m + S - B} x_m - \frac{Bx_B}{L_m + S - B} \tag{4.93}$$

where S is the rate of steam input to the column. The liquid obtained by condensing the overhead vapors, and returned to the top of the column is called the reflux. When all of the overhead products are condensed and returned to the top of the column we will have the total reflux. The internal reflux ratio is defined as L/V where L and V are the molar flow rates of liquid and vapor streams, respectively; while designing a column using the McCabe and Thiele Method the molar liquid and vapor flow rates are assumed to be constant.

Example 4.52. Distillation column design with McCabe and Thiele method for separation of ethanol from a fermentation product The operating conditions of the rectifying column described in this example are very similar to the ones used in the distilled beverage industry to improve the ethanol concentration of the product. An ethanol-water mixture available at its bubble point with 0.3 mole fraction ethanol will be distilled to produce 0.80 mole fraction ethanol with negligible ethanol in the bottom. The feed rate to the column is 910 kg/h, the column will contain 76 cm diameter cross-flow sieve trays (Fig. E.4.52.1) and operate under atmospheric pressure (101.3 kPa). The net tray area for the flow of the vapor phase is 0.417 m^2. Determine a suitable reflux ratio, the number of the trays needed and the location of the feed plate needed for this separation. Saturated open steam at 169 kPa will be used for heating.

Solution The feed composition corresponds to 52.27% (wt/wt) ethanol, and the feed contains (910)(0.5227)/(46.05) = 10.33 kmol ethanol and

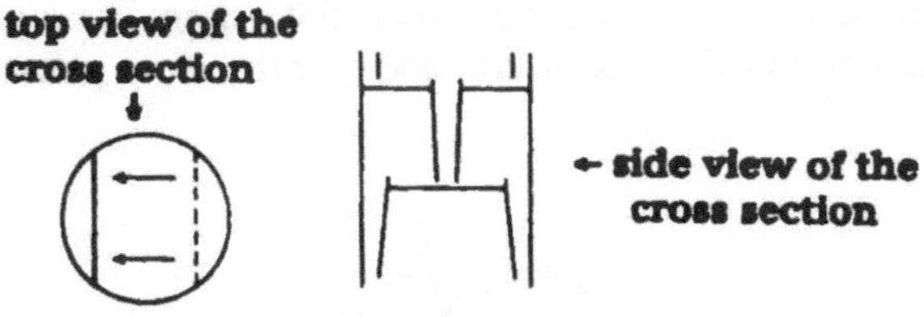

FIGURE E.4.52.1 A cross-flow tray arrangement. Bubbe-cap trays were used almost in every distillation column during 1920–1950, but they have been abandoned for new installation because of their high cost.

(910)(1−0.5227)/(18.02) = 24.10 kmol water, where 46.05 and 18.02 are the molecular weights of ethanol and water, respectively. If essentially all the ethanol is removed from the residue, the distillate $D = 10.33/0.80 = 12.91$ kmol/h = (12.91)(0.80)(46.05) + (12.91)(0.20)(18.02) = 522.2 kg/h. Equilibrium data for ethanol-water systems is depicted in Table E.4.52.1.

Table E.4.52.1 Equilibrium mass fractions for ethanol-water systems at 101.3 kPa

T (°C)	x	y	T (°C)	x	y
100.0	0	0	81.0	0.60	0.794
98.1	0.02	0.192	80.1	0.70	0.822
95.2	0.05	0.377	79.1	0.80	0.858
91.8	0.10	0.527	78.3	0.90	0.912
87.3	0.20	0.656	78.2	0.94	0.942
84.7	0.30	0.713	78.1	0.96	0.959
83.2	0.40	0.746	78.2	0.98	0.978
82.0	0.50	0.771	78.3	1.00	1.00

At the top of the column mole fraction of ethanol is 0.80 corresponding to a mass fraction $y = (0.80)(46.05)/((0.80)(46.05) + (0.20)(18.02)) = 0.91$, temperature of this plate may be estimated from Table E.4.52.1 as 78.28°C.

The diameters of the distillation columns are chosen to accommodate the pertinent liquid and vapor flow rates. The detailed design is made near flooding conditions. With large pressure drops in the space between the trays, the level of liquid leaving a tray at relatively low pressure and entering one of high pressure must necessarily assume an elevated position in the downcomers. The liquid is led from one tray to the next by means of the downcomers. As the pressure differences increase further due to the increased rate of either vapor or liquid, the level in the downcomers will increase further, and eventually the liquid level will reach that of the tray above, then an increase in either flow rate will force the liquid to occupy the entire space between the trays, this is called flooding. For a given type of a column at flooding the superficial velocity of the vapor phase is (Treybal, 1980):

$$V_F = C_F \sqrt{\frac{\rho_L - \rho_V}{\rho_V}} \tag{E.4.52.1}$$

where ρ_L is the density of the liquid phase and ρ_V is the density of the vapor phase, C_F is flooding constant of the tray tower and may be expressed empirically as (Treybal, 1980):

$$C_F = \alpha \log\left\{\frac{1}{(L_s/V_s)\sqrt{\rho_V/\rho_L}} + \beta\right\}\left(\frac{\sigma}{0.02}\right)^{0.2} \qquad \text{(E.4.52.2)}$$

where L_s and V_s are the superficial mass flow rates of liquid and vapor phases, respectively; σ is the surface tension of the liquid phase; α and β are constants and under the operating conditions given as $\alpha = 0.0452$ and $\beta = 0.0287$. At the temperature of the top of the column (tray number 20) $\rho_V = 1.405\ \text{kg/m}^3$, $\rho_L = 744.9\ \text{kg/m}^3$ and $\sigma = 0.021$ N/m. If we should tentatively take $(L_s/V_s)(\rho_V/\rho_L)^{0.5} = 0.1$, (E.4.52.2) will provide $C_F = 0.0746$; then we may use (E.4.52.1) to calculate $V_F = 1.72$ m/s. If we should choose a reflux ratio of 3 then we will have $V = D(3+1) = 12.91\ (3+1) = 51.64$ kmol/h vapor at the top, which (after using the ideal gas law) may be shown to correspond $0.414\ \text{m}^3/\text{s}$ volumetric vapor flow rate. The velocity of the vapor phase is $0.414/0.417 = 0.993$ m/s (where 0.417 = the net tray area for the flow of the vapor phase). This net velocity of the vapor phase corresponds to only 58 % of the flooding velocity of the tower and indicates that the choice of the reflux ratio is appropriate for the operation of the column.

The flow rate of the liquid phase is $L = 3D = (3)\ (12.91) = 38.73$ kmol/h corresponding to the mass flow rate of 38.73 ((0.80) (46.05) + (0.20) (18.02)) = 1567 kg/h, and $V = (51.64)\ ((0.80)\ (46.05) + (0.20)\ (18.02)) = 2089$ kg/h. We may calculate $(L_s/V_s)\ (\rho_V/\rho_L)^{0.5} = (1567/2089)\ (1.405/744.9)^{0.5} = 0.033$. When $(L_s/V_s)\ (\rho_V/\rho_L)^{0.5} \leqslant 0.1$, (E.4.52.1) uses the C_F calculated for $(L_s/V_s)\ (\rho_V/\rho_L)^{0.5} = 0.1$, therefore the calculation of V_F is correct.

All the feed will join the liquid coming from the enriching section in the feed plate, then the flow rate of the liquid and the vapor phases in the stripping section will be:

$$L_{\text{stripping}} = L + F \qquad \text{(E.4.52.3)}$$

$$V_{\text{stripping}} = V \qquad \text{(E.4.52.4)}$$

Therefore $L_{\text{stripping}} = 38.73 + 34.43 = 73.2$ kmol/h and $V_{\text{stripping}} = 51.64$ kmol/h. It should be noticed that (E.4.52.3) and (E.4.52.4) are valid only with liquid feed at its bubble point, partitioning of the feed among the phases depends on its enthalpy.

Enthalpy of the saturated steam at 169 kPa is 2699 kJ/kg (reference: enthalpy of liquid water at 0°C is 0 kJ/kg). It will adiabatically expand to the column conditions, where enthalpy of the saturated steam under 101.3 kPa is 2676 kJ/kg and the latent heat ΔH_{vap} is 2257 kJ/kg. We will use the following equation to relate the molar flow rates of steam and vapor phase at the bottom of the column:

$$V = S\left(1 + \frac{H_N - H_{\text{saturated}}}{\Delta H_{\text{vap}}}\right) \tag{E.4.52.5}$$

After substituting the numbers in (E.4.52.5) we will have: $51.64 = S(1 + (2699-2676)/2557)$, then we can easily calculate $S =$ 51.17 kmol steam/h. The molar material balance between the phases requires:

$$L = V - S + B \tag{E.4.52.7}$$

After substituting the numbers $73.2 = 51.64 - 51.1 + B$, we calculate $B = 72.7$ kmol/h.

We assumed that there is saturated steam at the bottom plate of the column at 101.3 kPa, therefore the temperature of this plate may be assumed as 100°C.

The locus of the intersection of the stripping and enriching operating lines is (Treybal, 1980):

$$y = \frac{q}{q-1}x - \frac{x_F}{q-1} \tag{E.4.52.3}$$

where

$$q = \frac{L_{\text{stripping}} - L_{\text{enriching}}}{F} = \frac{H_V - h_F}{H_V - h_L} \tag{E.4.52.4}$$

$L_{\text{stripping}}$ = molar flow rate of the liquid phase in the stripping section, $L_{\text{enriching}}$ = molar flow rate of the liquid phase in the enriching section and F = molar feed rate to the column, H_V = molar enthalpy of the vapor phase, h_L = molar enthalpy of the liquid phase and h_F = molar enthalpy of the feed and x_F is the average of the mole fraction of ethanol in both phases of the feed stream. Equation (E.4.52.3) describes the q line. In the y versus x diagram (Fig. E.4.52.2) (of $y = x$ (45° line) with $y = z_F$ line) found, then a line passing through the intercepte with slope of $q/(1-q)$ is plotted to obtain the q line. The value of q for a mixture of ethanol-water feed at its bubble point is 1.

After substituting the numbers in (4.90) and (4.93) the operating lines for the enriching and the stripping sections will be $y = 0.75x + 0.2$ and $y = 1.42x$, respectively. In Figure E.4.52.2 the operating lines of the enriching and the stripping sections were plotted using these equations, it was also confirmed that the intercept of these operating lines are on the q line. The equilibrium curve was plotted after converting the equilibrium data in mole fractions (Tab. E.4.52.2).

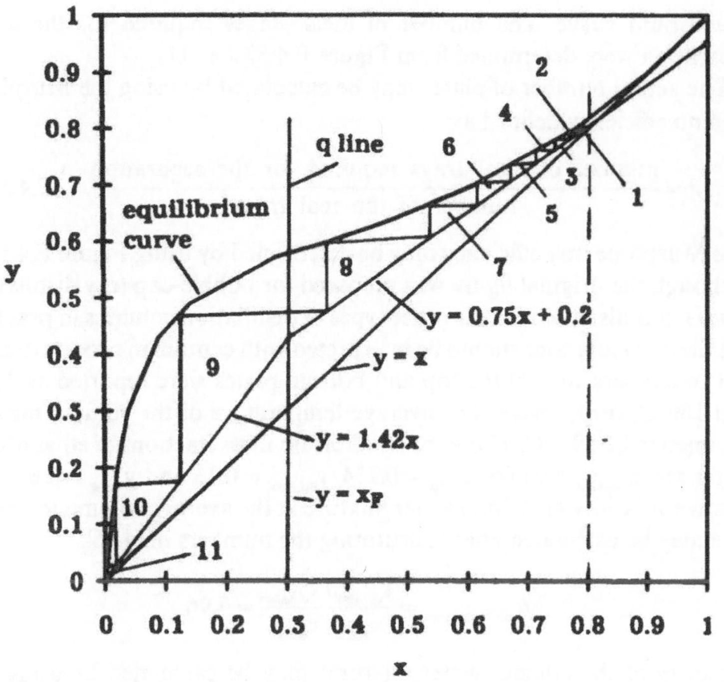

FIGURE E.4.52.2 Determination of the number of the ideal stages to achieve the required separation.

Table E.4.52.2 Equilibrium mole fractions for ethanol-water systems at 101.3 kPa

T (°C)	*x*	*y*	*T* (°C)	*x*	*y*
100.0	0	0	81.0	0.37	0.60
98.1	0.01	0.09	80.1	0.48	0.64
95.2	0.02	0.19	79.1	0.61	0.70
91.8	0.04	0.30	78.3	0.78	0.80
87.3	0.09	0.43	78.1	0.90	0.90
84.7	0.14	0.49	78.2	0.95	0.95
83.2	0.21	0.53	78.3	1.00	1.00
82.0	0.28	0.57			

The number of the ideal plates is the same as the number of the horizontal lines shown in Figure E.4.52.2 connecting the operating line to the equilibrium line. The horizontal stage lines of Figure E.4.52.2 indicate that when liquid and vapor streams enter into a tray their compositions are related by the operating line; after remaining in contact they attain equilibrium in an ideal plate, then the relation between their compositions is described by the equilibrium curve. The number of ideal stages required for the given separation were determined from Figure E.4.52.2 as 11.

The actual number of plates may be calculated by using the Murphree column efficiency defined as:

$$\eta = \frac{\text{number of ideal trays required for the separation}}{\text{number of the real trays}} \qquad \text{(E.4.52.5)}$$

The Murphree tray efficiency may be determined by using Figure E.4.52.3. Although the original figure was prepared for bubble-cap tray distillation towers, it is also used for the other types of distillation columns in practice, but these calculations should be interpreted with caution in process design.

The temperatures of the top and bottom plates were reported as 78.28 and 100°C, respectively. The average temperature of the column may be assumed to be 89.5°C, where the equilibrium mass fractions of ethanol and water are: $x_{ethanol} = 0.066$, $x_{water} = 0.934$, $y_{ethanol} = 0.36$ and $y_{water} = 0.64$. The relative volatility of ethanol water mixture at the average column temperature may be calculated after substituting the numbers in (4.85):

$$\alpha_{ethanol/water} = \frac{y_{ethanol}/x_{ethanol}}{y_{water}/x_{water}} = 7.96$$

Viscosity of the ethanol water mixtures may be calculated by using the Kendall-Monroe equation:

$$\mu_{mixture}^{1/3} = x_{ethanol}\,\mu_{ethanol}^{1/3} + x_{water}\,\mu_{water}^{1/3} \qquad \text{(E.4.52.6)}$$

where we substitute the mass fractions of ethanol and water in the feed stream for x_{water} and $x_{ethanol}$. After substituting $x_{water} = 0.447$, $x_{ethanol} = 0.523$

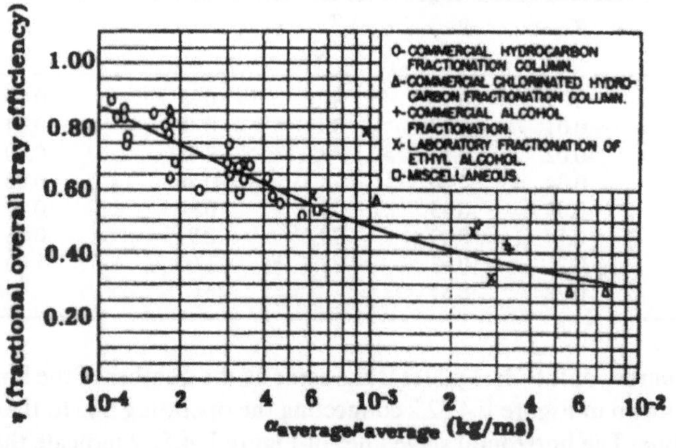

FIGURE E.4.52.3 **Overall tray efficiencies of bubble-cap distillation columns separating hydrocarbons and similar mixtures (O'Connel, 1946). Reproduced with permission of the American Institute of Chemical Engineers. © 1946 AIChE. All rights reserved.**

and the viscosities at 89.5°C $\mu_{ethanol} = 0.95 \times 10^{-3}$ kg/ms and $\mu_{water} = 0.8 \times 10^{-3}$ kg/ms in (E.4.52.6) we will calculate $\mu_{mixture} = 0.8 \times 10^{-3}$ kg/ms. Then $\alpha_{average}\,\mu_{average} = 6.4 \times 10^{-3}$. The Murphree plate efficiency may be read from Figure E.4.52.3 as $\eta = 0.57$. After substituting the numbers in (E.4.52.5) we will determine the number of real trays required for this separation as 20. Figure E.4.52.2 shows that feed enters to the ideal stage number 9. We may calculate the corresponding real tray number for the entrance of the feed as 16, after using the Murphree plate efficiency.

Example 4.53. Estimation of the tray number for maximum accumulation of congeners during brandy production (Guymon, 1974) The data given in Table E.4.53 is taken in a column with an internal reflux ratio of 0.7.

Table E.4.53 Concentration of ethanol and higher alcohols in the plates of a distillation column during production of brandy (Guymon, 1974; Table 2)

Tray number	°proof	n-propyl alcohol[a]	isobutyl alcohol[a]	amyls[a]
Over head	189.8	–	2.3	–
14	189.2	–	7.1	2.4
13	188.6	11.8	11.8	5.9
12	187.5	19.9	16.4	14.0
product	186.0	25.4	25.4	25.4
11	186.6	25.7	22.2	23.4
10	184.0	42.5	42.5	64.0
9	181.9	62.3	65.6	158
8	179.2	65.0	85.0	207.5
7	176.8	88.0	110.0	354
6	171.9	107.5	150.5	688
5	166.4	120.5	177	1000
4	160.2	135	200	1500
3	148.5	116	190	2150
2	128.5	128	208	3210
1	94.4	53	71	1178

[a] grams per 100 liters at existing proof

The proof strength at which the volatility curve of a particular component intersects that of ethyl alcohol indicates that proof at which the minor constituent will be concentrated in a fractionating column at the limiting condition of total reflux. For practical conditions, the maximum concentration of a particular congener or minor component may be assumed to occur at a proof in the column at which its volatility is approximately equal to the internal reflux ratio. The volatility curves of aliphatic alcohols at various ethyl alcohol-water concentrations are given in Figure E.4.53. Reading the curves of Figure E.4.53 at 0.7 volatility indicate that the maximum concentrations should be obtained as follows: *n*-propyl alcohol 170° proof, isobutyl alcohol 158° proof and amyls 128° proof. It was shown in Table

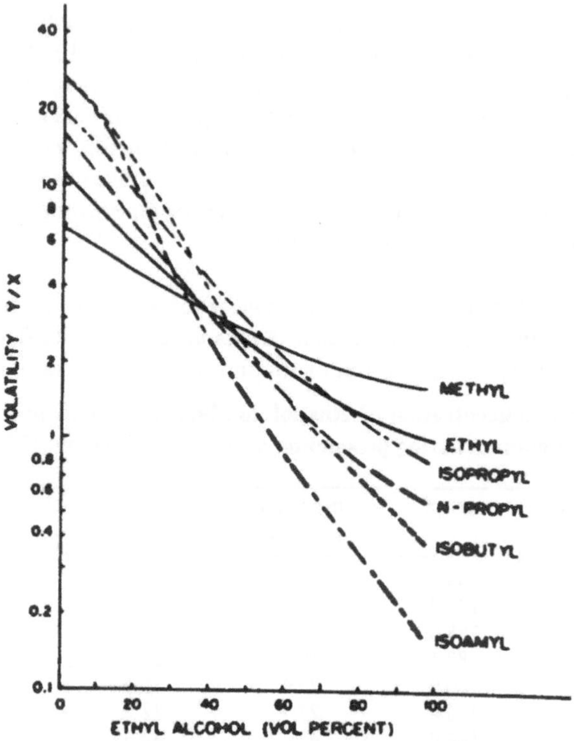

FIGURE E.4.53 The volatility curves of aliphatic alcohols at various ethyl alcohol-water concentrations (Reprinted with permission from Guymon, ©1974 ACS).

E.4.53 that the maximum *n*-propyl alcohol concentration was observed in tray number 4, where °proof was 160.2, the maximum isobutyl alcohol concentration was observed in tray number 2, where °proof was 128.5 and the maximum amyls concentration was observed in tray number 2, where °proof was 128.5. The experimental results are in good agreement with the estimates obtained from Figure E.4.53, with exception of the isobutyl alcohol. The second greatest accumulation of isobutyl alcohol was in plate number 4, where the °proof was 160.2 and this is in excellent agreement with the prediction from Figure E.4.53.

REFERENCES

Aiba, S., Humphrey, A. E. and Millis, N. F. (1973) *Biochemical Engineering*, Second Edition. University of Tokyo Press, Tokyo, pp. 258–264.

Andrieu, J. and Stamatopoulos, A. (1986) Moisture and heat transfer modeling during drum wheat pasta drying. In *Drying '86*. Volume 2, Mujumdar, A. S. (ed.) Hemisphere, New York, 1986, pp. 492–498.

Andrieu, J, Jallut, C., Stamatopoulos, A. and Zafiropoulos, M. (1988) Identification of water apparent diffusivities for drying of corn based extruded pasta. *Proceedings of the 6th International Drying Symposium, IDS'88*, Versailes, September 5–8.

Anonymous, (1942) A modern distillary, *Chemical and Metallurgical Engineering*, 49(11), 126.

Bhaskar, A. R., Rizvi, S. S. H. and Harriot, P. (1993) Performance of a packed column for continuous supercritical carbon dioxide processing of anhydrous milk fat. *Biotechnology Progress*, 9, 70–74.

Balaban, M and Pigott, G. M. (1988) Mathematical model of simultaneous heat and mass transfer in foods with dimensional changes and variable transport parameters. *Journal of Food Science* 53, 935–939.

Ball, C. O. (1923) Thermal process time for canned food. Bull. 37, National Research Council, Washington, D. C.

Bayindirli, L., Özilgen, M. and Ungan, S. (1989) Modeling of apple juice filtrations. *Journal of Food Science*, 54, 1003–1006.

Bluhm, L. (1983) Distilled beverages. In *Biotechnology*, Reed, G. (ed.) Verlag Chemie, Weinheim, Germany.

Brandt, D. A. (1982) Distilled beverage alcohol. In *Precott and Dunn's Industrial Microbiology*, Reed, G. (ed.), Avi, Westport, Connecticut.

Biswal, R. N. and Bozorgmehr, K. (1992) Mass transfer in mixed solute osmotic dehydration of apple rings. *Transactions of the ASAE*, 35, 257–262.

Bomben, J. L., Durkee, E. L., Lowe, E. and Secor, G. E. (1974) A laboratory study on countercurrent desalting of pickles, *Journal of Food Science*, 39, 260–263.

Brogle, H. (1982) CO_2 as a solvent: its properties and applications. *Chemistry and Industry*, (19 June) 385–390.

Brose, D. J., Childlaw, M. B., Friesen, D. T., LaChapelle, E. D. and van Eikeren, P. (1995) Fractionation of citrus oils by using a membrane based extraction process. *Biotechnology Progress*, 11, 214–220.

Buckle, E. R. (1961) Studies on the freezing of pure liquids II. The kinetics of homogeneous nucleation in supercooled liquids. *Proceedings of the Royal Society* (London) A261, 189–196.

Carslaw, H. S. and Jaeger, J. C. (1959) Conduction of Heat in Solids. 2nd ed. Clarendon Press, Oxford, England.

Chang, D. J. and Hwang, S. Y. (1994) Unsteady-state permeate flux of crossflow microfiltration, *Separation Science and Technology*, 29, 1593–1608.

Chang, S. Y., Toledo, R. T. and Lillard, H. S. (1989) Filtrate flow through filter cakes which exibit time dependent resistance, *Journal of Food Processing and Preservation*, 12, 253–269.

Charoenrein, S and Reid, D.S. (1989) The use of DSC to study the kinetics of heterogeneous nucleation of ice in aqueous systems. *Thermochimica Acta*, 156, 373–381.

Cheryan, M. (1992) Concentration of liquid foods by reverse osmosis, in *Handbook of Food Engineering*, Heldman, D. and Lund, D. B. (eds.) Marcel Dekker, Inc. New York.

Chiheb, A., Debray, E., Le Jean, G. and Piar, G. (1994) Linear model for predicting transient temperature during sterilization of a food product. *Journal of Food Science*, **59**, 441–446.

Cleland, A. C. and Earle, R. L. (1982a) Freezing time prediction for foods – a simplified procedure. *International Journal of Refrigeration*, **5**, 134–140.

Cleland, A. C. and Earle, R. L. (1982b) A simple method for prediction of heating and cooling rates in solids of various shapes, *International Journal of Refrigeration*, **5**, 98–106.

Cleland, A. C. and Earle, R. L. (1984) Assessment of freezing time prediction methods, *Journal of Food Science*, **49**, 1034–1042.

Cleland, A. C. (1990) *Food Refrigeration Process, Analysis, Design and Simulation*, Elsevier Applied Science, New York.

Cleland, D. J., Cleland, A. C. and Earle, R. L. (1987) Prediction of freezing and thawing times for multi-dimensional shapes by simple formulae: I. Regular shapes, *International Journal of Refrigeration*, **10**, 156–164.

Coulson, J. M. and Richardson, J. F. (1968) *Chemical Engineering*; 2nd ed. vol. 2, Pergamon Press, Oxford.

Crank, J. (1956) *Mathematics of Diffusion*, University Press, Oxford.

De Michelis, A. and Calvelo, A., (1982) Mathematical models for nonsymmetric freezing of beef, *Journal of Food Science*, **47**, 1211–1217.

De Filippi, R. P. (1982) CO_2 as a solvent: application to fats, oils and other materials. *Chemistry and Industry*, (19 June) 390–394.

Dunsford, P. and Boulton, R. (1981) The kinetics of potassium bitartrate crystallization from table wines. I. Effect of particle size, surface area and agitation. *American Journal of Enology and Vitticulture*, **32**, 100–105.

Ede, A. J. (1949) The calculation of freezing and thawing of foods. *Modern Refrigeration*, **52**, 52–55.

Farkas, B. E. (1994) *Modeling immersion frying as a moving boundary problem*. Ph.D. Dissertation. University of California at Davis.

Franks, F. (1982) The properties of aqueous solutions at subzero temperatures. In *Water – A comprehensive Treatise*, vol. 7. ed. F. Franks, New York: Plenum.

Gasson-Lara, J. H. (1993) Drying of nixtamal, in *Food Dehydration*, Canovas, G. V. B., and Okos, M. R. (eds.) IAChE syposium series, pp. 85–89, Volume 89.

Geankoplis, C. J. (1983) *Transport Processes and Unit Operations*, 2nd ed. Allyn and Bacon, Inc. MA.

Giner, S. A. and Calvelo, A. (1987) Modeling of wheat drying in fluidized beds. *Journal of Food Science*, **52**, 1358–1363.

Guymon, J. F. (1974) Chemical aspects of distilling wines into brandy. In *Chemistry of Winemaking*, Advances of Chemistry Series, **137**, American Chemical Society, Washington, D.C.

Hallström, B., Skjöldebrand, C. and Trägårdh, C. (1988) *Heat Transfer & Food Products*. Elsevier Applied Science.

Hallström, B. (1992) Mass transfer in foods, in *Handbook of Food Engineering*, (Heldman, D. R. and Lund, D. B., Eds.) Marcel Dekker Inc. New York.

Harvey, M. A., Bridger, K. and Tiller, F. M. (1988) Apparatus for studying incompressible and moderately compressible cake filtration, *Filtration and Separation* (Jan./Feb.) 21–29.

Hayakawa, K. and Hwang, P. M. (1981) Apparent thermophysical constants for thermal and mass exchanges of cookies undergoing commerical baking processes. *Lebensmittlel-Wissesnschaft und Technologie*, **14**, 336–345.

Hierro, M. T. G. and Santa-Maria, G. (1992) Supercritical fluid extraction of vegetable and fats with CO_2– A minireview. *Food Chemistry*, **45**, 189–192.

Heldman, D. R. (1983) Factors influencing food freezing rates. *Food Technology*, 37(4), 103–109.

Hiemenz, P. C. (1977) *Principles of Colloid and Surface Chemistry*, Marcel Dekker, Inc. New York.

Hoffman, J. D. (1958) Thermodynamic driving force in nucleation and growth processes. *Journal of Chemical Physics*, **29**, 1192–1193.

Hurst, A. (1977) Bacterial injury: A review. *Canadian Journal of Microbiology*, **23**, 935–944.

Ingmanson, W. L. Filtration resistance of compressible materials, *Chemical Engineering Progress*, **49**, 577–584.

Jackson, J. M. and Shinn, B. M. (1979) *Fundamentals of Food Canning Technology*, Avi Pub. Co., Westport, Connecticut.

Jay, M. J. (1978) *Modern Food Microbiology*, 2nd. ed. Van Nostrand Reinhold, New York.

Jen, Y., Manson, J. R., Stumbo, C. R. and Zahrandik, J. W. (1971) A procedure for estimating sterilization and quality factor degradation in thermally processed foods. *Journal of Food Science*, **36**, 692–698.

Kadem, O. and Katchalsky, A. (1958) Thermodynamic analysis of the permeability of biological membrances to non-electrolytes. *Biochimica Biophysica Acta*, **27**, 229–246.

Kaiser, J. M. and Glatz, C. E. (1988) Use of precipitation of alter flux and fouling performance in cheese whey ultrafiltration. *Biotechnology Progress*, **4**, 242–247.

Karel, M. (1975a) Concentration of foods. In *Principles of Food Science*, Part II: *Physical Principles of Food Preservation*, Karel, M., Fennema, O. R. and Lund, D. B. (eds.) Marcel Dekker Inc. New York.

Karel, M. (1975b) Heat and mass transfer in freeze drying. In *Freeze Dying and Advanced Food Technology*. Goldblith, S. A., Rey. L. and Rothmayr, W. W. (eds) Academic Press, New York.

King. C. J. (1968) Rates of moisture sorption and desorption in porous dried foodstuffs. *Food Technology*, **22**(4), 165–171.

Kozempel, M. F., Sullivan, J. F. and Craig, J. C. (1981) Model for blanching potatoes and other vegetables. *Lebensmittel-Wissenschaft und Technologie* **14**, 331–335.

Krukonis, V. J. (1989) Supercritical fluid processing of fish oils: Extraction of polychlorinated biphenyls. *Journal of the American Oil Chemists Society*, **66**, 818–821.

Labuza, T. P. and Simon, I. B. (1970) Surface tension effects during dehydration. *Food Technology*, **24**(6) 712–715.

Lenz, M. K. and Lund, B. (1977) The lethality-Fourier number method: Experimental verification of a model for calculating temperature profiles and lethality in conduction-heating canned foods. *Journal of Food Science*, **42**, 989–996, 1001.

Lima-Hon, V. M., Chen, C. S. and Marsaioli, A. (1979) Computer simulation of dynamic behavior in vacuum evaporation of tomato paste. *Transactions of the ASAE*, **22**, 215–218, 224.

Longsdale, H. K. (1972) Theory and practice of reverse osmosis and ultrafiltration. In *Industrial processing with membranes*. Lacey, R. E. and Loeb, S. (eds.) Wiley-Interscience, New York.

Mannapperuma, J. D. and Singh, P. (1988) Prediction of freezing and thawing times of foods using a numerical method based on enthalpy formulation. *Journal of Food Science*, **53**, 626–630.

Martino, M. N. and Zaritzky, N. E. (1988) Ice crystal size modifications during frozen beef storage. *Journal of Food Science*, **53**, 1631–1637, 1649.

Mascheroni, R. H. and Calvelo, A. (1982) A simplified model for freezing time calculations of foods. *Journal of Food Science*, **47**, 1201–1207.

McCabe, W. L., Smith, J. C. and Harriot, P. (1985) *Unit Operations of Chemical Engineering*. 4th ed. McGraw-Hill.

Merson, R. L., Singh, R. P. and Carroad, P. A. (1978) An evaluation of Ball's formula method of thermal process calculations. *Food Technology*, **32**(3), 66–72, 75.

Merten, U. (1966) Transport properties of osmotic membranes. In *Desalination by Reverse Osmosis*. Merten, U. (ed.), MIT Press, Cambridge, Massachusetts.

Moyler, D. A. (1988) Liquid CO_2 extraction in the flavour and fragrance industries. *Chemistry and Industry*, (17 October) 660–662.

National Canners Assiciation, Laboratory Manual For Food Canners & Processors, Vol. 1, fourth printing, Avi Pub. Co. Westport, Connecticut, USA, 1980, pp. 235–238.

Newman, A. E. (1936) Heating and cooling rectangular and cylindrical solids. *Industrial and Engineering Chemistry*, **28**, 545–548.

O'Connell, H. E. (1946) Plate efficiency of fractionating columns and absorbers *Transactions of the AIChE*. **42**, 741–755.

Oolman, T. and Liu, T. C. (1991) Filtration of mycelial broths. *Biotechnology Progress*, **7**, 534–539.

Özilgen, S. and Reid, D. S. (1993) The use of DSC to study the effects of solutes on heterogeneous ice nucleation kinetics in model food emulsions. *Lebensmittel-Wissenschaft und Technologie*, **26**, 116–120.

Pham, Q. T. and Willix, J. (1989) Thermal conductivity of fresh lamb meat, offals and fat in the range −40 to +30°C: Measurements and correlations. *Journal of Food Science*, **54**, 508–515.

Plank, R. (1941) Bietrage zur berechnung und bewertung der gefriergesch windigkeit von lebensmitteln. *Z. ges Kalteind.*, **10**(3), Beih Reihe 1–16.

Porter, H. F., McCormick, P. Y., Lucas, R. L. and Wells, D. F. (1973) Gas-solid systems, in *Chemical Engineer's Handbook*, McGraw-Hill Kogakusha, Japan.

Pusch, W. (1977) Determination of transport parameters of synthetic membrances by hyperfiltration experiments. I. Derivation of transport relationship from the linear relations from thermodynamics of irreversible processes. *Berichte der Bunsen Gesellschaft für Chemie Physikalische* **81**(2), 269–276.

Radovic, L. R., Tasic, A. Z., Grozdanic, D. K., Djordjevic, B. D. and Valent, V. J. (1979) Computer design and analysis of operation of a multiple-effect evaporator system in the sugar industry. *Industrial and Engineering Chemistry, Process Design and Development* **18**, 318–323.

Schwartzberg, H. G. (1990) Food freeze concentration, in *Biotechnology and Food Process Engineering*, Schwartzberg, H. G. and Rao, M. A. (eds.), Marcel Dekker, Inc., New York, pp. 127–202.

Schoeber, W. J. A. H. *Regular Regime in Sorption Processes.* Ph.D. Thesis, Tehcnical University of Eindhoven, The Netherlands.

Shi, Y., Liang, B. and Hartel, R. W. (1990) Crystallization kinetics of alpha-lactose monohydrate in a continuous crystallizer. *Journal of Food Science*, **55**, 817–820.

Singh, R. P. (1995) Heat and mass transfer in foods during deep-fat frying. *Food Technology*, **49**(4), 134–137.

Skjöldebrand. C. (1980) Convective oven frying. Heat and mass transfer between air and the product. *Journal of Food Science*, **45**, 1354–1358, 1362.

Skjöldebrand. C. and Hallström, B. (1980) Convective oven frying. Heat and mass transport in the product. *Journal of Food Science*, **45**, 1347–1362.

Svarovsky, L. (1985) Solid-liquid separation processes, in *Scaleup of Chemical Processes*, Bisio, A. and Kabel, R. L. (eds.) Wiley-Interscience, New York.

Thomas, R. L., Westfall, P. H., Louvieri, Z. A. and Ellis, N. D. (1986) Production of applie juice by single pass metallic membrane ultrafiltration. *Journal of Food Science*, **51**, 559–563.

Thomas, R. L., Gaddis, J. L., Westfall, P. H., Titus, T. C. and Ellis, N. D. (1987) Optimization of apple juice production by single pass metallic membrane ultrafiltration. *Journal of Food Science*, **52**, 1263–1266.

Tong, C. H. and Lund, D. B. (1990) Effective moisture diffusivity in porous materials as a function of temperature and moisture content. *Biotechnology Progress*, **6**, 67–75.

Treybal, R. E. (1980) *Mass Transfer Operations*, 3rd ed. McGraw-Hill Kogakusha Ltd, Tokyo.

Turhan, M. and Özilgen, M. (1991) Effect of oven temperature variations upon the drying behavior of thin biscuits, *Acta Alimentaria*, **20**, 197–203.

Ward, D. R., Pierson, M. D. and Minnick, M. S. (1984) Determination of equivalent process for the pasteurization of crabmeat in cans and flexible pouches, *Journal of Food Science*, **49**, 1003–1004, 1017.

Wehrle, J. C. (1980) *The effect of headspace on the rate of heat penetration in coduction-heated canned foods.* M.S. Thesis, University of California at Davis.

Yang, H. H. and Brier, J. C. (1980) Extraction of sugar from beets, *AIChE Journal*, **4**, 453–459.

Yang, B. B., Nunes, R. V. and Swartzel, K. R. (1992) Lethality distribution in the holding section of an aseptic processing system. *Journal of Food Science*, **57**, 1258–1265.

Yazicioğlu, T. (1974) *Ankara Bira Fabrikasinda Yapilan Viski İmal Denemeleri ve Elde Olunan Viskiler Üzerinde Araştirmalar* (Tukish) Publication of the Faculty of Agriculture of the University of Ankara, Publication Number 497.

CHAPTER 5

Statistical Process Analysis and Quality Control

5.1. STATISTICAL QUALITY CONTROL

Application of statistical techniques to maintain the desired quality level during all stages of production or storage is called *statistical quality control.* Quality of the commodities may be evaluated after comparing the measurements or attributes of the quality factor with the desired standard. If their difference is greater than the *tolerable limits* than the process inputs are re-adjusted to decrease or eliminate the difference (Sidebottom, 1986). This is called the *closed loop concept* (Fig. 5.1).

Usually a process is divided into stages and the principles of statistical quality control applied to the individual stages separately (Fig. 5.2). The raw materials, intermediary or final products are not permitted to pass to the next stage if they fail to satisfy the pre-specified quality standards. The *stagewise quality control* scheme helps to maintain the desired quality level at all levels of the process, and makes it possible to react rapidly if anything is going wrong. Corrective actions, including re-processing, may be applied on the rejected items. If it is not possible to correct the cause of the rejection these commodities may be discarded.

Example 5.1. Stagewise quality control in winemaking (*Peterson,* 1974) The common stagewise quality control scheme practiced in California wineries may be summarized as follows:

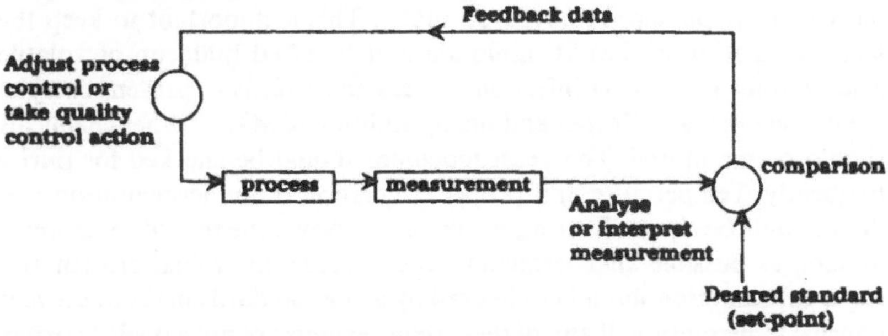

FIGURE 5.1 An outline of the control loop approach. Reprinted from *Food Processing*, volume 55(3), Sidebottom, B. Quality control by design, pages, 39–42, © 1986, with kind permission from Techpress Publishing Co. Ltd.

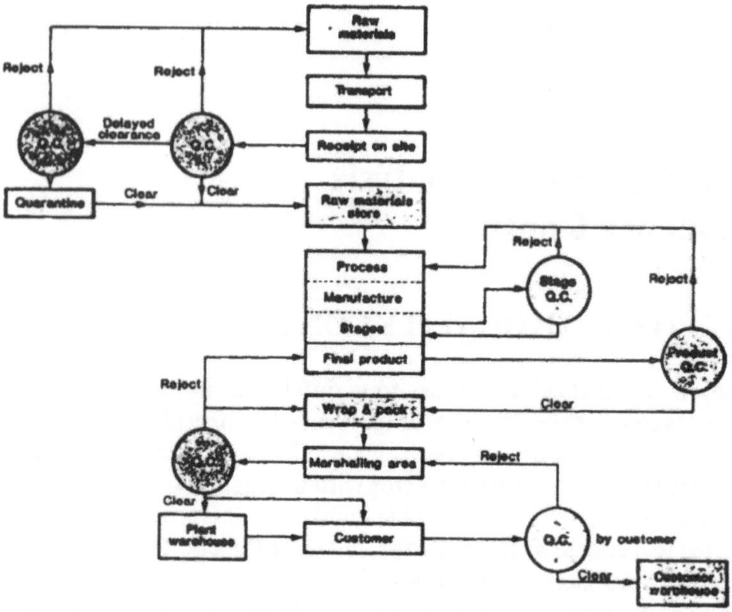

FIGURE 5.2 A typical quality control flow diagram, showing how the system is incorporated into the plant operations. Reprinted from *Food Processing*, volume 55(3), Sidebottom, B. Quality control by design, pages, 39–42, © 1986, with kind permission from Techpress Publishing Co. Ltd.

i) *Receiving the grapes* Maturity, sugar level, mildew, rot, foreign matter, spray residue, juice color, and firmness of the berries are checked. The variety is confirmed. The grapes must be picked early in the morning and rushed to the winery as soon as possible. Temperature of the grapes is an indicator of the time the grapes are picked. If these requirements are not satisfied the grapes may be rejected and not used for winemaking.

ii) *Crushing and fermentation* Check the crushing equipment is cleaned daily and the pomace removed promptly. This is important to keep the local population of flies at minimum and to avoid build up of volatile acidity from *Acetobacter* infections. Check the screens to prevent entry of the flies are in place. Proper and timely addition of SO_2 or other chemicals should be monitored. The yeast inoculum should be checked for purity frequently. Temperature and proper circulation of the fermentation medium should be checked during fermentation. New wine should be removed as soon as possible after fermentation lees settle. Microbial growth and sugar consumption should be checked by using standard curves to prevent unusual fermentation. If any of these requirements are not satisfied corrective actions may be taken to prevent quality loss.

iii) *Processing and aging* Wine tanks should be checked regularly for unusual oxygen entry or heating. With oxygen in the head space and unusually low SO_2 levels, acetone-like flavors can develop rapidly. Visual and taste inspection of the wines are carried out every two months. These analyses are done every two months with the wine in the barrels. Malic acid and lactic acid measurements are made once a week until the wine is bottled. Volatile acidity, SO_2, alcohol, pH and total acidity are also monitored. Organoleptic analysis is done to obtain the right blends. Sanitary condition and temperature of the equipment, and the volume and composition of the head space of the storage tanks are also monitored. If any requirements are not satisfied corrective actions may be taken to prevent quality loss.

iv) *Bottling* The right packaging material should be used. The bottling room and the bottles should be clean. Input and output of the filters are continuously sampled and microbiological analysis are made.

v.) *Warehousing and shipping* Precautions are taken not to confuse different wines. Temperature and humidity of the warehouse is monitored. Cork finish bottles are stored with cork pointed down to prevent drying of the cork. Wine should be handled delicately by the fork lift operators to prevent physical damaging. Stock rotation and chronology of bottling date should be carefully controlled. The loading practices are controlled, and the follow-up information to the distributors warehouse personnel is carefully analyzed.

Total quality management of a food plant is achieved in three levels: Quality control by line employees, assurance by mid-management to be certain that the control job is done effectively, and improvement of safer and more effective new procedures by top management. Procedures of quality control and product safety are integrated with the production process. The statistical techniques explained in Chapter 5.2–5.7 are usually employed by the line employees.

ISO 9000 is the world wide accepted quality system standard for all industries. Certification to ISO 9000 standards may be regarded as evidence that a food company is taking the required precautions for the production of safe food. An ISO 9000 certification shows that the company has an internationally recognized standard in managing its quality program, therefore many multi-national food producers, large distribution chains and government agencies insist that their suppliers have ISO 9000 registration. Regulatory agencies may also require an approved Quality management system for product certification (Thornton, 1992).

ISO 9000 gives a general description of the quality system standards and helps to select the appropriate part of the standards for the company seeking registration. ISO 9000 series has five elements: ISO 9001 is the most comprehensive. It covers the quality system applicable to companies whose activities include the design and development, production, installation, and servicing of products when the requirements of how those products must perform are specified by the customer and then provided by the supplier.

These specifications may be applied to some sectors of the food industry, i.e., flavor production. ISO 9002 sets out the quality system requirements for a company which is manufacturing goods to specifications which they publish, as part of their technical information and data service, or manufacturing goods to a customer's specifications but no significant element of the design is included in meeting these specifications. An example might be a company supplying glucose syrups, sold to freely available specifications, but may also supply glucose syrups which conform the customer's specific requirements, i.e., pH, color, etc. ISO 9003 is a standard which specifies a quality system which only applies to final inspection and test procedures for finished goods. Having registration for ISO 9003 may be sufficient for a company which imports and distributes foods. ISO 9004 provides extensive quality management guidance, assists the supplier organization to develop and implement quality systems and gives guidelines which help determine the extent to which each quality system element is applicable to the company concerned (Thornton, 1992).

5.2. STATISTICAL PROCESS ANALYSIS

Quality control may be done via measuring the quality factors, or on the basis of more practical judgments not involving measurements. Three of the major statistical distribution models commonly used in Food and Bioprocess Engineering analysis are depicted in Table 5.1. A typical histogram was used in Figure E.4.9 to visualize variation of the size distribution of ice crystals in meat tissue during storage. When a data set is described with a histogram the details are lost, i.e., crystal size is expressed within a 10 μm range in Figure E.4.9, but no information is available concerning the distribution within these 10 μm intervals. In an infinite data set if the variations in the measured quantity x are random, the distribution of the values of x around the population mean μ may be described by a normal (Gaussian) distribution model ((5.1) in Tab. 5.1). Major properties of the normal distribution curve are depicted in Table 5.2. We may classify by apples according to their size measuring them individually, or we may roll them over a slanted surface with holes, such that the apples smaller than a certain diameter pass through, but the larger ones are retained. In such analysis, if the apples smaller than the hole size are detrimental, they may be classified as non-conforming units (failure) while the ones retained on the screen are called the conforming ones (success). In some cases visual observations may be more practical and less costly than measuring the quality aspects. We may use binomial or poisson distribution models when we determine the quality attributes on the basis of the counts or fractions of the conforming or non-conforming units (Tab. 5.1). In order to be able to use the binomial distribution model, analysis should be done with a constant sample size n, we must have two possible outcomes of an observation i.e., conforming or non-conforming. There should be a constant probability of each outcome. If we know that a process produces 0.1% defective chocolate bars, this constant probability level is $p = 0.001$. If $p = 0.5$ the binomial distribution curve is symmetrical, for other values of p it is skew. For a large number of observations, i.e., $np \geqslant 10$, the binomial distribution curve approaches the normal distribution curve, and the properties of the normal distribution may be applied to the binomial distribution. Binomial distribution is not practical to use with large data sets. Poisson distribution is a good approximation of the Binomial distribution to analyze large data sets with rare defects, especially when $n \geqslant 20$ and $p \leqslant 0.05$ or $n \geqslant 100$ and $np \leqslant 10$. For large values of λ, the poisson distribution curve approaches the normal distribution curve then the properties of the normal distribution may be applied to the Poisson distribution. The distribution models may be used to find out the confidence limits of a statistical parameter (Tab. 5.3) or to test the validity of a hypothesis (Tab. 5.4).

Example 5.2. Cell size distribution in corn extrudates (Barrett and Peleg, 1992) Many distributions may be converted into a normal distribution after appropriate transformation. We may make the transformation $\zeta = \log x$, where x = cell area, $\zeta = \log$ (cell area) with the cell size distribution data of the corn extrudates, then the cell area distribution may be described as:

$$f(\zeta) = \frac{1}{\sigma_\zeta \sqrt{2\pi}} \exp\left\{ -\frac{1}{2}\left(\frac{\zeta - \mu}{\sigma_\zeta}\right)^2 \right\} \tag{E.5.2.1}$$

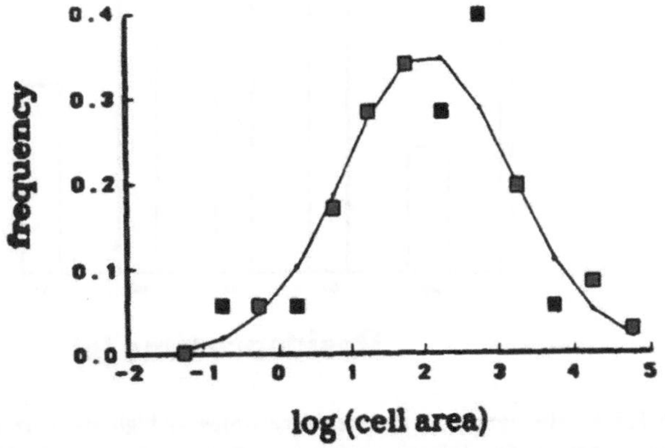

FIGURE E.5.2 A typical cell size distribution of the corn extrudates (Barrett and Peleg, 1992; © IFT, reproduced by permission).

where σ_ζ is the population standard deviation obtained after transformation of the data into new variables. A typical cell size distribution of the corn extrudates is given in Figure E.5.2.

Example 5.3. Residence time distribution in non-Newtonian flow during thermal processing (*Sandeep and Zuritz*, 1994) Residence time distribution of food in a continuous thermal processing equipment is needed for process calculations (Chapter 4.1). Residence times of neutrally bouyant multiple particles were measured in aqueous solutions of carboxymethylcellulose flowing through a holding tube arrangement (length = 13.18 m, internal diameter = 0.0471 m). A typical histogram of the residence times is depicted in Figure E.5.3.

The residence time distribution function $E(t)$ applicable to lethality calculations in aseptic processing, was expressed with normal distribution as:

$$E(t) = \frac{1}{\sqrt{2\pi\sigma^2}} \exp\left[-\frac{(t_R - t_{R_{mean}})^2}{2\sigma^2} \right] \tag{E.5.3.1}$$

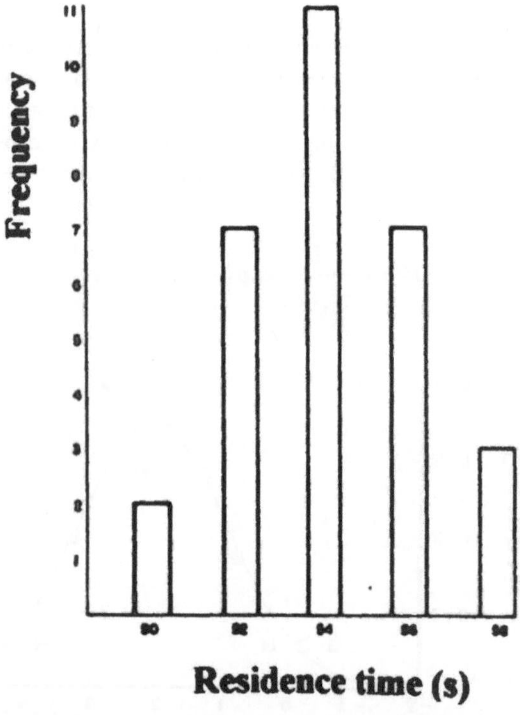

FIGURE E.5.3 Histogram of the residence times at high viscosity (consistency coefficient = 1.4 Pa s^n, flow behavior index n = 0.70), low flow rate (0.44 kg/s), high particle size (diameter = 12 mm) and low particle particle concentration (4% by volume) (Sandeep and Zuritz, 1994; © IFT, reproduced by permission).

where t_R and t_{Rmean} are the residence and mean residence times of the particles in the holding tube, respectively; σ is the standard deviation of the residence times. The following correlations were obtained from the experimental data:

$$t_R = 90.6 - 3.73\,\mu_{app} - 75.1\,Q + 0.363\,c_p \tag{E.5.3.2}$$

$$t_{Rmin} = 86.7 - 4.67\,\mu_{app} - 73.1\,Q + 0.427\,c_p \tag{E.5.3.3}$$

$$s_R = 3.33 - 2.42\,Q - 0.102\,d_p - 0.025\,c_p + 0.759\,\mu_{app}\,Q + 0.101\,Q\,d_p \tag{E.5.3.4}$$

where μ_{app} = apparent viscosity (Pa s^n), Q = flow rate (kg/s), c_p = particle concentration (%), d_p = particle size (mm); t_{Rmin} and s_R are the minimum residence time in the holding tube, and the standard deviation of the residence times, respectively.

Example 5.4. Probability of having more than a limiting weight of contents in a randomly selected can a) What is the probability of having *exactly* six cans with more than 500 g of contents when 8 cans are taken randomly from a population where 80% of the whole cans have more than 500 g of contents?

Solution The quality factor has been reported on the basis of conforming/non conforming units, and the conditions of the problem implies the binomial distribution as $f(x,n,p) = n!/(x!(n-x)!)(p^x(1-p)^{(n-x)})$ ((5.3) in Tab. 5.1) when $n = 8, x = 6, p = 0.80 f(6,8,0.80) = 0.29$. Exactly six cans will have at least 500 g contents with 29% probability.

b) What is the probability of getting at least 6 cans with 500 g contents?

Solution The total probability of getting six, seven or eight cans with 500 g contents is $f(6,8,0.80) + f(7,8,0.80) + f(8,8,0.80) = 0.29 + 0.34 + 0.17 = 80\%$.

Example 5.5. Probability of having a certain fraction of loss in fruit packing A fruit packing company discards, on average, 5% of the incoming produce to confirm to a standard minimum quality level. In a lot of 120 kg, what is the probability of having 10 kg of fruits discarded?

Solution The quality factor has been reported on the basis of conforming/non conforming units; and the defect is a rare defect, therefore Poisson distribution may be used. Problem statement implies that $n = 120, p = 0.05$ and $x = 10$ after substituting the numbers in (5.7) we will get $\lambda = np = 6$. Equation (5.6) requires that probability of having 10 kg defectives is

$$f(x,\lambda) = e^{-\lambda}\frac{\lambda^x}{x!} = 0.042 = 4.2\%$$

A normal distribution curve is characterized by the parameters μ and σ. Parameter μ is the mean around the data distributed; parameter σ determines the width of the distribution curve (Fig. 5.3.a). Distributions of the

Table 5.1 Statistical distribution models commonly used in Food and Bioprocess Engineering analysis

Normal (Gaussian) distribution	Probability (relative frequency) $f(\mu,\sigma)$ of obtaining the measurement x from a population with mean μ and standard deviation σ. $$f(x)=\frac{1}{\sigma\sqrt{2\pi}}\exp\left\{-\frac{1}{2}\left(\frac{x-\mu}{\sigma}\right)^2\right\} \quad (5.1)$$ where $$f=\frac{\text{group frequency}}{\text{total number of the data points}} \quad (5.2)$$
Binomial distribution	Probability $f(x,n,p)$ of getting x failures in n trials from a population where the overall probability of failure is p $$f(x,n,p)=\frac{n!}{x!(n-x)!}p^x(1-p)^{(n-x)} \quad (5.3)$$ The mean and the standard deviation $$\mu=np \quad (5.4)$$ $$\sigma=\sqrt{np(1-p)} \quad (5.5)$$
Poisson distribution	probability $f(x,\lambda)$ of getting x failures in n trials from a population where the overall probability of failure is p $$f(x,\lambda)=e^{-\lambda}\frac{\lambda^x}{x!} \quad (5.6)$$ where $\lambda=np$ (5.7) The mean and the standard deviation: $$\mu=\lambda \quad (5.8)$$ $$\sigma=\sqrt{\lambda} \quad (5.9)$$

variables of most natural phenomena are approximately normal. In a normal distribution curve 68.26, 95.46 and 99.73% of the total area is enclosed within the limits of $\mu\pm\sigma$, $\mu\pm2\sigma$ and $\mu\pm3\sigma$, respectively (Fig. 5.3.b). The fraction of the data set which is smaller than an arbitrarily chosen x value (shaded area in Fig. 5.3.b) may be calculated by dividing the shaded area by the total area under the curve. Parameter z (standard normal variable) defined in (5.33) (Tab. 5.4) is used in these calculations. The normal distribution curve is plotted by using variable z in Figure 5.4. It should be noticed that the curve is symmetrical and $z=0$, i.e., at the origin. Values of p corresponding to z_p for the normal curve are listed in Table 5.5. The normality of a data set may be assessed the skewness or kurtosis tests (Tab. 5.4; Levinson, 1990). The skewness (5.17) measures the symmetry of the data; the kurtosis (5.18) measures the tendency of the data to have long "tail" regions and a high center. When a population is formed by additive contributions from different sources, i.e., weight of a packaged food = weight of the package material + weight of the food; protein content of a food = sum of the protein contents of each individual making up the

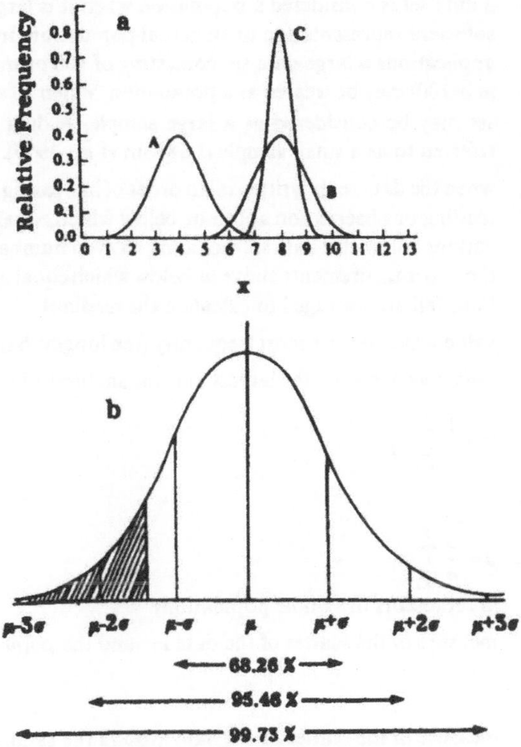

FIGURE 5.3 (a) Normal distribution curves with A: $\mu = 4$, $\sigma = 1$; B: $\mu = 8$, $\sigma = 1$; C: $\mu = 8$, $\sigma = 0.5$. (b) Fractions of the area enclosed between different ranges of the normal distribution curve.

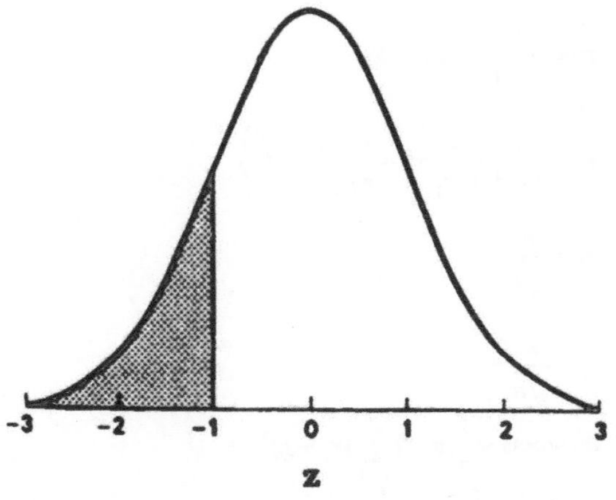

FIGURE 5.4 The z transform of the normal distribution curve. Fraction of the shaded area corresponds to the probability of obtaining a standardized score of less than -1.

Table 5.2 Common statistical terms used to describe a normally distributed data set

size of the data sets	a data set is considered a population when it is large enough for sufficient representation of an actual population. In practical applications a large data set consisting of 100 or more measurements ($n \geqslant 100$) may be treated as a population. When $30 \leqslant n \leqslant 100$ the data set may be considered as a large sample. A data set with $n \leqslant 30$ is referred to as a small sample (LaMont *et al.*, 1977).
median	when the data set is written in an order of increasing measurements the reading or observation above or below which equal number of observations fall (with a data set including an even number of measurements the two measurements above or below which equal number of observations fall are averaged to calculate the median)
mode	value which occurs most frequently (the longest bar in the histogram)
range	difference between the largest and the smallest observations
sample mean (called average in daily language)	$$\bar{x} = \frac{1}{n}\sum_{i=1}^{n} x_i \qquad (5.10)$$
population mean	$$\mu = \frac{1}{n}\sum_{i=1}^{n} x_i \qquad (5.11)$$ (n represents the whole population)
population standard deviation (σ)	measure of the scatter of the data around the population mean
sample standard deviation (s)	measure of the scatter of the data around the sample mean
population variance (σ^2)	$$\sigma^2 = \frac{1}{n}\sum_{i=1}^{n} (x_i - \mu)^2 \qquad (5.12)$$
sample variance (s^2)	$$s^2 = \frac{1}{n-1}\sum_{i=1}^{n} (x_i - \bar{x})^2 \qquad (5.13)$$ or $$s^2 = \frac{1}{n-1}\left\{\sum_{i=1}^{n} x_i^2 - \frac{1}{n}\left[\sum_{i=1}^{n} x_i\right]^2\right\} \qquad (5.14)$$
population variance of the sample means	$$\sigma_{\bar{x}} = \frac{\sigma}{\sqrt{n}} \qquad (5.15)$$
variance of the sample means	$$s_{\bar{x}} = \frac{s}{\sqrt{n}} \qquad (5.16)$$
skewness	$$\text{skewness} = \frac{1}{n}(\sigma^{-3})\sum_{i=1}^{n} (x_i - \mu)^3 \qquad (5.17)$$
kurtosis	$$\text{kurtosis} = \frac{1}{n}(\sigma^{-4})\sum_{i=1}^{n} (x_i - \mu)^4 \qquad (5.18)$$

Table 5.2 (Continued)

means of the added properties	$\mu = \sum_{i=1}^{k} \mu_i$	(5.19)
	where μ_i = population mean of the added property i	
	$\mu = \sum_{i=1}^{k} \omega_i \mu_i$	(5.20)
	where ω_i = weight or the arithmetic fraction	
	$\omega_i = \dfrac{n_i}{\sum_{j=1}^{k} n_j}$	(5.21)
variance of the added properties	$\sigma^2 = \sum_{i=1}^{k} \sigma_i^2$	(5.22)
	where σ_i = populaiton standard deviation of the added property i	
	$\sigma^2 = \sum_{i=1}^{k} \omega_i \sigma_i^2$	(5.23)

food, etc., the composite nature of the population should be considered in the statistical analysis. With such a population the mean and the variance of the added properties are calculated with (5.19)–(5.23) (Tab. 5.2). It should be noticed that the difference of the population means and the variances are calculated as: $\mu_2 = \mu - \mu_1$ and $\sigma_2^2 = \sigma^2 + \sigma_1^2$.

Example 5.6. Fraction of the total production falling between given limits A fruit juice is produced with sugar content of $\mu = 8\%$ and $\sigma = 0.5\%$. What percentage of the products contain between 6.7 and 9.15% sugar?

Solution After using (5.33) we will determine $z_1 = 9.15 - 8/0.5 = 2.3$, $p_{2.3} = 0.9893$ (Tab. 5.5) and $z_2 = 6.7 - 8/0.5 = -2.6$. The probability corresponding to $z = -2.6$ is not given in Table 5.5. The z distribution is symmetrical (Figure 5.4). The fraction of the area in z distribution curve between $z = -\infty$ and $z = -2.6$ is the same as the area between $z = 2.6$ and $z = \infty$, therefore $p_{-2.6} = 1 - p_{2.6}$ or after substituting the numbers $p_{-2.6} = 1 - 0.9953 = 0.0047$, indicating that 0.47% of the whole production contains less than 6.7% sugar. The required interval of probability is $p(-2.6 \leqslant z \leqslant 2.3) = 0.9893 - 0.0047 = 0.9846$, indicating that 98.46% of the total prodduct, contains 6.7% to 9.15% sugar.

Example 5.7. The maximum tolerable standard deviation a) Concentration of a certain food additive is required to be $0.2 \pm 0.02\%$. If the population mean of the additive concentration in the food is $\mu = 0.2$, what is the standard deviation?

Solution Additive contents of the food should be $\mu - 3\sigma \leqslant x \leqslant \mu + 3\sigma$ ($p = 0.9973$) (Fig. 5.3), since $\mu = 0.2\%$, the tolerance limits should correspond to $6\sigma = (2)(0.02\%)$, therefore $\sigma = 0.007\%$

b) If the additive is added with $\mu = 0.19\%$ and $\sigma = 0.021\%$, what percentage of the food satisfies the required tolerance limits?

Solution Allowed limits of the additive concentration:

$$x_{lower} = (0.2) - (0.02) = 0.18\% \quad \text{and} \quad x_{upper} = (0.2) + (0.02) = 0.22\%.$$

The z values corresponding to x_{lower} and x_{upper} are:

$$z_{lower} = \frac{x_{lower} - \mu}{\sigma} = \frac{0.18 - 0.19}{0.021} = -0.48 \quad \text{and} \quad z_{upper} = \frac{x_{upper} - \mu}{\sigma}$$

$$= \frac{0.22 - 0.19}{0.021} = 1.43$$

Fraction of the product with additive less than 0.18% is: $p_{-0.48} = 1 - p_{0.48} = 1 - 0.6844 = 31.56\%$. Fraction of the product with additive less than 0.22% is: $p_{1.43} = 92.36\%$. Amounts of product with allowed amount of additive $= (92.36 - 31.56)\% = 60.80\%$.

c) If only 3% of the product is permitted to contain less than 0.2% additive, $\sigma = 0.02\%$ and no upper limit is specified, what should be the population mean additive concentration?

Solution $p = 0.03, x = 0.2\%$, (5.33) requires $z_{0.03} = (0.2 - \mu)/0.02$. Value of $z_{0.03}$ is not given in Table 5.5, therefore we should read $z_{0.97} = 0.8340$, then calculate $z_{0.03} = -z_{0.97} = -0.8340$. After substituting value of $z_{0.03}$ in (5.33) we will calculate $\mu = 0.217\%$.

Example 5.8. Probability of having a sample mean smaller than a given value A margarine packaging machine was adjusted to pack 125 g of margarine ($\mu = 125$ g) with $\sigma = 5$ g. What is the probability of having less than 122 g of a sample mean with a sample of 25 packages?

Solution Population standard deviation is available, therefore we will use the z values in the test. The standard normal variable was defined as:

$$z = \frac{x - \mu}{\sigma} \tag{5.33}$$

Since the distribution of the sample means around the population mean follows normal behavior, the same expression may be stated for the distribution of the sample means as: $z = \bar{x} - \mu/\sigma_{\bar{x}}$, after substituting the numerical values $z = (122 - 125)/(5/\sqrt{25}) = -3$. Probability corresponding $z = -3$ is $p_{-3} = 1 - p_{+3} = 1 - 0.9987 = 0.0013$ (Tab. 5.5). Therefore the probability of having less than 122 g of a sample mean with a sample of 25 packages is 0.13%.

Example 5.9. Use of the standard normal variable due to the similarity of the normal, Poisson and Gaussian distribution curves If 5% of a packaged product is produced underweight, what is the probability that a random sample of 100 packages will contain 7 or more underweight packages?

Solution It is given in the problem statement that $p=0.05$, $n=100$ and $\lambda=\mu=np=5$. Since $\lambda<10$, $n\geqslant 100$ and $p=$ constant, we may use the poisson distribution with $\sigma=\sqrt{\lambda}=\sqrt{5}$. Similarity of Poisson and Gaussian distribution curves allows us to use the standard normal variable $z=(x-\mu)/\sigma=(7-5)/\sqrt{5}=0.8$

Probability of having 7 or less underweight packages is 81.3% (Tab. 5.5) implying that the probability of having 7 or more underweight packages is 100–81.3 = 18.7%.

Example 5.10. Statistical properties of broiler feeds and their raw materials (Alti and Özilgen, 1994) Statistical properties related to distribution of protein in raw materials and three different types of broiler feeds (starter, grower, finisher) are:

raw material or feed	μ (%)	σ (%)	$3-\dfrac{14.7}{\sqrt{N}}$	kurtosis	$3\pm\dfrac{14.7}{\sqrt{N}}$	skewness	$\pm\dfrac{7.35}{\sqrt{N}}$
rendering meal	55.0	1.7	−2.08	2.19	6.21	−0.05	±1.60
meat and bone meal	39.4	1.2	−0.68	2.59	6.68	−0.52	±1.84
sorghum	9.3	0.4	−0.29	2.72	6.29	−0.21	±1.64
soybean meal	45.7	1.7	1.31	3.20	4.69	−0.75	±0.84
tapioca	2.7	0.3	−0.13	3.83	6.13	0.71	±1.57
corn	7.5	0.6	1.05	4.63	4.95	0.04	±0.98
wheat	12.6	0.4	0.22	3.53	5.78	0.93	±1.39
starter	23.6	0.8	1.19	3.87	4.89	−0.56	±0.94
grower	21.6	0.9	1.49	4.28	4.51	−0.24	±0.75
finisher	20.4	0.9	1.60	5.12	4.40	−0.94	±0.70

where $N=$ total number of the measurements in a data set. These data shows that proteins are distributed normally in the raw materials (rendering meal, meat and bone meal, sorghum, soybean meal, tapioca, corn, wheat) and the feeds (starter, grower and finisher) with one exception: According to the kurtosis and skewness tests ((5.40) and (5.41) in Tab. 5.4) proteins were not distributed normally in the finisher.

Example 5.11. Estimatation of the population mean and variance from those of the additive contributing factors Population mean and standard deviation of the contents of fruit juice containers is 500 g and 10 g, respectively. There are 24 containers packed in a box. An empty container weighs 30 g with standard deviation of 2 g. The mean weight of an empty box is 200 g with standard deviation of 20 g. Find the limits of the weight of a single box filled with juice.

Solution Equations (5.19) and (5.22) requires

$$\mu = \sum_{i=1}^{k} \mu_i = (24)(500) + (24)(30) + 200 = 12920\,\mathrm{g}$$

$$\sigma^2 = \sum_{i=1}^{k} \sigma_i^2 = (24)(10)^2 + (24)(2)^2 + (20)^2 = 2896\,\mathrm{g},\ \sigma = 53\,\mathrm{g}$$

Limits of x were given in Figure 5.3 as:

$$\mu - 3\sigma \leqslant x \leqslant \mu + 3\sigma$$

After substituting the numbers we will obtain $12761 \leqslant x \leqslant 13079$ implying that with $p = 99.73\%$ the weight of the single box filled with juice will be between these limits.

Example 5.12. Estimate of the population mean and variance from food formulation, and those of the contributing factors Recipe of a certain food includes three protein-containing ingredients. Fraction of the ingredients in the recipe with their mean and standard deviation of protein contents are:

ingredient	ω_i	μ_i (%)	σ_i^2 (%)
1	0.10	30	6
2	0.15	7	1
3	0.05	6	0.5
others	0.70	0	0

Determine the confidence limits of the protein contents of the food.

Solution Equations (5.20) and (5.23) require

$$\mu = \sum_{i=1}^{k} \omega_i \mu_i = (0.10)(30) + (0.15)(7) + (0.05)(6) + (0.70)(0) = 4.35\%$$

$$\sigma^2 = \sum_{i=1}^{k} = \omega_i \sigma_i^2 = (0.10)(6) + (0.15)(1) + (0.05)(0.5) + (0.70)(0) = 0.775$$

Limits of x were given in Figure 5.3 as:

$\mu - 3\sigma \leqslant x \leqslant \mu + 3\sigma$, or $1.71(\%) \leqslant x \leqslant 6.99(\%)$ implying that with $p = 99.73\%$ the protein content of a serving will be between these limits.

Example 5.13. Additive property of population means and variances A vegetable with 85% population mean and 2% variance water contents is dried in three successive dryers. The following percentages of the initial moisture content were removed in each dryer:

drier	μ_i (%)	σ_i^2 (%)
1	35	2
2	21	1
3	12	0.5

Calculate the confidence limits of the final moisture content of the product.

Solution $\mu = \mu_{\text{initial}} - (\mu_1 + \mu_2 + \mu_3) = 85 - (35 + 21 + 12) = 17$

$$\sigma^2 = \sigma^2_{\text{initial}} + \sigma_1^2 + \sigma_2^2 + \sigma_3^2 = 2 + 2 + 1 + 0.5 = 5.5, \ \sigma = 2.34$$

It should be noticed here that although the population means are subtracted, the variances are added. The limits of the moisture of the product are calculated according to Figure 5.3 as:

$$\mu - 3\sigma \leqslant x \leqslant \mu + 3\sigma, \text{ or } 9.88 \leqslant x \leqslant 24.08(\%)(p = 99.73\%)$$

Production of a single commodity usually consists of numerous stages with many control loops involving multiple measurements. In most processes intermediate or finished products pass through these stages at very high rates. When quality assurance is based on the individual items, due to the complex nature of this system, extremely large data sets are formed. These large data sets are usually impossible to analyze totally, therefore sampling is made. Each sample is made of different members, therefore the sample means show a range of variation depending on the properties of the original population. The *Central Limit Theorem* states that *regardless of the distribution behavior of the individual values within their own population, distribution of the sample means will approach the normal distribution as the sample size increases.* In most cases the sample size, n, required to achieve an acceptable near-normal distribution of the sample mean around the population mean is four or five (Jacobs, 1990). The variance of the sample means decreases with the square root of the sample size ((5.15) in Tab. 5.2). In a typical normal distribution curve of the sample means the area confined between $\mu \pm i\sigma\bar{x}$ or $\mu \pm is\bar{x}$ $(i = 1, 2, 3)$ corresponds to 68.26%, 95.46% and 99.73% of the total area under the curve when i is 1, 2 and 3, respectively; implying that if a very large number of samples is drawn from the population, we may expect 68.26% of the sample means to fall between $\mu \pm \sigma\bar{x}$ or $\mu \pm s\bar{x}$, etc. (Fig. 5.5).

When the number of items in a data set (n) is less than about 30 and values of μ and σ are not known we may drive information from the sample to discuss the properties of the population by using t-distribution. Definition of parameter t is given with (5.27) in Table 5.3. The shape of the t-curves depend on the degrees of freedom of the data set (Fig. 5.6). Probability p of having population mean μ within the limits of $-t$ and $+t$ is listed in Table 5.6.

When the actual value of a statistical parameter is not known, we may use statistical techniques (Tab. 5.3) and estimate the interval where the exact value may fall with a given probability level. The limits of this interval are called the *confidence limits.* When we do not know the population mean and the population variance we may estimate the confidence limits of the population mean with (5.26) and the confidence limits of the population variance with (5.30). The possibility of having actual value of the statistical parameter beyond these limits is only $1 - p$.

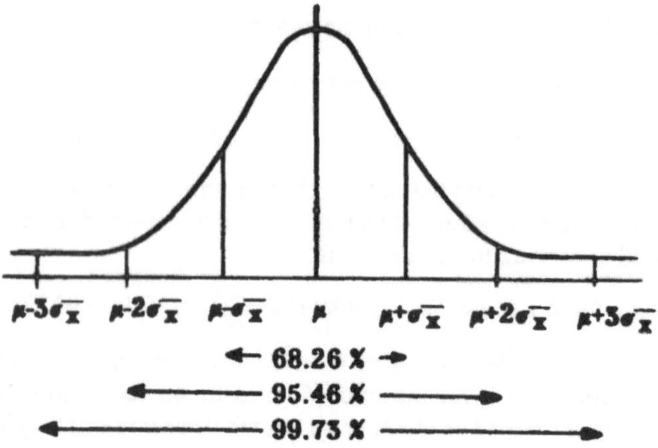

FIGURE 5.5 **Distribution curve of the sample means around the population mean.**

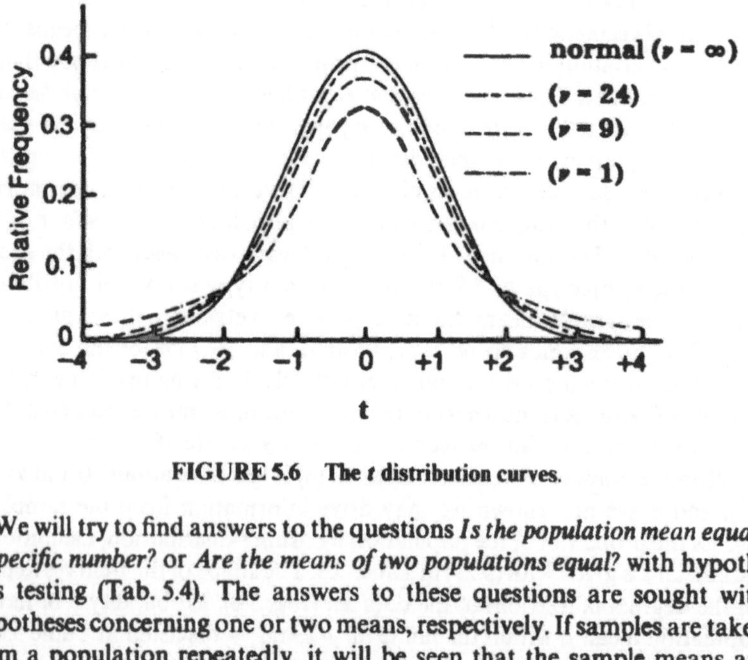

FIGURE 5.6 **The *t* distribution curves.**

We will try to find answers to the questions *Is the population mean equals a specific number?* or *Are the means of two populations equal?* with hypothesis testing (Tab. 5.4). The answers to these questions are sought with hypotheses concerning one or two means, respectively. If samples are taken from a population repeatedly, it will be seen that the sample means are scattered around the population mean over a range determined by the variance of the data. With hypothesis testing concerning one mean, we are actually trying to see if the mean of the sample with unknown origin falls into this range. If the answer to this question is yes, the sample may be a member of the suggested distribution curve. In Figure 5.7 the acceptance region is the range that the samples taken from a population with mean μ and variance σ fall with probability p. The possibility of having a sample

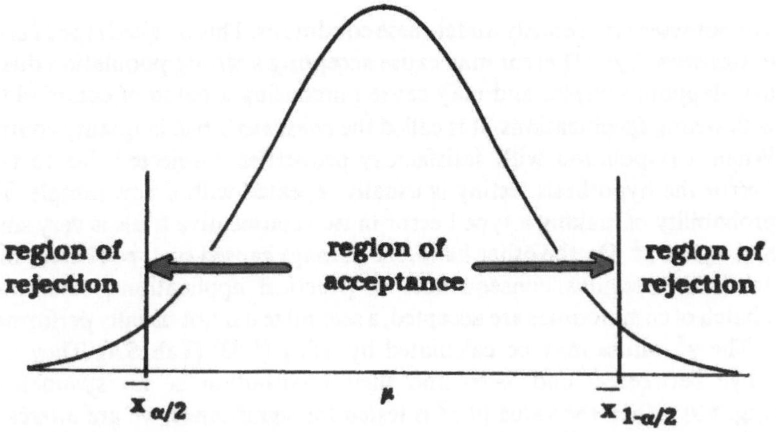

FIGURE 5.7 Schematic description of the acceptance and rejection regions of a normal distribution curve.

from that population with mean outside the given limits is $1 - p = \alpha$. Since the distribution curve is symmetrical, having a sample from this population with mean in either rejection region is only $\alpha/2$. A sample may actually belong to the population, but its mean may fall into the rejection range with probability α (Fig. 5.7). We may erroneously conclude from these data that the sample is not a member of the population described with μ and σ. This is called *type I error* in statistics. In quality control making a type *I* error means rejecting a batch of commodities because of information based on a bad sample. Rejection of a good population with such an unlucky sample usually hurts the producers and it is called the *producer's risk*. The way that we make our judgements in hypothesis testing is weak in the sense that it is based on the criterion whether a sample mean falls into a pre-determined range or not. It is possible that a sample may actually belong to another population but the two populations may overlap (Fig. 5.8). We may accept

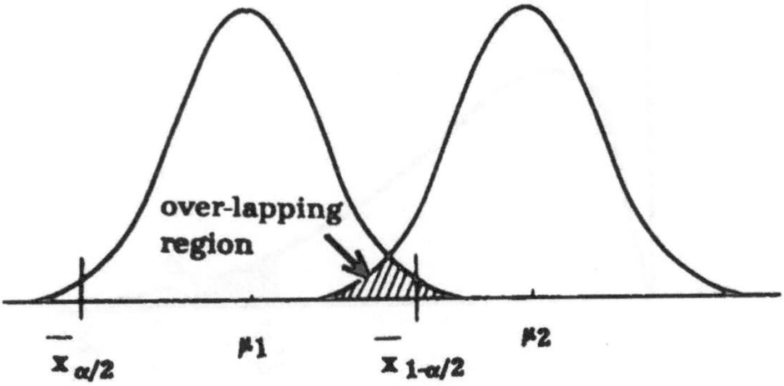

FIGURE 5.8 Over-lapping population distribution curves with means μ_1 and μ_2.

a hypotheses erroneously under these conditions. This is called *type II* error in statistics. Type II error may cause accepting a wrong population due to over-lapping samples and may cause purchasing a batch of commodities with wrong specifications. It is called the *consumer's risk* in quality control. When a population with satisfactory properties is rejected due to type I error the hypothesis testing is usually repeated with a new sample. The probability of making a type I error in two consecutive trials is very small and equals α^2. On the other hand, the damage caused by type II error may have more serious consequences in practical applications, since once a batch of commodities are accepted, a second test is not usually performed.

The χ^2 values may be calculated by using (5.43) (Tab. 5.4). They may range between 0 and $+\infty$ and their distribution is not symmetrical (Fig. 5.9). When the value of χ^2 is tested for significance, we are interested only in the probability of exceeding the observed value as a result of random errors. Standard χ^2 tables (Tab. 5.7) are provided to give χ^2 values reached or exceeded with given probability. The χ^2 values may also be used to determine the confidence limits of σ^2 (Tab. 5.3).

The F values may be calculated with (5.45) or (5.46) (Tab. 5.4). Distribution of the F values is not symmetrical (Fig. 5.10). Standard F tables (Tabs. 5.8 and 5.9) are provided to give F values reached or exceeded with given probability. The F values may be used to test if the variances of two populations are the same (Tab. 5.4) or for analysis of variance.

Example 5.14. Dependence of the confidence limits on the sample size a) The sample mean salt content of a sodium free diet food was detected as 0.10 g/serving after making 25 replicate measurements. There is 0.05 g/serving of population standard deviation due to variations in pro-

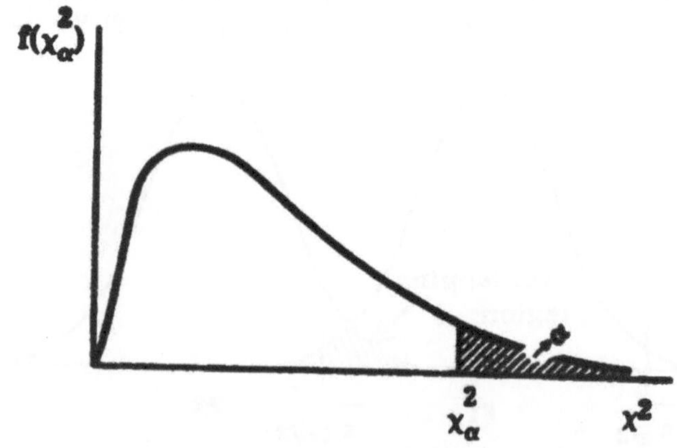

FIGURE 5.9 A χ^2 distribution curve (α = probability of having χ^2 in the shaded area).

Table 5.3 Confidence limits

Confidence limits of μ with normal distribution (σ *is* available, $p=0.9973$)	$\bar{x}-3\,\sigma_{\bar{x}}\leqslant\mu\leqslant\bar{x}+3\,\sigma_{\bar{x}}$	(5.24)
Confidence limits of μ with normal distribution (σ is available, $p=0.9973$)	$\bar{x}-3\dfrac{\sigma}{\sqrt{n}}\leqslant\mu\leqslant\bar{x}+3\dfrac{\sigma}{\sqrt{n}}$	(5.25)
Confidence limits of μ with normal distribution (*variable* p, σ is not known)	$\bar{x}-t\dfrac{s}{\sqrt{n}}\leqslant\mu\leqslant\bar{x}+t\dfrac{s}{\sqrt{n}}$	(5.26)
	where $t=\dfrac{\bar{x}-\mu}{s/\sqrt{n}}$	(5.27)
Confidence limits with binomial distribution ($p=0.9973$)	$np-3\sqrt{np(1-p)}\leqslant(np)_{\exp}\leqslant np+\sqrt{np(1-p)}$ ($p=0.9973$)	(5.28)
Confidence limits with Poisson distribution ($p=0.9973$)	$np-3\sqrt{\lambda}\leqslant(np)_{\exp}\leqslant np+3\sqrt{\lambda}$	(5.29)
Confidence limits of σ^2 with $p=\gamma-\beta$	$\dfrac{s^2v}{x_\beta^2}\leqslant\sigma^2\leqslant\dfrac{s^2v}{x_\gamma^2}$	(5.30)
	where $x^2=\dfrac{(n-1)s^2}{\sigma^2}$	(5.31)

Table 5.4 Statistical tests

Sample mean $\bar{x}_A$ is obtained from a normally distributed population. Check if μ_A = a specific value? Population standard deviation σ_A is known	$\mu-z_{1-\alpha/2}\dfrac{\sigma}{\sqrt{n}}\leqslant\bar{x}\leqslant\mu+z_{1-\alpha/2}\dfrac{\sigma}{\sqrt{n}}$	(5.32)
	where $z=\dfrac{x-\mu}{\sigma}$	(5.33)
Sample mean $\bar{x}_A$ is obtained from a normally distributed population. Check if μ_A = a specific value? Population standard deviation σ_A is not known	$\mu-t_{1-\alpha/2}\dfrac{s}{\sqrt{n}}\leqslant\bar{x}\leqslant\mu+t_{1-\alpha/2}\dfrac{s}{\sqrt{n}}$	(5.34)

Table 5.4 (Continued)

sample means $\bar{x}_A$ and $\bar{x}_B$ are obtained from normally distributed population. Check if $\mu_A=\mu_B$? Values of σ_A and σ_B are available	$$-z_{1-\alpha/2}\left(\frac{\sigma_A^2}{n_A}+\frac{\sigma_B^2}{n_B}\right)^{1/2} \leqslant \bar{x}_A-\bar{x}_B \leqslant z_{1-\alpha/2}\left(\frac{\sigma_A^2}{n_A}+\frac{\sigma_B^2}{n_B}\right)^{1/2} \quad (5.35)$$ where $$z=\frac{\bar{x}_A-\bar{x}_B}{\left(\frac{\sigma_A^2}{n_A}+\frac{\sigma_B^2}{n_B}\right)^{1/2}} \quad (5.36)$$
Sample means $\bar{x}_A$ and x_B are are obtained from normally distributed populations Check if $\mu_A=\mu_B$? It is assumed that $\sigma_A=\sigma_B$, but their values are not available	$$-t_{1-\alpha/2}s_p\left(\frac{1}{n_A}+\frac{1}{n_B}\right)^{1/2} \leqslant \bar{x}_A-\bar{x}_B \leqslant t_{1-\alpha/2}s_p\left(\frac{1}{n_A}+\frac{1}{n_B}\right)^{1/2} \quad (5.37)$$ where $s_p^2=v_A s_A^2+v_B s_B^2$ (5.38) s_p = pooled variance (degrees of freedom $v=v_A+v_B$) s_A = sample variance of the data set A (degrees of freedom of v_A) s_B = sample variance of the data set A (degrees of freedom of v_B) $$t=\frac{\bar{x}_A-\bar{x}_B}{s_p\left(\frac{1}{n_A}+\frac{1}{n_B}\right)^{1/2}} \quad (5.39)$$
Is the data set distributed normally	$$-\frac{7.35}{\sqrt{N}} \leqslant \text{skewness} \leqslant +\frac{7.35}{\sqrt{N}} \quad (5.40)$$ $$3-\frac{14.7}{\sqrt{N}} \leqslant \text{skewness} \leqslant 3+\frac{14.7}{\sqrt{N}} \quad (5.41)$$ where N = total number of the data poitns in the set
Check if the assumed distribution model is correct	$$\chi^2_{\text{observed}} \leqslant \chi^2_{\text{Table}} \quad (5.42)$$ where $$\chi^2_{\text{observed}}=\sum_{i=1}^{n}\left(\frac{f_0-f_e}{f_e}\right)^2 \quad (5.43)$$ where f_0 = observed frequencies f_e = frequencies estimated with a distribution model

Table 5.4 (Continued)

Check if $\sigma_A = \sigma_B$?	$\frac{1}{F_{\alpha/2}(\nu_A,\nu_B)} \leqslant (s_A/s_B)^2 \leqslant F_{\alpha/2}(\nu_A, \nu_B)$	(5.44)
	where $F(\nu_A, \nu_B) = \frac{s_A^2/\sigma_A^2}{s_B^2/\sigma_B^2}$	(5.45)
	when $\sigma_A = \sigma_B$, then $F(\nu_A, \nu_B) = s_A^2/s_B^2$	(5.46)
	important property $F_{1-\gamma}(\nu_A, \nu_B) = \frac{1}{F_\gamma(\nu_A, \nu_B)}$	(5.47)

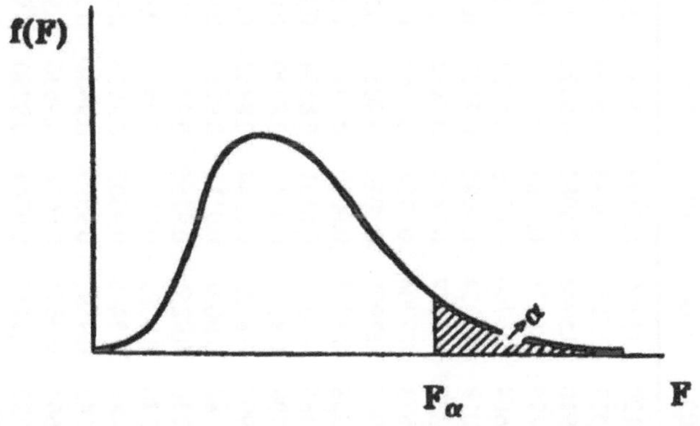

FIGURE 5.10 A typical F distribution curve (α = probability of having F in the shaded area).

duction. What are the confidence limits of the population mean salt content of the food based on this sample?

Solution The sample standard deviation according to (5.15) (Tab. 5.2) is $\sigma_{\bar{x}} = \sigma/\sqrt{n} = 0.05/\sqrt{25} = 0.01$. Since the population standard deviation is known when $p = 0.9973$ the confidence limits are calculated as $\bar{x} - 3\sigma_{\bar{x}} \leqslant \mu \leqslant \bar{x} + 3\sigma_{\bar{x}}$ ((5.25) in Tab. 5.3), or after substituting the numbers $0.07 \leqslant \mu \leqslant 0.13$ g.

b) What are the confidence limits, if the same result had been obtained with sample size of $n = 10$?

Solution The sample size $n = 10$, therefore $\sigma_{\bar{x}} = 0.016$; then the confidence limits become $0.05 \leqslant \mu \leqslant 0.15$ g. It should be noticed that decreasing the sample size n increases the confidence interval of μ.

c) What is the sample size, if the difference between the confidence limits of μ is required to be less than 0.04 g/serving?

Solution The difference between the confidence limits of μ is $6\sigma_{\bar{x}}$, therefore $\sigma_{\bar{x}} = (0.04/6) = 0.0067$. After substituting the values of σ and $\sigma_{\bar{x}}$ in (5.15) we will obtain $n = 56$.

Table 5.5 The normal distribution function $F(z) = \frac{1}{\sqrt{2\pi}} \int_{-\infty}^{z} \exp\left(-\frac{1}{2}\xi^2\right) d\xi$

z_p	0.00	0.01	0.02	0.03	0.04	0.05	0.06	0.07	0.08	0.09
0.0	0.5000	0.5040	0.5080	0.5120	0.5160	0.5199	0.5239	0.5279	0.5319	0.5359
0.1	0.5398	0.5438	0.5478	0.5517	0.5557	0.5596	0.5636	0.5675	0.5714	0.5753
0.2	0.5793	0.5832	0.5871	0.5910	0.5948	0.5987	0.6026	0.6064	0.6103	0.6141
0.3	0.6179	0.6217	0.6255	0.6293	0.6331	0.6368	0.6406	0.6443	0.6480	0.6517
0.4	0.6554	0.6591	0.6628	0.6664	0.6700	0.6736	0.6772	0.6808	0.6844	0.6879
0.5	0.6915	0.6950	0.6985	0.7019	0.7054	0.7088	0.7123	0.7157	0.7190	0.7224
0.6	0.7257	0.7291	0.7324	0.7357	0.7389	0.7422	0.7454	0.7486	0.7517	0.7549
0.7	0.7580	0.7611	0.7642	0.7673	0.7704	0.7734	0.7764	0.7794	0.7823	0.7852
0.8	0.7881	0.7910	0.7939	0.7967	0.7995	0.8023	0.8051	0.8078	0.8106	0.8133
0.9	0.8159	0.8186	0.8212	0.8238	0.8264	0.8289	0.8315	0.8340	0.8365	0.8389
1.0	0.8413	0.8438	0.8461	0.8485	0.8508	0.8531	0.8554	0.8577	0.8599	0.8621
1.1	0.8643	0.8665	0.8686	0.8708	0.8729	0.8749	0.8770	0.8790	0.8810	0.8830
1.2	0.8849	0.8869	0.8888	0.8907	0.8925	0.8944	0.8962	0.8980	0.8997	0.9015
1.3	0.9032	0.9049	0.9066	0.9082	0.9091	0.9115	0.9131	0.9147	0.9162	0.9177
1.4	0.9192	0.9207	0.9222	0.9236	0.9251	0.9265	0.9279	0.9292	0.9306	0.9319
1.5	0.9332	0.9345	0.9357	0.9370	0.9382	0.9394	0.9406	0.9418	0.9429	0.9441
1.6	0.9452	0.9463	0.9474	0.9484	0.9495	0.9505	0.9515	0.9525	0.9535	0.9545
1.7	0.9554	0.9564	0.9573	0.9582	0.9591	0.9599	0.9608	0.9616	0.9625	0.9633
1.8	0.9641	0.9649	0.9656	0.9664	0.9671	0.9678	0.9686	0.9693	0.9699	0.9706
1.9	0.9713	0.9719	0.9726	0.9732	0.9738	0.9744	0.9750	0.9756	0.9761	0.9767
2.0	0.9772	0.9778	0.9783	0.9788	0.9793	0.9798	0.9803	0.9808	0.9812	0.9817

2.1	0.9821	0.9826	0.9830	0.9834	0.9838	0.9842	0.9846	0.9850	0.9854	0.9857
2.2	0.9861	0.9864	0.9868	0.9871	0.9875	0.9878	0.9881	0.9884	0.9887	0.9890
2.3	0.9893	0.9896	0.9898	0.9901	0.9904	0.9906	0.9909	0.9911	0.9913	0.9916
2.4	0.9918	0.9920	0.9922	0.9925	0.9927	0.9929	0.9931	0.9932	0.9934	0.9936
2.5	0.9938	0.9940	0.9941	0.9943	0.9945	0.9946	0.9948	0.9949	0.9951	0.9952
2.6	0.9953	0.9955	0.9956	0.9957	0.9959	0.9960	0.9961	0.9962	0.9963	0.9964
2.7	0.9965	0.9966	0.9967	0.9968	0.9969	0.9970	0.9971	0.9972	0.9973	0.9974
2.8	0.9974	0.9975	0.9976	0.9977	0.9977	0.9978	0.9979	0.9979	0.9980	0.9981
2.9	0.9981	0.9982	0.9982	0.9983	0.9984	0.9984	0.9985	0.9985	0.9986	0.9986

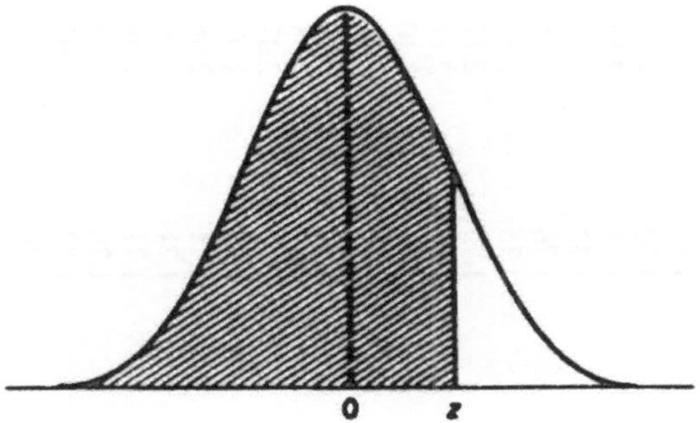

Table 5.6 Two-tailed critical values for student's t distribution (Adapted from Merrington, M, Critical values of student's t distribution, *Biometrika*, 32, 300 (1942) with permission of *Biometrika* trustees)

p↓/v→	0.9	0.95	0.975	0.99
1	6.31	12.71	25.45	63.66
2	2.92	4.30	6.20	9.93
3	2.35	3.18	4.18	5.84
4	2.13	2.78	3.50	4.60
5	2.02	2.57	3.16	4.03
6	1.94	2.45	2.97	3.71
7	1.90	2.37	2.84	3.50
8	1.86	2.31	2.75	3.36
9	1.83	2.26	2.69	3.25
10	1.81	2.23	2.63	3.17
11	1.80	2.20	2.59	3.11
12	1.78	2.18	2.56	3.06
13	1.77	2.16	2.53	3.01
14	1.76	2.15	2.51	2.98
15	1.75	2.13	2.49	2.95
16	1.75	2.12	2.47	2.92
17	1.74	2.11	2.46	2.81
18	1.73	2.10	2.45	2.88
19	1.73	2.09	2.43	2.86
20	1.73	2.09	2.42	2.85
21	1.72	2.08	2.41	2.83
22	1.72	2.07	2.41	2.82
23	1.71	2.07	2.40	2.81
24	1.71	2.06	2.39	2.80
25	1.71	2.06	2.39	2.79
26	1.71	2.06	2.38	2.78
27	1.70	2.05	2.37	2.77
28	1.70	2.05	2.37	2.76
29	1.70	2.05	2.36	2.76
30	1.70	2.04	2.36	2.75
∞	1.65	1.96	2.24	2.58

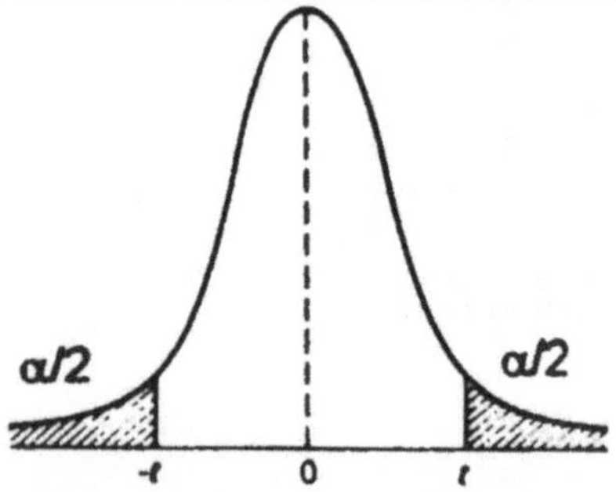

Example 5.15. Hypothesis testing concerning one mean when the population standard deviation is known A fruit juice producer is supposed to supply 200 g of concentrated juice in the containers. A set of 25 cans has a sample mean of 192 g. Standard deviation of the filling operation is $\sigma = 3$ g. Is the machine filling less than 200 g of concentrate into the containers?

Table 5.7 Values of χ^2_α (Adapted from Thompson, C.M, Table of percentage points of the χ^2 distribution, *Biometrika*, 32, 187–191 (1941) with permission of *Biometrika* trustees) values of a α

ν	0.995	0.99	0.975	0.95	0.05	0.025	0.01	0.005
1	0.00	0.00	0.00	0.00	3.84	5.02	6.64	7.88
2	0.01	0.02	0.05	0.10	5.99	7.38	9.21	10.60
3	0.07	0.12	0.22	0.35	7.82	9.35	11.35	12.84
4	0.21	0.30	0.48	0.71	9.49	11.14	13.28	14.86
5	0.41	0.55	0.83	1.15	11.07	12.83	15.09	16.75
6	0.68	0.87	1.24	1.64	12.59	14.45	16.81	18.55
7	0.99	1.24	1.69	2.17	14.07	16.01	18.48	20.28
8	1.34	1.65	2.18	2.73	15.51	17.54	20.09	21.96
9	1.74	2.09	2.70	3.33	16.92	19.02	21.67	23.59
10	2.16	2.56	3.25	3.94	18.30	20.48	23.21	25.19
11	2.60	3.05	3.82	4.58	19.68	21.92	24.73	26.76
12	3.07	3.57	4.40	5.23	21.03	23.34	26.22	28.30
13	3.57	4.11	5.01	5.89	22.36	24.74	27.69	29.82
14	4.08	4.66	5.63	6.57	23.69	26.12	29.14	31.32
15	4.60	5.23	6.26	7.26	25.00	27.49	30.58	32.80
16	5.14	5.81	6.91	7.96	26.30	28.85	32.00	34.27
17	5.70	6.41	7.56	8.67	27.59	30.19	33.41	35.72
18	6.27	7.02	8.23	9.39	28.87	31.53	34.81	37.16
19	6.84	7.63	8.91	10.12	30.14	32.85	36.19	38.53
20	7.43	8.26	9.59	10.85	31.41	34.17	37.57	40.00
21	8.03	8.90	10.28	11.59	32.67	35.48	38.93	41.40
22	8.64	9.54	10.98	12.34	33.92	36.78	40.29	42.80
23	9.26	10.20	11.69	13.09	35.17	38.08	41.64	44.18
24	9.89	10.86	12.40	13.85	36.42	39.36	42.98	45.56
25	10.52	11.52	13.12	14.61	37.65	40.65	44.31	46.93
26	11.16	12.20	13.84	15.38	38.89	41.92	45.64	48.29
27	11.81	12.88	14.57	16.15	40.11	43.19	46.96	49.65
28	12.46	13.57	15.31	16.93	41.34	44.46	48.28	50.99
29	13.12	14.26	16.05	17.71	42.56	45.72	49.59	52.34
30	13.79	14.95	16.79	18.49	43.77	46.98	50.89	53.67

Solution In this example σ is known. We use z values since σ is known. The hypotheses is $\mu = 200$ g, and the test is:

$$\mu - z_{1-\alpha/2}\frac{\sigma}{\sqrt{n}} \leqslant \bar{x} \leqslant \mu + z_{1-\alpha/2}\frac{\sigma}{\sqrt{n}} \qquad \text{((5.32) in Tab. 5.4)}$$

$z_{1-\alpha/2} = 2.57\,(\alpha = 0.01)$ (Tab. 5.5), $n = 25$ and $\sigma = 3$, after substituting the numbers in the test we will find $198.5 \leqslant \bar{x} \leqslant 201.5$. The experimentally determined value of the sample mean is 192 g and it is not in the given range, therefore the hypotheses is rejected. The manufacturer is not supplying the concentrate with $\mu = 200$ g.

Example 5.16. Hypothesis testing concerning two means when the population standard deviations are known 105 samples of a 1994 vintage wine had 0.72 g/100 ml of tartaric acid content. 110 samples of a 1996 vintage of the same wine had 0.61 g/100 ml of tartaric acid. Standard deviation of tartaric acid contents of these samples were 0.10 g/100 ml in 1994 and 0.06 g/100 ml

Table 5.8 Values of $F_{0.01}(v_n, v_d)$, where v_n = degrees of freedom for numerator, v_d = degrees of freedom for denominator (Adapted from Merrington, M. and Thompson, C.M. Tables of Percentage Points of the Inverted Beta (F) Distribution, *Biometrika*, 33, 73–88 (1943) with permission of *Biometrika* trustees)

$v_d \downarrow$ $v_n \rightarrow$ 1		2	3	4	5	6	7	8	9	10	12	15	20	24	30	40	60	120	∞
1	4052	5000	5403	5625	5744	5859	5928	5982	6023	6056	6106	6157	6209	6235	6261	6287	6313	6339	6366
2	98.5	99.0	99.2	99.2	99.3	99.3	99.4	99.4	99.4	99.4	99.4	99.4	99.4	99.5	99.5	99.5	99.5	99.5	99.5
3	34.1	30.8	29.5	28.7	28.2	27.9	27.7	27.5	27.3	27.2	27.1	26.9	26.7	26.6	26.5	26.4	26.3	26.2	26.1
4	21.2	18.0	16.7	16.0	15.5	15.2	15.0	14.8	14.7	14.5	14.4	14.2	14.0	13.9	13.8	13.7	13.7	13.6	13.5
5	16.3	13.3	12.1	11.4	11.0	10.7	10.5	10.3	10.2	10.1	9.89	9.72	9.55	9.47	9.38	9.29	9.20	9.11	9.02
6	13.7	10.9	9.78	9.15	8.75	8.47	8.26	8.10	7.98	7.87	7.72	7.56	7.40	7.31	7.23	7.14	7.06	6.97	6.88
7	12.2	9.55	8.45	7.85	7.46	7.19	6.99	6.84	6.72	6.62	6.47	6.31	6.16	6.06	5.99	5.91	5.82	5.74	5.65
8	11.3	8.65	7.59	7.01	6.63	6.37	6.18	6.03	5.91	5.81	5.67	5.52	5.36	5.28	5.20	5.12	5.03	4.95	4.83
9	10.6	8.02	6.99	6.42	6.06	5.80	5.61	5.47	5.35	5.26	5.11	4.96	4.81	4.73	4.65	4.57	4.48	4.40	4.31
10	10.0	7.56	6.55	5.99	5.64	5.39	5.20	5.06	4.94	4.85	4.71	4.56	4.41	4.33	4.25	4.17	4.08	4.00	3.91
11	9.65	7.21	6.22	5.67	5.32	5.07	4.89	4.74	4.63	4.54	4.40	4.25	4.10	4.02	3.94	3.86	3.78	3.69	3.60
12	9.33	6.93	5.95	5.41	5.06	4.82	4.64	4.50	4.39	4.30	4.16	4.01	3.86	3.78	3.70	3.62	3.54	3.45	3.36
13	9.07	6.70	5.74	5.21	4.86	4.62	4.44	4.30	4.19	4.10	3.96	3.82	3.66	3.59	3.51	3.43	3.34	3.25	3.17
14	8.86	6.51	5.56	5.04	4.70	4.46	4.28	4.14	4.03	3.94	3.80	3.66	3.51	3.43	3.35	3.27	3.18	3.09	3.00
15	8.68	6.36	5.42	4.89	4.56	4.32	4.14	4.00	3.89	3.80	3.67	3.52	3.37	3.29	3.21	3.13	3.05	2.96	2.87
16	8.53	6.23	5.29	4.77	4.44	4.20	4.03	3.89	3.78	3.69	3.55	3.41	3.26	3.18	3.10	3.02	2.93	2.84	2.75
17	8.40	6.11	5.19	4.67	4.34	4.10	3.93	3.79	3.68	3.59	3.46	3.31	3.16	3.08	3.00	2.92	2.83	2.75	2.65
18	8.29	6.01	5.09	4.58	4.25	4.01	3.84	3.71	3.60	3.51	3.37	3.23	3.08	3.00	2.92	2.84	2.75	2.66	2.57
19	8.19	5.93	5.01	4.50	4.17	3.94	3.77	3.63	3.52	3.43	3.30	3.15	3.00	2.92	2.84	2.76	2.67	2.58	2.49
20	8.10	5.85	4.94	4.43	4.10	3.87	3.70	3.56	3.46	3.37	3.23	3.09	2.94	2.86	2.78	2.69	2.61	2.52	2.42

21	8.02	5.78	4.87	4.37	4.04	3.81	3.64	3.51	3.40	3.31	3.17	3.03	2.88	2.80	2.72	2.64	2.55	2.46	2.36
22	7.95	5.72	4.82	4.31	3.99	3.76	3.59	3.45	3.35	3.26	3.12	2.98	2.83	2.75	2.67	2.58	2.50	2.40	2.31
23	7.88	5.66	4.76	4.26	3.94	3.71	3.54	3.41	3.30	3.21	3.07	2.93	2.78	2.70	2.62	2.54	2.45	2.35	2.26
24	7.82	5.61	4.72	4.22	3.90	3.67	3.50	3.36	3.26	3.17	3.03	2.89	2.74	2.66	2.58	2.49	2.40	2.31	2.21
25	7.77	5.57	4.68	4.18	3.86	3.63	3.46	3.32	3.22	3.13	2.99	2.85	2.70	2.62	2.53	2.45	2.36	2.27	2.17
30	7.56	5.39	4.51	4.02	3.70	3.47	3.30	3.17	3.07	2.98	2.84	2.70	2.55	2.47	2.39	2.30	2.21	2.11	2.01
40	7.31	5.18	4.31	3.83	3.51	3.29	3.12	2.99	2.89	2.80	2.66	2.52	2.37	2.29	2.20	2.11	2.02	1.92	1.80
60	7.08	4.98	4.13	3.65	3.34	3.12	2.95	2.82	2.72	2.63	2.50	2.35	2.20	2.12	2.03	1.94	1.84	1.73	1.60
120	6.65	4.79	3.95	3.48	3.17	2.96	2.79	2.66	2.56	2.47	2.34	2.19	2.03	1.95	1.86	1.76	1.66	1.53	1.38
∞	6.63	4.61	3.78	3.32	3.02	2.80	2.64	2.51	2.41	2.32	2.18	2.04	1.88	1.79	1.70	1.59	1.47	1.32	1.00

Table 5.9 Values of $F_{0.05}(\nu_n, \nu_d)$ where ν_n = degrees of freedom for numerator, ν_d = degrees of freedom for denominator (Adapted from Merrington, M. and Thompson, C. M. Tables of Percentage Points of the Inverted Beta (F) Distribution, *Biometrika*, 33, 73–88 (1943) with permission of *Biometrika* trustees)

$\nu_d \downarrow$ $\nu_n \rightarrow$ 1		2	3	4	5	6	7	8	9	10	12	15	20	24	30	40	60	120	∞
1	161	200	216	225	230	234	237	239	241	242	244	246	248	249	250	251	252	253	254
2	18.5	19.0	19.2	19.3	19.3	19.3	19.4	19.4	19.4	19.4	19.4	19.4	19.4	19.4	19.5	19.5	19.5	19.5	19.5
3	10.1	9.55	9.28	9.12	9.01	8.94	8.89	8.85	8.81	8.79	8.74	8.70	8.66	8.66	8.62	8.59	8.57	8.55	8.63
4	7.71	6.94	6.59	6.39	6.26	6.16	6.09	6.04	6.00	5.96	5.91	5.86	5.80	5.80	5.75	5.72	5.69	5.66	5.63
5	6.61	5.79	5.41	5.19	5.05	4.95	4.88	4.82	4.77	4.74	4.68	4.62	4.56	4.56	4.50	4.46	4.43	4.40	4.37
6	5.99	5.14	4.76	4.53	4.39	4.28	4.21	4.15	4.10	4.06	4.00	3.94	3.87	3.84	3.81	3.77	3.74	3.70	3.67
7	5.59	4.74	4.35	4.12	3.97	3.87	3.79	3.73	3.68	3.64	3.57	3.51	3.44	3.41	3.38	3.34	3.30	3.27	3.23
8	5.32	4.46	4.07	3.84	3.69	3.58	3.50	3.44	3.39	3.35	3.28	3.22	3.15	3.12	3.08	3.04	3.01	2.97	2.93
9	5.12	4.26	3.86	3.63	3.48	3.37	3.29	3.23	3.18	3.14	3.07	3.01	2.94	2.90	2.86	2.83	2.79	2.75	2.71
10	4.96	4.10	3.71	3.48	3.33	3.22	3.14	3.07	3.02	2.98	2.91	2.85	2.77	2.74	2.70	2.66	2.62	2.58	2.54
11	4.84	3.98	3.59	3.36	3.20	3.09	3.01	2.95	2.90	2.85	2.79	2.72	2.65	2.61	2.57	2.53	2.49	2.45	2.40
12	4.75	3.89	3.49	3.26	3.11	3.00	2.91	2.85	2.80	2.75	2.69	2.62	2.54	2.51	2.47	2.43	2.38	2.34	2.30
13	4.67	3.81	3.41	3.18	3.03	2.92	2.83	2.77	2.71	2.67	2.60	2.53	2.46	2.42	2.38	2.34	2.30	2.25	2.21
14	4.60	3.74	3.34	3.11	2.96	2.85	2.76	2.70	2.65	2.60	2.53	2.46	2.39	2.35	2.31	2.27	2.22	2.18	2.13
15	4.54	3.68	3.29	3.06	2.90	2.79	2.71	2.64	2.59	2.54	2.48	2.40	2.33	2.29	2.25	2.20	2.16	2.11	2.07
16	4.49	3.63	3.24	3.01	2.85	2.74	2.66	2.59	2.54	2.49	2.42	2.35	2.28	2.24	2.19	2.15	2.11	2.06	2.01
17	4.45	3.59	3.20	2.96	2.81	2.70	2.61	2.55	2.49	2.45	2.38	2.31	2.23	2.19	2.15	2.10	2.06	2.01	1.96
18	4.41	3.55	3.16	2.93	2.77	2.66	2.58	2.51	2.46	2.41	2.34	2.27	2.19	2.15	2.11	2.06	2.02	1.97	1.92
19	4.38	3.52	3.13	2.90	2.74	2.63	2.54	2.48	2.42	2.38	2.31	2.23	2.16	2.11	2.07	2.03	1.98	1.93	1.88
20	4.35	3.49	3.10	2.87	2.71	2.60	2.51	2.45	2.39	2.35	2.28	2.20	2.12	2.08	2.04	1.99	1.95	1.90	1.84

21	4.32	3.47	3.07	2.84	2.68	2.57	2.49	2.42	2.37	2.32	2.25	2.18	2.10	2.05	2.01	1.96	1.92	1.87	1.81
22	4.30	3.44	3.05	2.82	2.66	2.55	2.46	2.40	2.34	2.30	2.23	2.15	2.07	2.03	1.98	1.94	1.89	1.84	1.78
23	4.28	3.42	3.03	2.80	2.64	2.53	2.44	2.37	2.32	2.27	2.20	2.13	2.05	2.01	1.96	1.91	1.86	1.81	1.76
24	4.26	3.40	3.01	2.78	2.62	2.51	2.42	2.36	2.30	2.25	2.18	2.11	2.03	1.98	1.94	1.89	1.84	1.79	1.73
25	4.24	3.39	2.99	2.76	2.60	2.49	2.40	2.34	2.28	2.24	2.16	2.09	2.01	1.96	1.92	1.87	1.82	1.77	1.71
30	4.17	3.32	2.92	2.69	2.53	2.42	2.33	2.27	2.21	2.16	2.09	2.01	1.93	1.89	1.84	1.79	1.74	1.68	1.62
40	4.08	3.23	2.84	2.61	2.45	2.34	2.25	2.18	2.12	2.08	2.00	1.92	1.84	1.79	1.74	1.69	1.64	1.58	1.51
60	4.00	3.15	2.76	2.53	2.37	2.25	2.17	2.10	2.04	1.99	1.92	1.84	1.75	1.70	1.65	1.59	1.53	1.47	1.39
120	3.92	3.07	2.68	2.45	2.29	2.18	2.09	2.02	1.96	1.91	1.83	1.75	1.66	1.61	1.55	1.50	1.43	1.35	1.25
∞	3.84	3.00	2.60	2.37	2.21	2.10	2.01	1.94	1.88	1.83	1.75	1.67	1.57	1.52	1.46	1.39	1.32	1.22	1.00

in 1996. Is there a significant difference between tartaric acid contents of these vintages?

Solution The hypotheses is $\mu_A = \mu_B$, and the test is:

$$-z_{1-\alpha/2}\left(\frac{\sigma_A^2}{n_A}+\frac{\sigma_B^2}{n_B}\right)^{1/2} \leqslant \bar{x}_A - \bar{x}_B \leqslant z_{1-\alpha/2}\left(\frac{\sigma_A^2}{n_A}+\frac{\sigma_B^2}{n_B}\right)^{1/2} \quad ((5.35) \text{ in Tab. } 5.4)$$

Since the population standard deviation is known we use the z values. $z_{1-\alpha/2} = 1.88$ ($\alpha = 0.06$, Tab. 5.5), $\sigma_A = 0.10$ g/100 ml, $n_A = 105$, $\sigma_B = 0.06$ g/100 ml, $n_B = 110$, after substituting the numbers into the test we will get $-0.02 \leqslant \bar{x}_A - \bar{x}_B \leqslant 0.02$. Experimentally determined $\bar{x}_A - \bar{x}_B = 0.11$, and it is not in the limits, therefore the hypotheses has been rejected. The population mean tartaric acid contents of the vintages are different.

Example 5.17. Hypothesis testing concerning one and two means when the population standard deviations are not known a) In a tea processing plant, processed tea is packed with two sets of equipment. Grocery store representatives claimed that each machine was packing different amounts of tea. How would you respond to their claims after analyzing the following data?

Equipment A							
number of packages	3	5	9	11	8	6	1
package weight (g)	367	372	380	382	385	393	413

Equipment B							
number of packages	3	4	8	7	6	2	1
package weight (g)	380	382	390	392	394	400	401

Solution The hypotheses is $\mu_A = \mu_B$, since the population standard deviation is not known we use the t values (Tab. 5.6) and the test is:

$$-t_{1-\alpha/2}\, s_p\left(\frac{1}{n_A}+\frac{1}{n_B}\right)^{1/2} \leqslant \bar{x}_A - \bar{x}_B \leqslant t_{1-\alpha/2}\, s_p\left(\frac{1}{n_A}+\frac{1}{n_B}\right)^{1/2}$$

((5.37) in Tab. 5.4)

The sample means are: $\bar{x}_A = (1/43)\sum_{i=1}^{43} x_{Ai} = 382.2$, $\bar{x}_B = (1/31)\sum_{i=1}^{31} x_{Bi} = 390.2$. The sample variances are: $s_A^2 = 1/(43-1)\sum_{i=1}^{43}(x_{Ai} - \bar{x}_A)^2 = 64.64$ and $s_B^2 = 1/(31-1)\sum_{i=1}^{31}(x_{Bi} - \bar{x}_B)^2 = 33.31$. The pooled variance ((5.38) in Tab. 5.4) is $s_p^2 = (v_A s_A^2 + v_B s_B^2)/(v_A + v_B) = 51.58$, $s_p = 7.2$, $t_{1-\alpha/2} = 2.24$ ($\alpha = 0.05$, $v = v_A + v_B = 72$) after substituting the numbers into the test we will get $-3.79 \leqslant \bar{x}_A - \bar{x}_B \leqslant 3.79$. Experimentally determined $\bar{x}_A - \bar{x}_B = -8.2$, and it is not in the limits, therefore the hypotheses has been rejected. The two machines are packing different amounts of tea.

b) Determine i) Mode, ii) Median and iii) Range of these data sets.

Solution By using Table 5.1 we can determine the followings:

Equipment A i) Mode = 382 (value which occurs most frequently) ii) Median = 382 (when we write down the data in increasing order, it is the value that appears in the middle, i.e., the value above or below which equal number of observations fall), iii) Range = 413 − 367 = 46 (difference between the largest and the smallest observations)

Equipment B i) Mode = 390, ii) Median = 392, iii) Range = 401 − 380 = 21

c) Equipment A and B were supposed to pack 390 g of tea (i.e., $\mu_A = 390$ g, $\mu_B = 390$ g). Are they packing the right amounts?

Solution i) The hypotheses is $\mu_A = 390$ g, and the test is:

$$\mu - t_{1-\alpha/2}\frac{s}{\sqrt{n_A}} \leqslant \bar{x}_A \leqslant \mu + t_{1-\alpha/2}\frac{s}{\sqrt{n_A}} \qquad \text{((5.34) in Tab. 5.4)}$$

We have $t_{1-\alpha/2} = 2.24$ ($\alpha = 0.05$, $\nu_A = 42$, Tab. 5.6), $s_A = 8.4$ and $n_A = 43$ after substituting the numbers in the test we will find $387 \leqslant \bar{x}_A \leqslant 393$. The experimentally determined value of the sample mean is 382.2 and it is not in the given range, therefore the hypotheses is rejected. Equipment set A is not packing the right amount.

ii) The hypotheses is $\mu_B = 390$ g, and the test is:

$$\mu - t_{1-\alpha/2}\frac{s}{\sqrt{n_B}} \leqslant \bar{x}_B \leqslant \mu + t_{1-\alpha/2}\frac{s}{\sqrt{n_B}} \qquad \text{((5.34) in Tab. 5.4)}$$

$t_{1-\alpha/2} = 2.36$ ($\alpha = 0.05$, $\nu_B = 30$, Tab. 5.6), $s_B = 5.8$ and $n_B = 32$ after substituting the numbers in the test we will find $387.5 \leqslant \bar{x}_B \leqslant 392.5$. The experimentally determined value of the sample mean is 390.2 and it is in the given range, therefore the hypotheses is accepted. Equipment set B is packing the right amount.

Example 5.18. Effect of the probability level on the results of the hypothesis testing In a given linear wine taste scale scores range between 80 and 100 where 80 means ordinary, 100 means perfect. The previous vintage of wine sample was evaluated by a taste panel of 23 members. The mean taste score given by the panel members was 98 with standard deviation of 5. A new vintage of the same wine was also tasted by the same panel, where the average taste score was found 95 with standard deviation of 4. Is there a significant deviation between the taste of the two wine samples? Solve the problem by using $\alpha = 0.02$ and $\alpha = 0.20$, then compare the results (why are the limits obtained with $\alpha = 0.02$ and $\alpha = 0.20$ different, what does it mean?).

Solution The hypotheses is $\mu_A = \mu_B$. The pooled variance ((5.38) in Tab. 5.4) is $s_p^2 = (\nu_A s_A^2 + \nu_B s_B^2)/(\nu_A + \nu_B) = ((22)(5)^2 + (22)(4)^2)/((22) + (22)) = 20.5$, $s_p = 4.53$, and the test is:

$$-t_{1-\alpha/2}\, s_p\left(\frac{1}{n_A} + \frac{1}{n_B}\right)^{1/2} \leqslant \bar{x}_A - \bar{x}_B \leqslant s_p\left(\frac{1}{n_A} + \frac{1}{n_B}\right)^{1/2}. \qquad (5.37)$$

i) choose $\alpha = 0.02$ $t_{1-\alpha/2} = 2.58$ ($\nu = \nu_A + \nu_B = 44$, Tab. 5.6) after substituting the numbers into the test we will get $-3.51 \leqslant \bar{x}_A - \bar{x}_B \leqslant 3.51$. Experimentally determined $\bar{x}_A - \bar{x}_B = 98 - 95 = 3$, and it is in the limits, therefore the hypotheses has been accepted, the taste of the wines are not different.

ii) choose $\alpha = 0.2$ $t_{1-\alpha/2} = 1.65$ ($\nu = \nu_A + \nu_B = 44$, Tab. 5.6) after substituting the numbers into the test we will get $-2.24 \leqslant \bar{x}_A - \bar{x}_B \leqslant 2.24$. Experimentally determined $\bar{x}_A - \bar{x}_B = 3$, and it is not in the limits, therefore the hypotheses has been rejected, and the taste of the wines are different.

iii) choosing the level $\alpha = 0.02$ implies that if the experiments should be replicated 100 times in only 2 of them $\bar{x}_A - \bar{x}_B$ may be out of the limits, although the fact is that the samples are not actually different. The probability of such an occurence is ten times larger when $\alpha = 0.20$. The range of limits are larger with smaller α.

Example 5.19. Confidence limits with binomial distribution A recipe of canned mixed vegetables consist of 30% ground beef and 70% other ingredients. What are the limits of the ground beef contents of a 1000 g can?

Solution The total weight of the sample is $n = 1000$ g. The probability of having meat in any servings is $p = 0.30$, the standard deviation is $\sigma = \sqrt{np(1-p)} = 14.5$. The confidence limits of the meat contents of a can is: $np - 3\sigma \leqslant (np)_{exp} \leqslant np + 3\sigma$ ((5.28) in Tab. 5.3), after substituting the numbers $256.5 \leqslant (np)_{exp} \leqslant 343.5$. In any can we may expect to have 256.5 to 343.5 g ground beef.

Example 5.20. Confidence limits with Poisson distribution A food processing company, which returns the money of unsatisfied consumers, produces 1,000,000 cans of soup in a year. Approximately 500 cans are returned annually. After changing one of the major ingredients 820 cans were returned in one year. Has this ingredient contributed to the increase in number of unsatisfied consumers?

Solution It is given in the problem statement that $p = 500/1{,}000{,}000 = 0.0005$ and $\lambda = np = (1{,}000{,}000)(0.0005) = 500$. Since $n > 20$ and $p < 0.05$ and constant we may use the poisson distribution with $\sigma = \sqrt{\lambda} = 22.36$. The confidence limits are: $np - 3\sigma \leqslant (np)_{exp} \leqslant np + 3\sigma$ ((5.29) in Tab. 5.3), or after substituting the numbers $432 \leqslant (np)_{exp} \leqslant 567$. The observed number of returns $= (np)_{exp} = 820$ and it is not in the given limits, therefore we may conclude that the ingredient contributed to the rise in unsatisfied consumers.

Example 5.21. Testing the validity of normal distribution assumption A cereal company started a promotion targeting elementary school students. During the campaign surprise toys were placed in the cereal boxes. Most of these gifts were simple and inexpensive. A smaller fraction were more attractive and sophisticated. The campaign is expected to be more successful if the gifts are distributed normally. To check the success of the distribution process 500 different school districts were surveyed and number of sophisticated gifts received by the school districts were determined. By using this data discuss if the sophisticated gifts are distributed normally?

Solution Since $n = 500$ the population mean and the population variance may be calculated as $\mu = (1/n)\Sigma_{i=1}^{n} x_i = 4.48$ and $\sigma^2 = (1/n)\Sigma_{i=1}^{n}(x-\mu)^2 = 4.98$, the standard deviation is $\sigma = 2.23$. Expected frequency of the distribution of the gifts may be calculated as $f_e = 1/(\sigma\sqrt{2\pi})\exp\{-(1/2)((x-\mu)/\sigma)^2\}$. Expected frequency of obtaining no gifts is $f_e = 1/(2.23\sqrt{(2)(3.14)})$ $\exp\{-(1/2)((0-4.48)/2.23)^2\} = 0.024$, then the expected number of the districts receiving no gifts is $(0.024)(500) = 12$. We may calculate the expected frequency and the numbers of the gifts received by each district as:

number of gifts received	experimentally determined number of the districts received the gift	expected number of the districts to receive the gift
0	18	12
1	22	26.49
2	50	48.22
3	81	71.81
4	100	87.45
5	79	87.10
6	55	70.94
7	45	47.26
8	25	25.75
9	15	11.47
10	10	4.18
TOTAL	500	492.67

$\chi^2_{observed} = \Sigma_{i=1}^{10}(f_0 - f_e)^2/f_e = 13.65$. We have eleven terms in the equation, therefore $v = 11 - 1 = 10$, we choose $\alpha = 0.005$ then $\chi^2_{table} = 25.19$ (Tab. 5.7). Since $\chi^2_{table} > \chi^2_{observed}$ we may conclude that the variations between f_e and f_0 are random and the distribution is normal.

Example 5.22. Confidence limits associated with triangular tasting experiments and testing the validity of a distribution model In a *triangular* tasting test two identical and one different samples are tasted by panel members, who are required to distinguish the different one. A panel of 49 individuals were given 12 tasting tests and required to identify an imitation cheese. In each set there was one imitation and two real cheese samples. Number of the times that the panel members distinguished the imitation cheese is given in the following table:

number of times	0	1	2	3	4	5	6	7	8	9	10	11	12
number of the panel members	0	1	2	7	9	17	5	6	1	1	0	0	0

Is the imitation cheese actually distinguished by the panel members?

Solution The total number of the samples tasted is $n = (49)(12) = 588$. The probability of having a different sample in any trial is $p =$ number of different samples/total number of samples = 1/3, the standard deviation is $\sigma = \sqrt{np(1-p)} = 11.4$, the test is $np - 3\sigma \leqslant (np)_{exp} \leqslant np + 3\sigma$, after substituting the numbers $161.8 \leqslant (np)_{exp} \leqslant 230.2$, number of the correct replicates observed $= (np)_{exp} = (0)(0) + (1)(1) + (2)(2) + (3)(7) + (4)(9) + \cdots = 236$ and it is not in the given interval, therefore the panel members actually distinguished the imitation cheese.

b) Is the assumption of binomial distribution appropriate?

Solution Calculate the expected number of panel members distinguishing the imitation cheese 0, 1, 2, ..., 11, 12 times. The binomial formula is: $f(x, n, p) = n!/(x!(n-x))!\, p^x(1-p)^{(n-x)}$ with $n = 12$ = number of replications and $p = 1/3$ = probability of guessing the imitation cheese in any trial. The probability of one person detecting the imitation cheese exactly 0 times in 12 trials $= f(0, 12, 1/3) = 12!/((0)!\,(12)!)(1/3)^0(1-1/3)^{(12-0)} = 0.0079$. The probability of 49 people detecting the imitation cheese exactly 0 times in 12 trials = (0.0079)(49) = 0.4. The same procedure may be repeated for the other occasions:

number of times the imitation cheese distinguished	number of the panel members made the distinction	expected number of panel members to make distinction
0	0	0.4
1	1	2.3
2	2	6.3
3	7	10.4
4	9	11.6
5	17	9.3
6	5	5.4
7	6	2.4
8	1	0.7
9	1	0.2
10	0	0
11	0	0
12	0	0
TOTAL	49	49

$\chi^2_{observed} = \Sigma_{i=1}^{12}((f_0 - f_e)^2/f_e) = 20.88$. We have thirteen terms in the equation, therefore $v = 13 - 1 = 12$, we choose $\alpha = 0.01$ then $\chi^2_{table} = 26.22$ (Tab. 5.7). Since $\chi^2_{table} > \chi^2_{observed}$ we may conclude that the assumption was appropriate.

Example 5.23. Confidence limits associated with duo-trio tasting experiments and testing the validity of a distribution model In a *duo-trio* tasting test a standard is presented to a panel, then they are asked to taste two coded samples (one of them is the same as the standard, the other is different), and find the identical one. A food producer developed an inexpensive ingredient, and was concerned about the consumer response to the new formulation. With a taste panel of 70 individuals, 8 replica experiments were performed, and the following results were obtained:

number of times	0	1	2	3	4	5	6	7	8
number of the panel members	1	2	5	10	15	22	11	4	0

Has the additive affected the taste of the product?

Solution The total number of the samples tasted is $n = (70)(8) = 560$. The probability of having a different sample in any trial is p = number of different samples/total number of samples = 1/2, the standard deviation is $\sigma = \sqrt{np(1-p)} = 11.8$, the test is $np - 3\sigma \leqslant (np)_{exp} \leqslant np + 3\sigma$, after substituting the numbers $245 \leqslant (np)_{exp} \leqslant 315$, number of the correct replicates observed $= (np)_{exp} = (0)(2) + (1)(2) + (2)(5) + \cdots = 306$ and it is in the given interval, therefore the additive did not actually affect the taste of the product.

b) Is the assumption of binomial distribution appropriate?

Solution Calculate the expected number of panel members distinguished the food with inexpensive additive 0, 1, 2,..., 7, 8 times. The binomial formula is: $f(x, n, p) = (n!/(x!(n-x)!))\, p^x(1-p)^{(n-x)}$ with $n = 8$ = number of replications and $p = 1/2$ = probability of guessing the right sample in any trial. The probability of one person detecting the inexpensive additive exactly 0 times in 8 trials $= f(0, 8, 0.5) = (8!/((0!)(8)!))\,(0.5)^0\,(1-0.5)^{(8-0)} = 0.0039$. The probability of 70 people detecting the inexpensive additive exactly 0 times in 8 trials $= (0.0039)(70) = 0.27$. The same procedure may be repeated for the other occasions:

number of times the inexpensive additive distinguished	number of the panel members made the distinction	expected number of panel members to make distinction
0	1	0.27
1	2	2.19
2	5	7.66
3	10	15.31
4	15	19.14
5	22	15.31
6	11	7.66
7	4	2.19
8	0	0.27
TOTAL	70	70.00

$\chi^2_{observed} = \Sigma_{i=1}^{8}(f_0 - f_e)^2/f_e = 11.78$. We have nine terms in the equation, therefore $v = 9 - 1 = 8$, we choose $\alpha = 0.01$ then $\chi^2_{table} = 20.09$ (Tab. 5.7). Since $\chi^2_{table} > \chi^2_{observed}$ we may conclude that the assumption was appropriate.

Example 5.24. Probability of obtaining the required value of sample standard deviation A food processing company wants to keep sample standard deviation of the packages small to maintain uniform shipments. Population standard deviation of the product is 1.26 g. Assume that the samples were taken randomly from a normally distributed population and calculate the probability of having a shipment of 20 packages with more than 1.7 of sample standard deviation.

Solution It is given in the problem statement that $n = 20$, $s = 1.7$ and $\sigma = 1.26$. After substituting the numbers $\chi^2 = ((n-1)s^2)/\sigma^2 = 34.6$. With $v = 20 - 1 = 19$, and $\chi^2 = 34.6$ we determine with interpolation from Table 5.7 that $\alpha = 0.018$, implying that there is 1.8% probability for a shipment with $s \geqslant 1.7$.

b) Calculate the probability of having a sample standard deviation more than 1.7 in a shipment of 15 packages in the previous example.

Solution It is given in the problem statement that $n = 15$, $s = 1.7$ and $\sigma = 1.26$. After substituting the numbers $\chi^2 = ((n-1)s^2)/\sigma^2 = 25.5$, with $v = 14$, and $\chi^2 = 25.5$. We determine with interpolation from the χ^2 table that $\alpha = 0.038$, implying that there is 3.8% probability for a shipment with $s \geqslant 1.7$. It should be noted that the probability of the shipment of the packages with $s \geqslant 1.7$ increases as the sample size decreases.

Example 5.25. Confidence limits of population means and standard deviations A Turkish newspaper made a survey concerning the actual volumes of the so called "one liter" fruit juice and carbonated beverages. After measuring one box or bottle of the each of the given brand names they published the following data (Hürriyet, March 23, 1995; page 9):

fruit juices	actual volume measured (mL)	carbonated beverages	actual volume measured (mL)
Tamek (sour cherry)	1000	Coca Cola	1025
Aroma (orange)	1000	Pepsi Cola	1025
Dimes (orange)	1000	Uludağ	950
Tamek (peach)	1025	Seven-Up	1000
Meysu (apricot)	1010	Çamlica (lemon)	1000
Tikveşli (apple)	1030	Çamlica (orange)	1000
Cappy (apricot)	1000	Yedigün (orange)	1000
Aroma (apricot)	1000	Schweppes (tangarine)	1000
		Fanta (orange)	1025

All the juices and the carbonated beverages were reported to be purchased from the same market, therefore we may assume that they were subjected the same treatment during storage and handling. Although measuring the volume of a single container is not sufficient to make any judgement concerning any brandnames, the data may be regarded as stratified samples and used to make the following analysis:

a) Determine the confidence limits of μ_{juice}, $\mu_{beverage}$, σ_{juice} and $\sigma_{beverage}$.

Solution The sample means and standard deviations are:

$$\bar{x} = \frac{1}{n}\sum_{i=1}^{n} x_i \qquad \text{((5.10) in Tab. 5.2)}$$

$$s^2 = \frac{1}{n-1}\sum_{i=1}^{n}(x_i - \bar{x})^2 \qquad \text{((5.13) in Tab. 5.2)}$$

After substituting the numbers in (5.10) and (5.13) we will obtain $\bar{x}_{juice} =$ 1008.1 mL, $s_{juice} = 12.5$ mL, $\bar{x}_{beverage} = 1002.8$ mL and $s_{beverage} = 23.2$ mL

$$\text{confidence limits of } \mu: \bar{x} - t\frac{s}{\sqrt{n}} \leqslant \mu \leqslant \bar{x} + t\frac{s}{\sqrt{n}} \qquad \text{((5.26) in Tab. 5.3)}$$

$$\text{confidence limits of } \sigma: \frac{s^2 v}{\chi^2_\beta} \leqslant \sigma^2 \leqslant \frac{s^2 v}{\chi^2_\gamma} \qquad \text{((5.30) in Tab. 5.3)}$$

We have $t_{1-\alpha/2} = 3.50$ ($v_{juice} = 7$, $\alpha = 0.02$, Tab. 5.6) for the fruit juices and $t_{1-\alpha/2} = 3.36$ ($v_{beverage} = 8$, $\alpha = 0.02$, Tab. 5.6) for the beverages. The chi square values are: $\chi^2_\beta = 16.01$ with ($\beta = 0.025$ and $v = 7$, Tab. 5.7) and $\chi^2_\gamma = 1.69$ with ($\gamma = 0.975$ and $v = 7$) for the fruit juice and $\chi^2_\beta = 17.54$ with ($\beta = 0.025$ and $v = 8$, Tab. 5.7) and $\chi^2_\gamma = 2.18$ with ($\gamma = 0.975$ and $v = 8$) for the beverage. After substituting the numbers in (5.26) and (5.30) we will obtain the confidence limits as: $992.6 \leqslant \mu_{juice} \leqslant 1023.6$, $8.3 \leqslant \sigma_{juice} \leqslant 25.4$, $976.9 \leqslant \mu_{beverage} \leqslant 1028.7$ and $15.7 \leqslant \sigma_{beverage} \leqslant 44.4$.

b) Determine if $\mu_{juice} = \mu_{beverage}$?

Solution The hypothesis is $\mu_{juice} = \mu_{beverage}$ and the test is:

$$-t_{1-\alpha/2}\, s_p\left(\frac{1}{n_A} + \frac{1}{n_B}\right)^{1/2} \leqslant \bar{x}_A - \bar{x}_B \leqslant t_{1-\alpha/2}\, s_p\left(\frac{1}{n_A} + \frac{1}{n_B}\right)^{1/2}$$

((5.37) in Tab. 5.4)

where

$$s_p^2 = \frac{v_A s_A^2 + v_B s_B^2}{v_A + v_B} = 360 \qquad \text{((5.38) in Tab. 5.4)}$$

$t_{1-\alpha/2} = 2.13$ ($v = v_A + v_B = 15$, $\alpha = 0.10$, Tab. 5.6). After substituting the numbers in (5.37) we will obtain: $-9.94 \leqslant \bar{x}_{juice} - \bar{x}_{beverage} \leqslant 9.94$, since

$\bar{x}_{juice} - \bar{x}_{beverage} = 5.3$ mL and it is within the given limits we may conclude that $\mu_{juice} = \mu_{beverage}$.

c) Determine if $\sigma_{juice} = \sigma_{beverage}$?

Solution The hypothesis is $\sigma^2_{juice} = \sigma^2_{beverage}$, and the test is

$$1/(F_{\alpha/2}(v_{juice}, v_{beverage})) \leqslant (s^2_{juice}/s^2_{beverage}) \leqslant F_{\alpha/2}(v_{juice}, v_{beverage}) \quad ((5.44) \text{ in Tab. 5.4})$$

$F_{table} = 6.18$ ($\alpha = 0.02$, $v_{juice} = 7$, $v_{beverage} = 8$, Tab. 5.8), after substituting F_{table} in the test: $0.161 \leqslant (s^2_{juice}/s^2_{beverage})_{exp} \leqslant 6.18$, since $(s^2_{juice}/s^2_{beverage})_{exp} = 0.29$ is within the given limits, we may conclude that $\sigma_{juice} = \sigma_{beverage}$.

d) Are these distribution curves similar?

Solution Since both $\mu_{juice} = \mu_{beverage}$ and $\sigma_{juice} = \sigma_{beverage}$ we may conclude that the relative frequency versus the volume distribution curves of the juice and the beverages are the same.

e) What are the smallest fruit juice and beverage volumes possible to find in the containers according to this data?

Solution Figure 5.3 indicates that: $\mu - 3\sigma \leqslant x \leqslant \mu + 3\sigma (p = 99.73\%)$. After substituting the lower confidence limits of μ and the upper confidence limits of σ in this equation we will obtain: $916.4 \leqslant x_{juice} \leqslant 1053.1$ and $843.7 \leqslant x_{beverage} \leqslant 1110.1$, implying that the smallest juice and beverage volumes available in the marked would be 916.4 mL and 843.7 mL, respectively.

Example 5.26. One way analysis of variance to test consumer preference of different cake formulations Cost of cake with formulations A and B is the same, but substantially less expensive with C. Consumer preference of these formulations was assessed with test scores between +2 (the highest preference) and −2 (strongest rejection). The test scores for each cake are listed as n_A, n_B and n_C. Is the consumer preference of these formulations the same?

Data and solution

C1	C2	C3	C4	C5	C6	C7	C8	C9	C10
SCORE	n_A	C1 × C2	C1² × C2	n_B	C1 × C5	C1² × C5	n_C	C1 × C8	C1² × C8
+2	20	40	80	40	80	160	5	10	20
+1	75	75	75	70	70	70	40	40	40
0	35	0	0	20	0	0	70	0	0
−1	17	−17	17	14	−14	14	25	−25	25
−2	3	−6	12	6	−12	24	10	−20	40
SUMS	150	92	184	150	124	268	150	5	125

Total sum of responses $n = \Sigma n_A + \Sigma n_B + \Sigma n_C = 150 + 150 + 150 = 450$

Total sum of scores = Sum of columns $C3$, $C6$ and $C9 = \Sigma x = 92 + 124 + 5 = 221$

Total sum of squares = Sum of columns $C4$, $C7$, $C10 = \Sigma x^2 = 184 + 268 + 125 = 577$
Total variance $= \sigma_T^2 = \Sigma x^2 - (\Sigma x)^2/n = 577 - (221^2/450) = 468.5$
Variance of the formulations $= \sigma_F^2 = (\Sigma x_A)^2/n_A + (\Sigma x_B)^2/n_B + (\Sigma x_C)^2/n_C - (\Sigma x)^2/n = (92^2/150) + (124^2/150) + (5^2/150) - (221^2/450) = 50.6$
Residual variance $= \sigma_R^2 = \sigma_T^2 - \sigma_F^2 = 468.5 - 50.6 = 417.9$

Analysis of variance table

Variance	v	σ^2	σ^2/v	F
Total	449	468.5		
Formulations	2	50.6	25.3	
Residual	447	417.9	0.94	26.9

where v_T = Total degrees of freedom $= n - 1 = 450 - 1 = 449$; v_F = Formulations degrees of freedom = number of formulations $- 1 = 3 - 1 = 2$; v_R = Residuals degrees of freedom $= v_T - v_F$

$$F_{\text{calculated}} = (\sigma_F^2/v_F)/(\sigma_R^2/v_R) = 26.9$$

F value associated with this data may be obtained from Table 5.8 $F_{\text{table}} = 4.61$ ($\alpha = 0.02$ with $v_F = 2$, $v_R = 447$). Since $F_{\text{calculated}}$ is larger than F_{table} we may conclude that there is noticeable difference in the consumer preference of the formulations.

Example 5.27. Two way analysis of variance to check the difference between the scores of different panel members, and the difference in the scatter of the test scores Three different sets (A, B, C) of fermented olive samples were tasted by twelve panel members. Panel members scored (x) the samples between zero and five according to their bitterness, where zero means acceptable and five means too bitter. Is there an actual difference in bitterness of the olive samples? Is there an actual difference between the scores of the different panel members? Is there an actual difference in the scatter of the test scores of the samples A and B?

Data and solution

	C1	C2	C3	C4	C5	C6
panel members	x_A	x_A^2	x_B	x_B^2	x_C	x_C^2
1	3	9	0	0	1	1
2	2	4	2	4	2	4
3	3	9	1	1	2	4
4	1	1	1	1	0	0
5	3	9	1	1	3	9

panel members	C1 x_A	C2 x_A^2	C3 x_B	C4 x_B^2	C5 x_C	C6 x_C^2
6	2	4	1	1	1	1
7	3	9	2	4	2	4
8	2	4	0	0	1	1
9	3	9	1	1	2	4
10	4	16	2	4	3	9
11	1	1	1	1	0	0
12	2	4	2	4	2	4
SUMS:	29	79	14	22	19	41

Total sum of x values = Sum of columns $C1$, $C3$ and $C5 = \Sigma x = 29 + 14 + 19 = 62$
Total sum of squares = Sum of columns $C2$, $C4$ and $C6 = \Sigma x^2 = 79 + 22 + 41 = 142$
Total number of the data points $= n = 12 + 12 + 12 = 36$

$$\text{Total variance} = \sigma_T^2 = \Sigma x^2 - \frac{(\Sigma x)^2}{n} = 142 - \frac{62^2}{36} = 35.2$$

$$\text{Samples variance} = \sigma_S^2 = \frac{(\Sigma x_A)^2}{n_A} + \frac{(\Sigma x_B)^2}{n_B} + \frac{(\Sigma x_C)^2}{n_C} - \frac{(\Sigma x)^2}{n}$$

$$= \frac{29^2}{12} + \frac{14^2}{12} + \frac{19^2}{12} - \frac{62^2}{36} = 9.7$$

Sum of the first replicates $= \Sigma_1 = 3 + 0 + 1 = 4$
Sum of the second replicates $= \Sigma_2 = 2 + 2 + 2 = 6$
Sum of the third replicates $= \Sigma_3 = 3 + 1 + 2 = 6$
Sum of the fourth replicates $= \Sigma_4 = 1 + 1 + 0 = 2$
Sum of the fifth replicates $= \Sigma_5 = 3 + 1 + 3 = 7$
Sum of the sixth replicates $= \Sigma_6 = 2 + 1 + 1 = 4$
Sum of the seventh replicates $= \Sigma_7 = 3 + 2 + 2 = 7$
Sum of the eight replicates $= \Sigma_8 = 2 + 0 + 1 = 3$
Sum of the ninth replicates $= \Sigma_9 = 3 + 1 + 2 = 6$
Sum of the tenth replicates $= \Sigma_{10} = 4 + 2 + 3 = 9$
Sum of the eleventh replicates $= \Sigma_{11} = 1 + 1 + 0 = 2$
Sum of the twelfth replicates $= \Sigma_{12} = 2 + 2 + 2 = 6$

$$\text{Members variance} = \sigma_{Rep}^2 = \frac{\Sigma_1^2}{n_1} + \frac{\Sigma_2^2}{n_2} + \frac{\Sigma_3^2}{n_3} + \frac{\Sigma_4^2}{n_4} + \frac{\Sigma_5^2}{n_5} + \frac{\Sigma_6^2}{n_6} + \frac{\Sigma_7^2}{n_7} + \frac{\Sigma_8^2}{n_8}$$

$$+ \frac{\Sigma_9^2}{n_9} + \frac{\Sigma_{10}^2}{n_{10}} + \frac{\Sigma_{11}^2}{n_{11}} + \frac{\Sigma_{12}^2}{n_{12}} - \frac{(\Sigma x)^2}{n} = 17.2$$

Residual variance $= \sigma_R^2 = \sigma_T^2 - \sigma_S^2 - \sigma_{Rep}^2 = 35.2 - 9.7 - 17.2 = 8.3$

Analysis of variance table

variance	ν	σ^2	σ^2/ν	F
Total	35	35.2	1.01	
Samples	2	9.7	4.85	12.8
Replicate	11	17.2	1.56	4.1
Residual	22	8.3	0.38	

where $\nu_{residual} = \nu_{total} - \nu_{samples} - \nu_{replicate} = 35 - 2 - 11 = 22$, $F_{samples/table} = 5.72$ ($\alpha = 0.02$; $\nu_{samples} = 2$, $\nu_{residual} = 22$, Tab. 5.8), $F_{samples/calculated} = (\sigma_B^2/\nu_{batch})/(\sigma_R^2/\nu_{residual}) = 12.8$. Since $F_{samples/calculated}$ is larger than $F_{batch/table}$ we may conclude that there is an actual difference in the bitterness of the fermented olives. $F_{replicate/table} = 3.19$ ($\alpha = 0.02$, $\nu_{replicates} = 11$, $\nu_{residuals} = 22$, Tab. 5.8) and $F_{replicate/calculated} = 4.1$. Since $F_{replicate/table}$ is smaller than $F_{replicate/calculated}$ we may conclude that there is an actual difference between the scores of the different panel members.

Scatter of the test scores may be compared with hypothesis test if: $\sigma_A^2 = \sigma_B^2$,

$$\text{Test:} \frac{1}{F_{\alpha/2}(n_A, n_B)} \leqslant (s_A^2/s_B^2) \leqslant F_{\alpha/2}(n_A, n_B) \qquad \text{((5.44) in Table 5.4)}$$

$$\bar{x}_A = \frac{1}{n_A}\sum_{i=1}^{n} x_A = 2.42, \quad \bar{x}_B = \frac{1}{n_B}\sum_{i=1}^{n_B} x_B = 1.17$$

$$s_A^2 = \frac{1}{n_A - 1}\sum_{i=1}^{n} (x - \bar{x})^2 = 0.82, \quad s_B^2 = \frac{1}{n_B - 1}\sum_{i=1}^{n} (x - \bar{x})^2 = 0.52$$

$F_{table} = 4.47$ ($\alpha = 0.02$, $\nu_A = 11$, $\nu_B = 11$, Tab. 5.8), after substituting F_{table} in the test: $0.22 \leqslant (s_A^2/s_B^2)_{exp} \leqslant 4.47$, since the $(s_A^2/s_B^2)_{exp} = 1.54$ is within the given limits, we conclude that there is no actual difference in the scatter of the scores of the samples A and B.

5.3. EMPIRICAL MODELS AND LINEAR REGRESSION

Representation of large amounts of experimental data by means of empirical equations is a practical necessity in science and engineering. The empirical models are easy to use in mathematical operations over a continuous range. The form of the empirical models may be suggested by theoretical or dimensional analysis or by intuition. The simplest empirical model is a line:

$$y = ax + b \qquad (5.48)$$

where y is the dependent variable, x is the independent variable. In a plot of y versus x parameter a is slope and b is the intercept with $x = 0$ axis. If it is possible to linearize an equation parameters a and b may be evaluated with linear regression. Procedures of linearization to evaluate the slope and the intercept of some common simple models may be summarized as:

model	linearization procedure
$y = a/x + b$	plot y versus $1/x$, slope $= a$, intercept $= b$
$y = \dfrac{x}{a + bx}$	rewrite the equation as $1/y = a/x + b$, plot $1/y$ versus $1/x$, slope $= a$, intercept $= b$
$y = be^{ax}$	rewrite the equation as $\ln y = \ln b + ax$, plot $\ln y$ versus x, slope $= a$, intercept $= \ln b$

In a plot of y versus x where almost all the data points fall on the line, values of parameters a and b may be easily determined from the slope and the intercept of the graph. When the data points are scattered it may be possible to draw many lines passing through them. The *best line* is the one with smallest sum of squares of difference defined as $\Sigma d^2 = \Sigma_{i=1}^{n} [y_i - (ax_i + b)]^2$ where $ax_i + b$ is the value of parameter y predicted by (5.48) and y_i is the experimentally determined value of y corresponding to x_i. The term $d = y_i - (ax_i + b)$ is the difference between the experimentally determined and predicted values of y at the point x_i. This difference might be either negative or positive. Regardless of the sign, the magnitude of the difference describes the deviation of the line from the data. When the differences are added up over the entire data set, the negative and positive differences may cancel each other and cause an erroneous conclusion. Working with the squares of the differences eliminates the cause of the erroneous conclusion. The minimum value of the sum of the squares difference is obtained with the following best line parameters:

$$a = \frac{n \sum_{i=1}^{n} x_i y_i - \left(\sum_{i=1}^{n} x_i\right)\left(\sum_{i=1}^{n} y_i\right)}{n \sum_{i=1}^{n} x_i^2 - \left(\sum_{i=1}^{n} x_i\right)^2} \qquad (5.49)$$

$$b = \frac{\sum_{i=1}^{n} x_i^2 \left(\sum_{i=1}^{n} y_i\right) - \left(\sum_{i=1}^{n} x_i y_i\right)\left(\sum_{i=1}^{n} x_i\right)}{n \sum_{i=1}^{n} x_i^2 - \left(\sum_{i=1}^{n} x_i\right)^2} \qquad (5.50)$$

Although (5.49) and (5.50) gives the values of a and b of the best fitting line, it does not mean that the best fitting line will always describe the correct relation between variables y and x. It is always possible that the relation between y and x may not be actually described with (5.48). The success of the best fitting line to describe n sets of data points may be evaluated with standard error of estimate:

$$s_e = \sqrt{\frac{\Sigma d^2}{n}} \tag{5.51}$$

Standard error of estimate is a measure of the degree of association between the data points and the regression line. The larger the standard error of estimate, the larger is the scatter of the data points around the fitted line. If all the data points are scattered by approximately normal distribution around the regression line, 99.73% of all the data points are scattered within the range of $\pm 3s_e$.

The correlation coefficient of the data and the best line is:

$$r = \sqrt{1 - \frac{s_e^2}{s_y^2}} \tag{5.52}$$

where s_y is the standard deviation of values of y around its mean value:

$$s_y = \sqrt{\frac{\Sigma y^2}{n} - \left(\frac{\Sigma y}{n}\right)^2} \tag{5.53}$$

The term s_e^2/s_y^2 is actually the ratio of the variance of the data points around the fitted line to their variance around the mean value of y. Having s_e much smaller than s_y implies that the scatter of the data points around the fitted line is almost negligible when compared to the scatter around the average value of y. At the limiting condition of this case the ratio $s_e^2/s_y^2 = 0$, thus $r = 1$ implying a perfect fit. Having s_e the same as s_y implies that the scatter of the data points around the fitted line is almost the same as the scatter around the average value of y, implying that the fitted line does not represent the data.

At the limiting condition of this case the ratio $s_e^2/s_y^2 = 1$, thus $r = 0$ implying no fit. Values of the correlation coefficient r varies between the limits 0 and 1. The correlation coefficient may also be defined as:

$$r = \frac{s_{xy}}{s_x s_y} \tag{5.54}$$

where s_x is the standard deviation of values of x around their mean value:

$$s_x = \sqrt{\frac{\Sigma x^2}{n} - \left(\frac{\Sigma x}{n}\right)^2} \tag{5.55}$$

and s_{xy} is defined as:

$$s_{xy} = \frac{\Sigma xy}{n} - \left(\frac{\Sigma x}{n}\right)\left(\frac{\Sigma y}{n}\right) \tag{5.56}$$

Equation (5.54) also gives the sign of the correlation, whereas (5.52) gives its absolute value only. Having $r = -1$ implies that there is perfect correlation between the two parameters, but while one of them is increasing the other one is decreasing. Square of the correlation coefficient, r^2, may be preferred to describe the correlation between the variables, since it magnifies the deviation from the perfect correlation when r is close to 1.

Parameters a, b and r as described by (5.49), (5.50) and (5.54) respectively, may be easily calculated with simple scientific calculators containing the related built in functions by entering the data only. Common spread sheet computer software also serve the same purpose. Non-linear regression is more sophisticated, but may be easily done with the common commercially available statistics and mathematics software.

Example 5.28. Survival kinetics of freeze dried lactic acid bacteria The number of viable microorganisms per gram of a freeze dried preparation is the major quality factor of the freeze dried cultures. Variation of the number of freeze dried viable microorganisms with time in storage may be expressed as:

$$\frac{dx}{dt} = -k_d x \tag{3.67}$$

Equation (3.67) may be integrated as:

$$\log x = \log x_0 - \frac{k_d}{2.303} t \tag{E.5.28.1}$$

where x_0 is the number of the viable microorganisms at $t = 0$. The following counts (number of viable microorganisms/ml) of the freeze dried lactic acid starter culture microorganisms were reported by Alaeddinoğlu *et al.* (1988):

t (days)	0	15	30	45	60	90
$\log x$	7.8	7.0	6.2	5.4	4.8	4.4

a) Calculate values of the constants $\log x_0$ and k_d and the correlation coefficient.

Solution We may change the notation as $y = \log x$, $b = \log x_0$, $a = -k_d/2.303$, then the above equation may be expressed as $y = ax + b$. It may also be shown that $\Sigma x_i = 240$, $\Sigma y_i = 35.6$, $\Sigma x_i y_i = 1218$, $\Sigma x_i^2 = 14\,850$, $(\Sigma x_i^2)^2 = 57\,600$ and also $n = 6$. After substituting them into (5.49) and (5.50) $a = -0.039$, $b = 7.5$. Since $a = -k_d/2.303$ and $b = \log x_0$, we may calculate $k_d = 0.09$ day^{-1} and $\log x_0 = 7.5$.

The fitted equation is $\log x_{reg} = 7.5 - 0.039t$. The squares of the differences between $\log x$ and $\log x_{reg}$ may be calculated as

t (days)	$\log x$	$\log x_{reg}$	d^2
0	7.8	7.5	0.09
15	7.0	6.9	0.01
30	6.2	6.3	0.01
45	5.4	5.7	0.09
60	4.8	5.1	0.09
90	4.4	4.0	0.16
			$\Sigma d^2 = 0.45$

$$s_e = \sqrt{\frac{\Sigma d^2}{n}} = 0.27, \quad s_y = \sqrt{\frac{\Sigma y^2}{n} - \left(\frac{\Sigma y}{n}\right)^2} = 1.2 \quad \text{and} \quad r = \sqrt{1 - \frac{s_e^2}{s_y^2}} =$$

0.95, (5.52) was used here and implied an almost perfect fit. Correlation coefficient may also be calculated using (5.54):

$$s_x = \sqrt{\frac{\Sigma x^2}{n} - \left(\frac{\Sigma x}{n}\right)^2} = 29.6, \quad s_{xy} = \frac{\Sigma xy}{n} - \left(\frac{\Sigma x}{n}\right)\left(\frac{\Sigma y}{n}\right) = -34.3$$

$$\text{and consequently } r = \frac{s_{xy}}{s_x s_y} = -0.97.$$

Equation (5.54) implies a perfect in agreement with the previous result. It gives additional information that there is negative correlation between $\log x$ and t, i.e., log number of the viable freeze dried microorganisms decrease as time passes by. Comparison of the best line with the experimental data (symbols) are shown in Figure E.5.28.

b) How many microorganisms/ml will remain viable on the seventy-fifth day of storage? State with 99.73% probability.

Solution When $t = 75$ we may calculate $\log x_{reg} = 7.5 - 0.039t = 4.56$. Since with $p = 0.9973$ experimental data are scattered within $\pm 3\,s_e$ range of the best line, the confidence limits are $\log x_{reg} - 3\,s_e \leqslant \log x \leqslant \log x_{reg} + 3\,s_e$ after substituting the numbers $3.75 \leqslant \log x \leqslant 5.37$.

5.4. QUALITY CONTROL CHARTS FOR MEASUREMENTS

There is always a certain amount of variability in any production process. It is mostly random and cannot be completely eliminated. A manufacturing process is called under *statistical control* when the variability is confined in the *random variability limits*. If the process variability goes out of these

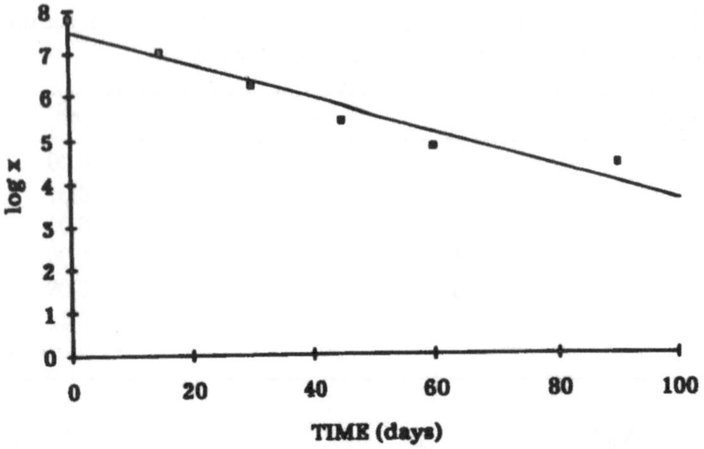

FIGURE E.5.28 Comparison of the best fitted line (—) with the experimental data (symbols).

limits, there must be something wrong with the process and corrective action is needed (Fig. 5. 11).

While constructing the statistical quality control charts we are actually preparing the graphical representation of the confidence limits, presented in the previous discussion. When we consider the analogy of the two concepts CL, UCL and LCL corresponds to the population mean, upper confidence limit and the lower confidence limit, respectively. The quality factor may be evaluated either by measurements or by the counts of the non-conforming units, i.e., attributes. Quality control charts are based on different statistical distribution models for each case.

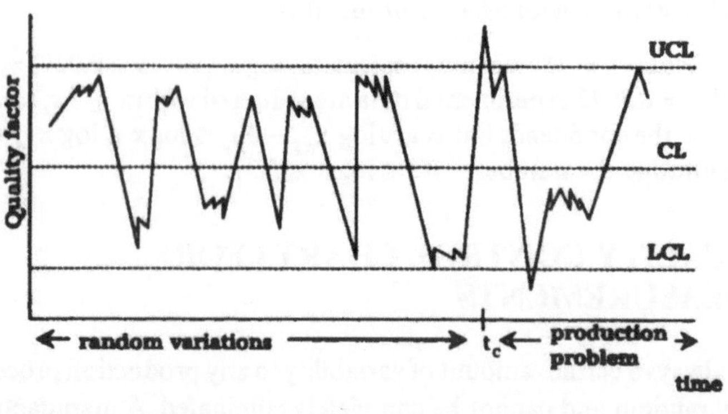

FIGURE 5.11 Variations in a quality factor during processing. The process goes out of statistical control at time $t = t_c$

Quality control techniques are generally applied to commodities consisting of a large number of individuals. It is not usually feasible to measure the quality factor of each item, therefore the tests are applied on the samples. The samples may be destroyed, i.e., pressed or ground and the measurements are made with the juice or powder. The sample size is usually chosen much smaller than the total size of the population to reduce the cost of quality control. The Shewhart control charts consist of means and range charts and are used in quality control with measured properties. The means chart aims to confine the variation of the sample means between the predetermined acceptable limits. The range chart aims to limit the range within individual samples. The mean of the replicate measurements may be calculated as:

$$\bar{x} = \frac{1}{n} \sum_{i=1}^{n} x_i \tag{5.10}$$

Where $\bar{x}$ is the sample mean, x_i's are the values of the individual items constituting the sample, and n is the number of the items in the sample set. The means charts are constructed to control the variation of the means of the replicate samples within the required UCL and LCL. The range of a sample is defined as:

$$R_i = x_{i,\max} - x_{i,\min} \tag{5.57}$$

where $x_{i,\max}$ and $x_{i,\min}$ corresponds to the maximum and the minimum measurements of the quality factor in the ith sample set. The mean value of the ranges may be calculated as:

$$\bar{R} = \frac{1}{k} \sum_{i=1}^{k} R_i \tag{5.58}$$

where k is the number of the sample sets. In such a quality control plan the central line (CL_x) is

$$CL_x = \frac{1}{k} \sum_{i=1}^{k} \bar{x}_i \tag{5.59}$$

The upper (UCL_x) and lower (LCL_x) control limits of the means chart may be calculated as:

$$UCL_x = CL_x + \frac{3\sigma}{\sqrt{n}} \tag{5.60}$$

and

$$LCL_x = CL_x - \frac{3\sigma}{\sqrt{n}} \tag{5.61}$$

If σ is not known it may be estimated as:

$$\sigma = \frac{\bar{R}}{d_2} \tag{5.62}$$

values of d_2 are given in Table 5.10. After substituting (5.62) in (5.60) and (5.61):

$$UCL_x = CL_x + \frac{3\bar{R}}{d_2\sqrt{n}} \tag{5.63}$$

$$LCL_x = CL_x - \frac{3\bar{R}}{d_2\sqrt{n}} \tag{5.64}$$

The range charts are constructed to ensure that the range, i.e., the variance, of the individual samples are confined within pre-determined limits. The central line (CL_R) and the upper (UCL_R) and lower (LCL_R) control limits of the range chart may be calculated as:

$$CL_R = \bar{R} \tag{5.65}$$

$$UCL_R = \bar{R} + 3d_3\sigma \tag{5.66}$$

$$LCL_R = \bar{R} - 3d_3\sigma \tag{5.67}$$

Table 5.10 Control chart constants

n	d_2	D_1	D_2	D_3	D_4
2	1.128	0	3.686	0	3.267
3	1.693	0	4.358	0	2.575
4	2.059	0	4.698	0	2.282
5	2.326	0	4.918	0	2.115
6	2.534	0	5.078	0	2.004
7	2.704	0.205	5.203	0.076	1.924
8	2.847	0.387	5.307	0.136	1.864
9	2.970	0.546	5.394	0.184	1.816
10	3.078	0.687	5.496	0.223	1.777
11	3.173	0.812	5.534	0.256	1.744
12	3.258	0.924	5.592	0.284	1.716
13	3.336	1.026	5.646	0.308	1.692
14	3.407	1.121	5.693	0.329	1.671
15	3.472	1.207	5.737	0.348	1.652

After substituting the equivalent of $\bar{R}$ from (5.62) and letting $D_1 = d_2 - 3\,d_3$ and $D_2 = d_2 + 3\,d_3$ we may obtain

$$UCL_R = D_2\sigma \tag{5.68}$$

$$LCL_R = D_1\sigma \tag{5.69}$$

When σ is not known its value may be estimated from (5.62) and substituted in (5.68) and (5.69) as:

$$UCL_R = D_4\bar{R} \tag{5.70}$$

$$LCL_R = D_3\bar{R} \tag{5.71}$$

where $D_3 = D_1/d_2$ and $D_4 = D_2/d_2$. Equations (5.70) and (5.71) may be used when σ is not known. Values of d_3, D_1, D_2, D_3 and D_4 are given in Table 5.10.

In any statistical quality control plan the ranges and the means of the samples are tried to be kept within the pre-determined control limits. Violation of these limits warns the operators that the quality will deteriorate if the process conditions are not corrected, and the operators should readjust the process parameters to prevent quality loss.

The central limit theorem states that regardless of the distribution behavior of the individual values around the population mean, the sampling distribution of the means will approach normality as the sample size increases, i.e., $n \geqslant 4$. Therefore the means and range charts may be used to control a process even when the distribution of individual measurements around the population mean is not Gaussian.

Example 5.29. The means and range charts for a packaging process Five consecutive packages were taken from a production line and weighed for ten days while a filling machine was operating. Sample weights were reported in grams as:

X_A	X_B	X_C	X_D	X_E	X_F	X_G	X_H	X_I	X_J
103	104	102	103	102	101	103	104	102	100
101	103	102	102	101	97	104	106	104	102
104	105	104	99	104	100	102	102	99	98
100	99	101	102	101	102	101	103	102	102
99	103	102	98	100	100	102	99	102	102

Prepare control charts for means and ranges.

Solution The sample means and the ranges were calculated as

$$\bar{x} = \frac{1}{n}\sum_{i=1}^{n} x_i \quad \text{and} \quad R_i = x_{i,max} - x_{i,min}$$

	A	B	C	D	E	F	G	H	I	J
$\bar{x}_i$	101.4	102.8	102.2	100.8	101.6	100.0	102.4	102.8	101.8	100.8
R_i	5	6	3	5	4	5	3	7	5	4

These data may be used to calculate

$$CL_x = \frac{1}{k}\sum_{i=1}^{k} \bar{x}_i = 101.7 \text{ and } \bar{R} = \frac{1}{k}\sum_{i=1}^{k} R_i = 4.7.$$

Since σ is not known, we may calculate UCL_x, LCL_x, UCL_R and LCL_R with $n = 5$, $d_2 = 2.326$, $D_3 = 0$ and $D_4 = 2.115$ as $UCL_x = CL_x + 3\bar{R}/d_2\sqrt{n} = 104.4$, $LCL_x = CL_x - 3\bar{R}/(d_2\sqrt{n}) = 98.9$, $LCL_R = D_3\bar{R} = 0$, $UCL_R = D_4\bar{R} = 9.9$ The means and the range chart are shown in Figure E.5.29.

b) Determine the control chart limits if the minimum sample mean fill is required to be 100 g.

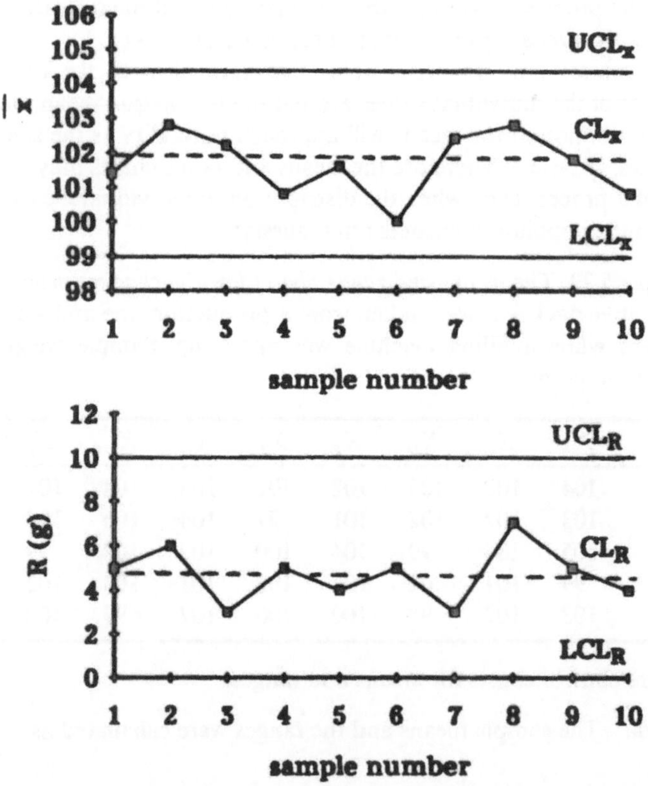

FIGURE E.5.29 The means and range charts for weight control of the packages.

Solution $LCL_x = 100$ g. Equation (5.64) requires $CL_x = LCL_x + 3\bar{R}/(d_2\sqrt{n}) = 102.7$, therefore $UCL_x = CL_x + 3\bar{R}/(d_2\sqrt{n}) = 105.4$. The range chart will remain the same.

c) Readjust the control limits if only 10% of the average package weights are tolerated to be less than 100 g.

Solution Standard normal variable is $z_{0.10} = (\bar{x} - CL_x)/(\sigma/\sqrt{n})$, $z_{0.10} = -z_{0.90} = -1.29$ (Tab. 5.5), $\bar{x} = 100$, $n = 5$ and $\sigma = \bar{R}/d_2 = 2.02$. After filling the numbers in the above equation we obtain, $-1.29 = (100 - CL_x)/(2.02/\sqrt{5})$, $CL_x = 101.2$, consequently $UCL_x = CL_x + 3\bar{R}/(d_2\sqrt{n}) = 103.9$ and $LCL_x = CL_x - 3\bar{R}/(d_2\sqrt{n}) = 98.5$. The range chart will remain the same.

Equations (5.59)–(5.71) are used in processes where quality factors are maintained within pre-specified constant limits, and are not suitable for use in quality control of some foods, i.e., apricots, apples and eggs, in storage, where deterioration is inevitable and acceptable at reasonably slow rates (Şumnu *et al.*, 1994a and 1994b; Kahraman-Doğan *et al.*, 1994). The central line of the means and the range charts may be determined as:

$$CL_x = CL_{x0} - \alpha t \tag{5.72}$$

and

$$CL_R = CL_{R0} - \beta t \tag{5.73}$$

where CL_x and CL_R represent the fitted sample means, and the ranges, respectively. CL_{x0} and CL_{R0} are the initial average weight and range of the sample, α and β are the slopes. Equations of the lines given in (5.72) and (5.73) may be determined with linear regression. Equation (5.72) may be substituted for CL_x and (5.73) for CL_R in (5.63)–(5.71) to construct the means and the range charts for the processes with time dependent CL_x and CL_R.

Example 5.30. Means and range charts for storage of eggs (*Kahraman-Doğan et al.*, 1994) Variation of the mean and range of the internal quality of untreated eggs in storage were evaluated in Haugh Units (HU) as $CL_x = 76.7 - 1.15t$ and $CL_R = 12.7$. By using the following sample mean and ranges ($n = 3$) determine if the process was under statistical control?

t (days)	0	3	6	10	20	25	28	31	38	45
$\bar{x}$ (HU)	89.4	73.4	67.7	66.8	43.2	38.3	40.4	38.3	36.4	37.1
R (HU)	4.6	23.7	24.4	20.0	3.5	7.6	7.7	4.8	17.2	9.3

Solution After substituting $CL_x = 76.7 - 1.15t$ and $\bar{R} = CL_R = 12.7$ in (5.63), (5.64), (5.70) and (5.71) with $d_2 = 1.693$; $D_3 = 0$ and $D_4 = 2.575$ we will obtain $LCL_x = 63.7 - 1.15t$, $UCL_x = 89.7 - 1.15t$, $LCL_R = 0$ and

$UCL_R = 32.7$. The control charts are shown in Figure E.5.30. Since all the measurements are confined within the limits we may conclude that the process is under statistical control.

In most processes *USL* (upper specification limit) and *LSL* (the lower specification limit) are different from the control limits. Usually LCL_x ($LCL_x > LSL$) CL_x and UCL_x ($UCL_x < USL$) are established to assure a tolerance band for process fluctuations. When the control limits are constant, *process capability index* C_{pk} measures centering of the data in the control charts and defined as the smallest of

$$C_{pk} = \frac{CL_x - LSL}{3\sigma} \tag{5.74}$$

and

$$C_{pk} = \frac{USL - CL_x}{3\sigma} \tag{5.75}$$

The C_{pk} as defined in equations (5.74) and (5.75) measures the distance between the process mean and *LSL* or *USL* and expresses this as a ratio of half of the width of the normal distribution curve (Oakland and Followell, 1990). A C_{pk} of 1 or less means that the width of the distribution curve and its

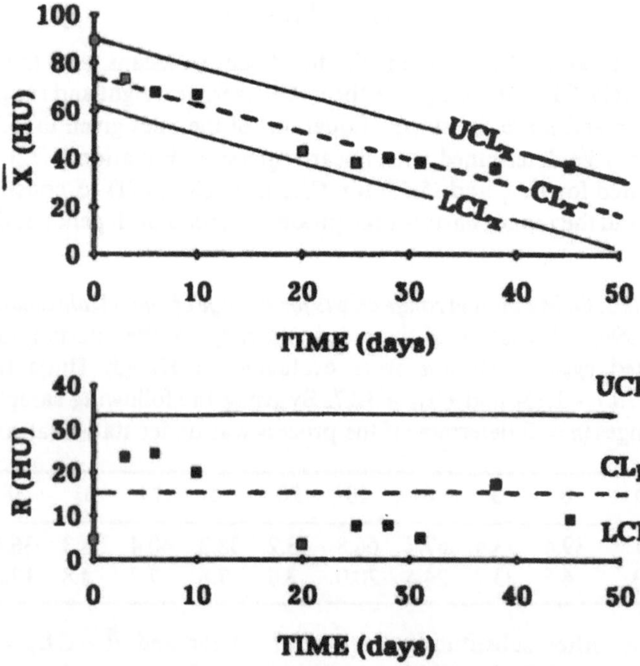

FIGURE E.5.30 Control charts for the Haugh Units of the eggs.

centering is such that it infringes the tolerance limit and the process needs improvement.

Example 5.31. Process capability index as a measure of the goodness of the quality control chart a) A set of packaging equipment operates with the following CL_x and σ at different settings. If USL and the LSL of the packages are 505 and 495 g, respectively, which one of these settings should be preferred?

Setting	CL_x (g)	σ (g)
I	498	1.2
II	500	2.0
III	501	0.5

Solution Process capability index is defined as the smaller one of $C_{pk} = (CL_x - LSL)/3\sigma$ and $C_{pk} = (USL - CL_x)/3\sigma$ calculated as

Setting	I	II	III
C_{pk}	0.83	0.83	2.67

Settings I and II yields $C_{pk} = 0.83$, implying that the data are badly scattered and the process does not satisfy the required standards. Having $C_{pk} = 2.67$ with setting III shows that the data are scattered among the process limits satisfactorily.

b) What is the probability of a sample mean to violate USL or LSL at setting III with sample size $n = 5$?

Solution Standard normal variable is $z = (USL - CL_x)/(\sigma/\sqrt{n}) = 16$ or $z = (LSL - CL_x)/(\sigma/\sqrt{n}) = -12$ indicating that the probability of violating USL or LSL with random process fluctuations is practically zero.

It was shown in Example 5.2 that by taking the logarithm of a variable it was possible to convert a population from non-normal to normal distribution. There are also a number of other conversion methods as depicted in Table 5.11.

Example 5.32. Quality control charts for alcohol content of bottled beer (*Özilgen* 1998) Sample means of the alcohol content $\bar{x}_i$ in bottled beer are non-normally distributed around μ, therefore the conventional control charts are not satisfactory; natural logarithmic transformation is applied (Tab. 5.11) as $y_i = \ln(x_i - \theta)$ (where $\theta = 0.5$). The original (x_i) and the transformed (y_i) data are shown in Table E.5.32. Prepare the means and range control charts for the process.

Table 5.11 The common transformation techniques from non-normal to normal distribution, the associated standard variable and control limits ($p = 0.9973$) (Jacobs, 1990)

transformation	transformed variable	standard variable	LCL_x *and* UCL_x
natural logarithm	$y_i = \ln(x_i - \theta)$ where $(\theta < x_{min})$	$z = \dfrac{\ln(x-\theta) - \bar{y}}{\sigma_v}$	$UCL_x = \theta + \exp(\bar{y} + 3\sigma_y)$ $UCL_x = \theta + \exp(\bar{y} + 3\sigma_y)$
square root	$y_i = \sqrt{(x_i - \theta)}$ where $(\theta \leqslant x_{min})$	$z = \dfrac{\sqrt{(x-\theta)} - \bar{y}}{\sigma_v}$	$LCL_x = \theta + (\bar{y} - 3\sigma_y)^2$ $UCL_x = \theta + (\bar{y} + 3\sigma_y)^2$
arcsine	$y_i = \arcsin\left[\sqrt{\dfrac{x_i}{\theta}}\right]$ where $\theta \leqslant x_{max}$	$z = \dfrac{\arcsin\left[\sqrt{\dfrac{x_i}{\theta}}\right] - \bar{y}}{\sigma_v}$	$LCL_x = \theta \sin(\bar{y} - 3\sigma_y)^2$ $UCL_x = \theta \sin(\bar{y} + 3\sigma_y)^2$

where x_i = non-normal parameter, y_i = normally distributed parameter, x_{min} = minimum attainable value of x_i in the data set, x_{max} = maximum attainable value of x_i in the data set, θ = threshold parameter, $\bar{y}$ = average of the values of y and σ_y = standard deviation of the y values.

Table E.5.32 The original and the transformed alcohol contents of the beer

Sample	x_i (%)	$\bar{x}$	y_i	$\bar{y}$	R_x
1	4.3, 4.3, 4.2	4.27	1.34, 1.34, 1.31	1.33	0.1
2	4.3, 4.3, 4.4	4.33	1.34, 1.34, 1.36	1.35	0.1
3	4.7, 4.7, 4.7	4.7	1.44, 1.44, 1.44	1.44	0
4	4.6, 4.5, 4.6	4.56	1.41, 1.39, 1.41	1.40	0.1
5	4.4, 4.6, 4.4	4.47	1.36, 1.41, 1.36	1.38	0.2
6	4.2, 4.4, 4.3	4.3	1.31, 1.36, 1.34	1.34	0.2
7	4.8, 4.8, 4.6	4.73	1.46, 1.46, 1.41	1.44	0.2
8	4.3, 4.4, 4.6	4.43	1.34, 1.36, 1.41	1.37	0.3
9	4.1, 4.3, 4.3	4.23	1.28, 1.34, 1.34	1.32	0.2
10	4.0, 4.0, 4.0	4.0	1.25, 1.25, 1.25	1.25	0
11	4.1, 4.1, 4.0	4.07	1.28, 1.28, 1.25	1.27	0.1
12	3.9, 3.8, 4.2	3.97	1.22, 1.20, 1.31	1.24	0.4
13	4.3, 4.0, 4.2	4.17	1.34, 1.25, 1.31	1.30	0.3
14	4.1, 4.1, 4.0	4.07	1.28, 1.28, 1.25	1.27	0.2
15	4.0, 4.2, 4.2	4.13	1.25, 1.31, 1.31	1.29	0.2
16	4.0, 4.1, 4.1	4.07	1.25, 1.28, 1.28	1.27	0.1
17	3.9, 3.9, 3.9	3.9	1.22, 1.22, 1.22	1.22	0
18	4.0, 3.9, 3.9	3.93	1.25, 1.22, 1.22	1.23	0.1
19	4.4, 4.3, 4.3	4.33	1.36, 1.34, 1.34	1.35	0.1
20	4.3, 4.3, 4.4	4.33	1.34, 1.34, 1.36	1.35	0.1
21	4.1, 4.1, 4.0	4.07	1.28, 1.28, 1.25	1.27	0.1
22	3.9, 3.9, 4.0	3.93	1.22, 1.22, 1.25	1.23	0.1
23	4.1, 4.0, 4.1	4.07	1.28, 1.25, 1.28	1.27	0.1
24	3.9, 3.9, 4.0	3.93	1.22, 1.22, 1.25	1.23	0.1
25	4.1, 4.1, 4.1	4.1	1.28, 1.28, 1.28	1.28	0
26	3.7, 3.8, 3.8	3.77	1.16, 1.20, 1.20	1.19	0.1

We have

$$\bar{y} = \frac{1}{k}\sum_{i=1}^{k} y_i = 1.30, \quad \sigma_y = \sqrt{\frac{1}{k-1}\sum_{i=1}^{k} (y_i - \bar{y})^2} = 0.07,$$

$LCL_x = \theta + \exp(\bar{y} - 3\sigma_y) = 3.47$, $UCL_x = \theta + \exp(\bar{y} + 3\sigma_y) = 5.03$, $\bar{R}_x = 1/k\,\Sigma_{i=1}^{k} R_i = 0.12$; therefore $CL_R = \bar{R} = 0.12$, since $D_3 = 0$ and $D_4 = 2.575$ with $n = 3$ (Tab. 5.10) we will have $LCL_R = D_3\bar{R} = 0$ and $UCL_R = D_4\bar{R} = 0.3$. The means and the range charts (Fig. E.5.32) indicate that there are no data points outside the control limits indicating that the process is under control.

Example 5.33. Analysis of sour dough bread-making process (*Durukan et al.*, 1992) In a typical sour dough bread-making process (Fig. E.5.33.1) dough is cut into small portions with the divider, then these portions are put into a spherical shape in a rounder. The dough undergoes the fermentation process in the a resting chamber. The loaves are given their final elongated shape in a long moulder. The loaves are fermented further in the final proofer and baked in the oven. The hot bread is cooled for about five hours before being distributed to the markets.

The distribution of the relative frequency of the dough portion and the loaf weights at the sampling locations A-E indicated two superimposed populations (Fig. E.5.33.2). The first population was initiated by the dough portions with exactly 400 g weight, and the remaining dough portions initiated the second population. There were four knives in the divider and apparently one of them was malfunctioning. Since the malfunctioning did not cause considerable commercial effect on the operation of the plant it was not replaced, but created the superimposed populations.

Each sample contained 15 dough portions (or loaves). Measurement of each sample set were separated intuitively into two groups. Each group was

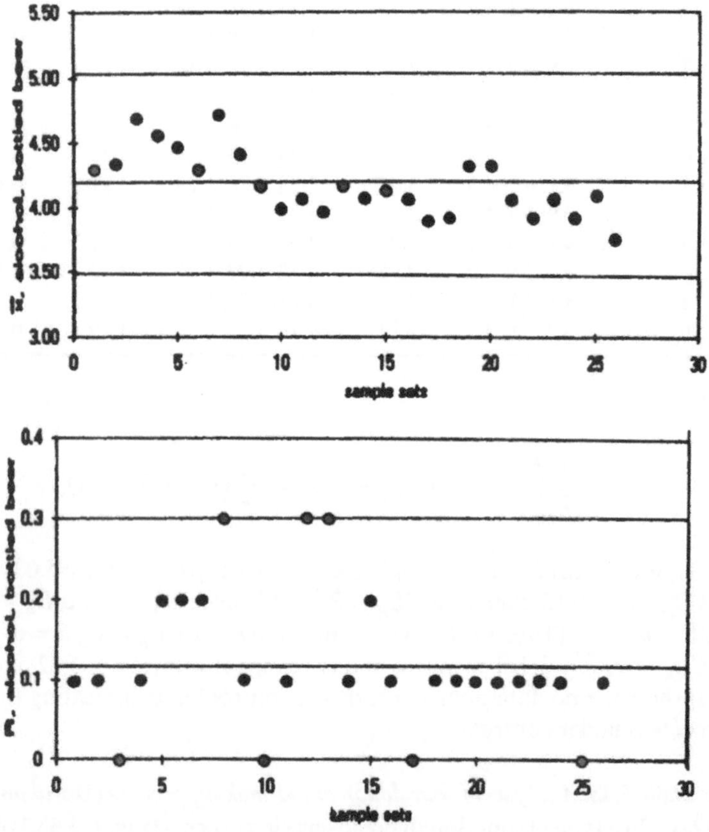

FIGURE E.5.32 The means and the range charts for the percentage of alcohol in the bottled beer (Özilgen 1998).

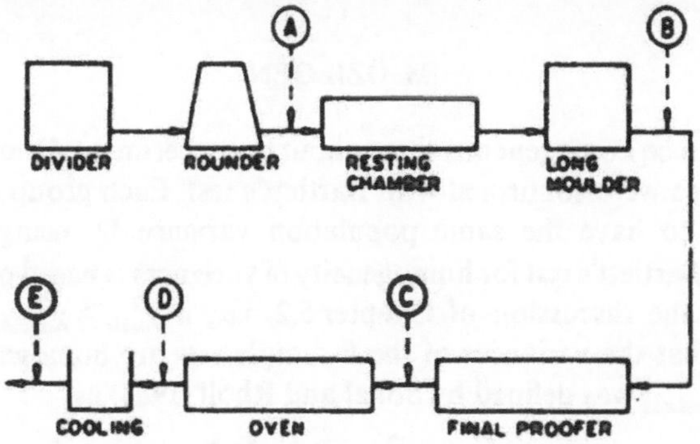

FIGURE E.5.33.1 Flow chart of the sour dough bread making process. A, B, C, D and E are the sampling locations (Durukan *et al.*, 1992; © Elsevier Science Publishers B.V., reproduced by permission).

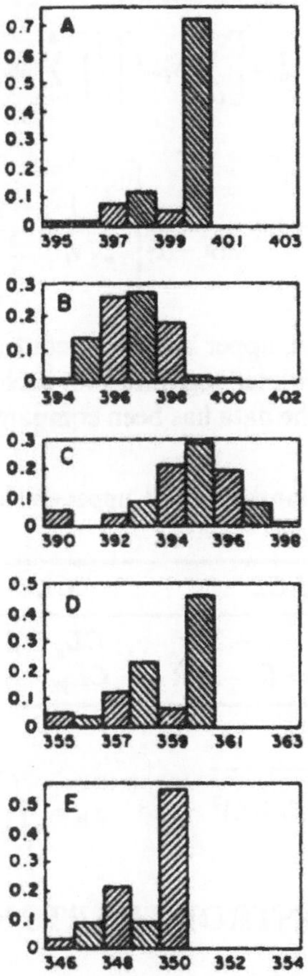

FIGURE E.5.33.2 Relative frequency of the dough portion (loaf) weights at the sampling locations A-E (Durukan *et al.*, 1992; © Elsevier Science Publishers B.V., reproduced by permission).

supposed to be homogeneous throughout the experiment. Homogeneity of these groups were confirmed with Bartlett's test. Each group of data was confirmed to have the same population variance by using the F-test (Tab. 5.4). Bartlett's test for homogeneity of variances is based on the χ^2 test similar to the discussion of Chapter 5.2, i.e., if $\chi^2_{table} \geq \chi^2_{calculated}$ we may conclude that the variances of the k sample sets are homogeneous. Parameter $\chi^2_{calculated}$ was defined by Sokal and Rholf (1981) as:

$$\chi^2_{calculated} = \frac{\left[\sum_{i=1}^{k} v_i\right] \ln(s_p^2) - \left[\sum_{i=1}^{k} v_i \ln(s_p^2)\right]}{C}$$ (E.5.33.1)

where k is the number of the sample sets, s_p^2 and C are the generalized pooled variance and correction factors:

$$s_p^2 = \left[\sum_{i=1}^{k} v_i s_i^2\right] \Big/ \left[\sum_{i=1}^{k} v_i\right]$$ (E.5.33.2)

and

$$C = 1 + \frac{1}{3(k-1)}\left[\sum_{i=1}^{k} \frac{1}{v_i} - \frac{1}{\sum_{i=1}^{k} v_i}\right]$$ (E.5.33.3)

Equations for the central, upper and lower control lines of the means and the range charts are calculated separately for each group with the formulas given in Table E.5.33. The data has been compared with the control limits in Figure E.5.33.3.

Table E.5.33. Equations for the central, upper and lower control limits for the means and the range charts

	LCL	CL	UCL
Means chart	$LCL_x = \mu - 3\sigma\bar{x}$	$CL_x = \mu$	$UCL_x = \mu + 3\sigma\bar{x}$
Range chart	$LCL_R = \bar{R} - 3\sigma\bar{R}$	$CL_R = \bar{R}$	$UCL_R = \bar{R} + 3\sigma\bar{R}$

where $\sigma_x = \sqrt{\dfrac{1}{k-1}\sum_{i=1}^{k}(\bar{x}_i - \mu)^2}$ and $\sigma_R = \sqrt{\dfrac{1}{k-1}\sum_{i=1}^{k}(R_i - \bar{R})^2}$

5.5. QUALITY CONTROL CHARTS FOR ATTRIBUTES

Attributes are characteristics used for quality control when actual measurements of the quality factors are not available. In quality control with attributes, simpler methods, i.e., visual observation of the defects, etc., may substitute for measurements. Each unit in a sample set is characterized on

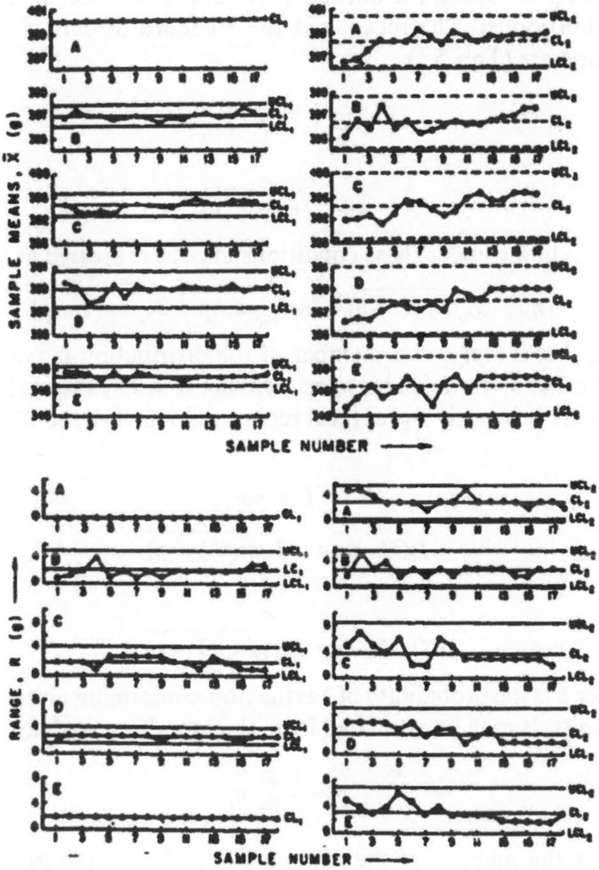

FIGURE E.5.33.3 The means and range charts for the samples taken at locations A-E. The charts appearing on the left are pertaining to the 400 g dough portions. The charts appearing on the right are pertaining to the remaining dough portions (Durukan *et al.*, 1992; © Elsevier Science Publishers B.V., reproduced by permission).

the basis of either *conforming* or *non-conforming* the required standards. Quality control with attributes saves time and money when used effectively. Weighing apples one by one is a way of obtaining the measurements to construct the quality control charts. The same purpose may be achieved if apples are rolled over a slanted surface with constant diameter holes. Apples larger than the hole size are the conforming ones and will remain on the top while the others, i.e., non-confirming small apples, are collected under the separator. This procedure helps to obtain counts of the non-conforming apples rapidly. The latter procedure may be preferred to the former because it requires less labor and costs less. The counts of the non-conforming units in a sample set of constant size are used to construct the *np charts* following the normal curve approximation of the binomial distribution model. The mean and the standard deviation of binomial distribution are (Tab. 5.1):

$$\mu = np \tag{5.4}$$

and

$$\sigma = \sqrt{np(1-p)} \tag{5.5}$$

Confidence limits under these conditions may be expressed as (Tab. 5.3):

$$np - 3\sqrt{np(1-p)} \leqslant (np)_{exp} \leqslant np + 3\sqrt{np(1-p)} \tag{5.28}$$

Implying that if $(np)_{exp}$ is a member of the distribution curve with given values of μ and σ, it will fall between these limits in 99.73% of all occasions. The *np* chart is actually a graphical representation of the confidence limits with

$$CL = np \tag{5.76}$$

$$LCL = np - 3\sqrt{np(1-p)} \tag{5.77}$$

and

$$UCL = np + 3\sqrt{np(1-p)} \tag{5.78}$$

Parameter p is the probability of having non-conforming units in a sample set of n units. It may be estimated from the experimental data:

$$p = \frac{1}{k}\sum_{i=1}^{k} p_i \tag{5.79}$$

where k is the number of the sample sets and p_i is the fraction of the non-confirming units in the ith sample set.

When the fractions of defectives, instead of their counts, are determined with constant sample size n, the *p chart* may be prepared. Control chart parameters of the *p chart with constant n* may be obtained after dividing (5.76)–(5.78) into n:

$$CL = p \tag{5.80}$$

$$LCL = p - 3\left(\sqrt{\frac{p(1-p)}{n}}\right) \tag{5.81}$$

and

$$UCL = p + 3\left(\sqrt{\frac{p(1-p)}{n}}\right) \tag{5.82}$$

Example 5.34. The np and p charts for quality control of confectionery products In a confectionery process 200 bars were sampled after coating for 10 days and the following number (np_i) and fraction (p_i) of defectives were found:

day	np_i	p_i
1	5	0.025
2	4	0.020
3	4	0.020
4	7	0.035
5	3	0.015
6	2	0.010
7	6	0.030
8	4	0.020
9	6	0.030
10	7	0.035

a) Construct an np chart for maintaining the present level of operation.

Solution $k = 10$ and $p = (1/k)\,\Sigma_{i=1}^{k} p_i = 0.024$. The control line parameters are $CL = np = 4.8$, $LCL = np - 3\sqrt{np(1-p)} = -1.69$ and $UCL = np + 3\sqrt{np(1-p)} = 11.3$. Since it is not possible to have a negative number of defective bars the lower control limit will be re-established as $LSL = 0$. The control chart is given in Figure E.5.34.1.

b) Construct a p chart with the same data.

Solution The p chart control parameters are $CL = p = 0.024$, $LCL = p - 3\sqrt{p(1-p)/n} = -0.008$. Since it is not possible to have a negative fraction of defective bars lower control limit will be re-established as $LCL = 0.$ and $UCL = p + 3\sqrt{p(1-p)/n} = -0.056$. Comparison of the data with the control limits is given in Figure E.5.34.2.

Equations (5.80)–(5.82) may also be used with variable size samples (n is not constant). In this case p may be evaluated as

$$p = \frac{\sum_{i=1}^{k} p_i n_i}{\sum_{i=1}^{k} n_i} \tag{5.83}$$

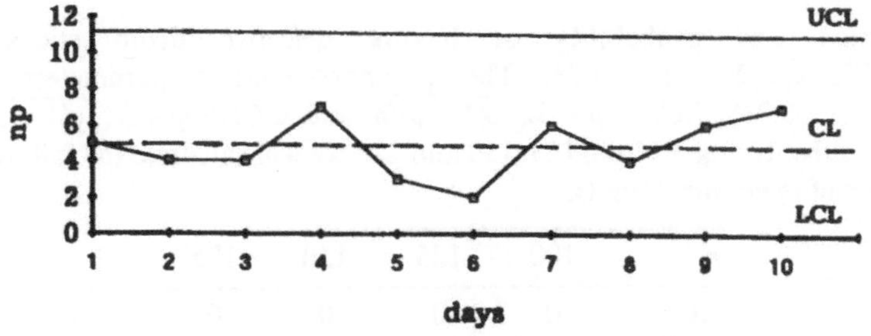

FIGURE E.5.34.1 The *np* chart for maintaining the present level of the operation.

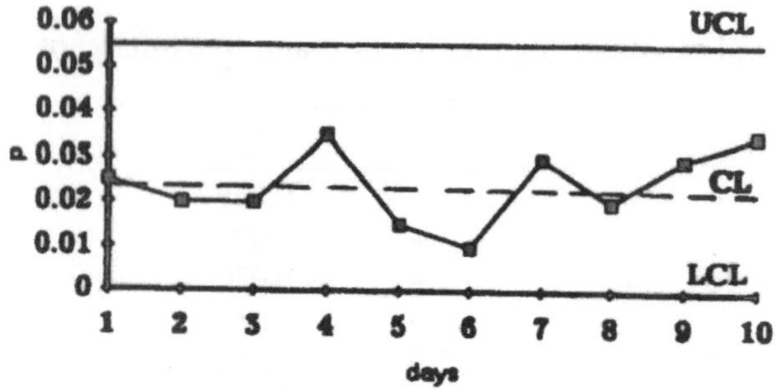

FIGURE E.5.34.2 **The *p* chart for maintaining the present level of the operation.**

where n_i is the size and p_i is the fraction of non-conforming units in the ith sample set. Several control limits LCL and UCL are determined for a whole range of sample sizes and the appropriate ones are used with each n.

Example 5.35. The p charts with variable sample size Prior to packaging broiler drum sticks, quality control is made to ascertain an acceptable level of feather is removed. The following samples with the given number of defectives were accounted during one week of operation.

	Sample size (n)	number of defectives (np)
Monday	175	5
Tuesday	150	7
Wednesday	100	3
Thursday	125	7
Friday	175	8
Saturday	125	6
Sunday	150	2

a) Calculate the limits of the p chart for variable sample size to maintain the present level of operation.

Solution The probability of having defective drum sticks is $p=\Sigma_{i=1}^{k} p_i n_i / \Sigma_{i=1}^{k} n_i = 0.038$. The p chart control parameters are $CL=p=0.038$, $LCL=p-3\sqrt{p(1-p)/n}$ and $UCL=p+3\sqrt{p(1-p)/n}$. After substituting $n=100$, 125, 150 and 175 we will calculate the following values of the control limits:

n	100	125	150	175
LCL	0	0	0	0
UCL	0.095	0.089	0.085	0.081

b) In a sample with $n = 175$ the number of the defectives was $np = 16$. Is this sample within the previously established acceptable limits?

Solution The data shows that $p = 0.091$. The control limits when $n = 175$ are $LCL = 0$ and $UCL = 0.081$, since p is not within the given range the sample violates the previously established limits.

The *c charts* are based on the Poisson distribution model and used in processes with *rare defects* to control the number of the defects. It should be noticed that an individual defective unit may have more than one defect. An apple with bruises at three different locations has three defects, but it is a single defective unit. The mean and the standard deviation of poisson distribution are (Tab. 5.1):

$$\mu = \lambda \tag{5.8}$$

and

$$\sigma = \sqrt{\lambda} \tag{5.9}$$

For large values of λ, the poisson distribution curve approaches the normal distribution curve and the properties of the normal distribution may be applied to the poisson distribution. Confidence limits under these conditions ($p = 0.9973$) with constant sample size n are (Tab. 5.3):

$$np - 3\sqrt{\lambda} \leqslant (np)_{exp} \leqslant np + 3\sqrt{\lambda n} \tag{5.29}$$

Parameter c is substituted for λ while constructing the c charts

$$c = \frac{1}{k}\sum_{i=1}^{k} c_i \tag{5.84}$$

where k and c_i are the number of the sample sets and counts of the defects in the ith sample set, respectively. The c chart is actually the graphical representation of the confidence limits with

$$CL = c \tag{5.85}$$

$$LCL = c - 3\sqrt{c} \tag{5.86}$$

and

$$UCL = c + 3\sqrt{c} \tag{5.87}$$

Example 5.36. c chart for more than one type of defect Visual quality control of canned dry beans is done by observing five different groups of defects. One can is opened in every batch and each can contains about 1000 beans. The following data were taken with 15 consecutive batches of acceptable quality level:

NUMBER OF DEFECTS/CAN

Batch number	split beans	cracked beans	discolored beans	wrinkled beans	pealed beans	total defects
1	0	2	0	3	0	5
2	2	0	1	5	0	8
3	3	3	0	2	0	8
4	3	1	2	1	1	8
5	0	3	2	1	0	6
6	1	0	4	0	1	5
7	1	4	2	1	1	9
8	4	2	1	1	1	9
9	1	3	5	1	1	11
10	1	2	1	2	2	8
11	1	6	0	0	2	9
12	1	4	4	1	1	11
13	0	0	5	1	3	9
14	2	2	2	2	2	10
15	2	1	3	1	3	10

With new batches of beans the following data was obtained:

NUMBER OF DEFECTS/CAN

Batch number	split beans	cracked beans	discolored beans	wrinkled beans	pealed beans	total defects
1	3	2	1	3	0	9
2	2	7	1	5	0	15
3	0	3	4	2	0	9
4	4	1	4	1	1	11
5	0	0	2	0	3	5

Is the quality of the new beans under control to the standard of the previous ones?

Solution The total number of defects are used to construct the *c* charts. Control limits will be determined with the previous data: $CL = c = 1/k\,\Sigma_{i=1}^{k} c_i = 8.4$, $LCL = c - 3\sqrt{c} = -0.29$, since it is not possible to have negative numbers of defective beans, the lower control limits will be reestablished as $LCL = 0$ and $UCL = c + 3\sqrt{c} = 17$. Since no data points fall outside the control llimits in Figure E.5.36, we may conclude that the process is under statistical control.

5.6. ACCEPTANCE SAMPLING BY ATTRIBUTES

Although inspecting each member of a population may be desirable to assure a pre-determined level of quality, this is not usually feasible. *Accept-*

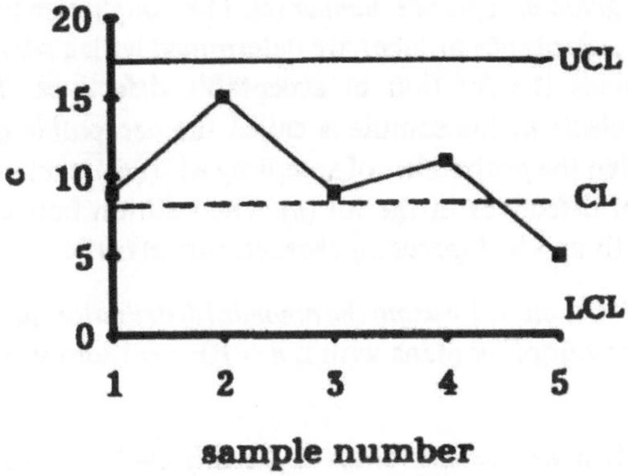

FIGURE E.5.36. Comparison of the data with the control limits.

ance sampling is preferred over total analysis of the population for many reasons: Chemical and microbiological tests may require homogenization or destruction of the sample and make total inspection of the population impossible. Accuracy of inspection may diminish due to fatigue or boredom of the inspector when the same tests are repeated too many times. Cost of inspection increases with the number of the tests due to need for more staff, laboratory space, chemicals, etc. It is usually practical to do quality control with samples and extend the results to the whole population.

A sample is supposed to represent the whole population. Different protocols may be used for sampling. A sample may be taken from a production line at pre-specified time intervals, i.e., sampling may be done at two hour intervals and a certain number, or weight of a sample may be taken. The same procedure may be applied also with pre-specified unit intervals, i.e., five consecutive items may be sampled after every 100 items. Sampling interval may be decided by using random number tables. Most statistics books include random number tables, and some hand calculators have built in programs to generate these numbers. Throwing dice may also be used for the same purpose. From trucks or store houses samples are taken to represent all locations including the sides, center and layers. When the commodities are in boxes or cases, samples, should be taken from different containers. When the sampling plan does not indicate the number of containers to be opened, total sample is collected by opening the boxes or cases according to the square root or cubic root principle, i.e., 9 out of 81 cases or 100 out of 1 000 000 cases may be opened and equal number of items collected from each. One method to randomly select these cases is to place individually numbered slips for each case in a box, shake and withdraw one slip of paper, record the number, return the slip into the box, shake and repeat until the required sample size has been selected (FAO, 1988). A lot may be accepted if the number of defectives in a sample set do not exceed a given *acceptance number* (c). The *sample size* to be taken from a lot and the acceptance number are determined with a *sampling plan*. This plan determines the fraction of acceptable defectives. The acceptable number of defects in the sample is called the *acceptable quality level*. In a sampling plan the probability of accepting a lot (P_a) is related to the actual percentage of defectives in the lot (p). The relation between P_a and p is visualized with an OC (*operating characteristics*) *curve*.

Example 5.37. OC curve by using the binomial distribution model Construct OC curves for sampling plans with i) $n = 10$, $c = 1$ and ii) $n = 8$, $c = 1$ and compare.

Solution When we use attributes to classify each unit as confirming or non-confirming with constant n and p we use the binomial distribution

$$f(x, n, p) = \frac{n!}{x!\,(n-x)!} p^x (1-p)^{(n-x)} \tag{5.3}$$

where $x \leqslant c$ ($x = 0$ and $x = 1$). Actual fraction of defectives in the lot is p. We may use (5.3) to determine the probability of accepting a lot (P_a) with varying values of p. When $n = 10$ and $p = 0\%$ with $x = 1$, probability of accepting the lot is $f(0, 10, 0) = 1$, with $x = 0$ the probability of accepting the lot is $f(1, 10, 0) = 0$. Probability P_a of accepting a sample with no defectives ($p = 0$) is $P_a = f(0, 10, 0) + f(1, 10, 0) = 1$. We repeat the same calculation with different levels of p:

p	0	0.1	0.2	0.3	0.4	0.5
P_a	1.0	0.74	0.38	0.15	0.04	0.01

The same procedure is followed to construct the OC curve with $n = 8, c = 1$:

p	0	0.1	0.2	0.3	0.4	0.5	0.6
P_a	1.0	0.81	0.50	0.26	0.11	0.03	0.01

The OC curves for both sampling plans are given in Figure E.5.37. When $n = 10$, $c = 1$ probability of accepting a lot (P_a) with 10% defectives ($p = 10\%$) is 74% ($P_a = 0.74$); there is $1 - 0.74 = 26\%$ probability for rejecting the lot. With the same plan the probability of accepting a lot with 40% defectives is drastically smaller, i.e., 4%($p = 0.40$, $P_a = 0.04$). It is also shown that when n is constant, P_a at constant p increases with acceptance number.

Example 5.38. OC curve by using the Poisson distribution model The cumulative probability distribution curve is given in Figure 5.12 for Poisson distribution. We will choose p, calculate np and read P_a from Figure 5.12 to construct the OC curves for i) $n = 100$, $c = 3$ and ii) $n = 100$, $c = 2$.

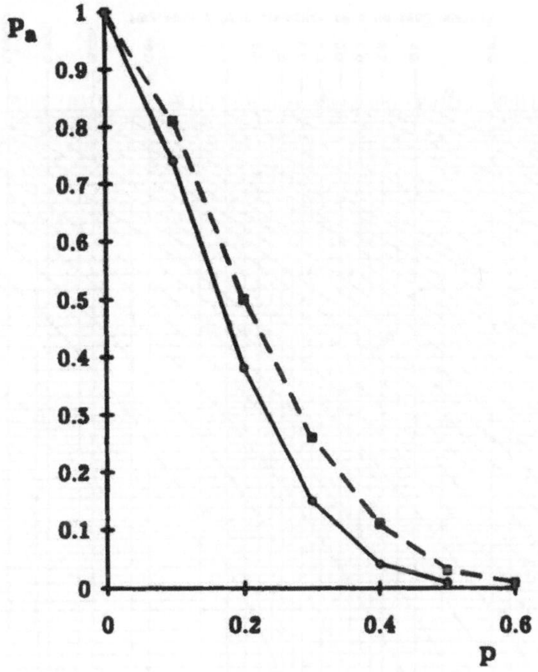

FIGURE E.5.37 OC curves for sampling plans of $n=10$, $c=1$ (—) and $n=8$, $c=1$ (-).

i) $n=100$, $c=3$:

p	0.001	0.01	0.02	0.03	0.04	0.05	0.06	0.07	0.08	0.09	0.10
np	0.1	1	2	3	4	5	6	7	8	9	10
P_a	1.00	0.98	0.86	0.65	0.44	0.26	0.15	0.08	0.04	0.02	0.01

ii) $n=100$, $c=2$:

p	0.001	0.01	0.02	0.03	0.04	0.05	0.06	0.07	0.08
np	0.1	1	2	3	4	5	6	7	8
P_a	1.00	0.93	0.67	0.45	0.28	0.12	0.06	0.02	0.005

The OC curves with $n=100$, $c=3$ and $n=100$, $c=2$ are compared in Figure E.5.38, where it should be noticed that when $n=$constant P_a decreases with c at the same p.

Consumer and producer interests conflict in a sampling plan. The consumers want to reduce the probability of accepting lots which include

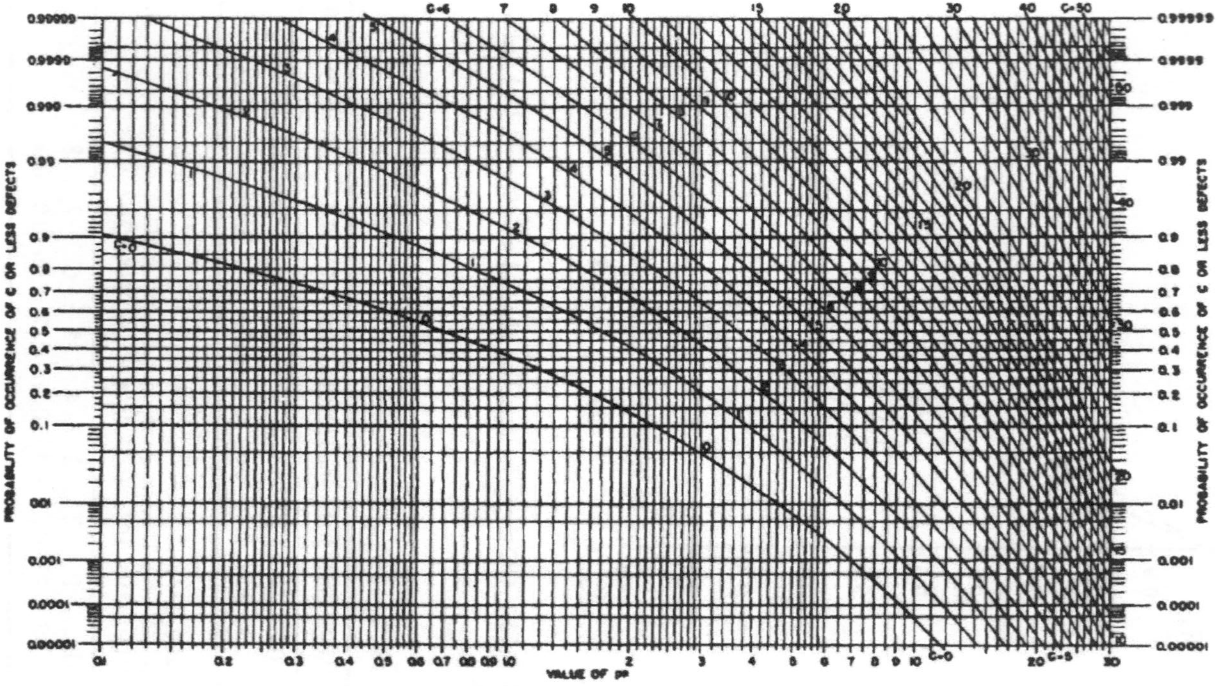

FIGURE 5.12 Cumulative poisson distribution curves for probability of occurrence of c or less defects in a sample size of n selected from a population where the defective fraction is p (Reproduced by permission of John Wiley & Sons Inc. from Dodge, H. F. and Romig, H. G. *Sapling Inspection Tables*. Second ed. © 1959 John Wiley & Sons Inc., and AT&T Bell Labs.).

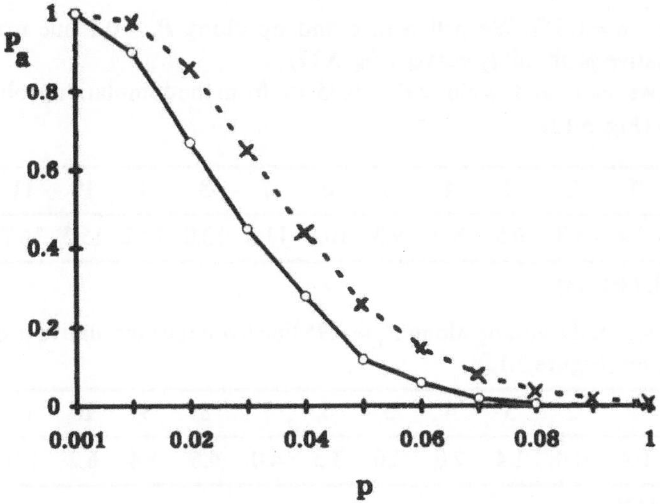

FIGURE E.5.38 The OC curves with (-x-) $n = 100$, $c = 3$ and (-o-) $n = 100$, $c = 2$.

too many defectives, while the producers want to minimize the probability of rejecting lots with an acceptable number of defectives. In any sampling plan there is a risk of accepting a lot with an unacceptable level of defectives. This is called *type II error* in statistics (Chapter 5.2) or *consumer's risk* in quality control literature. There is also a risk of rejecting a lot with less than unacceptable level of defectives, this is called *type I error* in statistics (Chapter 5.2) or *producer's risk* in quality control literature. The desired quality level at which probability of acceptance should be high is called the *acceptable quality level* (AQL) and the quality level below which lots are considered unacceptable is called the *lot tolerance percent defective* (LTPD). An OC curve may be constructed to compromise the interests of the consumer and the producer and is required to pass through the points defined by (AQL, α) and (LTPD, β) which describe the locations of $p = \text{AQL}$ and $p = \text{LTPD}$ in P_a versus p plots, with the corresponding values of $1 - \alpha$ and β of P_a. This may be achieved by using a trial and error procedure with binomial distribution or with cumulative probability curves (Fig. 5.12) when the requirements of poisson model are met.

Example 5.39. An OC curve to satisfy the required LTPD and AQL values Use the cumulative probability curves (Fig. 5.12) to construct an OC curve which (as nearly as possible) satisfies the following requirements: LTPD = 5%, $\beta = 10\%$ (consumer's risk) and AQL = 2%, $\alpha = 5\%$ (producer's risk).

Solution The OC is required to pass through the points (LTPD = 5%, $\beta = 10\%$) i.e., ($p = 0.05$, $P_a = 0.1$) and AQL = 2% $\alpha = 5\%$) i.e., ($p = 0.02$,

$P_a = 1 - \alpha = 0.95$). We will read c and np along $P_a = 0.1$ line from the cumulative probability curves (Fig. 5.12):
When we read *c* and *np* along $P_a = 0.95$ line from the cumulative probability curves (Fig. 5.12):

c	1	2	3	4	5	6	7	8	9	10	11
np	3.9	5.3	6.5	8.0	9.3	10.5	11.8	13.0	14.2	15.5	16.7

(np = LTPDxn)

When we read *c* and np along $P_a = 0.95$ line from the cumulative probability curves (Figure 5.12):

c	1	2	3	4	5	6	7	8	9	10	11
np	0.4	0.8	1.4	2.0	2.6	3.3	4.0	4.6	5.4	6.2	7.0

(np = AQLxn)

The required ratio of LTPD/AQL is 2.5. LTPD/AQL may be calculated for the individual values of *c* (LTPD/AQL = LTPD × *n*/AQL × *n*):

c	1	2	3	4	5	6	7	8	9	10	11
LTPD/AQL	9.75	6.7	4.7	4.1	3.6	3.2	3.0	2.8	2.6	2.5	2.4

LTPD/AQL = 2.5 when $c = 10$. The above data shows that LTPDxn = 15.5 or AQLxn = 6.2 when $c = 10$, therefore we may easily calculate $n = 310$, implying that we may compromise the producer and the consumer interests when $n = 310$ and $c = 10$. The required OC curve is shown in Figure E.5.39.

Probability for defectives to pass a sampling plan is $P_a x p$. *Average out going quality percentage* (*AOQ%*) of a lot is defined as:

$$AOQ\% = \left(\frac{\text{number of defective items in an accepted lot}}{\text{initial lot size}}\right)100 \quad (5.88)$$

When we destroy a sample set of *n* items from a lot of initial size *N* we will have a remaining batch of size $N - n$.

Example 5.40. Construction of the AOQ% curves Construct an AOQ% curve for the OC of $n = 10$, $c = 1$ with $N = 200$. The sample is destroyed in the analysis.

Solution Read values of *p* and P_a from Fig. E.5.37 with the same sampling plan. AOQ% may be calculated as:

$$AOQ\% = \frac{(P_a)(p)(N - n)}{N}100 \quad (5.88)$$

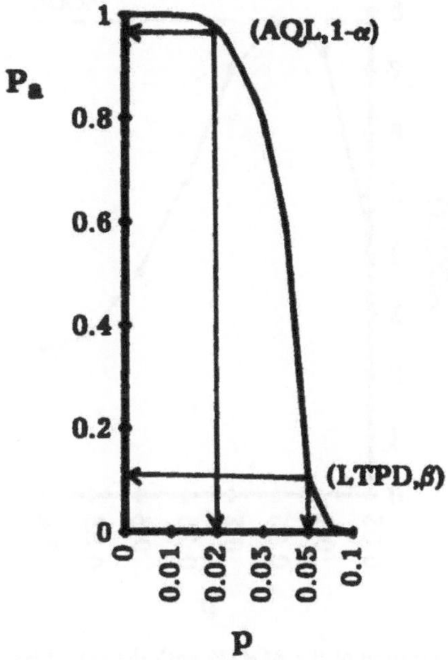

FIGURE E.5.39 Sketch of the required OC curve with LTPD = 5%, β = 10% and AQL = 2%, α = 5%.

p	0.05	0.10	0.15	0.20	0.25	0.30	0.35	0.40
P_a	0.90	0.77	0.55	0.37	0.22	0.13	0.08	0.06
AOQ%	4.3	7.3	7.8	7.0	5.2	3.7	2.7	2.4

Variation of the AOQ% with the actual level of the defectives in the product is shown in Figure E.5.40.

Average outgoing quality limit (AOQL) is a built in limit for the maximum probability of the defectives which might be accepted with the sampling plan.

Example 5.41. Average out going quality limits from the OC curves Determine the AOQL from the OC curves of the sampling plans $n = 10, c = 1$ and $n = 8, c = 1$ for a sample lot size of $N = 200$. Which one of these plans would permit acceptance of a lower number of defectives?

Solution We may read P_a versus p data from the OC curves of Example 5.37 AOQ% may be calculated as:

$$\text{AOQ\%} = \frac{(P_a)(p)(N-n)}{N} 100 \tag{5.88}$$

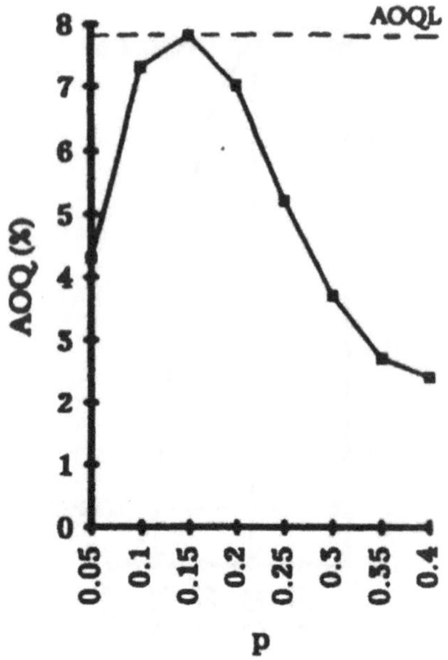

FIGURE E.5.40 Variation of the AOQ% with the actual level of the defectives in the product.

i) $n = 10, c = 1$

p	0	0.1	0.2	0.3	0.4	0.5
P_a	1.0	0.74	0.38	0.15	0.04	0.01
AOQ%	0	7.0	7.2	4.3	1.5	0.5

ii) $n = 8, c = 1$

p	0.	0.1	0.2	0.3	0.4	0.5	0.6
P_a	1.0	0.81	0.50	0.26	0.11	0.03	0.01
AOQ%	0	7.8	9.6	7.5	4.2	1.4	0.6

Variation of the AOQ% with the actual level of the defectives p is shown in Figure E.5.41.

AOQL is a built-in limit of a sampling plan for the maximum level of defectives which may be accepted. Sampling plan with $n = 10, c = 1$ permits acceptance of lower numbers of defectives than the other one.

The sampling calculations described here are time consuming. A computer flow chart (Fig. 5.13) has been recommended by Pappas and Rao (1987) to determine sample size and acceptance number for a single sampling plan and may make calculations easier with the aid of a computer.

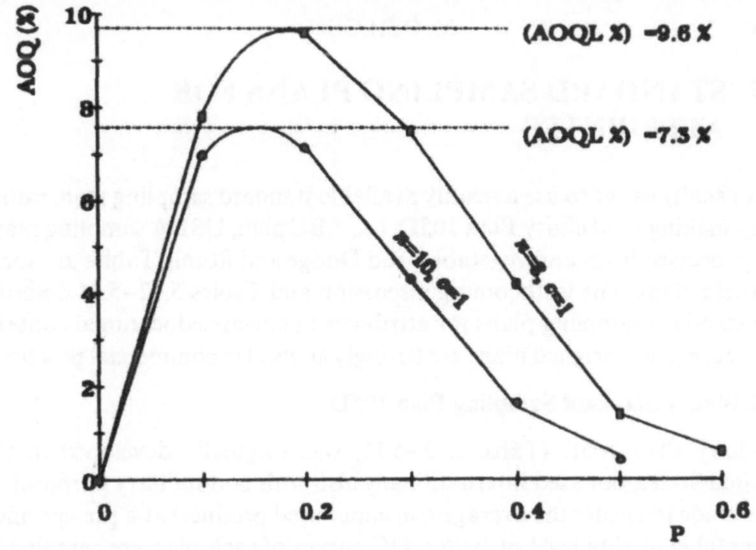

FIGURE E.5.41 Variation of the AOQ% with the actual level of the defectives.

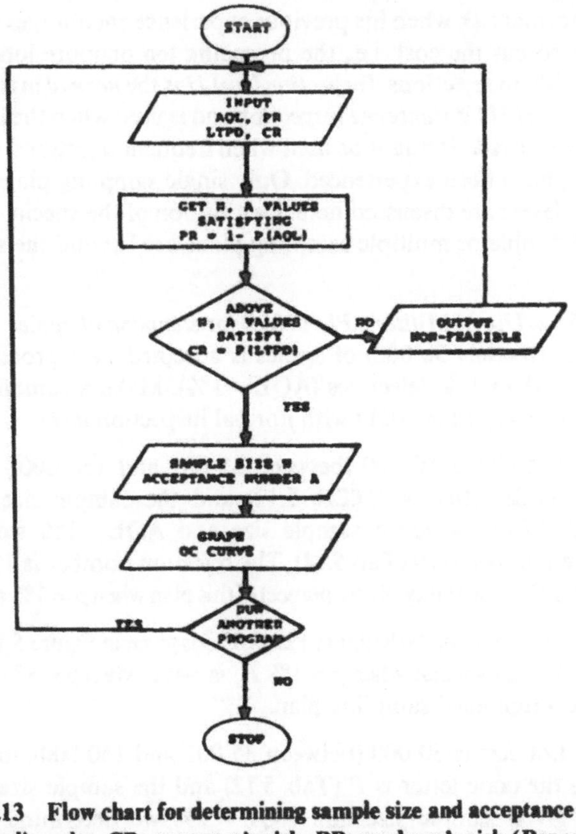

FIGURE 5.13 Flow chart for determining sample size and acceptance number for a single sampling plan CR: consumer's risk, PR: producer's risk (Pappas and Rao, 1987; © Food & Nutrition Press, Inc., reproduced by permission).

5.7. STANDARD SAMPLING PLANS FOR ATTRIBUTES

It is usually easier to use a readily available standard sampling plan, rather than making it. Military Plan 105D, i.e., ABC plan, USDA sampling plans for processed fruits and vegetables and Dodge and Romig Tables are some of these plans. The forthcoming discussion and Tables 5.12–5.24 describe the standard sampling plans for attributes in concise educational context; reference to the original plans are strongly advised in commercial practices.

i) Military Standard Sampling Plan 105D

Military Plan 105D (Tabs. 5.12–5.15) was originally developed in the United States, but used internationally also with non-military purposes. It was made to ensure the average consumer used products at a pre-specified acceptable quality level or better. OC curves of each plan are supplied in Figure 5.14. The plan has three inspection levels. *Inspection level I is reduced inspection* and is applied when a consumer accepts to take greater than the normal risk when his previous experience encourages the reduced inspection to cut the cost, i.e., the preceding ten or more lots have been accepted with no rejections. *Inspection level II* is the *normal inspection* level. *Inspection level III* is *tightened inspection* and is used when the buyer wants to minimize his risk. It might be used when frequent rejections, i.e., two lots out of five, have been experienced. Only single sampling plans at general inspection levels are discussed here. Description of the special application levels and double or multiple sampling plans are beyond the scope of the text.

Example 5.42. Use of Military Plan 105D for sampling of apples a) A truck load (approximately 50 000) of apples is accepted by a processor if they contain less than 1% defectives (AQL = 1%). Make a sampling plan by using the Military Plan 105D with normal inspection level.

Solution Lot size is 50 000 (between 35 001 and 150 000), for normal inspection code letter is N (Tab. 5.12) and the sample size is $n = 500$ (Tab. 5.14). With the given sample size and AQL = 1% we determine acceptance number = 10 (Tab 5.14). The rejection number is 11

b) What is P_a (probability of acceptance) of this plan when $p = 1\%$ and $p = 8\%$?

Solution OC curves of the Military Plan 105D is given in Figure 5.14. The curve for $AQL = 1\%$ shows that when $p = 1\%$ P_a is 94%; when $p = 8\%$ P_a is 4%.

c) Make a tightened sampling plan.

Solution Lot size is 50 000 (between 35 001 and 150 000), for tightened inspection the code letter is P (Tab. 5.12) and the sample size is $n = 800$. With the given sample size and AQL = 1% we determine acceptance number = 12 (Tab. 5.15).

d) Make a sampling plan if the apples are shipped in crates (100 crates, 500 apples in each crate).

Solution Since no levels are specified we may use the previous sampling plan at normal level: $n_{\text{apples}} = 500$, acceptance number = 10. We should also determine how many crates should be opened for sampling. There are 100 crates, the sample size is between 91 and 150 corresponding to code letter F at normal inspection level (Tab. 5.12) with $n_{\text{crates}} = 20$ (Tab. 5.14). $n_{\text{apples}}/n_{\text{crates}} = 25$, implying that we should open 20 crates and take 25 apples from each. If we should find 10 or less defectives in the previous sample with 500 apples we may accept the lot, we will reject it otherwise.

Example 5.43. Use of Military Plan 105D for sampling of individually wrapped sliced cheese A cheese producer markets its products in boxes of 100 packages. Each package contains 18 individually wrapped slices. Make an acceptance sampling plan for a purchase of 100 boxes of cheese. Find out how many slices should be analyzed. How many boxes and packages should be opened? The supplier has been doing business with this company for 6 months and it is known that none of their shipments has ever been rejected. AQL of the sampling plan is required to be 0.065%.

Solution We may use Military Plan 105D with *reduced inspection level* since the supplier has been doing business with this company for 6 months and non of their shipments has ever been rejected. Lot size of boxes is 100 (between 91 and 150), for reduced inspection code letter is D (Tab. 5.12) and $n_{\text{box}} = 3$ (Tab. 5.13).

Lot size of the packages is (100)(100) = 10 000, for reduced inspection code letter is J (Tab. 5.12) and $n_{\text{packages}} = 32$ (Tab. 5.13).

Lot size of the slices (100) (100) (18) = 180 000. For the reduced inspection code letter is M (Tab. 5.12) and $n_{\text{slices}} = 125$, with AQL = 0.065% acceptance number = 0 and rejection number = 1 (Tab. 5.13).

The sampling plan implies that 3 boxes should be opened, 11 packages should be taken from two of them and 10 packages should be taken from the third box. Since 125/32 = 3.9 (not an integer) we may make the partitioning of the slices as 125 = 29x4 + 3x3, therefore 4 slices should be taken from each of the first 29 packages and 3 slices should be taken from the remaining 3 packages. If any of the slices is defective the lot should be rejected.

Example 5.44. Use of Military Plan 105D for sampling of products with seasonally changing risk of spoilage One thousand cases of eggs are inspected by the purchasing department of a market at each delivery. Each case contains 20 boxes. There are 12 eggs in each box. It is known that the eggs may spoil fast in the summer and there is very small possibility for spoilage in winter. Make separate sampling plans for each season, i.e., fall/spring, winter and summer. Determine how many eggs should be taken from how many boxes and cases; also determine the acceptance number. AQL is required to be maintained as small as possible.

Solution We may use the normal inspection level in spring and fall, reduced inspection level in winter and tightened inspection level in the summer. The lot size is 1000 with cases, (1000)(20) = 20 000 with boxes, (1000)(20)(12) = 240 000 with eggs. The following sampling plans may be suggested at different levels:

	cases	boxes	eggs (AQL = 0.01%)
level I	code letter = G	code letter = K	code letter = M
	$n_{cases} = 13$	$n_{boxes} = 50$	$n_{eggs} = 125$, c = 0
letter II	code letter = J	code letter = M	code letter = P
	$n_{cases} = 80$	$n_{boxes} = 315$	$n_{eggs} = 800$, c = 0
level III	code letter = K	code letter = N	code letter = Q
	$n_{cases} = 125$	$n_{boxes} = 500$	$n_{eggs} = 1250$, c = 0

Code letters were determined from Table 5.12. Sample size and acceptance numbers were determined from Tables 5.13, 5.14 and 5.15 for levels I, II and III, respectively.

	n_{boxes}/n_{cases}	n_{eggs}/n_{boxes}
level I	3.84	2.5
level II	3.93	2.53
level III	4	2.5

Table 5.12 Military plan 105D Sample size code letters general inspection levels

Lot or batch size	I (Reduced)	II (normal)	III (tightened)
2–8	A	A	B
9–15	A	B	C
16–25	B	C	D
26–50	C	D	E
51–90	C	E	F
91–150	D	F	G
151–280	E	G	H
281–500	F	H	J
501–1 200	G	J	K
1 201–3 200	H	K	L
3 201–10 000	J	L	M
10 001–35 000	K	M	N
35 001–150 000	L	N	P
150 001–500 000	M	P	Q
500 001 and over	N	Q	R

Table 5.13 Military plan 105D Master table at reduced inspection level

Sample size code letter	Sample size	AQL (%) 0.010	0.015	0.025	0.040	0.065	0.10	0.15	0.25	0.40	0.65	1.0	1.5	2.5	4.0	6.5	10
		Ac Re	Ac Re	Ac Re	Ac Re	Ac Re	Ac Re	Ac Re	Ac Re	Ac Re	Ac Re	Ac Re	Ac Re	Ac Re	Ac Re	Ac Re	Ac Re
A	2	↓	↓	↓	↓	↓	↓	↓	↓	↓	↓	↓	↓	↓	↓	0 1	↓
B	2	↓	↓	↓	↓	↓	↓	↓	↓	↓	↓	↓	↓	↓	0 1	↑	↓
C	2	↓	↓	↓	↓	↓	↓	↓	↓	↓	↓	↓	↓	0 1	↑	↓	0 2
D	3	↓	↓	↓	↓	↓	↓	↓	↓	↓	↓	↓	0 1	↑	↓	0 2	1 3
E	5	↓	↓	↓	↓	↓	↓	↓	↓	↓	↓	0 1	↑	↓	0 2	1 3	1 4
F	8	↓	↓	↓	↓	↓	↓	↓	↓	↓	0 1	↑	↓	0 2	1 3	1 4	2 5
G	13	↓	↓	↓	↓	↓	↓	↓	↓	0 1	↑	↓	0 2	1 3	1 4	2 5	3 6
H	20	↓	↓	↓	↓	↓	↓	↓	0 1	↑	↓	0 2	1 3	1 4	2 5	3 6	5 8
J	32	↓	↓	↓	↓	↓	↓	0 1	↑	↓	0 2	1 3	1 4	2 5	3 6	5 8	7 10
K	50	↓	↓	↓	↓	↓	0 1	↑	↓	0 2	1 3	1 4	2 5	3 6	5 8	7 10	10 13
L	80	↓	↓	↓	↓	0 1	↑	↓	0 2	1 3	1 4	2 5	3 6	5 8	7 10	10 13	↑
M	125	↓	↓	↓	0 1	↑	↓	0 2	1 3	1 4	2 5	3 6	5 8	7 10	10 13	↑	↑
N	200	↓	↓	0 1	↑	↓	0 2	1 3	1 4	2 5	3 6	5 8	7 10	10 13	↑	↑	↑
P	315	↓	0 1	↑	↓	0 2	1 3	1 4	2 5	3 6	5 8	7 10	10 13	↑	↑	↑	↑
Q	500	0 1	↑	↓	0 2	1 3	1 4	2 5	3 6	5 8	7 10	10 13	↑	↑	↑	↑	↑
R	800	↑	↑	0 2	1 3	1 4	2 5	3 6	5 8	7 10	10 13	↑	↑	↑	↑	↑	↑

↓ = Use for sampling plan below arrow.
↑ = Use for sampling plan above arrow.
Ac = Acceptance number.
Re = Rejection number.

Table 5.14 Military plan 105D master table at normal inspection level (see Tab. 5.13 for legends)

Sample size code letter	Sample size	AQL (%) 0.010	0.015	0.025	0.040	0.065	0.10	0.15	0.25	0.40	0.65	1.0	1.5	2.5	4.0	6.5	10
		Ac Re	Ac Re	Ac Re	Ac Re	Ac Re	Ac Re	Ac Re	Ac Re	Ac Re	Ac Re	Ac Re	Ac Re	Ac Re	Ac Re	Ac Re	Ac Re
A	2	↓	↓	↓	↓	↓	↓	↓	↓	↓	↓	↓	↓	↓	↓	0 1	↓
B	3	↓	↓	↓	↓	↓	↓	↓	↓	↓	↓	↓	↓	↓	0 1	↑	↓
C	5	↓	↓	↓	↓	↓	↓	↓	↓	↓	↓	↓	↓	0 1	↑	↓	1 2
D	8	↓	↓	↓	↓	↓	↓	↓	↓	↓	↓	↓	0 1	↑	↓	1 2	2 3
E	13	↓	↓	↓	↓	↓	↓	↓	↓	↓	↓	0 1	↑	↓	1 2	2 3	3 4
F	20	↓	↓	↓	↓	↓	↓	↓	↓	↓	0 1	↑	↓	1 2	2 3	3 4	5 6
G	32	↓	↓	↓	↓	↓	↓	↓	↓	0 1	↑	↓	1 2	2 3	3 4	5 6	7 8
H	50	↓	↓	↓	↓	↓	↓	↓	0 1	↑	↓	1 2	2 3	3 4	5 6	7 8	10 11
J	80	↓	↓	↓	↓	↓	↓	0 1	↑	↓	1 2	2 3	3 4	5 6	7 8	10 11	14 15
K	125	↓	↓	↓	↓	↓	0 1	↑	↓	1 2	2 3	3 4	5 6	7 8	10 11	14 15	21 22
L	200	↓	↓	↓	↓	0 1	↑	↓	1 2	2 3	3 4	5 6	7 8	10 11	14 15	21 22	↑
M	315	↓	↓	↓	0 1	↑	↓	1 2	2 3	3 4	5 6	7 8	10 11	14 15	21 22	↑	↑
N	500	↓	↓	0 1	↑	↓	1 2	2 3	3 4	5 6	7 8	10 11	14 15	21 22	↑	↑	↑
P	800	↓	0 1	↑	↓	1 2	2 3	3 4	5 6	7 8	10 11	14 15	21 22	↑	↑	↑	↑
Q	1250	0 1	↑	↓	1 2	2 3	3 4	5 6	7 8	10 11	14 15	21 22	↑	↑	↑	↑	↑
R	2000	↑	↑	1 2	2 3	3 4	5 6	7 8	10 11	14 15	21 22	↑	↑	↑	↑	↑	↑

Table 5.15 Military plan 105D master table at tightened inspection level (see Tab. 5.13 for legends)

Sample size code letter	Sample size	AQL (%) 0.010	0.015	0.025	0.040	0.065	0.10	0.15	0.25	0.40	0.65	1.0	1.5	2.5	4.0	6.5	10
		Ac Re	Ac Re	Ac Re	Ac Re	Ac Re	Ac Re	Ac Re	Ac Re	Ac Re	Ac Re	Ac Re	Ac Re	Ac Re	Ac Re	Ac Re	Ac Re
A	2	↓	↓	↓	↓	↓	↓	↓	↓	↓	↓	↓	↓	↓	↓	↓	↓
B	3	↓	↓	↓	↓	↓	↓	↓	↓	↓	↓	↓	↓	↓	↓	0 1	↓
C	5	↓	↓	↓	↓	↓	↓	↓	↓	↓	↓	↓	↓	↓	0 1	↓	↓
D	8	↓	↓	↓	↓	↓	↓	↓	↓	↓	↓	↓	↓	0 1	↓	↓	1 2
E	13	↓	↓	↓	↓	↓	↓	↓	↓	↓	↓	↓	0 1	↓	↓	1 2	2 3
F	20	↓	↓	↓	↓	↓	↓	↓	↓	↓	↓	0 1	↓	↓	1 2	2 3	3 4
G	32	↓	↓	↓	↓	↓	↓	↓	↓	↓	0 1	↓	↓	1 2	2 3	3 4	5 6
H	50	↓	↓	↓	↓	↓	↓	↓	↓	0 1	↓	↓	1 2	2 3	3 4	5 6	8 9
J	80	↓	↓	↓	↓	↓	↓	↓	0 1	↓	↓	1 2	2 3	3 4	5 6	8 9	12 13
K	125	↓	↓	↓	↓	↓	↓	0 1	↓	↓	1 2	2 3	3 4	5 6	8 9	12 13	18 19
L	200	↓	↓	↓	↓	↓	0 1	↓	↓	1 2	2 3	3 4	5 6	8 9	12 13	18 19	↑
M	315	↓	↓	↓	↓	0 1	↓	↓	1 2	2 3	3 4	5 6	8 9	12 13	18 19	↑	↑
N	500	↓	↓	↓	0 1	↓	↓	1 2	2 3	3 4	5 6	8 9	12 13	18 19	↑	↑	↑
P	800	↓	↓	0 1	↓	↓	1 2	2 3	3 4	5 6	8 9	12 13	18 19	↑	↑	↑	↑
Q	1250	↓	0 1	↓	↓	1 2	2 3	3 4	5 6	8 9	12 13	18 19	↑	↑	↑	↑	↑
R	2000	0 1	↑	↓	1 2	2 3	3 4	5 6	8 9	12 13	18 19	↑	↑	↑	↑	↑	↑
S	3150			1 2													

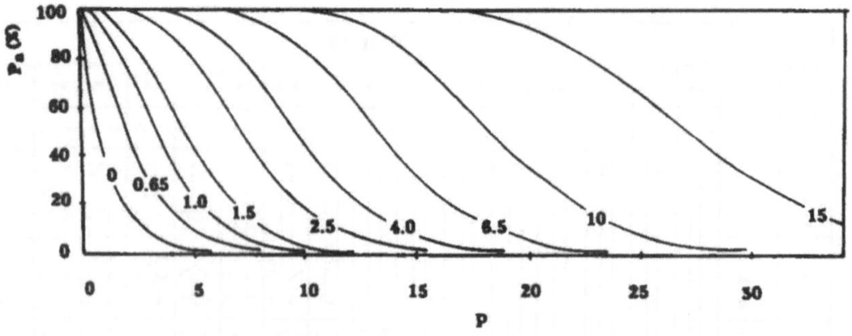

FIGURE 5.14 **Operating characteristics curves for Military Plan 105D. AQL (%) values are shown near the curves.**

	sampling plan
level I	open 13 cases, take 3 boxes from the first two and 4 boxes from the remaining 11 cases. Take 2 eggs from the first 25 boxes, take 3 eggs from the remaining 25 boxes.

level II open 80 cases, take 3 boxes from the first five and 4 boxes from the remaining 75 cases. Take 2 eggs from the first 145 boxes, take 3 eggs from the remaining 170 boxes.

level III open 125 cases, take 4 boxes from each. Take 2 eggs from the first 250 boxes, take 3 eggs from the remaining 250 boxes.

---

ii) USDA Sampling Plan

USDA sampling plans for processed fruits and vegetables are given in Tables 5.16–5.19 (Gould, 1983). The plan was established at 6% AQL. Lot inspection and continuous inspection may be made according to this plan. *Lot inspection* is inspection and grading of specific lots of processed fruits and vegetables located in plant warehouse, commercial storage, railway cars and trucks. The Inspector does not know the conditions or the practices under which the product is packed. *Continuous (in-plant, on-line)* inspection is made by inspectors in the plant at all times and stages, i.e., processing, storage, etc. of operation.

USDA sampling plan makes reference to certain standard size cans. No 303, No 3 cylinder and No 12 size cans have filling capacity of 621 g, 1464 g and 3919 g water, respectively.

Example 5.45. Sampling plan with USDA tables for canned beans A lot of beans consisting of 500 cases, each containing forty eight 200 g cans will be certified. Make i) a lot inspection and ii) *on-line, in-plant inspection* sampling plans by using the USDA tables.

Solution Can size is smaller than that of No 303 (621 g) can. Lot size = (500)(48) = 24 000. We may use Table 5.16, Group 1, column for 12 001-39 000.

i) Lot inspection: $n = 13, c = 2$
ii) On-line, in-plant inspection: $n = 6, c = 1$.

In lot inspection we should open 13 cans (each from a different case) we may accept the lot if two or less defectives are found. Since the inspector knows the conditions or the practices under which the product is packed, smaller sample size is employed with on-line, in-plant inspection.

Example 5.46. Sampling plan with USDA tables for packaged dried apricots 5000 packages of dried apricots are to be inspected. If the net weight of each package is between 500 g to 600 g make i) lot inspection, ii) on-line, in-plant inspection sampling plan by using the USDA tables.

Solution We may use the Table 5.19 for dehydrated (low moisture) fruits and vegetables. We may use Group 2 (container size is between 450 g and 2.7 kg), lot size between 2401 to 7800:

i) Lot inspection: $n = 13, c = 2$
ii) On-line, in-plant inspection: $n = 6, c = 1$.

Table 5.16 USDA sampling plan for canned or similarly processed fruits, vegetables and products containing units of such size and character as to be readily separable (based on the charts given by Gould (1983), converted into SI units with minor modification of the container sizes, published with kind permission of the copyright owner Chapman & Hall Inc.)

Container size group	Lot size (number of containers)*							
Group 1. Container volume less than that of No 303 size can	3 000 or less	3 001– 12 000	12 001– 39 000	39 001 84 000	84 001– 145 000	145 001– 228 000	228 001– 336 000	336 001– 480 000
Group 2. Container volumer between those of No 303 and No 3 cylinder size cans	1 500 or less	1 501 6 000	6 001– 19 500	19 501– 42 000	42 001– 72 500	72 501– 114 000	114 001– 168 000	168 001– 240 000
Group 3. Conatiner volume between those of No 3 cylinder and No 12 size cans	1500 or less	1 501– 6 000	6 001– 19 500	19 501– 42 000	42 001– 72 500	72 501– 114 000	114 001– 168 000	168 001– 240 000
Group 4. Container volume exceeding that of No 12 can	Convert to equivalent number of 2.750 kg containers and use group 3							

Lot inspection								
Sample size**	3	6	13	21	29	38	48	60
Acceptance number	0	1	2	3	4	5	6	7
On-line in-plant inspection								
Sample size**	3	6	6	13	21	29	38	48
Acceptance number	0	1	1	2	3	4	5	6

*Under on-line in-plant inspection a 5% overrun in number of containers may be permitted by the inspector before going to the next size

**Number of sample units. When a standard sample size unit is not specified in the US grade standards, the sample unit for the various container size groups are as follows: Groups 1, 2 and 3: One container and its entire content. Group 4: Approximately 1 kg of product. When determined by inspector that a 1 kg sample unit is inadequate a larger sample unit may be substituted.

Example 5.47. Sampling plan with USDA tables for soup in containers A truck load (15 000 cans) of ready to eat microwaveable vegetable soup (250 g containers) will be inspected at the receiving department of a supermarket chain. Make an acceptance sampling plan. Explain your reasoning in detail.

Solution USDA sampling plan for canned, comminuted, fluid or homogeneous foods (Tab. 5.18) will be used. Group = 1 (can size = 250 g), lot size is between 4 501-18 000. Inspection is made at the receiving department of the supermarket chain, processing conditions are not known, therefore lot inspection will be made. The sampling plan is $n = 6$, $c = 1$.

Example 5.48. Sampling plan with USDA tables for frozen mixed vegetables A food processor produces 45 000 kg of frozen mixed vegetables (diced carrots, cut beans and peas) in one month. These vegetables will be divided into three equal portions and marketed in 5 kg, 1 kg and 250 g bags. Make a lot inspection plan for each group of packages.

Solution Each lot will be 15 000 kg. USDA sampling plan (Tab. 5.17) will be used.

package weight	number of containers	group	effective lot size	n	c
250 g	60 000	1	60 000	21	3
1 kg	15 000	2	15 000	21	3
5 kg	3 000	3	13 274	13	2

Effective lot size was calculated after converting the lot size of 5 kg packages into an equivalent number of 1.130 kg containers

iii) The Dodge and Romig Tables

The Dodge and Romig Tables (1944) were constructed to assure the required LTPD (Tabs. 5.20 and 5.21) or AOQL (Tabs. 5.22 and 5.23).

Example 5.49. Sampling plan with Dodge and Romig Tables for apples a) A truck load (approximately 50 000) of apples is required to fail inspection if they contain more than i) 3% defectives (LTPD = 3%), ii) 2% defectives (LTPD = 2%). An average load is expected to have 0.61 to 0.80% defectives. Make a sampling plan. iii) Make the above sampling plans when an average load is expected to have 0 to 0.02% defectives

Solution We will use Dodge and Romig Tables (Tabs. 5.20–5.21).

lot size	p	LTPD	n	c	AOQL
50 000	0.61–0.90%	3%	520	10	1.2%
50 000	0.61–0.90%	2%	1060	15	0.93%
50 000	0–0.02%	3%	130	1	0.65%
50 000	0–0.02%	2%	200	1	0.42%

Table 5.17 USDA sampling plan for frozen or similarly processed fruits, vegetables and products containing units of such size and character as to be readily separable (based on the charts given by Gould (1983), converted into SI units with minor modification of the container sizes, published with kind permission of the copyright owner Chapman & Hall Inc.)

Container size group	Lot size (number of containers)*							
Group 1. Container weight less than 450 g	2 400 or less	2 401–9 600	9 601–31 200	31 201–67 200	67 201–116 000	116 001–182 400	182 401–268 800	268 801–384 000
Group 2. Container weight 450–1130 g	1 200 or less	1 201–4 800	4 801–15 600	15 601–33 600	33 601–58 000	58 001–91 200	91 201–134 400	134 401–192 000
Group 3. Container weight more than 1130 g	Convert to equivalent number of 1.130 kg containers and use group 3							
Lot inspection								
Sample size**	3	6	13	21	29	38	48	60
Acceptance number	0	1	2	3	4	5	6	7
On–line in–plant inspection								
Sample size**	3	6	6	13	21	29	38	48
Acceptance number	0	1	1	2	3	4	5	6

*Under on–line in–plant inspection a 5% overrun in number of containers may be permitted by the inspector before going to the next size

**Number of sample units. When a standard sample size unit is not specified in the US grade standards, the sample unit for the various container size groups are as follows: Groups 1 and 2: One container and its entire content. Group 3: With containers up to 4.5 kg, approximately 1130 g of product. When determined by inspector that a 1130 g sample unit is inadequate a larger sample unit of one of more containers and entire contents may be substituted for one or more sample units of each 1130 g.

Table 5.18 USDA sampling plan for canned, frozen or otherwise processed fruits, vegetables and products in comminuted, fluid or homegeneous state (based on the charts given by Gould (1983), converted into SI units with minor modification of the container sizes, published with kind permission of the copyright owner Chapman & Hall Inc.)

Container size group	Lot size (number of containers)*							
Group 1. Container weight less than 450 g	4 500 or less	4 501–18 000	18 001–58 000	58 001–126 000	126 001–217 000	217 001–342 000	342 001–504 000	504 001–720 000
Group 2. Container weight 450 g and 1.7 kg	3 000 or less	3 001–12 000	12 001–39 000	39 001–84 000	84 001–145 000	145 001–228 000	228 001–336 000	336 001–480 000
Group 3. Container weight between 1.7 and 4.5 kg	1 500 or less	1 501–6 000	6 001–19 500	19 501–42 000	42 001–72 500	72 501–114 000	114 001–168 000	168 001–240 000
Group 4. Container weight exceeding 4.5 kg	Convert to equivalent number of 2.7 kg containers and use group 3							
Lot inspection								
Sample size**	3	6	13	21	29	38	48	60
Acceptance number	0	1	2	3	4	5	6	7
On–line in–plant inspection								
Sample size**	3	6	6	13	21	2	38	48
Acceptance number	0	1	1	2	3	4	5	6

*Under on–line in–plant inspection a 5% overrun in number of containers may be permitted by the inspector before going to the next size
**Number of sample units. When a standard sample size unit is not specified in the US grade standards, the sample unit for the various container size groups are as follows: Groups 1, 2 and 3: One container and its entire content. A smaller sample unit may be substituted in group 3 at the inspector's discretion. Group 4: Approximately 450 g of product. When determined by inspector that a 450 g sample unit is inadequate a larger sample unit may be substituted.

Table 5.19 **USDA sampling plan for dehydrated fruits and vegetables (based on the charts given by Gould (1983), converted into SI units with minor modification of the container sizes, published with kind permission of the copyright owner Chapman & Hall Inc.)**

Container size group	Lot size (number of containers)*							
Group 1. Container weight less than 450 g	1 800 or less	1 801–7 200	7 201–23 400	23 401–50 400	50 401–87 000	87 001–136 000	136 001–201 600	201 601–288 000
Group 2. Container weight 450–2.7 kg	600 or less	601–2 400	2 401–7 800	7 801–16 800	16 801–29 000	29 001–45 600	45 601–67 200	67 201–96 000
Group 3. Container weight more than 2.7 kg	Convert to equivalent number of 2.7 kg containers and use group 2							
Lot inspection								
Sample size**	3	6	13	21	29	38	48	60
Acceptance number	0	1	2	3	4	5	6	7
On–line in–plant inspection								
Sample size**	3	6	6	13	21	29	38	48
Acceptance number	0	1	1	2	3	4	5	6

*Under on–line in–plant inspection a 5% overrun in number of containers may be permitted by the inspector before going to the next size

**Number of sample units. When a standard sample size unit is not specified in the US grade standards, the sample unit for the various container size groups are as follows: Groups 1 and 2: One container and its entire content. Group 3: With containers up to 4.5 kg, approximately 1.3 kg of product. When determined by inspector that a 1.3 kg sample unit is inadequate a larger sample unit of one of more containers and entire contents may be substituted for one or more sample units of each 1.3 kg.

Table 5.20 Single Sampling Table for LTPD = 2.0% (Reproduced by permission of John Wiley & Sons Inc. from Dodge, H. F. and Romig, H. G. *Sapling Inspection Tables*. Second ed. © 1959 John Wiley & Sons Inc., and AT&T Bell Labs.)

	process average 0 to 0.02%			process average 0.03 to 0.20%			process average 0.21 to 0.40%			process average 0.41 to 0.60%			process average 0.61 to 0.80%			process average 0.81 to 1.00%		
lot size	n	c	AOQL %	n	c	AOQL %	n	c	AOQL %	n	c	AOQL %	n	c	AOQL %	n	c	AOQL %
1–75	All	0	0	All	0	0	All	0	0	All	0	0	All	0	0	All	0	0
76–100	70	0	0.16	70	0	0.16	70	0	0.16	70	0	0.16	70	0	0.16	70	0	0.16
101–200	85	0	0.25	85	0	0.25	85	0	0.25	85	0	0.25	85	0	0.25	85	0	0.25
201–300	95	0	0.26	95	0	0.26	95	0	0.26	95	0	0.26	95	0	0.26	95	0	0.26
301–400	100	0	0.28	100	0	0.28	100	0	0.28	160	1	0.32	160	1	0.32	160	1	0.32
401–500	105	0	0.28	105	0	0.28	105	0	0.28	165	1	0.34	165	1	0.34	165	1	0.34
501–600	105	0	0.29	105	0	0.29	175	1	0.34	175	1	0.34	175	1	0.34	235	2	0.36
601–800	110	0	0.29	110	0	0.29	180	1	0.36	240	2	0.40	240	2	0.40	300	3	0.41
801–1 000	115	0	0.28	115	0	0.28	185	1	0.37	245	2	0.42	305	3	0.44	305	3	0.44
1 001–2 000	115	0	0.30	190	1	0.40	255	2	0.47	325	3	0.50	380	4	0.54	440	5	0.56
2 001–3 000	115	0	0.31	190	1	0.41	260	2	0.48	385	4	0.58	450	5	0.60	565	7	0.64
3 001–4 000	115	0	0.31	195	1	0.41	330	3	0.54	450	5	0.63	510	6	0.65	690	9	0.70
4 001–5 000	195	1	0.41	260	2	0.50	335	3	0.54	455	5	0.63	575	7	0.69	750	10	0.74
5 001–7 000	195	1	0.42	265	2	0.50	335	3	0.55	515	6	0.69	640	8	0.73	870	12	0.80
7 001–10 000	195	1	0.42	265	2	0.50	395	4	0.62	520	6	0.69	760	10	0.79	1050	15	0.86
10 001–20 000	200	1	0.42	265	2	0.51	460	5	0.67	650	8	0.77	885	12	0.86	1230	18	0.94
20 001–50 000	200	1	0.42	335	3	0.58	520	6	0.73	710	9	0.81	1060	15	0.93	1520	23	1.0
50 001–100 000	200	1	0.42	335	3	0.58	585	7	0.76	770	10	0.84	1180	17	0.97	1690	26	1.1

Table 5.21 **Single Sampling Table for LTPD = 3.0% (Reproduced by permission of John Wiley & Sons Inc. from Dodge, H. F. and Romig, H. G. *Sapling Inspection Tables*. Second ed. © 1959 John Wiley & Sons Inc., and AT&T Bell Labs.)**

	process average 0 to 0.03%			process average 0.04 to 0.30%			process average 0.31 to 0.60%			process average 0.61 to 0.90%			process average 0.91 to 1.20%			process average 1.21 to 1.50%		
lot size	*n*	*c*	AOQL %	*n*	*c*	AOQL %	*n*	*c*	AOQL %	*n*	*c*	AOQL %	*n*	*c*	AOQL %	*n*	*c*	AOQL %
1–40	All	0	0	All	0	0	All	0	0	All	0	0	All	0	0	All	0	0
41–55	40	0	0.18	40	0	0.18	40	0	0.18	40	0	0.18	40	0	0.18	40	0	0.18
56–100	55	0	0.30	55	0	0.30	55	0	0.30	55	0	0.30	55	0	0.30	55	0	0.30
101–200	65	0	0.38	65	0	0.38	65	0	0.38	65	0	0.38	65	0	0.38	65	0	0.38
201–300	70	0	0.40	70	0	0.40	70	0	0.40	110	1	0.48	110	1	0.48	110	1	0.48
301–400	70	0	0.43	70	0	0.43	115	1	0.52	115	1	0.52	115	1	0.52	155	2	0.54
401–500	70	0	0.45	70	0	0.45	120	1	0.53	120	1	0.53	160	2.	0.58	160	2	0.58
501–600	75	0	0.43	75	0	0.43	120	1	0.56	160	2	0.63	160	2	0.63	200	3	0.65
601–800	75	0	0.44	125	1	0.57	125	1	0.57	165	2	0.66	205	3	0.71	240	4	0.74
801–1 000	75	0	0.45	125	1	0.59	170	2	0.67	210	3	0.73	250	4	0.76	290	5	0.78
1 001–2 000	75	0	0.47	130	1	0.60	175	2	0.72	260	4	0.85	300	5	0.90	380	7	0.95
2 001–3 000	75	0	0.48	130	1	0.62	220	3	0.82	300	5	0.95	385	7	1.0	460	9	1.1
3 001–4 000	130	1	0.63	175	2	0.75	220	3	0.84	305	5	0.96	425	8	1.1	540	11	1.2
4 001–5 000	130	1	0.63	175	2	0.76	260	4	0.91	345	6	1.0	465	9	1.1	620	13	1.2
5 001–7 000	130	1	0.63	175	2	0.76	265	4	0.92	390	7	1.1	505	10	1.2	700	15	1.3
7 001–10 000	130	1	0.64	175	2	0.77	265	4	0.93	390	7	1.1	550	11	1.2	775	17	1.4
10 001–20 000	130	1	0.64	175	2	0.78	305	5	1.0	430	8	1.2	630	13	1.3	900	20	1.5
20 001–50 000	130	1	0.65	225	3	0.86	350	6	1.1	520	10	1.2	750	16	1.4	1090	25	1.6
50 001–100 000	130	1	0.65	265	4	0.96	390	7	1.1	590	12	1.3	830	18	1.5	1215	28	1.6

Table 5.22 Single Sampling Table for AOQL = 1.0% (Reproduced by permission of John Wiley & Sons Inc. from Dodge, H. F. and Romig, H. G. *Sapling Inspection Tables*. Second ed. © 1959 John Wiley & Sons Inc., and AT&T Bell Labs.)

	process average 0 to 0.02%			process average 0.03 to 0.20%			process average 0.21 to 0.40%			process average 0.41 to 0.60%			process average 0.61 to 0.80%			process average 0.81 to 1.00%		
lot size	*n*	*c*	%	*n*	*c*	%	*n*	*c*	%	*n*	*c*	%	*n*	*c*	%	*n*	*c*	%
1–25	All	0	–	All	0	–	All	0	–	All	0	–	All	0	–	All	0	–
26–50	22	0	7.7	22	0	7.7	22	0	7.7	22	0	7.7	22	0	7.7	22	0	7.7
51–100	27	0	7.1	27	0	7.1	27	0	7.1	27	0	7.1	27	0	7.1	27	0	7.1
101–200	32	0	6.4	32	0	6.4	32	0	6.4	32	0	6.4	32	0	6.4	32	0	6.4
201–300	33	0	6.3	33	0	6.3	33	0	6.3	33	0	6.3	33	0	6.3	65	1	5.0
301–400	34	0	6.1	34	0	6.1	34	0	6.1	70	1	4.6	70	1	4.6	70	1	4.6
401–500	35	0	6.1	35	0	6.1	35	0	6.1	70	1	4.7	70	1	4.7	70	1	4.7
501–600	35	0	6.1	35	0	6.1	75	1	4.4	75	1	4.4	75	1	4.4	75	1	4.4
601–800	35	0	6.2	35	0	6.2	75	1	4.4	75	1	4.4	75	1	4.4	120	2	4.2
801–1000	35	0	6.3	35	0	6.3	80	1	4.4	80	1	4.4	120	2	4.3	120	2	4.3
1001–2000	36	0	6.2	80	1	4.5	80	1	4.5	130	2	4.0	130	2	4.0	180	3	3.7
2 001–3000	36	0	6.2	80	1	4.6	80	1	4.6	130	2	4.0	185	3	3.6	235	4	3.3
3 001–4000	36	0	6.2	80	1	4.7	135	2	3.9	135	2	3.9	185	3	3.6	295	5	3.1
4 001–5000	36	0	6.2	85	1	4.6	135	2	3.9	190	3	3.5	245	4	3.2	300	5	3.1
5 001–7000	37	0	6.1	85	1	4.6	135	2	3.9	190	3	3.5	305	5	3.0	420	7	2.8
7 001–10000	37	0	6.2	85	1	4.6	135	2	3.9	245	4	3.2	310	5	3.0	430	7	2.7
10 001–20 000	85	1	4.6	135	2	3.9	195	3	3.4	250	4	3.2	435	7	2.7	635	10	2.4
20 001–50 000	85	1	4.6	135	2	3.9	255	4	3.1	380	6	2.8	575	9	2.5	990	15	2.1
50 001–100 000	85	1	4.6	135	2	3.9	255	4	3.1	445	7	2.6	790	12	2.3	1520	22	1.9

Table 5.23 Single Sampling Table for AOQL = 2.0% (Reproduced by permission of John Wiley & Sons Inc. from Dodge, H. F. and Romig, H. G. *Sapling Inspection Tables*. Second ed. © 1959 John Wiley & Sons Inc., and AT&T Bell Labs.)

	process average 0 to 0.04%			process average 0.05 to 0.40%			process average 0.41 to 0.80%			process average 0.81 to 1.20%			process average 1.21 to 1.60%			process average 1.61 to 2.00%		
lot size	*n*	*c*	%	*n*	*c*	%	*n*	*c*	%	*n*	*c*	%	*n*	*c*	%	*n*	*c*	%
1–15	All	0	–	All	0	–	All	0	–	All	0	–	All	0	–	All	0	–
16–50	14	0	13.6	14	0	13.6	14	0	13.6	14	0	13.6	14	0	13.6	14	0	13.6
51–100	16	0	12.4	16	0	12.4	16	0	12.4	16	0	12.4	16	0	12.4	16	0	12.4
101–200	17	0	12.2	17	0	12.2	17	0	12.2	17	0	12.2	35	1	10.5	35	1	10.5
201–300	17	0	12.3	17	0	12.3	17	0	12.3	37	1	10.2	37	1	10.2	37	1	10.2
301–400	18	0	11.8	18	0	11.8	38	1	10.0	38	1	10.0	38	1	10.0	60	2	8.5
401–500	18	0	11.9	18	0	11.9	39	1	9.8	39	1	9.8	60	2	8.6	60	2	8.6
501–600	18	0	11.9	18	0	11.9	39	1	9.8	39	1	9.8	60	2	8.6	60	2	8.6
601–800	18	0	11.9	40	1	9.6	40	1	9.6	65	2	8.0	65	2	8.0	85	3	7.5
801–1 000	18	0	12.0	40	1	9.6	40	1	9.6	65	2	8.1	65	2	8.1	90	3	7.4
1 001–2 000	18	0	12.0	41	1	9.4	65	2	8.2	65	2	8.2	95	3	7.0	120	4	6.5
2 001–3 000	18	0	12.0	41	1	9.4	65	2	8.2	95	3	7.0	120	4	6.5	180	6	5.8
3 001–4 000	18	0	12.0	42	1	9.3	65	2	8.2	95	3	7.0	155	5	6.0	210	7	5.5
4 001–5 000	18	0	12.0	42	1	9.3	70	2	7.5	125	4	6.4	155	5	6.0	245	8	5.3
5 001–7 000	18	0	12.0	42	1	9.3	95	3	7.0	125	4	6.4	185	6	5.6	280	9	5.1
7 001–10 000	42	1	9.3	70	2	7.5	95	3	7.0	155	5	6.0	220	7	5.4	350	11	4.8
10 001–20 000	42	1	9.3	70	2	7.6	95	3	7.0	190	6	5.6	290	9	4.9	460	14	4.4
20 001–50 000	42	1	9.3	70	2	7.6	125	4	6.4	220	7	5.4	395	12	4.5	720	21	3.9
50 001–100 000	42	1	9.3	95	3	7.0	160	5	5.9	290	9	4.9	505	15	4.2	955	27	3.7

Observations: 1) with constant lot size and p, the acceptance number/sample size ratio increases, AOQL decreases with decreasing LTPD. 2) With constant lot size and LTPD the acceptance number/sample size ratio and AOQL decreases with decreasing p.

b) Make a sampling plan for inspecting the same lot to assure i) AOQL = 1.0%, when the average percentage of defects is 0.61 to 0.80%. ii) AOQL = 2.0%, when the average percentage of defects are expected to be in the range of 0.61 to 0.80%. iii) Make the above sampling plans when an average load is expected to have 0 to 0.02% defectives

Solution We will use the Dodge and Romig Tables (Tabs. 5.22–5.23).

lot size	p	AOQL	n	c	LTPD
50 000	0.61–0.80%	1%	575	9	2.5
50 000	0.61–0.80%	2%	125	4	6.4
50 000	0–0.02%	1%	85	1	4.6
50 000	0–0.02%	2%	42	1	9.3

Observations: 1) with constant lot size and p, the acceptance number/sample size ratio and LTPD increases with AOQL. 2) With constant lot size and AOQL, the acceptance number/sample size ratio decreases, and LTPD increases with decreasing p.

iv) FDA Sampling Plan for Mycotoxin Analysis

Mycotoxins are usually found in extremely high concentrations when toxin-producing microorganisms invade a food. For small items, such as cottonseeds, either the entire item or a portion of the item will contain high mycotoxin concentrations. For larger items, such as apples or loaves of bread only a small portion of the item will contain high concentrations of the toxin. Variation in concentration of mycotoxin from one item to another within a lot requires that the product should be sampled as frequently as feasible, and the items taken each time should be small. The sample units are required to be collected from as many random sites as possible. The FDA (United States Food and Drug Administration) sampling procedures are given in Table 5.24. If the toxin level of the sample is detected to be greater than the acceptable levels the product is rejected. A wide variety of equipment is available for comminuting, blending and subsampling operations to form a thoroughly blended paste (Dickens and Whitaker, 1982). The expected error in the average mycotoxin analysis is directly proportional to the variance in the mycotoxin concentration among the lots and inversely proportional to the number of lots sampled.

Example 5.50. Aflotoxin sampling and analysis program for peanuts The aflotoxin sampling and analysis program for peanuts is better documented than any other mycotoxin testing program (Tab. E.5.50). The techniques used to develop this program may be used to develop sampling programs for mycotoxin analysis in other commodities. The OC curve of the program is given in Figure E.5.50.

The procedure explained in Table E.5.50 is a typical example of a multiple sampling plan.

Table E.5.50 Aflotoxin testing program employed for peanuts in the United States (Converted into SI units from Dickens and Whitaker, 1982)

STEP 1 comminute first 22 kg sample in subsampling mill	STEP 2 extract 1100 g subsample	STEP 3 make duplicate analysis of extract (1A and 1B)	STEP 4 let x = average mycotoxin content of 1A and 1B	STEP 5 accept if $x \leqslant 16 \mu g/kg$ reject if $x \geqslant 75 \mu g/kg$ go to STEP 6 if $16 \leqslant x \leqslant 75 \mu g/kg$
STEP 6 comminute second 22kg sample in subsampling mill	STEP 7 extract 1100 g subsample	STEP 8 make duplicate analysis of extract (2A and 2B)	STEP 9 let y = average mycotoxin content of 1A 1B 2A and 2B	STEP 10 accept if $y \leqslant 22 \mu g/kg$ reject if $y \geqslant 38 \mu g/kg$ go to STEP 11 if $22 \leqslant y \leqslant 38 \mu g/k$
STEP 11 comminute third 22 kg sample in subsampling mill	STEP 12 extract 1100 g subsample	STEP 13 make duplicate analysis of extract (3A and 3B)	STEP 14 let z = average mycotoxin content of 1A, 1B, 2A, 2B 3A and 3B	STEP 15 accept if $z < 25 \mu g/kg$ reject if $z \geqslant 25 \mu g/kg$

5.8. HACCP PRINCIPLES

In food processing quality control is applied to achieve the required quality level. HACCP (hazard analysis critical control point) is an additional preventive system designed to assure safety. HACCP principles originated in non-food applications. They were first used in the chemical processing industry and nuclear power plants in the 1950s, then in space missions by NASA in 1960s. NASA recommended the use of HACCP concepts to

Table 5.24 Sampling procedures for mycotoxin analysis (converted into SI units from FDA, 1979)

Product	Package	Lot size	Minimum number of sample units	Minimum unit size	Minimum total sample size
peanut butter (smooth)	consumer or bulk	NA	24	225 g	5.5 kg
			12	450 g	5.5 kg
peanut butter (crunchy), peanuts (shelled, roasted, or unroasted), peanuts (ground for topping)	consumer or bulk	NA	48	450 g	22.8 kg
				initial sample	
three nuts (except in shell Brasil nuts and all pistachio nuts in import status) shelled, in-shell, slices, pieces or flour	consumer or bulk	NA	10	450 g	4.5 kg
			as follow-up to positive analysis		
			50	500 g	22.7 kg
three nuts (paste)	consumer or bulk	NA	12	500 g	5.5 kg
Brasil nuts (in shell, in import status)	bulk (in bags)	0–200	20	500 g	9 kg
		201–800	40	500 g	18 kg
		801–2000	60	500 g	27 kg
Pistachio nuts (in shell, in import status)	bulk	lot portion = 340 000 kgs	20% of units	–	22.5 kg/lot portion
Pistachio nuts (shelled, in import status)	bulk	lot portion = 340 000 kgs	20% of units	–	11.5 kg/lot portion
corn (shelled, meal flour or grits)	consumer or bulk	NA	10	500 g	4.5 kg

cottonseed	bulk	NA	15	2 kg	27 kg
oil seed meals (peanut meal, cottonseed meal	bulk	NA	20	500 g	9 kg
edible seeds (melon, pumpkin, sesame, etc.)	bulk	NA	50	500 g	22.5 kg
ginger root: i) dried whole ii) ground	i) bulk ii) consumer	i) n units ii) NA	i) $\sqrt{n}$ ii) 16	i) – ii) 30 g	i) 7 kg ii) 4.5 kg
milk (whole, skim, low fat)	i) consumer ii) bulk	NA NA	10 –	500 g	4.5 kg 4.5 kg
small grains (wheat, sorghum, barley, etc.)	bulk	NA	10	500 g	4.5 kg
dried fruits (figs, etc.)	consumer or bulk	NA	50	500 g	22.5 kg
mixtures containing commodities susceptible to mycotoxin contamination with i) large particles ii) relatively small particles	consumer	NA	i) 50 ii) 10	i) 500 g ii) 500 g	i) 22.5 kg ii) 4.5 kg

NA = not applicable

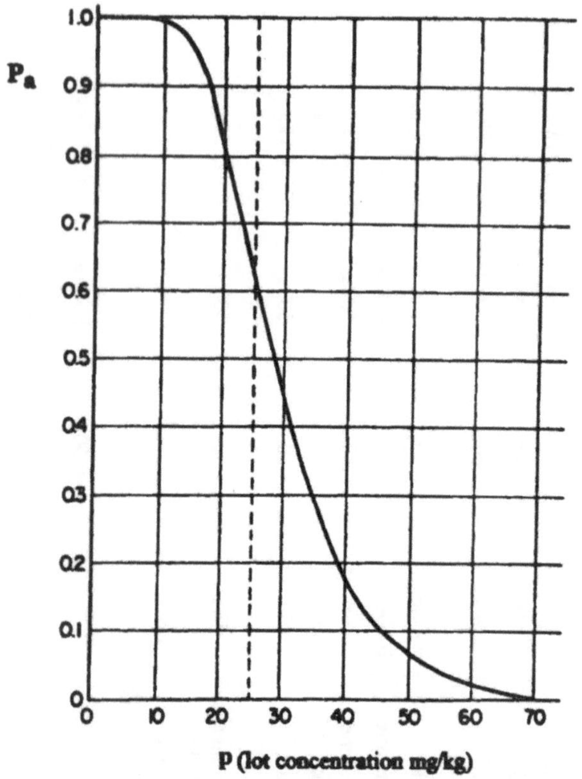

FIGURE E.5.50 The operating *OC* curve for the peanut sampling plan (Dickens and Whitaker 1982; © International Agency for Reseach on Cancer, reproduced by permission).

prevent food borne illness of space crews. US Army Natic Laboratories used these concepts to prepare the space food for NASA. Philsbury Company, a contractor of space food, adapted and expanded the concept for commercial applications (Snyder, 1992).

A hazard is defined as any biological, chemical or physical property that may cause an unacceptable health risk. Any point (or process) in a specific food system, where loss of control may result in unacceptable health risk is called a critical control point (CCP) (USDA, 1989). Control points (CP) are differentiated from CCPs on the basis that losing control in the CPs will not result in unacceptable health risk. Control points are generally non-safety points related to product quality or regulatory compliance. In yogurt production lactic acid bacteria decreases the pH and prevents growth of pathogenic microorganisms. Here the hazard is the pathogens and CCP is acid fermentation. Losing control in acid fermentation, i.e., failing to achieve safe pH may cause unacceptable health risk. Microorganisms that contaminate the *low acid foods* may form heat stable spores, therefore processing requires temperatures above 100 °C for a certain time (Chapter 4.1). In the process of canning low acid foods the hazard is *Clostridium botulinum*, and the CCP is the thermal process (temperature and duration). Failing to achieve a safe time-temperature combination may cause unacceptable health risk. HACCP principles are applied to assure food safety by monitoring critical control points during food processing and preservation. The critical points are continuously monitored to keep the critical parameters within tolerable limits. HACCP principles are usually not enforced by regulatory agencies, but applied voluntarily by food processors following their own HACCP plan.

The HACCP plan is a written document which has been prepared considering the hazards associated with growing or harvesting the raw materials and ingredients; and processing, distribution, marketing, preparation and consumption of the food. It shows the CCPs and the procedures to monitor and control the identified hazards, and establishes the limits which must be met at these points. Examples of critical limits include minimum processing time and temperature, maximum refrigeration holding temperature, maximum pesticide level, etc. The plan also establishes the corrective action to be taken when the required limits are not met. The companies also establish their own record keeping procedures to verify that the HACCP plan is working correctly.

Example 5.51. Philsbury's Hazard and quality control plan (Sperber, 1991) HACCP system is combined with statistical quality control. In the overall system there are CCPs (critical control points), CPs (control points) and MCPs (manufacturing control points). The CCPs are required for HACCP, hazards including *Salmonella*, aflotoxin, antibiotics, inadequate pasteurization and cross-contamination, etc., are monitored. CPs are required for regulatory and economic reasons; coliforms, insects, food color, fumigant, net weight and labeling controls are made. MCPs are required for statistical quality control purposes; total microbial counts, formulation controls, viscosity measurements are made.

Foods are ranked from 0 to IV according to the associated risks and assigned a risk category (Tab. 5.25). Processing and handling practices are conducted according to their category.

Category VI is assigned to for non-sterile foods designated for consumption by high risk groups. Categories V to 0 show the total number of hazard characteristics (Hazard B to Hazard F) associated with the food. The above considerations allow us to identify the areas in which hazards in the food system may be reduced. This analysis may result in changing the form of an ingredient, e.g., fresh to canned, or changing a step in the manufacturing process, e.g., chilled to frozen, to reduce the risk. If the supplier of food

Table 5.25 Hazard characteristics of the foods (USDA, 1989)

HAZARD A	food is non sterile and intended to be consumed by high-risk populations, i.e., infants, elderly, hospitalized people
HAZARD B	food contains microbiologically sensitive ingredients
HAZARD C	there is no control step to effectively destroy the microorganisms
HAZARD D	there is a significant risk of post processing contamination by microorganisms and toxins before packaging
HAZARD E	there is substantial potential for abusive handling in distribution or by consumer
HAZARD F	there is no terminal heat process after packaging or cooking at home

ingredients develops a HACCP program, lot acceptance tests may be replaced by reduced frequency audit programs to verify that the HACCP plan is working correctly (Microbiology and Food Safety Committee of the National Food Processors Association, 1993).

Example 5.52. HACCP flow diagram for cake mix HACCP flow diagram for cake mix is depicted in Figure E.5.52 and its work sheet is given in

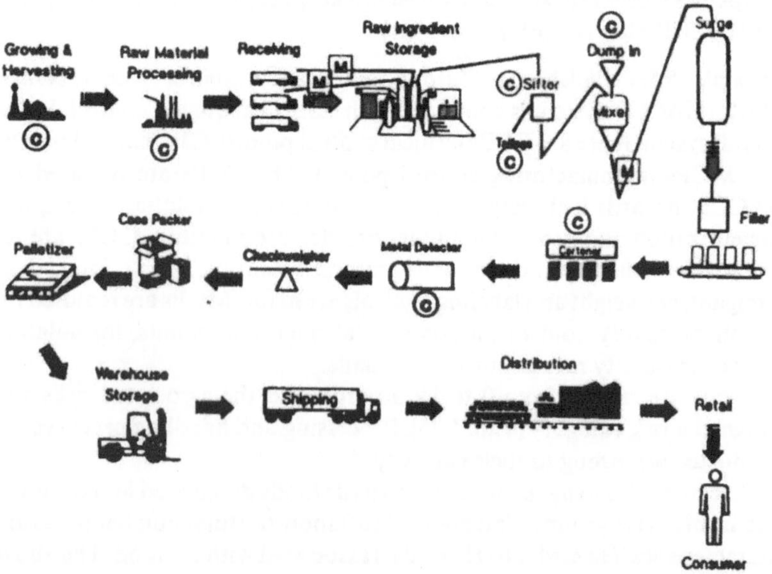

FIGURE E.5.52 HACCP flow diagram for cake mix. C: control point, M: magnet (Microbiology and Food Safety Committee of the National Food Processors Association, 1993, Reprinted from *Journal of Food Protection* with permission. © International Association of Milk Food and Environmental Sanitarians, Inc.).

Table E.5.52 (Microbiology and Food Safety Committee of the National Food Processors Association, 1993).

Example 5.53. Model based approach to automated hazard identification (*Catino and Ungar*, 1995) Flow diagram of a prototype hazard identification system, qualitative hazard identifier (QHI) is depicted in Figure E.5.53. The QHI works by positing possible faults, automatically building qualitative process models, simulating them, and checking for hazards. To automatically generate a mathematical model library of the physical and chemical phenomena, the process flow diagram and process conditions are required. Each phenomenon defined in the library consists of a set of preconditions required for that particular phenomenon to occur, and the equations contributing to model this phenomenon if it is applicable. The input of the physical description of the plant includes the substances and the equipment present, their connectivity and assumptions about operating conditions. First a qualitative differential equation (QDE) model of the plant as it was designed to operate is automatically generated from the library of phenomena and the physical description of the plant. The pre-processor determines if a fault is applicable. The faults may be described by perturbations of the original design model, or they may require a new model. In general, deviations in the process parameters, i.e., an increase or decrease of temperature, can be analyzed by using a single model, whereas deviations in process intentions, i.e., involvement of additional, reverse or different phenomena, require building a new model to describe the process. Qualitative process compiler (QPC) generates sets of equations that are abstractions of ordinary differential equations. These qualitative equations are then sent to QSIM algorithm for qualitative simulation and the possible behaviors (qualitative solutions of the model equations) are determined. The post processor determines if any of the behaviors are hazardous and prints out the possible hazards. The QHI has been originally developed for the chemical process industries. The same principles may be used for hazard identification in food processing if the required input files are provided.

Example 5.54. Predictive microbial modeling There is a strong and increasing demand for more convenient, fresher, more natural, less heavily processed, less heavily preserved (less salt, sugar and additives, etc.) foods. The microorganisms that can spoil these foods, limit their shelf lives and cause foodborne diseases need to be controlled. *Predictive microbial modeling* is a powerful tool that can underpin improved control in food processing and preservation. There are international collaborations toward developing the required databases of key microorganisms to processes, and conditions to which they are exposed in processing (COST 914, 1994; Whiting and Butchanan, 1994). Accumulation of relevant microbial data was standardized and coordinated and microbial models taking account of the pH value,

Table E.5.52 **Work sheet for HACCP analysis of cake mix, (Microbiology and Food Safety Committee of the National Food Processors Association, 1993; Reprinted from *Journal of Food Protection* with permission. ©International Association of Milk Food and Environmental Sanitarians, Inc.)**

item	hazard	control	limit	monitoring frequency/ documentation	action (for failure of CCP)	personnel responsible
1. Growing and harvesting	improper chemical application	grower records	EPA approved chemicals, specified tolerances	certificate of guarantee for each lot; QC random audit of records and chemical assay	reject ingredient lot to supplier. Increase frequency of audit/record review	shipment receiver; QC audit responsibility
2. Supplier HACCP programs	chemical, physical and microbiological hazards specifically identified	supplier HACCP programs QC audits	no hazardous foreign material in the lot intact	QC audits supplier HACCP program at least annually	failure of supplier HACCP will result in delisting as supplier	QC has audit responsibility for supplier audit programs
3. Sifter	hole in screen allowing physical hazards (wood, metal, glass, plastics, etc.) to pass thorough	routine monitoring of sifter screen	screens	operator check and record in processing log every shift	defective screen will result in all product run since last check placed on hold. System emptied and cleaned. Screen replaced. Lot rejected back to supplier	Line operator. QC notified and handles disposition of any product placed on hold, per standard operating procedures

Table E.5.52 (Continued)

item	hazard	control	limit	monitoring frequency documentation	action (for failure of CCP	personnel responsible
4. Tailings from sifter	physical hazards (wood, metal, glass, plastics, etc.)	tailings check from sifter	no hazardous material	operator checks and records findings in processing log every tow hours every tow hours	any hazardous findings. supervisor and QC notified. All product since last OK check placed on hold. Ingredient lot rejected back to supplier	Line operator. QC notified and handles disposition of held product and rejection
5. Dump in	physical hazards (wood, metal, glass, plastics, etc.)	visual observation by mixer during dumping of bagged ingredients	no hazardous material	operator dumps each ingredient through 4 mesh screen and observes for hazards	any hazardous finding reported to supervisor. Current batch diverted and placed on hold. Mixer cleaned. Lot rejected back to supplier	Mixer. QC notified of hazardous findings and material placed on hold. Handles disposition of product and rejection
6. Cartoner	improper labeling which may cause health hazard (e.g. Yellow 5 and 6, sulfites, etc.)	check off to ensure proper labels are used for product being produced	proper labels must be used	packaging operator reviews and records in packaging log that proper labels or cartons in use every two hours or at changeover	improper packages or labels must be reported to supervisor and QC. All product since last OK check placed on hold	Packaging operator. QC notified of any improper packaging or labels being used. Dispositions held product
7. Metal detector	metal	metal detector	no hazardous findings	calibration check every two hour with appropriate test piece; recorded in log	all kick-outs checked by QC. Any hazardous findings investigated. Action to follow QC policy	Line operator checks calibration, QC audits twice each shift. QC handles metal detector rejections

EPA: Environmental Protection Agency of the United States
QC: Qiality control personnel

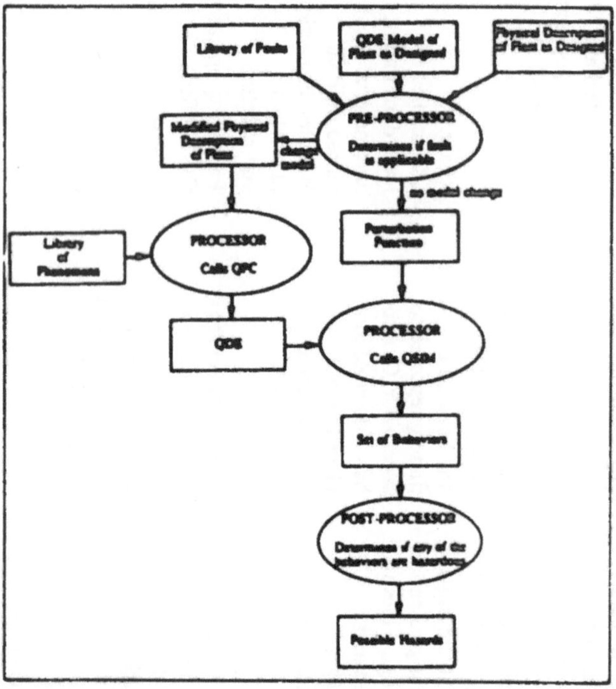

FIGURE E.5.53 Computer flow diagram of the qualitative hazard identifier (Catino and Ungar, 1995). Reproduced with permission of the American Institute of Chemical Engineers.

water activity and storage temperature have been developed so that survival or growth can be predicted using single simple computer software. The second generation of these models are expected to be used in HACCP procedures and lead to greater confidence in new and competitive product innovations.

An excellent example to predictive microbial modeling of food spoilage has been presented by Nicolai *et al.* (1993), who studied surface growth of lactic acid bacteria in vacuum-packed meat. Spoilage of the vacuum-packed meat at temperatures below 20°C is dominated by anaerobic growth of the lactic acid bacteria, which attain about $10^7 - 10^8$ cfu/cm^2 during 5 to 10 weeks of storage time and cause a sharp off-odor. Some hetero-fermentative species such as *Lactobacillus viridescens* may produce peroxides which react with meat pigments and cause green off-colors. It is desirable to predict the microbial spoilage level during storage of vacuum packed meat to control economic and hygienic aspects. The growth and lactic acid production may be simulated as:

$$\frac{dx}{dt} = \mu x \qquad \text{(E.5.54.1)}$$

and

$$\frac{d(\text{total lactic acid produced})}{dt} = \nu x \qquad \text{(E.5.54.2)}$$

where specific growth rate μ and the specific product formation rate ν are functions of the lactic acid concentration and pH of the medium:

$$\mu = \mu(\text{lactic acid, pH}) \qquad \text{(E.5.54.3)}$$

$$\nu = \nu(\text{lactic acid, pH}) \qquad \text{(E.5.54.4)}$$

Nicolai *et al.* (1993) assumed that the spoilage phenomena occurs in a surface liquid film of 10 µm thickness and is related the availability of the positively charged hydrogen atoms and other chemical species through chemical equilibria.

Post mortem glycolysis of the meat pH is usually depressed from about 7.4 to an ultimate level of 5.6. Cooked dark firm dry meat cuts usually have pH $\geqslant$ 6.0. The meat components contribute to the buffering capacity. In Figure E.5.54 the time course for biomass and pH is simulated after using (E.5.54.1) – (E.5.54.4) with appropriate functional relations and chemical equilibrium reactions. The initial pH of the meat is higher than the optimum pH for growth. In case 1, the initial buffer amount is large enough to induce a rather large lag time. Consequently growth will start slowly. When the produced lactic acid causes a pH shift towards the optimum value, the growth rate passes through a maximum. After a fast growth period the pH will decrease further and because of this and the corresponding high amount of undissociated lactic acid, the growth decreases and eventually stops. This results in the typical sigmoidal appearance. The smaller the buffer capacity the smaller the corresponding lag time as depicted in cases 2 and 3.

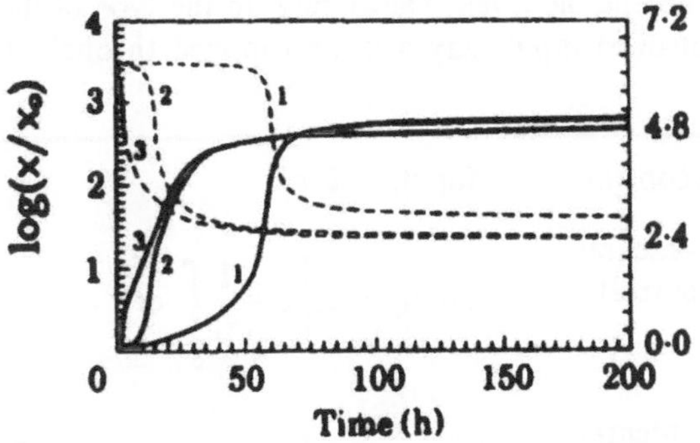

FIGURE E.5.54 Surface growth of *Lactobacillus delbrueckii* for different values of the initial buffer which is 10^{-8}, 10^{-9} and 10^{-10} mols in cases 1, 2 and 3, respectively (Nicolai *et al.*, 1993; © Academic Press, reproduced by permission).

5.9. PROCESS CONTROL THROUGH MATHEMATICAL MODELING

Process control is implemented to prevent deviations from constant processing conditions. The feed back control concept was depicted in Figure 5.1, where the measured value of an output from a process is compared with its set point value. If the difference is larger than the tolerable amounts, the inputs are re-adjusted to reduce the difference to acceptable levels. In traditional process control the requirement of constant process conditions is met by automatic controllers, which operate either pneumatically or electronically. These controllers form an integral part of a control loop, in which they receive a signal from a sensor measuring the parameter to be controlled (such as temperature or concentration), and transmit the signal to a control valve, causing it to open or close in accordance with the corrective action required to minimize the error signal being received from the sensor. The rate at which the controller responds to the error signal is a function of the on/off or proportional, integral and derivative components of the system response time relative to preset adjustments for these functions in the controllers themselves. Under steady state conditions the value of a measured output parameter remains at the system settings $\mathcal{P}_{set}$. When the system settings are changed or the steady state conditions are disturbed the measured parameter may attain an instant value of $\mathcal{P}$. The difference of the measured output of $\mathcal{P}$ and its set point value $\mathcal{P}_{set}$ is called the *error*, ε:

$$\varepsilon = \mathcal{P}_{set} - \mathcal{P} \tag{5.89}$$

The simplest type of control devices are the on/off controllers which turn the controlled input to the system (Input_c) on or off to maintain the error within acceptable levels. Depending on the type of the controllers used the controlled input may also be supplied through the following ways:

proportional control	$\text{Input}_c = K_c \varepsilon$	(5.90)
proportional-integral control (PI control)	$\text{Input}_c = K_c\left[\varepsilon + \frac{1}{\tau_I}\int_0^t \varepsilon\, dt\right]$	(5.91)
proportional-integral-derivative control (PID control)	$\text{Input}_c = K_c\left[\varepsilon + \frac{1}{\tau_I}\int_0^t \varepsilon\, dt + \tau_D \frac{d\varepsilon}{dt}\right]$	(5.92)

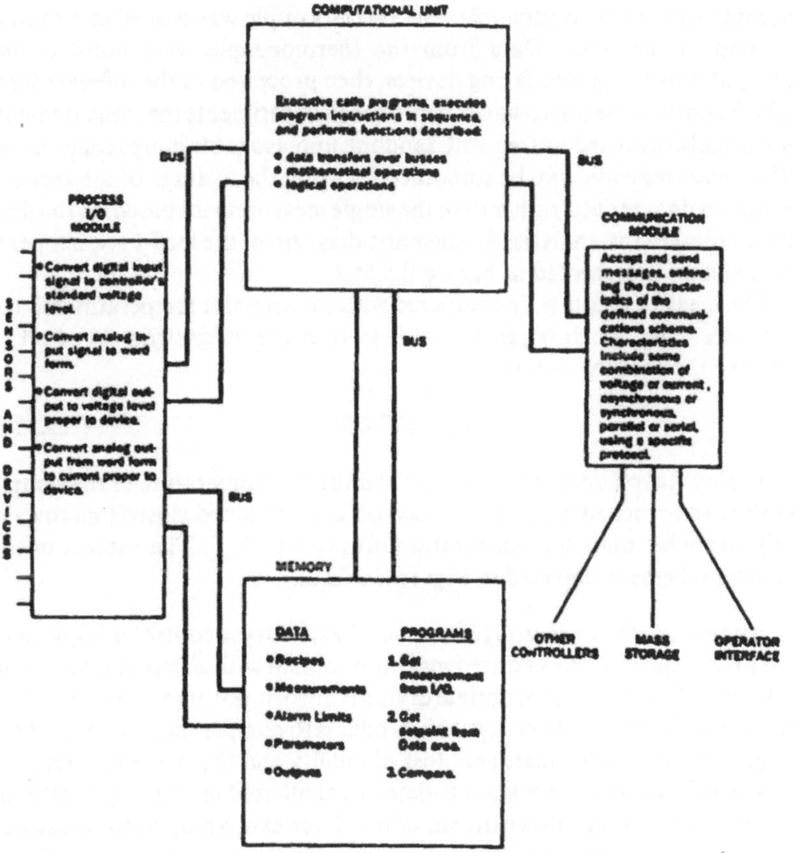

FIGURE 5.15 General configuration of a computer-based control system (Teixeira and Shoemaker, 1989; © Chapman & Hall. Reprinted by permission of Chapman & Hall, NY).

where K_c is called the gain or sensitivity. Parameters τ_I and τ_D are named as the integral and the derivative times, respectively.

Computers may be interfaced with instrument sensors and devices through I/O (input/output) process modules with A/D (analog/digital) or D/A (digital/analog) converters. General configuration of a computer-based control system is depicted in Figure 5.15, which also illustrates how the same computers could be connected to various memory storage devices and other communication modules for data storage and retrieval, operator interface, and communication with other computers and controllers. Applications of computer-based control to food and fermentation related systems are depicted in Examples 5.55–5.61.

Example 5.55. Feed-back temperature control system for microwave ovens Ramaswamy *et al.* (1991) modified a domestic microwave oven to accommodate a thermocouple. The thermocouple was placed in a sample located in the oven. Data from the thermocouple were input to the computer by using interfacing devices, then processed in the software logic which controls the microwave oven. In such experiments the measurements are usually distorted by frequent, random, unreasonable jumps called *noise*. The measurements may be smoothed by using the average of the recently acquired data points (rather than the single measurement based on the final data point) in the analysis. A schematic diagram of the feed-back temperature system is presented in Figure E.5.55.1.

The measurements were compared with the set point temperatures in the software logic which triggers a relay to turn the magnetron on or off to achieve the control such that:

$$T_{min} \leqslant T \leqslant T_{max} \tag{E.5.55.1}$$

The relay stayed open (microwave on) until the temperature of the sample in the oven reached T_{max}, then turned off and remained closed (microwave off) thereafter until the temperature dropped to T_{min}. The success of the control scheme is depicted in Figure E.5.55.2.

Example 5.56. Dryer control (*Robinson*, 1992) Dryer control is implemented to dry a product to a desired moisture content with acceptable variation (Fig. E.5.56.1). When appropriate drying control is not employed, manufacturers usually tend to over dry their products to compensate the effect of the large variance, with consequent loss of quality and thermal efficiency.

Manual feedback control is out-dated but still used in many applications, where at some point downstream of the dryer exit, an operator measures the moisture content and mentally compares the measurement with the desired value, then makes the adjustments to the energy input based on the comparison. Alternatively, the energy input rate may be held constant and the feed rate may be manipulated, or both energy input and feed rate may be manipulated to maintain the desired moisture content. *Closed-loop feedback control* (Fig. E.5.56.2.a) employs a sensor to measure the moisture content or a controlled variable at a point downstream from the dryer. This value is transmitted to a controller, which compares it to the desired set point and uses the error to calculate the change in energy input required to bring the moisture content of the product back to the set point value. If this adjustment in the manipulated value is automatic, the system loop is closed. If not, the system is open-loop and an operator must manually adjust the energy input value, as discussed previously. Closing the loop improves upon manual feed back control by speeding up the return of controlled variable data, upon which the control decision is made. *Feed-forward control systems* (Fig. E.5.56.2.b) include sensors for measuring the effect of disturbances to the drying operation not accounted for by feed back systems. Figure E.5.56.2.b illustrates feed-forward control added to the

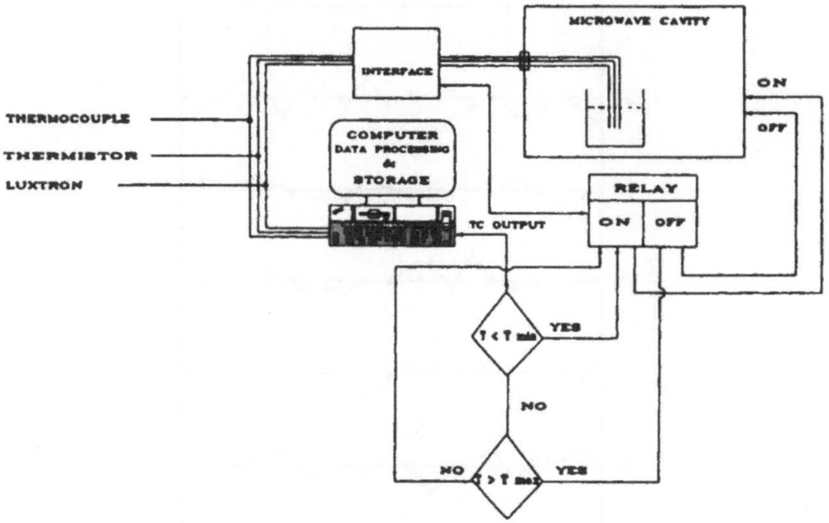

FIGURE E.5.55.1 Schematic diagram of the experimental set-up and description of the control logic (Ramaswamy *et al.*, 1991; © IFT, reproduced by permission). Luxron and thermistor are used for comparison with the thermocouple.

feedback system of Figure E.5.56.2.a to correct for variations in evaporative load on dryer. Here a disturbance is sensed and the necessary correction in the manipulated variable, i.e., the energy input rate, is made. Such systems require additional sensors, as well as a thorough knowledge of the relationship between disturbances and the manipulated control variables. To measure the disturbance caused by variations in entering a evaporative load, it is necessary to include a feed rate sensor and a transmitter at the front end of the dryer. A computer-managed material balance gives the theoretical amount of water to be evaporated and the amount of energy required, for comparison with the energy actually consumed. The difference is used to correct the manipulated variable. To meet higher quality and efficiency demands, dryer control systems became more sophisticated and costly. There is an additional moisture sensor and a transmitter, a feed rate sensor and transmitter and a process computer to improve the moisture distribution of the product in Figure E.5.56.2.b.

A temperature-drop model control system is used in Figure E.5.56.2.c. This system enables determination and control of the moisture content at any appropriate point along or inside batch or continuous dryers before the product leaves the dryer. The following model is used with the continuous dryers:

$$\text{product moisture content} = K_1(\Delta T)^Q - \frac{K_2}{S^P} \qquad \text{(E.5.56.1)}$$

where K_1, K_2, P and Q are constants; ΔT = (air temperature after contact-

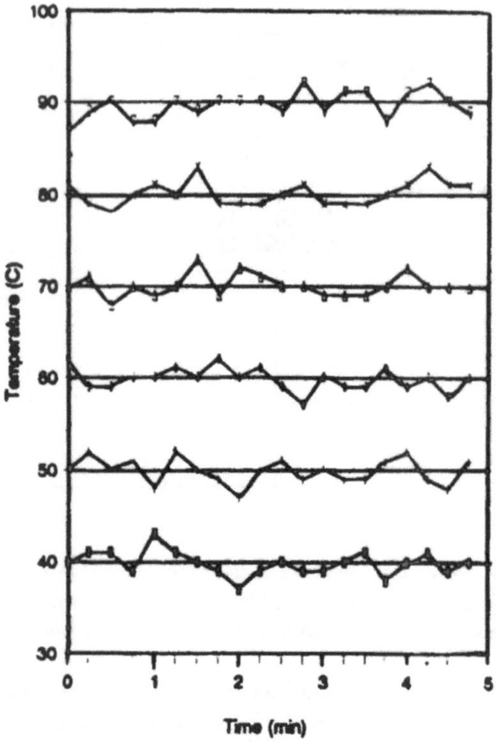

FIGURE E.5.55.2 Microwave temperature stability plots obtained during measurements with small sample volume (5 mL) in the presence of additional microwave sink. Obtaining the measurement from a small sample volume caused stability problems when there was no other object in the oven to absorb the microwaves (Ramaswamy *et al.*, 1991; © IFT, reproduced by permission).

ing with the product)-(air temperature before contacting with the product), S = dryer speed or production rate. With the batch dryers the model becomes:

$$\text{product moisture content} = K_1(\Delta T)^Q - K_3(t_D)^R \quad \text{(E.5.56.2)}$$

where K_3 and R are constants, t_D = drying time. These models lump all the variables affecting drying and drying rate into variables ΔT and S. If any of the input variables of the feed to the dryer varies for any reason, this will be sensed by the variation in ΔT and adjustments will be made in the system to maintain constant product moisture content. A two-loop cascaded control system is employed in Figure E.5.56.2.c. Comparison of the dryer control methods are given in Table E.5.56.

Table E.5.56 Comparison of the dryer control methods (Robinson, 1992; Table I)

control method	Tradeoffs	Relative cost	Distribution of moisture contents
manual feed back	i. simple to operate ii. relatively inexpensive iii. simple to maintain iv. wide range of product moisture content distribution	base	±(6 to 8) moisture content units
automatic feedback	i. simple to operate ii. relatively inexpensive iii. moderately easy to maintain iv. moderate range of product moisture content distribution	(1.5 to 2) × base	±(4 to 35) moisture content units
feed forward combined with feed back	i. complex to operate ii. relatively expensive iii. requires expertise to maintain iv. requires process information v. range of product moisture content is good	(3 to 4) × base	±(2 to 3.5) moisture content units
temperature drop	i. requires moderate expertise to maintain and operate ii. moderate cost iii. little process information is required iv. excellent range of product moisture content distribution	(1.5 to 2) × base	±(1 to 1.5) moisture content units

Example 5.57. Computer control of citrus juice evaporator (*Chen et al.,* 1981) In a thermally activated, short time evaporation process (TASTE) the feed juice normally at 11–13°Brix is concentrated to 60–65°Brix. Under manual control the steam flow is at constant pressure and the feed flow is set according to the desired output °Brix. Both flows are regulated by an operator, who measures the Brix of the concentrate, and adjusts the flow rate of either the steam or the feed or both. The operator usually needs to make calculations for these adjustments which takes up time, meanwhile concentrate with undesirable °Brix is produced. Such problems may be solved with feed-back control (Fig. E.5.57.1).

Real time temperature, °Brix, steam and juice flow rates were monitored in the process described in Figure E.5.57.1. Separate feed-back loops were needed for juice and steam control in each evaporator. It was possible to control several evaporators by using one computer. Temperatures and concentrate °Brix values were sensed electronically, then interpreted with the computer and appropriate signals were sent to the steam and feed

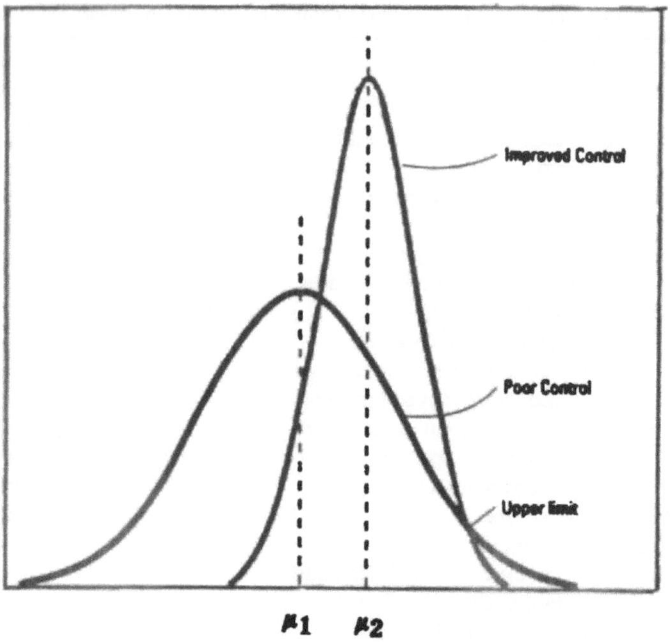

FIGURE E.5.56.1 Variation of the average moisture contents of a product at the exit of a dryer. Poor control is characterized with the large variance, while the average moisture distribution with improved control is confined within narrower limits (Robinson, 1992). Reproduced with permission of the American Institute of Chemical Engineers. © 1988 AIChE. All rights reserved.

control valves at five second intervals. Steam condensation temperature was used for steam control and second stage juice temperature was used for juice flow control. The steam control was achieved by PID control model:

$$V_n = K_c \varepsilon_n + \frac{K_c t}{\tau_I} \sum_{i=0}^{n} \varepsilon_i + \frac{\tau_D K_c}{t}(\varepsilon_n - \varepsilon_{n-1}) + V_r \qquad \text{(E.5.57.1)}$$

where t = sampling time, V_n = value of the manipulated variable at the nth sampling instant, V_r = reference value at which the control action is initialized, ε_n = value of the error at the nth sampling instant. Equation (E.5.57.1) is actually a different version of (5.92). If (E.5.57.1) is written for V_{n-1} and subtracted from (E.5.57.1) as given above, the result is:

$$\Delta V_n = K_c(\varepsilon_n - \varepsilon_{n-1}) + \frac{K_c t}{\tau_I}\varepsilon_n + \frac{\tau_D K_c}{t}(\varepsilon_n - 2\varepsilon_{n-1} + \varepsilon_{n-2}) \qquad \text{(E.5.57.2)}$$

where $\Delta V_n = V_n - V_{n-1}$ is equal to the change in the manipulated variable. Equation (E.5.57.1) was used at the start-up period, (E.5.57.2) is used later.

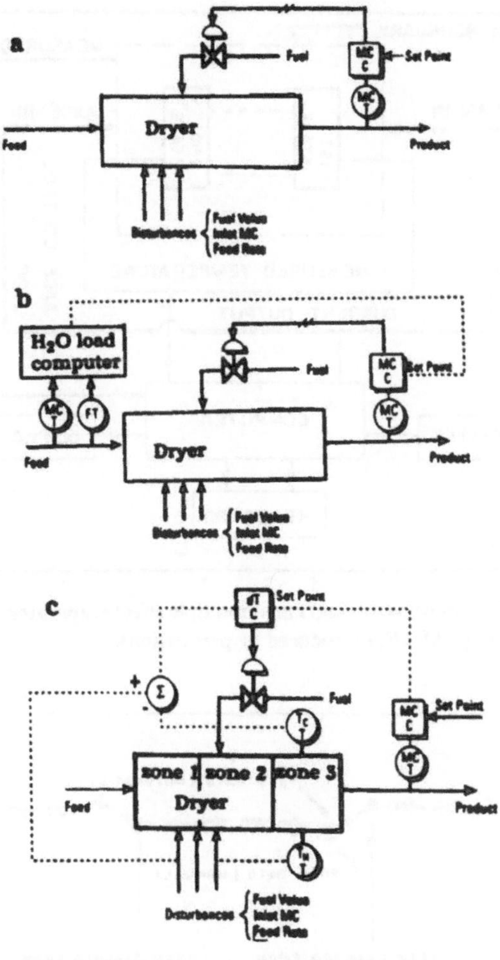

FIGURE E.5.56.2 (a) Closed-loope feed-back control scheme. (b) feed-forward control scheme. (c) two-loop cascaded control system based on temperature-drop drying model. MC: moisture content, C: controller, T: temperature, FT:feed temperature (Robinson, 1992). Reproduced with permission of the American Institute of Chemical Engineers. © 1992 AIChE. All rights reserved.

The control constants K_c, τ_I and τ_D were determined with the trial and error method. Figure E.5.57.2 shows the variations of the steam and feed rates with manual and feed-back control. It was discovered that the feed-back control achieved 17% savings over manual control for the production of the same amount of concentrated juice.

Example 5.58. Computer control of batch retort operations with on-line correction of process deviations (Teixeira and Manson, 1982) Heat transfer in

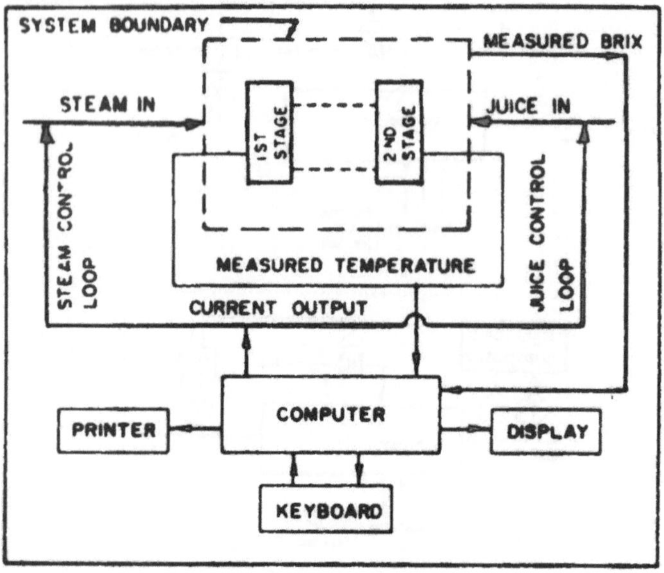

FIGURE E.5.57.1 **Schematic diagram of multiple effect evaporator control scheme (Chen *et al.*, 1981; © ASAE, reproduced by permission).**

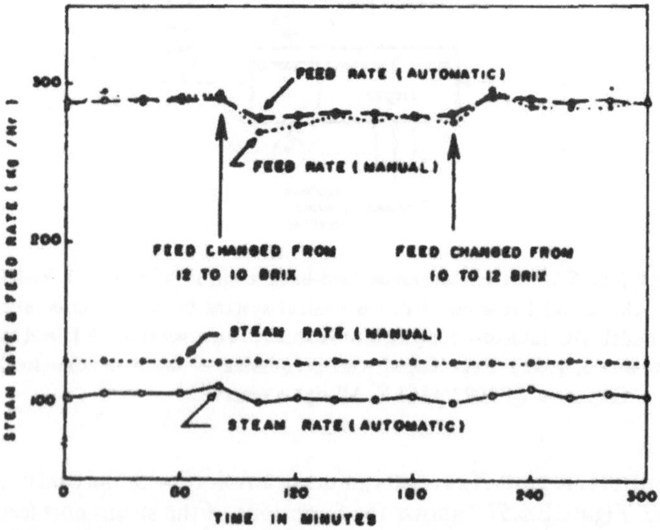

FIGURE E.5.57.2 **Variation of the steam and feed rates under manual and feed-back control (Chen *et al.*, 1981; © ASAE, reproduced by permission).**

a conduction heating food is described with the simplified form of (2.17) as:

$$\frac{\partial T}{\partial t}=\alpha\left[\frac{1}{r}\frac{\partial^2 T}{\partial r^2}+\frac{1}{r}\frac{\partial T}{\partial r}+\frac{\partial^2 T}{\partial z^2}\right] \tag{E.5.58.1}$$

where thermal diffusivity α may also be expressed after rearranging (E.4.5.13) as:

$$\alpha=\frac{0.398}{f_h\left[\frac{1}{R^2}+\frac{0.427}{H^2}\right]} \tag{E.5.58.2}$$

where $2H=$ height and $R=$ radius of the can. Equation (E.5.58.1) may be converted into a finite difference equation as:

$$T_{i,j}^{t+\Delta t}=T_{i,j}^{t}+\frac{\alpha\Delta t}{\Delta r^2}[T_{i-1,j}-2T_{i,j}+T_{i+1,j}]^t$$

$$+\frac{\alpha\Delta t}{2r\Delta r}[T_{i-1,j}-T_{i+1,j}]^t+\frac{\alpha\Delta t}{\Delta h^2}[T_{i,j-1}-2T_{i,j}+T_{i,j+1}]^t \tag{E.5.58.3}$$

Equation (E.5.58.3) was used to evaluate the temperature profiles along the internal volume elements of the can after setting the initial temperature as the initial product temperature and the surface temperature equal to the retort temperature. When the steam is shut off and cooling water is turned on, the retort temperature on the boundary condition is substituted with the cooling water temperature. Thermal processing received at the can center is:

$$F_0=\int_0^t 10^{(T-T_{ref})/z}\,dt \tag{E.5.58.4}$$

When the numerical computer model is used to calculate the process time required at a given retort temperature to achieve a specified sterilizing value, the computer follows a programmed search routine of assumed process times. It quickly converges on the precise time at which cooling should begin in order to achieve the specified F_0 value such that

$$F_{required}=F_0 \tag{E.5.58.5}$$

In this way, the model can be used to determine the process time required for any given set of constant or variable retort temperature conditions.

The schematic representation of the computer-based retort control system is described in Figure E.5.58.1. The retort is instrumented with sensors to measure temperature, pressure, and water levels in the retort, along with various valves and switches to control flow of air, water and steam into and out of retort. The controller in the system is a small digital computer (microprocessor). It is programmed to execute a specific sequence of control functions, as well as to read and interpret input signals, make

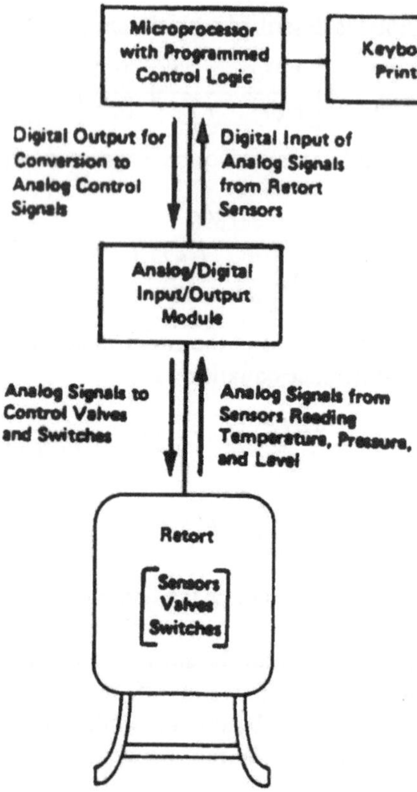

FIGURE E.5.81.1 Schematic representation of computer based retort control system (Teixeira and Manson, 1982; © IFT, reproduced by permission).

decisions, and send output signals to the system. The analog/digital input/output module converts the analog signals received from the sensing instruments in the retort into electronic digital signals required by the computer.

The flow diagram in Figure E.5.58.2. illustrates the type of control logic concept that could be used to evaluate and automatically correct a process deviation during on-line control. The first decision box in the flow diagram is used to assure that the process would not begin until the computer has compared all the operator input data with the established specifications on the control program specified for the product. Once processing begins, the control logic calls for the computer to read the actual retort temperature and compare it with the established retort temperature specified for that point in the programmed processing cycle. If these temperatures agree, the control logic moves to the left side of the flow diagram and the computer

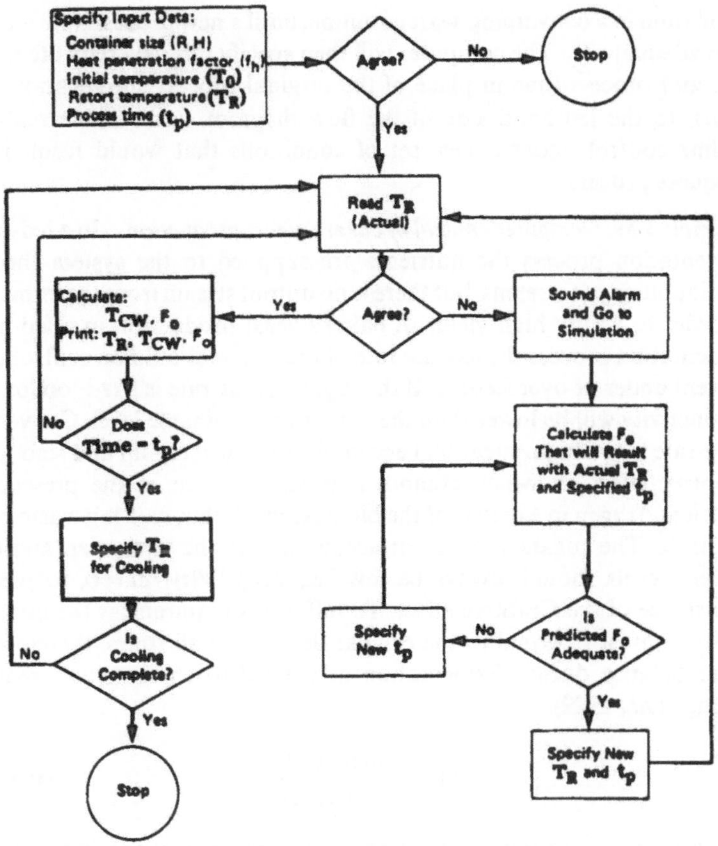

FIGURE E.5.58.2 Control logic for on-line correction of process deviations (Teixeira and Manson, 1982; © IFT, reproduced by permission). TC = cooling temperature.

will calculate the new temperature reached at the can center over that time interval and the resulting sterilization value accomplished. When the specified process time has been reached and cooling begins, the control logic will ,call for the computer to specify the new retort temperature established for the cooling cycle. The computer will continue to read, compare, and calculate the previously explained parameters until the cooling cycle ends. If a process deviation occurs in the system, then the control logic moves to the right side of the flow diagram. An alarm is sound to inform the operator, and the computer immediately uses the model to simulate completion of the process based on the actual retort temperature just read, and predict the final F_0 value that would result. If this is not adequate the computer will specify a new process time and repeat the simulation in a converging search routine, until a new process time is found with adequate F_0. The computer will then specify this new retort temperature and process time in place of the original process specifications and return to the left-hand side of the flow diagram, and resume real-time on-line control under a new set of conditions that would result in an adequate process.

Example 5.59. Computer control of bakers' yeast production In a fed-batch fermentation process the nutrients are supplied to the system through continuous input streams, but there is no output stream from the fermenter. In order to assure high yields of baker's yeast production in a fed-batch fermentation process the feeding rate of the sugars should be controlled to prevent under or over feeding. If the sugar feeding rate is fixed too low, the productivity will be lower than the maximum attainable level. Conversely, if the rate is too high, sugar will accumulate in the medium and lead to the Crabtree effect, in which ethanol is produced even in the presence of sufficient oxygen in expense of the biomass yield, thus causing waste of the substrate. The sugars should not accumulate in the fermenter, and their present levels should always be low i.e., $d[c_s(t)\,V(t)]/dt = 0$, to prevent occurrence of the Crabtree effect. To fulfill this requirement the molasses feed rate must be equal to the cellular demand at all times, therefore the sugar balance during fermentation in the fed-batch fermenter requires (Wang *et al.*, 1979):

$$F_s(t) = \frac{\mu(t)\,[x(t)\,V(t)]}{Y(t)\,c_s(t)} \qquad \text{(E.5.59.1)}$$

where $c_s(t)$ is the sugar concentration, $F_s(t)$ is the volumetric addition rate of molasses to the fermenter, $x(t)$ is the biomass concentration, $V(t)$ is the volume of the fermentation broth, $Y(t)$ is the cellular yield coefficient of the molasses and $\mu(t)$ is the specific growth rate of the yeast. Equation (E.5.59.1) was used as the initial estimate of $F_s(t)$ in the anticipatory feed-back control scheme depicted in Figure E.5.59.1. The actual value of $F_s(t)$ was calculated after including a proportionality constant α to (E.5.59.1) to achieve greater versatility in the control scheme:

$$F_s(t) = \alpha\frac{\mu(t)\,[x(t)\,V(t)]}{Y(t)\,c_s(t)} \qquad \text{(E.5.59.2)}$$

A control strategy based on anticipation of the required sugar demand is a form of feed-forward control. The controlled variable is not used as one of the inputs in a true feed-forward control. The volume of the fermentation broth volume $V(t)$ is calculated by using $F_s(t)$:

$$\frac{dV(t)}{dt} = F_s(t) + F_a(t) + F_n(t) \qquad \text{(E.5.59.3)}$$

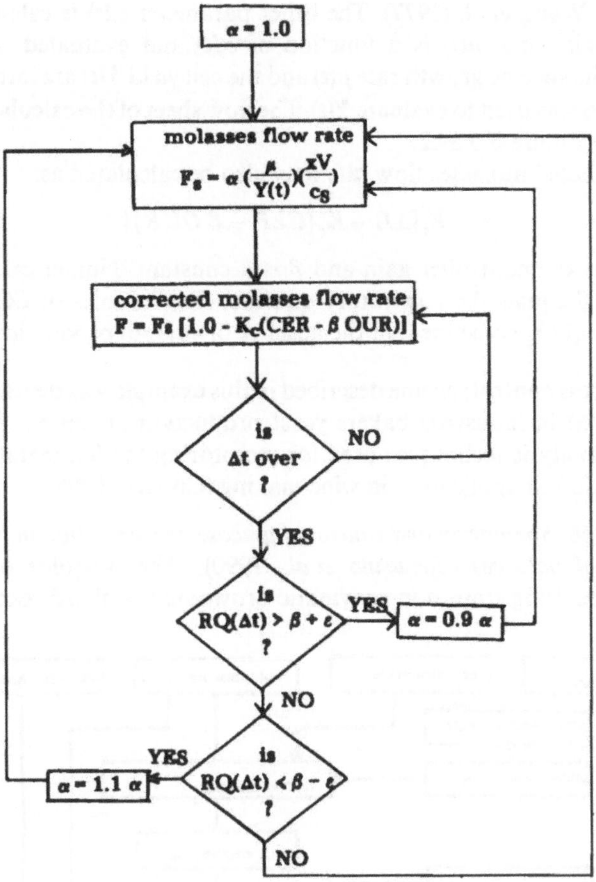

FIGURE E.5.59.1 The flow sheet to show the execution of the anticipatory feed-back control of the molasses addition during the yeast production (Reprinted by permission of John Wiley & Sons Inc. from Computer control of baker's yeast production, Wang *et al.*, *Biotechnology and Bioengineering* © 1979 John Wiley & Sons Inc.,). Δt = predetermined time difference between the steps of corrective action. The flow diagram indicates the F_s is reduced if the CO_2 production rate exceeds a predetermined value such that the respiratory quotient (RQ) becomes grater than $\beta + \varepsilon$; the control action is taken to increse F_s if the CO_2 production becomes less than a predetermined value such that the RQ becomes smaller than $\beta - \varepsilon$.

where $F_a(t)$ and $F_n(t)$ are the addition rates of ammonia and the other nutrients, respectively. Since $F_s(t)$ is used to calculate V(t), which is in turn used to estimate $F_s(t)$ the control scheme is not called a feed-forward control, but referred to as the anticipatory feed-back control (Wang *et al.* 1979). The other terms $x(t)$ and $V(t)$ which appear in (E.5.59.1) were calculated with material balancing as described in detail by Cooney *et al.*

(1977) and Wang *et al.* (1977). The other parameter $c_s(t)$ is calculated by material balancing, $\mu(t)$ is a function of $c_s(t)$ and evaluated from this relation. The specific growth rate $\mu(t)$ and the cell yield $Y(t)$ are interrelated, this relation was used to evaluate $Y(t)$. The flow sheet of the calculations are depicted in Figure E.5.59.2.

The corrected molasses flow rate may also be calculated as:

$$F = F_s[1.0 - K_c(CER - \beta\, OUR)] \tag{E.5.59.4}$$

where K_c is the controller gain and β is a constant. Numerical value of parameter β equals the respiratory quotient (RQ = moles of CO_2 evaluated/moles of O_2 consumed) in the absence of ethanol production.

The process control scheme described in this example was developed and is being used in industrial bakers yeast production processes. The head space gas analysis techniques used for monitoring the fermentation processes also found application in wine making (Corrieu, 1995).

Example 5.60. Simulation and control of glucose concentration in hot-water blanching of potatoes (*Tomasula et al.*, 1990) The off-color of potato products resulting from non-enzymatic browning (Maillard reaction) in-

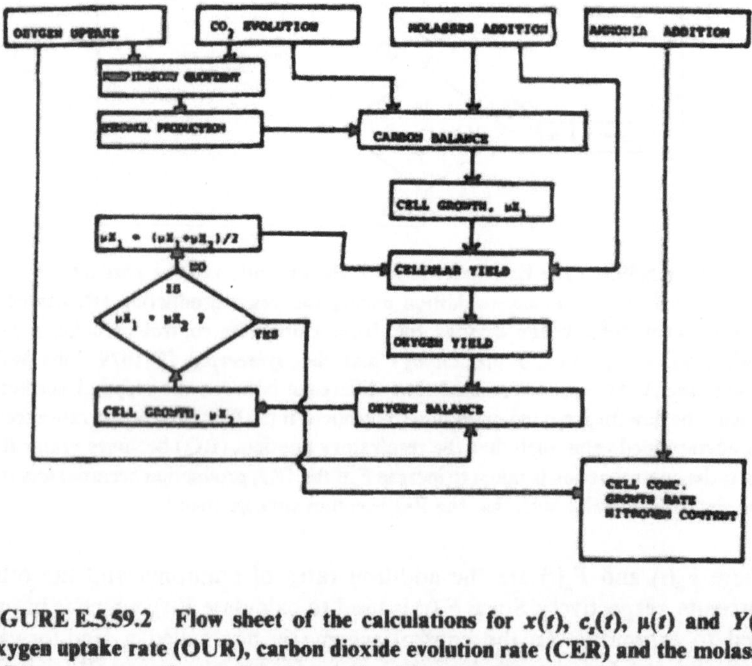

FIGURE E.5.59.2 Flow sheet of the calculations for $x(t)$, $c_s(t)$, $\mu(t)$ and $Y(t)$. Oxygen uptake rate (OUR), carbon dioxide evolution rate (CER) and the molasses addition rates were the monitored variables (Reprinted by permission of John Wiley & Sons Inc. from Computer-aided baker's yeast fermentations, Wang *et al.*, *Biotechnology and Bioengineering* © 1977 John Wiley & Sons Inc.,).

volves the reaction of reducing sugars with amino acids. High reducing sugar content enhances the Maillard reaction. Processed potatoes that result in a dark product are judged unacceptable. One strategy processors use to control the color is hot water blanching to extract the components that participate in the Maillard reaction. Blanchers are generally of the screw conveyor type continuous units with direct steam injection and are run at residence times of 2–30 minutes. The sugar concentration in the blanch water should be maintained at an appropriate level to yield potatoes leached to a sufficiently low sugar concentration. This means that the flow of water to the blancher must be continuously adjusted so that the acceptable concentration of reducing sugars in the potato is attained. An automatic control system that senses the reducing sugar concentration in water and adjusts the water flow rate to the blancher would aid in the processor to improve the color and lower the utility cost. The process control loop used by Tomasula *et al.* (1990) for glucose control is shown in the block diagram of Figure E.5.60.1, where concentration of glucose in the blancher is determined on-line by using a glucose analyzer. The glucose analyzer provides electrical current output in proportion with the glucose

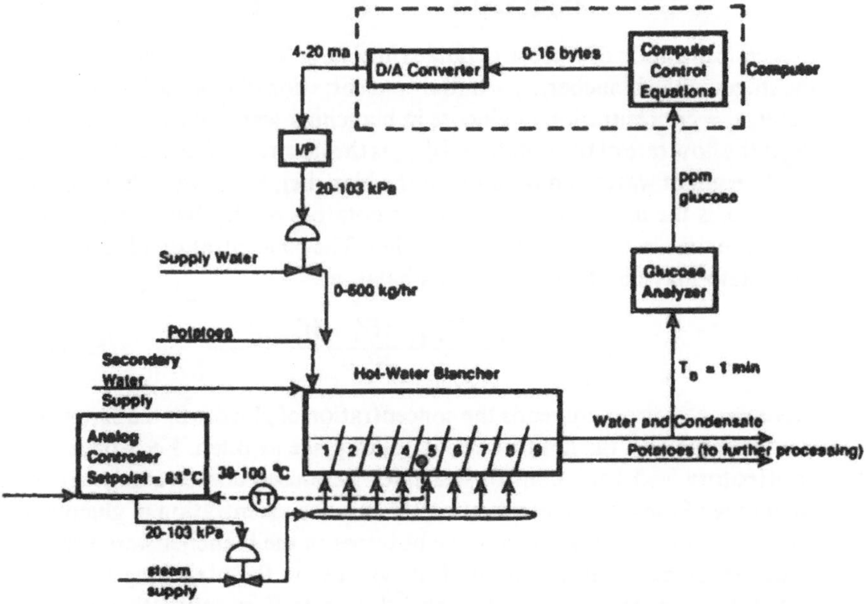

FIGURE E.5.60.1 Schematic drawing of the glucose analyzing and controlling system. The I/P unit converts the electrical current into the pressure signals which are used to turn on/off, the water supply to the hot water blancher. Steam temperature is controlled by means of a separate analog controller (Reprinted with permission from Tomasula *et al.*, © 1990 ACS).

concentration measurements and is interfaced to a computer with an analog output interface board. The computer program monitors the glucose concentration in the exit water of the blancher and compares it to the setpoint, and writes it to the D/A board, which provides a graphical and digital display on the monitor.

A mathematical model is needed to control the glucose concentration in the potatoes exiting the blancher, in terms of the glucose concentration in the blanching water:

$$s_{z+1} = \frac{W_p c_z E + W_{\text{steam}} s_z}{W_p x E + W_z + W_{\text{steam}}} + \frac{(W_p x E + W_z + W_{\text{steam}}) s_0 - (W_p c_z E + W_z s_z)}{W_p x E + W_z + W_{\text{steam}}} \exp\left[-\frac{W_p x E + W_z + W_{\text{steam}}}{W_{\text{total}}/N_{\text{total}} + W_p x \tau E} t \right] \quad \text{(E.5.60.1)}$$

where

$$E = 1 - \frac{8}{\pi^2} \sum_{m=0}^{\infty} \frac{1}{(2m+1)^2} \exp\left[-\frac{D(2m+1)^2 \pi^2 t}{L^2} \right] \quad \text{(E.5.60.2)}$$

c_z = concentration of glucose in potato in zone z, N_{total} is the total number of the stages in the blancher, s_0 = initial concentration of glucose in blanching water, s_z = concentration of glucose in blanching water in zone z, t is time, W_p is the flow rate of the potatoes, W_{steam} is the flow rate of steam, W_{total} is the total weight of water and potatoes in the blancher, W_z is water flow rate to zone z, x is the moisture content of the potatoes on dry basis and τ is the residence time in one zone of the blancher. The concentration of glucose in the potato may be obtained by mass balance:

$$c_{z+1} = \frac{W_p c_z + W_z s_z - (W_z + W_{\text{steam}}) s_{z+1}}{W_p} \quad \text{(E.5.60.3)}$$

The computer program reads the concentration of glucose in the exit water every minute and the control action takes place as often. Feed-back and anticipatory feed-back control strategies were used consequently. At the beginning of the control experiment, the initial concentration of glucose in the potato and the flow rates of the potatoes in the blancher were used to calculate the exit concentration of the glucose in the blanching water by (E.5.60.1), then an appropriate water flow rate is supplied through the control units to maintain the required exit glucose concentration. An error of $\varepsilon = \pm 30$ ppm was considered tolerable between the set point and the experimentally measured values of the glucose concentration in the exit water from the blancher. There was 18 minutes delay between the control action and sensing of glucose in the blancher water, and this was accounted for by the computer program. A process control run with proportional feed-back control indicates that once the setpoint is reached the glucose concentration in the potato has also reached the setpoint value, but glucose that has built up in the blancher was washed out in one residence time since the control valve opens to the value to keep the glucose concentration at that setpoint. One more hour can elapse until the set point is approached again. Since the proportional feed-back controller was found to overact, a control based on anticipation was used next. In order to anticipate the disturbance caused by washing out the glucose in the blancher, (E.5.60.1)–(E.5.60.3) were used. Value of the glucose concentration in water under steady state conditions was estimated with the model when there was no water flow initially. The flow rate of water which gives steady state values of glucose concentration in the exit water and potato streams and closely responds to the setpoints was chosen for a particular run and used until the setpoint is reached. To avoid washing out glucose after attaining the setpoint, the control valve was closed for one residence time. The results are depicted in Figure E.5.60.2.

Example 5.61. Grain dryer controllers Cereal grain is often harvested at moisture contents that are too high to permit storage of the grain without spoilage. Grain dryers are used to reduced this moisture content to reasonable levels so that the grain will not deteriorate in storage. Over-drying, on the other hand, results in shrinkage of the grains and introduces additional energy cost (Moreira and Bakker-Arkema, 1992). Manual control of a drying system is a complicated task. It requires extensive experience on the part of the dryer operator not to overdry or underdry the grain

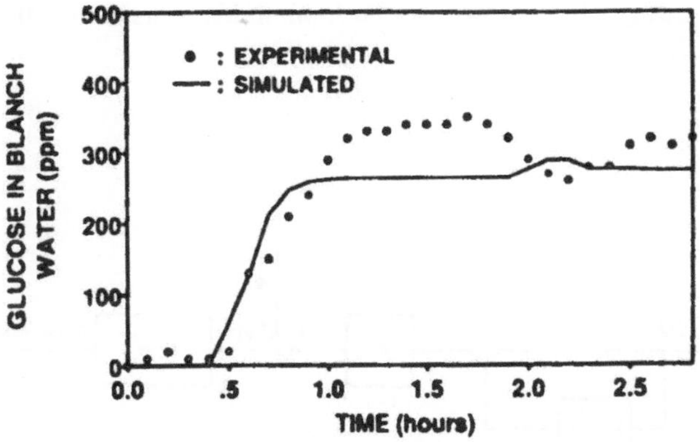

FIGURE E.5.60.2 Simulation of process control for glucose. The glucose concentrations were 1580 ppm and 300 ppm initially in the potatoes and for the setpoint of the blancher exit water, respectively (Reprinted with permission from Tomasula *et al.*, © 1990 ACS).

while operating the dryer under acceptable conditions with respect to grain quality deterioration and fuel consumption. A well designed controller substitutes for expertise and in fact outperforms an experienced operator (Moreira and Bakker-Arkema, 1992).

Whitfield (1986) suggested a PI (proportional-integral) algorithm for control of the counter-current flow grain dryers:

$$F(t) = K_c\left[\varepsilon(t) + \frac{1}{\tau_I}\int_0^t \varepsilon(t)dt + F_0\right] \tag{E.5.61.1}$$

Where F is the flow rate of the grain, subscript 0 indicates the initial conditions when the control action is started, $\varepsilon(t)$ was the error, i.e., difference between the set point and the measured values of the moisture content of the grain at the exit of the dryer. The proportional term gives quick responses to errors, the integral term is added to F_0 to adjust the mean level and to prevent steady state errors. The differential term of the PID controllers is employed for rapid response to the deviations from the set point value. Very rapid response is not needed for control of the packed bed grain dryers, therefore a differential control term is not included to (E.5.61.1). Block diagram of the suggested control scheme is depicted in Figure E.5.61.1.

Whitfield (1986) tested (E.5.61.1) after determining the "best" values for the parameters K_c and τ_I. It was noted that the best parameter values for a sudden increase in the inlet moisture content are different from those for a sudden decrease; and the best values depend on the amount of drying that is required. Thus, (E.5.61.1) was shown to be useful in a limited initial moisture range, but shown to be non-stable outside of this range (Fig. 5.61.2).

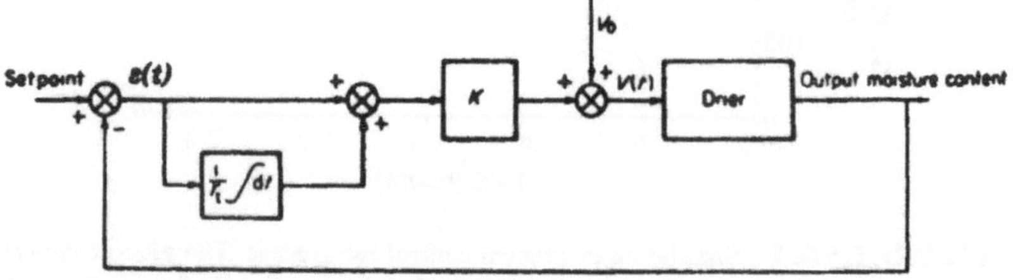

FIGURE E.5.61.1 Block diagram of the control scheme suggested by (E.5.61.1) (Whitfield, 1986; © Silsoe Research Institute, reproduced by permission).

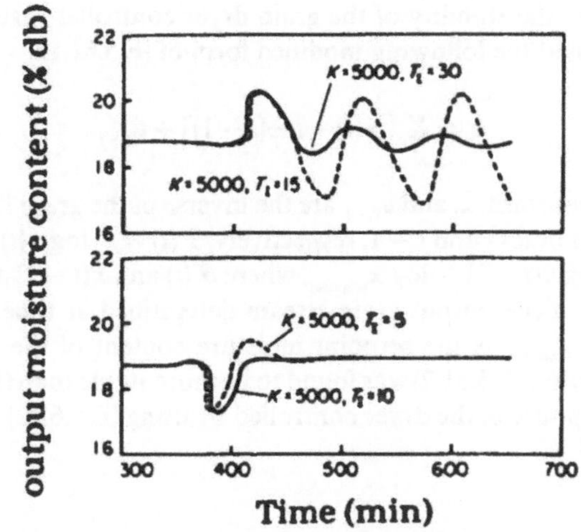

FIGURE E.5.61.2 Variation of the output moisture content from the continuous flow grain dryer controlled with the control scheme suggested by (E.5.61.2) in the stable initial moisture content range by employing different values of K_c and τ_I (Whitfield, 1986; © Silsoe Research Institute, reproduced by permission).

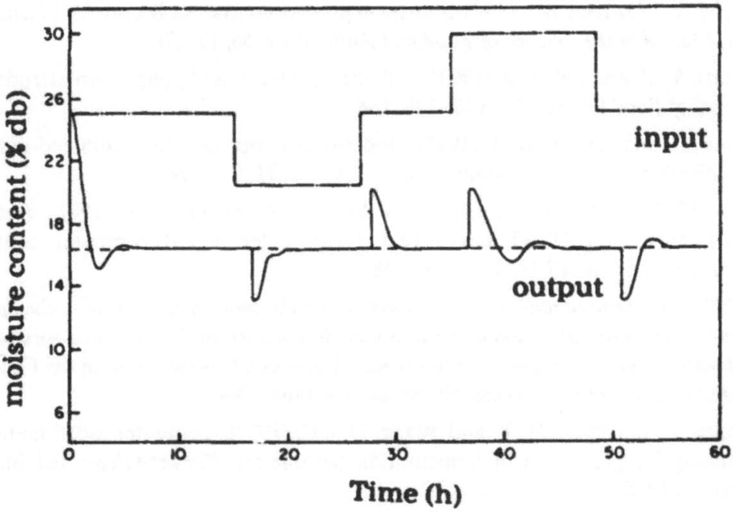

FIGURE E.5.61.3 The predicted response of the dryer controlled by using (E.5.61.2). Variations in the input (–.–.–) and the output (–) moisture contents of the grains and the setpoint (–––) are shown with lines. The model parameters were: $K_c = 6.3 \times 10^{-4}$, $a = 0.8$ (Whitfield, 1988; © Silsoe Research Institute, reproduced by permission).

To improve the stability of the grain dryer controller system Whitfield (1988) suggested the following modified form of (E.5.61.1):

$$u_t = K_c[\mathscr{X}(t) - a\mathscr{X}(t-1)] + u_{t-1} \qquad \text{(E.5.61.2)}$$

where a is a constant, u_t and u_{t-1} are the inverse of the grain flow rate (1/F) determined at times t and $t-1$, respectively; $\mathscr{X}(t) = -\log x(t) - \log x_{\text{setpoint}}$, $\mathscr{X}(t-1) = \log x(t-1) - \log x_{\text{setpoint}}$, where $\mathscr{X}(t)$ and $x(t-1)$ are the moisture contents of the output grain stream determined at times t and $t-1$, respectively; x_{setpoint} is the setpoint moisture content of the output grain stream. Equation (E.5.61.2) was found to be more stable than (E.5.61.1). The predicted response of the dryer controlled by using (E.5.61.2) is depicted in Figure E.5.61.3.

REFERENCES

Alaeddinoğlu, G., Güven, A. and Özilgen, M. (1988). Activity loss kinetics of freeze-dried lactic acid starter cultures. *Enzyme and Microbial Technology*, **11**, 765–769

Alti, M. and Özilgen, M. (1994). Statistical process analysis in Broiler feed formulation. *Journal of the Science of Food and Agriculture*, **66**, 13–20

Barrett, A. M. and Peleg, M. (1991). Cell size distributions of puffed corn extrudates, *Journal of Food Science*, **57**, 146–148, 154

Catino, C. and Ungar, L. H. (1995) Model-based approach to automated hazard identification of chemical plants. *AIChE Journal*, **41**, 97–109

Chen, C. S., Carter, R. D., Miller, W. M. and Wheaton, T. A. (1981). Energy performance of a HTST citrus evaporator under digital computer control, *Transactions of the ASAE*, **24**, 1678–1682

COST 914, *Memorandum of understanding for the implementation of a European research action on validation of predictive models of microbial growth and survival in food matrices and throughout the food chain.* European Co-operation in the Field of Scientific and Technical Research, Brussels, 8 June 1994.

Cooney, C. L., Wang, H. Y. and Wang, D. I. C. (1977). Computer-aided material balancing for prediction of fermentation parameters, *Biotechnology and Bioengineering*, **19**, 55–67

Corrieu, D. P. (1995). Advances in analytical techniques for food process control, *The World of Ingredients*, (March-April) 49–54

Dickens, J. W and Whitaker, T. B. (1982). Sampling and Sample preparation, in *Environmental Carcinogens Selected Methods of Analysis*, Volume 5, Egan, E. (Editor-in-Chief) International Agency for Research on Cancer, Lyon, France.

Dodge, H. F. and Romig, H. G. (1959). *Sapling Inspection Tables*. Second ed. John Wiley and Sons., New York.

Durukan, A., Özilgen, S. and Özilgen, M. (1992). Analysis of a baking process with aplication of means and range charts to samples coming from cobined populations. *Process Control and Quality*, **2**, 327–333

FDA, *Inspection Operation Manual of the Food and Drug Administration*, United States Department of Health, Education and Welfare (October 19, 1979)

FAO, Manuals of Food Quality Control 9. Introduction to Food Sampling. Food and Agriculture Organization of the United Nations, Rome, 1988

Gould, W. A. (1983). *Food Quality Assurance*, AVI, USA, pp. 31–34.

Jacobs, D. C. (1990). Watch out for nonnormal distributions, *Chemical Engineering Progress*, **86**(11) 19–27

Kahraman-Dogan, H., Bayindirli, L. and Özilgen, M. (1994). Quality control charts for storage of eggs. *Journal of Food Quality*, **17**, 495–501

LaMont, M. D., Douglas, L. L., Oliva, R. A., Hoffman, R. I., Zastrocky, M., Reed, J. F., Battaglia, S., Alderman, H., Staiert, P., Bardwell, G., Wise, R. and Poynes, J. (1977). Calculator Decision-Making Sourcebook. Texas Instruments Inc., USA

Levinson, W. (1990). Understand the basics of statistical quality control. *Chemical Engineering Progress*, **86**(11) 28–37

Microbiology and Food Safety Committee of the National Food Processors Association, Implementation of HACCP in food processing plant, *Journal of Food Protection*, **56**, 548–554

Moreira, R. G. and Bakker-Arkema, F. W. (1992). Grain dryer controls: A review, *Cereal Chemistry*, **69**, 390–366

Nicolai, B. M., Van Impe, J. F., Verlinden, B., Martens, T, Vandewalle, J. and De Baerdemaeker, (1993). J. Predictive modeling of surface growth of lactic acid bacteria in vacuum-packed meat. *Food Microbiology*, **10**, 229–238

Oakland J. S. and Followell R. F. (1990). *Statistical Process Control*, 2nd ed. Heinemann Newness, Great Britain

Özilgen, M. Construction of the quality control charts with sub-optimal size samples. Food control, in press (1998).

Pappas, G. and Rao, V. N. M. (1987). Development of a sampling plan for specified risks by microcomputers. *Journal of Food Processing and Preservation*, **11**, 339–345

Peterson, R. G. (1974). Wine quality control and evaluation, in *Chemistry of Winemaking*, Webb, A. D. (ed.) American Chemical Society, USA

Ramaswamy, H., van de Voort, F. R., Raghavan, G. S. V., Lightfoot, D. and Timbers, G. (1991). Feed-back temperature control system for microwave ovens using a shielded thermocouple. *Journal of Food Science*, **56**, 550–555

Robinson, J. W. (1992). Improve dryer control. *Chemical Engineering Progress*, **88**(12), 28–33

Sandeep, K. P. and Zuritz, C. A. (1994). Residence time distribution of multiple particles in non-Newtonian holding tube flow: Statistical analysis. *Journal of Food Science*, **59**, 1314–1317

Sidebottom, B. (1986). Quality control by design. *Food Processing*, **55**(3), 39–42

Snyder, O. P. HACCP- An industry food safety self-control program. Dairy Food and Environmental Sanitation, January 1992, pp. 26–27.

Sokal, R. R. and Rohlf, F. J. (1981). *Biometry, Freeman,* New York

Sperber, W. S. (1991). The modern HACCP system. *Food Technology,* **45**(6), 116–118, 120

Şumnu, G., Bayindirli, L. and Özilgen, M. (1994a). Quality control charts for storage of apricots. *Zeitschrift für Lebensmittel-Untersuchung und-Forschung,* **199**, 201–205

Şumnu, G., Bayindirli, L. and Özilgen, M. (1994b). Quality control charts for storage of apples. *Lebensmittel – Wissenschaft und Technologie,* **27**, 496–499

Teixeira, A. A. and Manson, J. E. (1982). Computer control of batch retort operations with on-line correction of process deviations. *Food Technology,* **36**(4) 85–90

Teixeira, A. A. and Shoemaker, C. F. (1989). *Computerized Food Processing Operations,* Avi, New York

Thornton, J. (1992). ISO 9000 & the food industry. *International Food Ingredients.* **5**(4), 8–14

Tomasula, P. M., Kozempel, M. F. and Craig, J. C. (1990). Simulation and control of glucose concentration in hot-water blanching of potatoes. *Biotechnology Progress,* **6**, 249–254

USDA, *HACCP Principles for Food Production* (November, 1989)

Wang, H. Y., Cooney, C. L. and Wang, D. I. C. (1977). Computer-aided baker's yeast fermentations, *Biotechnology and Bioengineering,* **19**, 69–86

Wang, H. Y., Cooney, C. L. and Wang, D. I. C. (1979). Computer control of bakers' yeast production. *Biotechnology and Bioengineering,* **21**, 975–995

Whiting, R. C. and Buchanan, R. L. (1994). Microbial modeling. *Food Technology* **48**(6) 113–120

Whitfield, R.D. (1986). An unsteady-state simulation to study the control of concurrent and counter-flow grain driers. *Journal of Agricultural Engineering Research,* **33**, 171–178

Whitfield, R. D. (1988). Control of a mixed-flow drier, Part I: Design of the control algorithm. *Journal of Agricultural Engineering Research,* **41**, 275–287

Author Index

Subject Index

* 9 7 8 9 0 5 6 9 9 1 4 3 2 *

T - #0448 - 101024 - C0 - 229/152/29 - PB - 9789056991432 - Gloss Lamination